Algebra – Representation Theory

NATO Science Series

A Series presenting the results of scientific meetings supported under the NATO Science Programme.

The Series is published by IOS Press, Amsterdam, and Kluwer Academic Publishers in conjunction with the NATO Scientific Affairs Division

Sub-Series

I. Life and Behavioural Sciences	IOS Press
II. Mathematics, Physics and Chemistry	Kluwer Academic Publishers
III. Computer and Systems Science	IOS Press
IV. Earth and Environmental Sciences	Kluwer Academic Publishers

The NATO Science Series continues the series of books published formerly as the NATO ASI Series.

The NATO Science Programme offers support for collaboration in civil science between scientists of countries of the Euro-Atlantic Partnership Council. The types of scientific meeting generally supported are "Advanced Study Institutes" and "Advanced Research Workshops", and the NATO Science Series collects together the results of these meetings. The meetings are co-organized bij scientists from NATO countries and scientists from NATO's Partner countries – countries of the CIS and Central and Eastern Europe.

Advanced Study Institutes are high-level tutorial courses offering in-depth study of latest advances in a field.
Advanced Research Workshops are expert meetings aimed at critical assessment of a field, and identification of directions for future action.

As a consequence of the restructuring of the NATO Science Programme in 1999, the NATO Science Series was re-organized to the four sub-series noted above. Please consult the following web sites for information on previous volumes published in the Series.

http://www.nato.int/science
http://www.wkap.nl
http://www.iospress.nl
http://www.wtv-books.de/nato-pco.htm

Series II: Mathematics, Physics and Chemistry – Vol. 28

Algebra – Representation Theory

edited by

I. Klaus W. Roggenkamp

Fakultät Mathematik,
Universität Stuttgart,
Stuttgart, Germany

and

Mirela Ştefănescu

Faculty of Mathematics,
Ovidius University,
Constanta, Romania

Kluwer Academic Publishers

Dordrecht / Boston / London

Published in cooperation with NATO Scientific Affairs Division

Proceedings of the NATO Advanced Study Institute on
Algebra – Representation Theory
Constanta, Romania
2–12 August 2000

A C.I.P. Catalogue record for this book is available from the Library of Congress.

ISBN 0-7923-7113-5

Published by Kluwer Academic Publishers,
P.O. Box 17, 3300 AA Dordrecht, The Netherlands.

Sold and distributed in North, Central and South America
by Kluwer Academic Publishers,
101 Philip Drive, Norwell, MA 02061, U.S.A.

In all other countries, sold and distributed
by Kluwer Academic Publishers,
P.O. Box 322, 3300 AH Dordrecht, The Netherlands.

Printed on acid-free paper

Printed in the Netherlands.

TABLE OF CONTENTS

PREFACE

Over the last 30 years representation theory of groups, Lie algebras and associative algebras has undergone a rapid development through the powerful tool of *almost split sequences and the Auslander-Reiten quiver*; further insight into *cohomology of finite groups* has eluminated their representation theory. The study of *quantum groups, Hopf-algebras and non-commutative geometry* is another new branch of representation theory, which pushes the classical theory further. All this can only be seen in connection with the understanding of the *structure of special classes of rings*. The considerable progress of these fields was accomplished by fruitful interaction among each other.

The aim of these proceedings is to introduce the interested algebraists (especially in representation theory and structure of rings) by internationally highly recognized specialists to some modern developments in

i. Lie algebras, quantum groups, Hopf algebras and algebraic groups,
ii. non-commutative algebraic geometry,
iii. representation theory of finite groups and cohomology,
iv. the structure of special classes of rings.

1. **Lie theory and algebraic groups.**

Let G be a reductive algebraic group defined over an algebraically closed field K of positive characteristic with maximal torus T and associated root system R. A finite dimensional rational module M has a "good filtration" if

it can be filtered such that the sub-factors are isomorphic to the module of global sections of the line bundle on the flag variety G/B of G modulo a Borel subgroup B induced from some character of T. A G-module M is a *tilting module* provided both M and its dual M^* have a good filtration. Tilting modules are closed under tensor products and can be characterized cohomologically. These results carry over to modules of the quantized enveloping algebra U_q at a complex root of unity q of odd order, associated to the root system R. One can apply tilting modules to recover in positive characteristic part of the classical Schur-Weyl duality for $G = GL(V)$. Tilting modules play also a crucial role in the study of the so-called quantum invariants of 3-manifolds associated to the above quantum group U_q. They can also be used for the difficult task of calculating the characters of irreducible G–modules (*H. H. Andersen*).

Given the action of a connected reductive complex algebraic group G on a complex algebraic variety X with only finitely many orbits, one can construct the equivariant extension algebra $Ext^*_G(X)$. This needs, in addition to an anti-holomorphic involution γ on G stabilizing a Borel subgroup, a character χ of the center of the universal enveloping algebra. The "modified Langlands parameter space" $X(\chi)$ on which the dual group $G^\vee$ acts was defined by Adams, Barbasch and Vogan. The "Langlands philosophy" centers around the conjecture

$$\oplus_{\delta \in H^1(Gal(\mathbb{C}/\mathbb{R}),G)} \mathcal{M}(G(\mathbb{R},\delta))_\chi \simeq Ext^*_{G^\vee}(X(\chi))\text{–Nil}\,.$$

Here $\mathcal{M}(G(\mathbb{R},\delta))_\chi$ is the category of Harish-Chandra modules for $G(\mathbb{R},\delta)$ (every δ as above gives rise to such a real form of G) killed by some power of χ, and for a graded ring $A = \oplus_{i \geq 0} A_i$ the category A–Nil consists of all finite dimensional A-modules killed by some A_i (*W. Soergel*).

In the representation theory of a finite dimensional semisimple complex Lie algebra with a triangular decomposition $\mathfrak{g} = \mathfrak{n}_- \oplus \mathfrak{h} \oplus \mathfrak{n}_+$ one is mainly interested in the Bernstein-Gelfand-Gelfand category $\mathcal{O}$, whose projective modules have a filtration by Verma-modules. The "blocks" $\mathcal{O}_\theta$ of $\mathcal{O}$ are equivalent to the finitely generated modules over a finite dimensional quasi-hereditary $\mathbb{C}$-algebra A_θ. The structure of the projective A_θ-modules is in close connection with the Kazhdan-Lusztig conjectures. The algebra $e \cdot A_\theta \cdot e$ – for a suitable primitive idempotent e – is the cohomology algebra of the flag manifold G/B for a Borel subgroup

B of the associated complex Lie group G, and A_θ is a self(Ringel)-dual quadratic Koszul algebra, which is QF-3 with dominant dimension at least 2 (*S. König*).

(Lie)Conformal algebras and the related vertex operator algebras – which are in a relation similar to Lie algebras and their enveloping algebras – have deep connections to physics and the Moonshine representation of the Monster groups. The main open problem in this theory is the classification of the simple Lie (associative) conformal algebras of Gelfand-Kirillov dimension one. In this direction one has the following results: Such a finitely generated *unital* associative F-algebra (F is the ground field of characteristic 0) is isomorphic to $M_n(W) := M_n(F) \otimes_F W$, where W is the conformal algebra generated by the formal distributions $J_k = \sum_{i\in\mathbb{Z}} t^i(\frac{d}{dt})^k z^{-i-1}$, $k \geq 0$. However, not all such algebras are unital. V. Kac has modified $M_n(W)$ in order to include also non-unital algebras and conjectured that in this way a full classification will be achieved. Moreover, there already exists a partial classification of simple finitely generated Lie conformal algebras, describing them in terms of the above $M_n(W)$ (*E. Zelmanov* [1]).

In connection with quantum groups the study of Hopf algebras has gained more and more interest. The development of a theory for finite dimensional semisimple Hopf algebras H over an algebraically closed field of characteristic 0 is guided by the strong similarity with classical representation theory of finite groups. One can thus use the character theory of H to derive a kind of class equation for H (G. I. Kac). The Grothendieck ring $K_0(H)$, which is isomorphic to the integral character ring, strongly influences the structure of H; $K_0(H_1) \simeq K_0(H_2)$ if and only if H_1 is a twist of H_2 by a (dual) pseudo-co-cycle (Nikshych; Drinfel'd). The Schur indicator is computed for the representations of certain Hopf algebras (*S. Montgomery*).

2. **Non-commutative Geometry.**

Classical geometry can be considered as the study of commutative Noetherian algebras and their modules, so non-commutative geometry is the study of non-commutative algebras and their modules; quantum groups can be thought of as non-commutative

[1] Fields medalist

analogues of algebraic groups, and non-commutative projective geometry corresponds to non-commutative graded rings.

In the last 30 years representation theory of algebras over a field K has made a considerable progress via the representation theory of hereditary algebras (as graph algebras of quivers) together with the "tilting theory", which transfers essential properties of hereditary algebras to tilted algebras since both are derived equivalent. The path algebra of the Kronecker quiver is derived equivalent to the category of coherent sheaves on the projective line; more general equivalences between suitable categories of modules and the category of coherent sheaves on the weighted projective line, an important source from (non-commutative) geometry, are known. Auslander-Reiten sequences give a central construction in the representation theory of finite dimensional algebras, and Serre duality is an important concept in algebraic geometry. Hereditary K–categories provide a common frame both for geometry and algebra; within them the concepts of Auslander-Reiten sequence and Serre duality coincide (*I. Reiten*).

On the representation space X to the dimension vector α of a finite directed graph Γ, a reductive linear group G acts as isomorphisms and the aim is to classify the orbits. Using Luna's stratification of X/G (in the representation theoretic fashion as given by Procesi and Le Bruyn) one can – knowing the simple dimension vectors – list all possible representation types and compute the dimensions of the strata. In this connection the extension groups of semisimple representations have a geometric interpretation as quotients of tangent spaces (*M. Domokos – A. N. Zubkov*).

Since commutative regular Noetherian rings correspond to smooth varieties there have been many generalizations to the non-commutative situation (Artin-Schelter, Tate, Van den Berg). Here the approach is taken via a generalization of Cohen-Macaulay modules. The latter is based upon a family $\mathcal{P} := \{P_1, \dots P_n\}$, which is a Cohen-Macaulay system, of invertible two-sided ideals, extending the classical definition. The localization of the universal enveloping algebra $U(\mathfrak{g})$ of a nilpotent Lie-algebra $\mathfrak{g}$ at the augmentation ideal $I(\mathfrak{g})$ is Cohen-Macaulay and Auslander regular. The indecomposable basic semi-perfect regular rings are described as matrix rings (*W. Rump*).

Let X be a projective scheme with an automorphism σ and let $\mathcal{L}$ be a σ-ample invertible sheaf, then the twisted homogeneous coordinate ring $\mathcal{B} = \mathcal{B}(X, \mathcal{L}, \sigma)$ is a right Noetherian connected graded ring such that the category of coherent $\mathcal{O}_X$-modules over the sheaf of regular functions over X is equivalent to the category $qgr(\mathcal{B})$ of finitely generated $\mathcal{B}$-modules modulo torsion (meaning here finite dimensional) modules. This is Artin-Van den Berg's non-commutative analogon of Serre duality; for example this can be applied to derive in a unified way both standard "quantum planes" from the theory of quantum groups: $K\langle x, y\rangle/(yx - yx - x^2)$ and $K\langle x, y\rangle/(yx - \lambda yx), \lambda \in K$. Another generalization is Artin-Zhang's theorem which characterizes certain K-linear abelian categories (with an ample pair) as $qgr(B)$ for a right Noetherian $\mathbb{N}$-graded ring. These techniques are used to "classify" non-commutative projective curves – equivalently non-commutative connected graded prime rings of Gelfand-Kirillov dimension two. A "Serre duality" for non-commutative varieties is derived. Finally surfaces – i.e. non-commutative graded domains of Gelfand-Kirillov dimension 3 – are discussed. Here one is still far away from a satisfactory classification (*T. Stafford*).

3. **Representations of finite groups and cohomology.**

One of the most outstanding problems in the representation theory of a finite group G over a complete discrete valuation ring R of characteristic 0 with residue field k of characteristic $p > 0$ is *Broué's conjecture*: A block B of RG with abelian defect group is derived equivalent to its Brauer correspondent b. Broué's conjecture has been proved for a multitude of groups; for even more groups a *perfect isometry* has been verified, which is a consequence of Broué's conjecture. Whenever a perfect isometry has been found, even the much stronger *isotypy* could be verified. This concept has been widely generalized into the context of derived categories, leading to so-called *splendid* equivalences (*J. Rickard*).

Since the triangulated stable category of B is a quotient of the triangulated derived category, Broué's conjecture would imply a stable equivalence between B and b. The strategy to attack Broué's conjecture is to "glue" Rickard's splendid equivalences for local subgroups to produce an equivalence of the stable module categories and then try to lift this to a derived equivalence. This is where non-commutative geometry enters via gradings and Koszul duality on Koszul algebras (e. g. the crossed product of an exterior algebra by a finite automorphism group of the underlying vector space). The invariance of automorphisms under derived or stable equivalences is applied to construct gradings, which allow in some cases to pass from the stable to the derived category (*R. Rouquier*).

Whenever in mathematics there is an obstruction, a small cohomology set will be the reason. So there is an intimate relation between modular representation theory of a finite group and group cohomology. The main observation and starting point is that for elementary abelian groups the rank variety of a finitely generated module coincides with the cohomological variety. Quillen's result that the components of the cohomological variety are determined by restrictions to elementary abelian subgroups is generalized. Subvarieties for infinitely generated modules, which arise naturally through a limiting process when studying finitely generated modules, can be defined as subset of the projective prime ideal spectrum of the cohomology ring; however, they are not really varieties, since they are not closed in the Zariski topology. Nevertheless they still conserve all the properties of the proper varieties for finitely generated modules that make them useful for applications to module theory (*J. Carlson*). There is also a cohomology theory for semi-groups (*B. V. Novikov*).

4. **Special classes of rings.**

Varieties of rings. The prime varieties of countably generated algebras over a field F were described in characteristic 0 by Kemer. The problem is still open in positive characteristic. The prime varieties in any characteristic fall into regular and irregular varieties. For the regular varieties there are some structural results. The multilinear components of the prime sub-varieties of the variety generated by 2×2 matrices over a field of characteristic $p > 0$ are described. The matrix type of the Grassmann algebra is linked to irregular varieties and a conjecture of Procesi (*A. Kemer*).

1-*dimensional orders.* Let R be a complete discrete valuation ring. Using a far-reaching generalization of Drozd–Kirichenko's rejection lemma one can completely characterize the Auslander–Reiten quivers (as translation quivers) of R–orders of finite representation type inductively. These techniques can be used to achieve important classification results: for example the classification of almost Bass orders is reduced to that of Bass orders; ring indecomposable almost Bass orders of rank ≥ 2 in a semisimple ambient ring can be described explicitly as matrix rings. The results allow generalizations to orders over complete regular local Noetherian domains (of any Krull dimension) as well as partial classifications in low dimensions. They also lead to new insight in the tame/wild dichotomy for R–orders (*O. Iyama*).

2-*dimensional orders, in particular integral Hecke orders.* In connection with Hecke algebras and quantum groups it is indicated to study orders over the arithmetical surface $\mathbb{Z}[q]$ or over $\mathbb{Z}[q, q^{-1}]$. Firstly some elementary facts are derived, since little seems to be known, as the Higman-ideal, which is the framework in which to prove Tits' deformation theorem. Blocks of cyclic defect for Hecke orders of Weyl-groups can be described as 2-dimensional tree orders. The indecomposable Cohen-Macaulay modules over 2-dimensional tree orders – though there are infinitely many of them – can be described. Finally it is proved that extension groups between the Specht modules of Hecke orders of the symmetric groups turn out to be finitely generated torsion-free abelian groups. (*K. W. Roggenkamp*).

Of the lectures by T. Stafford, there is only a short abstract included, since an extended version has already appeared in:

Stafford, J. T. and Van den Bergh, M.; Noncommutative projective curves and surfaces, Bull. Amer. Math. Soc., 38 (2001), 171-216.

Unfortunately some of the invited main speakers, who had committed themselves earlier did not participate in the end: M. van den Berg, F. van Oestayen and C. M. Ringel.

We thank Ms. Stefanie Ide very grateful for making the meeting run smoothly. We thank Joachim Simon for preparing carefully the manuscript. Finally our thank goes to the numerous helpful hands both from the universities of Constanţa and Stuttgart.

Mirela Ştefănescu, University of Constanţa and
Klaus W. Roggenkamp, University of Stuttgart

TILTING MODULES FOR ALGEBRAIC AND QUANTUM GROUPS

HENNING HAAHR ANDERSEN
(mathha@imf.au.dk)
Department of Mathematics
University of Aarhus
Building 503
Ny Munkegade
8000 Aarhus C, Denmark

Let G be a connected reductive algebraic group over an algebraically closed field k of characteristic $p > 0$. The concept of tilting modules for G was introduced by Donkin [15] based on some previous work by Ringel [26] on modules for quasi-hereditary algebras. Modules with analogous properties in the category $\mathcal{O}$ were studied earlier by Collingwood and Irving [12] (although they did not use the word tilting). It should also be added that tilting objects and complexes have a longer history in the theory of finite dimensional algebras, see e.g. the lectures by Reiten at this workshop.

It turns out that there is a straightforward analogue of tilting modules in the theory of quantum groups U_q at roots of unity. In our presentation we shall start out by describing in some details the definition and basic properties of tilting modules for G. Then we introduce the corresponding quantized modules (leaving most of the justifications to the reader). Our goal is to show both the close resemblance between the two cases and to point out some important differences. The interplay between the representation theory of G and the one for U_q was of crucial importance in [6] where the focus was on irreducible modules. As we shall emphasize there is an expected similar very close relationship between indecomposable tilting modules for G and their counterparts for U_q in the case

K.W. Roggenkamp and M. Ştefănescu (eds.), Algebra - Representation Theory, 1–21.

where the highest weights are in the lowest p^2-alcove. Outside this alcove, however, the situation for G is almost completely unknown whereas for U_q the characters of all indecomposable tilting modules have been determined by Soergel [28].

Apart from studying tilting modules for the purpose of comparing the situations for G and U_q we also want to give a couple of applications in other directions. First we look at the group $GL(V)$ with V a finite dimensional vector space over k. Here one may use Schur-Weyl duality to obtain an explicit relation between the multiplicities of the summands of $V^{\otimes r}, r \in \mathbb{N}$ and the dimensions of irreducible $k\Sigma_r$- modules. When combined with the fusion rule for the case where the highest weight in question belongs to the lowest p-alcove we deduce a formula for a special class of irreducible $k\Sigma_r$-modules, cf [23].

Another application is related to the Reshetikhin-Turaev construction of quantum invariants for 3-manifolds [27]. In order to make this theory work for U_q one needs a tensor product on U_q-modules with highest weights in the bottom l-alcove (l being the order of q). It turns out that by using the theory of tilting modules one can quite easily check the required property.

For both of the two applications it is essential that tensor products of tilting modules are again tilting. This is a non-trivial fact. After having dealt with the basic properties of tilting modules in Section 1 (the modular case) and Section 2 (the quantum case) we indicate in Section 3 how one may handle tensor products in both cases by appealing to canonical bases. The two applications mentioned above can be found in Section 4.

Acknowledgements. It is a pleasure to thank Joachim Simon and Harald Weber whose careful notes from my lectures in Constanta have formed the basis of this paper.

1. The modular case.

1.1 We fix a maximal torus T in G and let R be the corresponding root system. Then we choose a set $R^+ \subset R$ of positive roots and a Borel subgroup B in G containing T such that the roots of B are negative. The character group of T is denoted $X(T)$ or just X and the partial order on X induced by R^+ is written $\leq$. Moreover, the set of dominant weights $X^+ \subset X$ is given by

$$X^+ = \{\lambda \in X \mid \langle \lambda, \alpha^\vee \rangle \geq 0 \text{ for all } \alpha \in R^+\}.$$

Here $\alpha^\vee$ is the coroot of α.

The Weyl group for G (or for R) is denoted W. In addition to the standard action $\lambda \mapsto w\lambda$, $\lambda \in X$, $w \in W$ of W on X we shall also consider the "dot action" given by

$$\lambda \mapsto w \cdot \lambda = w(\lambda + \rho) - \rho, \lambda \in X, w \in W.$$

As usual ρ denotes half the sum of the positive roots.

We also have the affine Weyl group W_p. It is the subgroup of the group of permutations of X generated by the set of affine reflections $\{s_{\alpha,n} \mid \alpha \in R, n \in \mathbb{Z}\}$. Here $s_{\alpha,n}$ is the map which takes $\lambda \in X$ into $s_\alpha \cdot \lambda + np\alpha$.

The bottom alcove in X^+ is

$$C_p = \{\lambda \in X \mid 0 < \langle \lambda + \rho, \alpha^\vee \rangle < p \text{ for all } \alpha \in R^+\}.$$

Its closure $\overline{C}_p$ is the set obtained by replacing all $<$ by $\leq$. This is a fundamental domain for the action of W_p on X.

1.2 Unless otherwise said all modules considered in this paper will be finite dimensional. If M is a T-module then M splits into a direct sum of its weight spaces:

$$M = \bigoplus_{\lambda \in X} M_\lambda.$$

Here the λ-weight space M_λ in M is given by

$$M_\lambda = \{m \in M \mid tm = \lambda(t)m, \ t \in T\}.$$

We say that λ is a weight in M if $M_\lambda \neq 0$.

The character $\mathrm{ch}(M)$ of M is the following element of $\mathbb{Z}[X]$

$$\mathrm{ch}(M) = \Sigma_{\lambda \in X} \dim(M_\lambda) e^\lambda.$$

Clearly, ch is additive on exact sequences and we have $\mathrm{ch}(M \otimes_k M') = \mathrm{ch}(M)\,\mathrm{ch}(M')$ for any two T-modules M and M'.

1.3 Let M be a B-module. The G-module $\operatorname{Ind}_B^G M$ obtained by inducing M from B to G is defined by

$$\operatorname{Ind}_B^G M = \{f : G \to M \mid f(gb) = b^{-1}f(g), g \in G, b \in B\}.$$

In this definition it is tacitly agreed that the maps considered are maps of varieties (between the two affine varieties G and M).

The action of G on $\operatorname{Ind}_B^G M$ is given by

$$xf : g \mapsto f(x^{-1}g),\ f \in \operatorname{Ind}_B^G M, x, g \in G.$$

It is a wellknown fact that $\operatorname{Ind}_B^G M$ is a finite dimensional G-module. Moreover, the same is true for all the cohomology modules $H^i(M) = R^i \operatorname{Ind}_B^G M, i \geq 0$ (even though one needs to consider infinite dimensional (rational) B-modules in order to get an injective resolution of M).

The universal property of $\operatorname{Ind}_B^G M$ is the socalled Frobenius reciprocity law

$$\operatorname{Hom}_G(V, \operatorname{Ind}_B^G M) \simeq \operatorname{Hom}_B(V, M) \text{ for all } G\text{-modules } V.$$

In other word, Ind_B^G is left adjoint to the restriction functor from G-modules to B-modules. The isomorphism above is given by composing with the evaluation at $1 \in G$: $\operatorname{Ind}_B^G M \to M$.

1.4 Let $\lambda \in X$. Then λ is a 1-dimensional T-module and since T is a quotient of B (namely by the unipotent radical of B) we may regard λ also as a B-module. It has become fashionable to denote the G-module induced from λ by $\nabla(\lambda)$. Then we have

(1) $\nabla(\lambda) \neq 0$ if and only if $\lambda \in X^+$.

(2) If $\lambda \in X^+$ then $\nabla(\lambda)$ contains a unique simple G-module which we shall denote $L(\lambda)$. Each simple G-module is isomorphic to $L(\lambda)$ for some $\lambda \in X^+$.
Moreover, all other composition factors $L(\mu)$ of $\nabla(\lambda)$ have $\mu < \lambda$.

Note that (2) is the reason for the notation $\nabla(\lambda)$. The dual of $\nabla(\lambda)$ is then a module with a unique simple quotient. We shall use the symbol Δ to denote these modules. However, in order to ensure that λ becomes the highest weight we define

$$\Delta(\lambda) = \nabla(-w_0\lambda)^*. \tag{3}$$

Here w_0 is the element of maximal length in W.

Then we have

(4) $$\operatorname{ch}(\nabla(\lambda)) = \operatorname{ch}(\Delta(\lambda)) = J(e^{\lambda+\rho})/J(e^{\rho}),$$

where J is the operator $J = \Sigma_{w\in W}(-1)^{l(w)}w$. The equation (4) is the famous Weyl character formula (often $\Delta(\lambda)$ is called the Weyl module associated to λ).

1.5 An important result (which is in fact already required in 1.4 to prove (4)) is the Kempf vanishing theorem [21] (see also [2]). In the notation of of 1.3 it says

(5) If $\lambda \in X^+$ then $H^j(\lambda) = 0$ for $j > 0$.

Combining (1) with the Frobenius reciprocity law (see 1.3) we get

(6) $$\operatorname{Ext}^j_G(V, \nabla(\lambda)) \simeq \operatorname{Ext}^j_B(V, \lambda) \text{ for all } j \geq 0.$$

for any G-module V and any $\lambda \in X^+$.

The trivial B-module k (given by the 0-weight in X) has an injective B-resolution

$$0 \to k \to I_0 \to I_1 \to \cdots$$

in which all weights in the (infinite dimensional) B-modules I_j are ≤ 0, with strict inequality for $j > 0$. It follows from this together with (2)

(3) If $\operatorname{Ext}^j_G(V, \nabla(\lambda)) \neq 0$ for some G-module V, $\lambda \in X^+$ and $j \in \mathbb{N}$ then there exists $\mu \geq \lambda$ with $V_\mu \neq 0$. If $j > 0$ then $\mu > \lambda$.

1.6 We are now able to deduce the following result which will play a crucial role in our treatment of tilting modules.

THEOREM . *(Cline, Parshall, Scott and van der Kallen, [11])* *Let* $\lambda, \mu \in X^+$. *Then* $\operatorname{Ext}^j_G(\Delta(\lambda), \nabla(\mu)) = 0$ *for all* $j > 0$.

Proof: By (1.5)(3) the statement is clear if $\lambda \not\geq \mu$ (because 1.4(2) tells us that the weights of $\Delta(\lambda)$ are $\leq \lambda$). So assume $\lambda > \mu$. Then we observe that $\mathrm{Ext}^j_G(\Delta(\lambda), \nabla(\mu)) \simeq \mathrm{Ext}^j_G(\Delta(\mu), \nabla(\lambda))$ and apply 1.5(3) once more.

REMARK . *It is easy to include the case $j = 0$ in the considerations above: By 1.4(2) we see immediately that*

$$\mathrm{Hom}_G(\Delta(\lambda), \nabla(\mu)) = \delta_{\lambda,\mu} k \textit{ for all } \lambda, \mu \in X^+.$$

1.7 EXAMPLE . *Let V be a finite dimensional vector space over k. Then $G = GL(V)$ is a connected reductive algebraic group. If we choose a basis, say $\{v_1, \cdots, v_m\}$ of V then the set T of diagonal matrices in G forms a maximal torus in G. The character group $X = X(T)$ has basis $\{\epsilon_1, \cdots, \epsilon_m\}$ where $\epsilon_j \in X$ is defined as the character of T which takes a diagonal matrix into its $j-$th entry.*

The root system of the pair (G, T) is then $R = \{\epsilon_i - \epsilon_j \mid i \neq j\}$ and we choose $R^+ = \{\epsilon_i - \epsilon_j \mid i < j\}$. The Borel subgroup in G corresponding to the negative roots is then the subgroup consisting of the lower triangular matrices in G.

Note that in this case we have

$$X^+ = \{\Sigma_{i=1}^m a_i \epsilon_i \mid a_1 \geq \cdots \geq a_m\}$$

and

$$C_p = \{\Sigma_{i=1}^m a_i \epsilon_i \in X^+ \mid a_1 - a_m + m - 1 < p\}.$$

If $\lambda \in X^+$ is polynomial, i.e $\lambda = \Sigma_{i=1}^m a_i \epsilon_i$ for some integers $a_1 \geq \cdots \geq a_m \geq 0$, then we may view λ as a partition of $\mid \lambda \mid = \Sigma_{i=1}^m a_i$ into at most m parts.

Finally, observe that the G-module V has weights $\epsilon_1, \cdots, \epsilon_m$. In fact, $V_{\epsilon_j} = kv_j$ for all j. It is not hard to check that this implies that $V = \nabla(\epsilon_1)$. A little further work gives

(1) $$\Delta^j V = \nabla(\epsilon_1 + \cdots + \epsilon_j),\ j = 1, \cdots, m$$

and

(2) $$S^j V = \nabla(j\epsilon_1),\ j = 0, 1, \cdots .$$

1.8 DEFINITION . *Let V be a G-module. A good filtration of V is a sequence of G-submodules*

$$0 = F_0 \subset F_1 \subset \cdots \subset F_r = V$$

such that $F_j/F_{j-1} \simeq \nabla(\lambda_j)$ for some $\lambda_1, \cdots, \lambda_r \in X^+$.

Note that in the notation of this definition we get $\operatorname{ch} V = \Sigma_{j=1}^r \operatorname{ch} \nabla(\lambda_j)$. Since $\{\operatorname{ch} \nabla(\lambda) \mid \lambda \in X^+\}$ is a basis of $\mathbb{Z}[X]^W$ (see 1.4) we conclude that for each $\lambda \in X^+$ the number $\#\{j \mid \lambda_j = \lambda\}$ is uniquely determined. We shall denote this number $[V : \nabla(\lambda)]$. In fact, we may extend this notation to an arbitrary G-module M by letting $[M : \nabla(\lambda)]$ be the integer determined by the formula

$$\operatorname{ch} M = \Sigma_{\lambda \in X^+} [M : \nabla(\lambda)] \operatorname{ch} \nabla(\lambda).$$

If V has a good filtration then clearly the numbers $[V : \nabla(\lambda)]$ are non-negative for all $\lambda \in X^+$.

REMARK . *i) If in the above definition we replace ∇ by Δ then the corresponding filtrations are called Weyl filtrations. Clearly, V has a good filtration if and only if the dual module V^* has a Weyl filtration.*

ii) In characteristic zero we have $\nabla(\lambda) = L(\lambda) = \Delta(\lambda)$ for all $\lambda \in X^+$. Hence in this case a good filtration (and a Weyl filtration) is the same thing as a composition series. Thus in this case all G-modules have good filtrations (and are in fact direct sums of their composition factors as follows from Theorem 1.6).

iii) In characteristic p the equalities

$$\nabla(\lambda) = L(\lambda) = \Delta(\lambda)$$

hold only for very special λ. By the strong linkage principle [1] this is for instance the case if $\lambda \in \bar{C}_p$ or if $\lambda = (p-1)\rho$. As in ii) we conclude that modules whose dominant weights are all in $\bar{C}_p$ have good filtrations (and are in fact direct sums of their composition factors).

1.9 Observe that Theorem 1.6 shows that if the G-module V has a good filtration then $\operatorname{Ext}_G^j(\Delta(\lambda), V) = 0$ for all $\lambda \in X^+, j > 0$. We shall now see that this property characterizes modules with a good filtration

THEOREM . *(Donkin [13]). Let V be a G-module. Then the following condition on V are equivalent*

i) V has a good filtration.

ii) $\mathrm{Ext}_G^j(\Delta(\lambda), V) = 0$ *for all* $\lambda \in X^+, j > 0$.

iii) $\mathrm{Ext}_G^1(\Delta(\lambda), V) = 0$ *for all* $\lambda \in X^+$.

Proof: We already observed that i) implies ii). So the only step we have left is to see that iii) implies i).

To check this choose $\lambda \in X^+$ minimal such that $L(\lambda) \subset V$. We claim that this inclusion extends to $\nabla(\lambda)$. The obstruction for this is $\mathrm{Ext}_G^1(\nabla(\lambda)/L(\lambda), V$. We claim that in fact $\mathrm{Ext}_G^1(L(\mu), V) = 0$ for all composition factors $L(\mu)$ of $\nabla(\lambda)/L(\lambda)$. Note that for such $L(\mu)$ we have $\mu < \lambda$, see 1.4 (2). Consider now the short exact sequence

$$0 \to \mathrm{Rad}(\Delta(\mu)) \to \Delta(\mu) \to L(\mu) \to 0$$

where $\mathrm{Rad}(\Delta(\mu))$ denotes the radical of $\Delta(\mu)$. By assumption $\mathrm{Ext}_G^1(\Delta(\mu), V) = 0$ so we are done if we prove $\mathrm{Hom}_G(\mathrm{Rad}(\Delta(\mu)), V) = 0$. However, since all composition factors $L(\nu)$ of $\mathrm{Rad}(\Delta(\mu))$ have $\nu < \mu < \lambda$ we see that this follows by the minimality of λ.

So we have $\nabla(\lambda) \hookrightarrow V$. The quotient then also satisfies iii) (use Theorem 1.6) and i) follows via an easy induction on $\dim V$.

1.10 We are no ready to define the concept of tilting in our setup.

DEFINITION . *A G-module Q is called a tilting module if Q as well as Q^* have good filtrations.*

Clearly, an alternative way of stating this definition is to require Q to have both a good filtration

$$0 = F_0 \subset F_1 \subset \cdots \subset F_r = Q$$

and a Weyl filtration

$$0 = E_0 \subset E_1 \subset \cdots \subset E_r = Q.$$

Note that if $F_i/F_{i-1} \simeq \nabla(\lambda_i)$, $i = 1, \cdots, r$ and $E_j/E_{j-1} \simeq \Delta(\mu_j)$, $j = 1, \cdots, r$ then the sets $\{\lambda_1, \cdots, \lambda_r\}$ and

$\{\mu_1, \cdots, \mu_r\}$ coincide (with multiplicities). This is so because $\operatorname{ch}\nabla(\lambda) = \operatorname{ch}\Delta(\lambda)$ for all $\lambda \in X^+$ so that for each given λ we have $\#\{i \mid \lambda_i = \lambda\} = \#\{j \mid \mu_j = \lambda\}$. In particular, the lengths of the two filtrations are the same (as indicated in the notation).

By 1.5 (3) we can always arrange the filtrations above such that if λ is a maximal weight in Q then $E_1 = \Delta(\lambda)$, respectively $Q/F_{r-1} = \nabla(\lambda)$.

1.11 Theorem 1.9 immediately leads to the following characterization of tilting modules

COROLLARY . *Let Q be a G-module. Then the following conditions on Q are equivalent*

i) Q is a tilting module.

ii) $\operatorname{Ext}^j_G(\Delta(\lambda), Q) = \operatorname{Ext}^j_G(Q, \nabla(\lambda)) = 0$ *for all $\lambda \in X^+$ and all $j > 0$.*

iii) $\operatorname{Ext}^1_G(\Delta(\lambda), Q) = \operatorname{Ext}^1_G(Q, \nabla(\lambda)) = 0$ *for all $\lambda \in X^+$.*

1.12 All G-modules decompose into direct sums of indecomposable modules. The following result classifies all indecomposable tilting modules.

THEOREM . *(Ringel [26], Donkin [15])*

i) Let $\lambda \in X^+$. There exists an indecomposable tilting module $T(\lambda)$ with unique highest weight λ and $\dim T(\lambda)_\lambda = 1$.

ii) If Q is an indecomposable tilting module for G then there exists a unique $\lambda \in X^+$ such that $Q \simeq T(\lambda)$.

Proof: i) Consider first $\Delta(\lambda_0)$ where $\lambda_0 = \lambda$. Choose then $\lambda_1 \in X^+$ minimal such that $d_1 = \dim \operatorname{Ext}^1_G(\Delta(\lambda_1), \Delta(\lambda_0)) > 0$ and let $0 \to \Delta(\lambda_0) \to E_1 \to \Delta(\lambda_1)^{d_1} \to 0$ be the corresponding extension. According to 1.5 (3) we have $\lambda_1 < \lambda_0$. Next we choose λ_2 minimal such that $d_2 = \dim \operatorname{Ext}^1_G(\Delta(\lambda_2), E_1) > 0$ and we let $0 \to E_1 \to E_2 \to \Delta(\lambda_2)^{d_2} \to 0$ be the corresponding extension. Continuing in this way we obtain a filtration

$$0 = E_0 \subset E_1 \subset \cdots \subset E_r$$

with $E_i/E_{i-1} = \Delta(\lambda_i)^{d_i}$ for some $\lambda_i < \lambda$ and $d_i \in \mathbb{N}$. Since the set $\{\mu \in X^+ \mid \mu < \lambda\}$ is finite the process must stop at some point r, say. Then the G-module E_r has a Weyl filtration by construction and $\mathrm{Ext}^1_G(\Delta(\mu), E_r) = 0$ for all $\mu \in X^+$. The last property ensures by Theorem 1.9 that E_r has a good filtration. Hence E_r is tilting. By construction it has λ as its unique highest weight and its λ-weight space is 1-dimensional. So we may define $T(\lambda)$ to be the unique indecomposable summand of E_r which contains the λ-weight space (in fact, it is not hard to check that all E_j are indecomposable so that $T(\lambda) = E_r$, see [5], but we do not need this here).

ii) Let Q be a tilting module for G. Pick a maximal weight λ in Q. Then - as observed in 1.10 - we may arrange the Weyl filtration of Q to start with the term $\Delta(\lambda)$. Similarly, $\Delta(\lambda) \subset T(\lambda)$. We claim that there exists a G-homomorphism $\phi : T(\lambda) \to Q$ making the diagram

$$\begin{array}{ccc} \Delta(\lambda) & \hookrightarrow & T(\lambda) \\ \downarrow & \swarrow_{\phi} & \\ Q & & \end{array}$$

commutative. This follows because by Theorem 1.6 we have $\mathrm{Ext}^1_G(T(\lambda)/\Delta(\lambda), Q) = 0$. Similarly, we have a G-homomorphism $\phi' : Q \to T(\lambda)$ which also commutes with the inclusions. The endomorphism $\phi' \circ \phi \in \mathrm{End}_G(T(\lambda))$ is then non-zero on the 1-dimensional λ-weight space $T(\lambda)_\lambda$. Therefore, it cannot be nilpotent and since $T(\lambda)$ is indecomposable we conclude that it must be an isomorphism. This means that $T(\lambda)$ is a summand of Q and ii) follows.

2. The quantum case

In this section we shall indicate how the theory in Section 1 can we developed quite analogously for quantum groups at roots of unity. In fact, we shall see that for a complex root of unity the quantized tilting modules are much better understood than the modular ones studied above.

2.1 Associated to our root system R from Section 1 we have a semi-simple Lie algebra $\mathfrak{g}$. Letting v denote an indeterminate we set U_v equal to the quantized enveloping algebra over $\mathbb{Q}(v)$ associated with $\mathfrak{g}$. As a $\mathbb{Q}(v)$-algebra U_v has $4n$ generators

$$E_i, F_i, K_i, K_i^{-1}, i = 1, \cdots, n$$

where n is the rank of R. These generators satisfy a list of relations for which we refer to [8].

Moreover, U_v is a Hopf algebra with comultiplication Δ, counit ϵ and antipode S. These homomorphisms may be given by their values on the generators, see loc. cit.

2.2 From now on we shall for convenience assume that all roots of R have the same length. This will simplify the notation considerably (see [8] for the general case).

Set $A = \mathbb{Z}[v, v^{-1}]$. For $m \in \mathbb{Z}$ we set $[m] = \frac{v^m - v^{-m}}{v - v^{-1}} \in A$ and if $m \in \mathbb{N}$ then we let $[m]! = [m][m-1]\cdots[1]$.. If $d \in \mathbb{N}$ we denote by $\left[\begin{smallmatrix} m \\ d \end{smallmatrix}\right]$ the Guassian binomial coefficient $\frac{[m][m-1]\cdots[m-d+1]}{[d]!}$. It is not hard to see that $\left[\begin{smallmatrix} m \\ d \end{smallmatrix}\right] \in A$ for all $m \in \mathbb{Z}, d \in \mathbb{N}$ and that $\left[\begin{smallmatrix} m \\ d \end{smallmatrix}\right] = 0$ for $0 \leq m < d$.

The quantum group U_A over A is defined to be the A-subalgebra of U_v generated by $E_i^{(r)}, F_i^{(r)}, K_i, K_i^{-1},\ i = 1, \cdots, n;\ r \in \mathbb{N}$. Here $X^{(r)} = X^r/[r]!$. It can be checked that the Hopf algebra structure on U_v induces a Hopf algebra structure on U_A.

If Γ is a commutative A-algebra then we set $U_\Gamma = U_A \otimes_A \Gamma$ and call this the quantum group over Γ. In the following we shall focus on the case $\Gamma = \mathbb{C}$ made into an A-algebra by specializing v to $q \in \mathbb{C} \setminus \{0\}$. We write U_q for this complex quantum group.

If $q = 1 \in \mathbb{C}$ then U_1 is very closely related to the ordinary enveloping algebra $U(\mathfrak{g})$ of $\mathfrak{g}$. In fact,

$$U_1/(\{K_i - 1 \mid i = 1, \cdots, n\}) \simeq U(\mathfrak{g}).$$

Similarly, if we make the field k from Section 1 into an A-algebra by specializing v to $1 \in k$ then the corresponding quantum group U_k is related to the hyperalgebra $U_k(\mathfrak{g})$ of $\mathfrak{g}$ over k by

$$U_k/(\{K_i - 1 \mid i = 1, \cdots, n\}) \simeq U_k(\mathfrak{g}).$$

2.3 The quantum group U_v has a triangular decomposition $U_v = U_v^- U_v^0 U_v^+$ where U_v^- (resp. U_v^0, resp. U_v^+) is the subalgebra of U_v generated by F_i (resp. K_i, K_i^{-1}, resp. E_i), $i = 1, \cdots, n$. There is a similar result about U_A and U_q. Here we define U_A^- (resp. U_q^-) as the subalgebra of U_A (resp. U_q) generated by $\{F_i^{(r)} \mid i = 1, \cdots, n; r \in \mathbb{N}\}$ (resp. the image of the same elements (which we denote by the same symbols) in U_q). Similarly for U_A^+ and U_q^+. To define U_A^0 we need the following elements of U_A

$$\left[{K_i \atop t} \right] = \prod_{j=1}^{t} \frac{K_i v^{(1-j)} - K_i^{-1} v^{(j-1)}}{v^j - v^{-j}}$$

Then U_A^0 is the A-subalgebra of U_A generated by

$$\{K_i, K_i^{-1}, \left[{K_i \atop t} \right] \mid i = 1, \cdots, n;\ t \in \mathbb{N}\}.$$

Again U_q^0 is the $\mathbb{C}$-subalgebra of U_q generated by the images of these elements (once more we use the same symbols to denote these images).

We then have $U_A = U_A^- U_A^0 U_A^-$ and $U_q = U_q^- U_q^0 U_q^-$.

2.4 As in Section 1 we let X denote the set of weights which we identify with $\mathbb{Z}^n$. If $\lambda = (\lambda_1, \cdots, \lambda_n) \in X$ then $\lambda_i = \langle \lambda, \alpha_i^\vee \rangle$ where $\{\alpha_1, \cdots, \alpha_n\}$ is an enummeration of the set of simple roots in R^+. The set of dominant weights X^+ consists of all $\lambda \in X$ for which $\lambda_i \geq 0$ for all $i = 1, \cdots, n$.

Each $\lambda \in X$ defines a character of U_q^0 which we also denote by λ. It is given by

$$\lambda(K_i) = q^{\lambda_i},\ \lambda(\left[{K_i \atop t} \right]) = \left[{\lambda_i \atop t} \right]_{v=q},\ i = 1, \cdots, n; t \in \mathbb{N}.$$

This character extends to $U_q^- U_q^0$ by $\lambda(F_i^{(r)}) = 0, i = 1, \cdots, n, r \in \mathbb{N}$.

If M is a finite dimensional U_q-module (of type **1**) one can show [8] that as a U_q^0-module M splits into weight spaces

$$M = \bigoplus_{\lambda \in X} M_\lambda,$$

where the λ-weight space $M_\lambda = \{m \in M \mid um = \lambda(u)m,\ u \in U_q^0\}$. It therefore makes sense to define the character of M by

$$\operatorname{ch} M = \Sigma_{\lambda \in X} \dim(M_\lambda) e^\lambda \in \mathbb{Z}[X].$$

2.5 Let M be an arbitrary (i.e not necessarily finite dimensional) U_q-module. Then we set

$$\mathcal{F}(M) = \{m \in \bigoplus_{\lambda \in X} M_\lambda \mid E_i^{(r)} m = F_i^{(r)} m = 0, i = 1, \cdots, n,\ r \gg 0\}.$$

It is not hard to check that $\mathcal{F}(M)$ is a U_q-submodule of M. We use this to define induction from $U_q^- U_q^0$-modules to U_q-modules:

Let M be a $U_q^- U_q^0$-module. Then we set

$$\mathrm{Ind}_{U_q^- U_q^0}^{U_q}(M) = \mathcal{F}(\mathrm{Hom}_{U_q^- U_q^0}(U_q, M)).$$

Here U_q is a $U_q^- U_q^0$-module via left multiplication, whereas $\mathrm{Hom}_{U_q^- U_q^0}(U_q, M)$ is made into a (left) U_q-module via right multiplication on U_q (i.e $uf : x \mapsto f(xu),\ u, x \in U_q, f \in \mathrm{Hom}_{U_q^- U_q^0}(U_q, M)$.)

In the special case where M is the 1-dimensional $U_q^- U_q^0$-module given by $\lambda \in X$ (see 2.4) we write

$$\nabla_q(\lambda) = \mathrm{Ind}_{U_q^- U_q^0}^{U_q}(\lambda).$$

2.6 It is proved in [8] that $\nabla_q(\lambda)$ behaves in very much the same way as $\nabla(\lambda)$. In particular, we have

(1) $\nabla_q(\lambda) \neq 0$ if and only if $\lambda \in X^+$.
(2) If $\lambda \in X^+$ then $\nabla_q(\lambda)$ contains a unique simple submodule $L_q(\lambda)$. Any simple U_q-module (of type $\mathbf{1}$) is isomorphic to $L_q(\lambda)$ for some $\lambda \in X^+$. Moreover, all composition factors $L_q(\mu)$ of $\nabla_q(\lambda)/L_q(\lambda)$ have $\mu < \lambda$.

REMARK . *If $q \in \mathbb{C} \setminus \{0\}$ is not a root of unity we have $L_q(\lambda) = \nabla_q(\lambda)$ for all $\lambda \in X^+$ (and all U_q-modules are semisimple). For this reason we shall from now on assume that q is a primitive l-th root of 1. To avoid technical difficulties we shall also assume that l is odd and if R has a component of type G_2 that l is not divisible by 3.*

2.7 In analogy with the modular case we set $\Delta_q(\lambda)=\nabla_q(-w_0\lambda)^*$. Here the dual of a module is defined via the antipode S on U_q.

We can now define good filtrations and Weyl filtrations of a module just as in Section 1. Likewise the definition of a tilting module for U_q is completely analogous to the corresponding one for G.

Also Theorems 1.9 and 1.12 carry over verbatim. In particular, the indecomposable tilting modules for U_q are $T_q(\lambda)$, $\lambda \in X^+$.

2.8 As illustrated above the category $\mathcal{C}_q$ of finite dimensional U_q-modules behaves very similar to the category $\mathcal{C}$ of finite dimensional
G-modules considered in Section 1. However, we shall now point out an essential difference: Whereas $\mathcal{C}$ does not contain a single injective module we have

PROPOSITION . *The category $\mathcal{C}_q$ has enough injectives.*

Proof: (sketch, see [3] for details) We set $St_q = L_q((l-1)\rho)$ (the quantum Steinberg module). Then we claim

(1) $$St_q \text{ is injective in } \mathcal{C}_q.$$

The linkage principle for U_q (see [8]) ensures that $\mathrm{Ext}^1_{\mathcal{C}_q}(L_q(\lambda), St_q) = 0$ unless $\lambda = (l-1)\rho + l\lambda^1$ for some $\lambda^1 \in X^+$. However, for such a weight λ we have $L_q(\lambda) \simeq St_q \otimes_{\mathbb{C}} L_{\mathbb{C}}(\lambda^1)^{[q]}$ by the tensor product theorem for simple U_q-modules (see [9]). Here $L_{\mathbb{C}}(\lambda^1)$ denotes the simple $\mathfrak{g}$-module with highest weight λ^1. This module is made into a U_q-module $L_q(\lambda^1)^{[q]}$ via the quantum Frobenius homomorphism $U_q \to U_{\mathbb{C}}(\mathfrak{g})$, see loc. cit. Now

$$\mathrm{Ext}^1_{\mathcal{C}_q}(St_q \otimes_{\mathbb{C}} L_q(\lambda^1)^{[q]}, St_q) \simeq \mathrm{Ext}^1_{\mathfrak{g}}(L_{\mathbb{C}}(\lambda^1), \mathbb{C}) = 0$$

for all $\lambda^1 \in X^+$ and (1) follows.

Observe that St_q is selfdual and that there is a corresponding non-zero U_q-homomorphism $\mathbb{C} \to St_q \otimes_{\mathbb{C}} St_q$. Hence if $M_q \in \mathcal{C}_q$ we get an inclusion $M \subset St_q \otimes_{\mathbb{C}} St_q \otimes_{\mathbb{C}} M$. The last module is clearly injective because of (1) and the proposition follows.

2.9 It follows from the quantum analogue of Theorem 1.9 that injective modules in $\mathcal{C}_q$ are tilting. More precisely, we find

PROPOSITION . *i)* $T_q(\lambda)$ *is injective in* $\mathcal{C}_q$ *if and only if and only if* $\lambda \in (l-1)\rho + X^+$.

ii) If $\lambda = \lambda_0 + l\lambda_1$ *with* $l-1 \leq \lambda_{0j} < 2l-1$ *for all* $j = 1, \cdots, n$ *then* $T_q(\lambda)$ *is the injective envelope of* $L_q(2(l-1)\rho + w_0\lambda_0 + l\lambda_1)$.

Proof: (sketch, see [3] for details) If $\lambda = (l-1)\rho + \nu$ for some $\nu \in X^+$ then $T_q(\lambda)$ is clearly a direct summand of $St_q \otimes_{\mathbb{C}} L_q(\nu)$. Since $T_q(\lambda)$ is indecomposable it is therefore by 2.8(1) the injective envelope of some simple module. The rest is now a matter of easy weight considerations.

2.10 Consider $\lambda \in (l-1)\rho + X^+$. Using the q-analogue of Theorem 1.6 and Remark 1.6 we get

$$[T_q(\lambda) : \nabla_q(\mu)] = \dim_{\mathbb{C}} \operatorname{Hom}_{\mathcal{C}_q}(\Delta_q(\mu), T_q(\lambda)). \tag{1}$$

(This is in fact true for all $\lambda, \mu \in X^+$).

Writing now $\tilde{\lambda} = 2(l-1)\rho + w_0\lambda_0 + l\lambda_1$ (see 2.9) and using the fact that $T_q(\lambda)$ is the injective hull of $L_q(\tilde{\lambda})$ (see Theorem 2.9 ii)) we get on the other hand

$$\dim_{\mathbb{C}} \operatorname{Hom}_{\mathcal{C}_q}(\Delta_q(\mu), T_q(\lambda)) = [T_q(\mu) : L_q(\tilde{\lambda})].$$

It follows that the set of characters $\{\operatorname{ch} T_q(\lambda) \mid \lambda \in (l-1)\rho + X^+\}$ is determined by the irreducible characters $\{\operatorname{ch} L_q(\nu) \mid \nu \in X^+\}$ and vice versa. The latters have been determined (under some very mild restrictions on l) by the Kazhdan-Lusztig theory, see [20] and [18], [19].

Even better: All the indecomposable tilting characters have been determined by W. Soergel, [28] (with the same mild restrictions on l).

2.11 We close this section by stating the expected relation between modular tilting characters and their quantized analogues.

CONJECTURE . *Suppose* $l = p$. *If* $\lambda \in X^+$ *belongs to the lowest* p^2*-alcove* C_{p^2} *then* $\operatorname{ch} T_q(\lambda) = \operatorname{ch} T(\lambda)$.

REMARK . *Let us emphasize that a proof of this conjecture would only give a small beginning towards a solution of the problem of finding all indecomposable tilting characters in the modular case. There is no known algorithm for determining* $\operatorname{ch} T(\lambda)$ *for large* λ *from the knowledge of smaller ones (except for the special weights covered by Proposition 2.1 in [15]).*

3. Tensor products.

3.1 An important result in both of the two applications we mentioned in the introduction (see also Section 4) as well as in many other situations is the following

THEOREM . *If* M *and* N *are tilting modules then so is* $M \otimes N$.

This result holds both in $\mathcal{C}$ and in $\mathcal{C}_q$. Below we indicate a unified proof for both cases.

3.2 To prove Theorem 3.1 it is clearly enough to prove the corresponding statement for modules with good filtrations. Moreover, to handle this statement we can immediately reduce to the situation where the filtration of both modules have only one term. For this case we have

THEOREM . *Let* $\lambda, \mu \in X^+$. *The* U_A*-module* $\nabla_A(\lambda) \otimes_A \nabla_A(\mu)$ *has a good filtration.*

Here we have used the notation from Section 2 and ∇_A is defined just as ∇_q. The tensor product is made into a U_A-module via the comultiplication $\Delta : U_A \to U_A \otimes_A U_A$.

It is a (non-trivial) fact that $\nabla_A(\lambda) \otimes_A \mathbb{C} \simeq \nabla_q(\lambda)$ and $\nabla_A(\lambda) \otimes_A k \simeq \nabla(\lambda)$ for all $\lambda \in X^+$, see [8], [9]. Here the last isomorphism relies on the connection between U_k and the hyperalgebra $U_k(\mathfrak{g})$ mentioned in Section 2.2. Once one has this fact it is clear that the theorem gives the desired results both in $\mathcal{C}_q$ and in $\mathcal{C}$.

For a proof of the theorem we refer to [17] where the key is to explore the wonderful properties of the canonical basis for U_A^-, see in particular Section 27 in [22].

3.3 In the modular case the result in this section was proved by Wang [30], Donkin [14] and Mathieu [23] (in increasing generality).

4. Applications.

In this section we shall indicate two applications of the theory of tilting modules. The first concerns the modular case and the second the quantum case.

4.1 In the notation from Section 1 we consider the reductive algebraic group $G = GL(V)$ where V is a finite dimensional vector space over k. As observed in Example 1.7 the vector representation V of G is a module with good filtration. In fact, $V \simeq \nabla(\epsilon_1) \simeq \Delta(\epsilon_1)$, i.e V is also a tilting module for G. Therefore by Theorem 3.1 so is $V^{\otimes r}$ for all $r \in \mathbb{N}$.

By Theorem 1.12 we can then write

$$V^{\otimes r} \simeq \bigoplus_{\lambda \in X^+} T(\lambda)^{d_r(\lambda)} \tag{1}$$

for some $d_r(\lambda) \in \mathbb{N}$.

Easy weight considerations show that $d_r(\lambda) = 0$ unless $\lambda = \Sigma_{i=1}^m c_i\epsilon_i$ ($m = \dim V$) for some $c_1 \geq c_2 \geq \cdots \geq c_m \geq 0$, $\Sigma_{i=1}^m c_i = r$. In other words, the only summands that occur in (1) involve partitions of r into at most m parts.

4.2 Fix $r \in \mathbb{N}$ and let λ be a partition of r into at most m parts, say $\lambda = (c_1 \geq c_2 \geq \cdots \geq c_m \geq 0)$. Then the Specht module $S^\lambda(r)$ for $k\Sigma_r$ has a unique simple quotient $D^\lambda(r)$ provided λ is p-regular (i.e has no p rows of equal length). If λ is not p-regular we set $D^\lambda(r) = 0$. Then

THEOREM . *(Mathieu, [24]) If λ is a partition of r into at most m parts then $d_\lambda(r) = \dim_k D^\lambda(r)$.*

The proof of this theorem is obtained by noticing that Σ_r acts on $V^{\otimes r}$ (by permuting the factors). Since this action commutes with the G-action we get a homomorphism

$$k\Sigma_r \to \mathrm{End}_G(V^{\otimes r}). \tag{1}$$

If $r \leq m$ ($= \dim V$) then this homomorphism is an isomorphism. This can be seen by rather direct computations, see e.g [10] (note that by Theorem 1.6 and Remark 1.6 the dimension of the right hand side is independent of the characteristic of k). It is somewhat harder to see that (1) is always a surjection (see loc. cit). Once this is established, however, the theorem follows easily.

4.3 So far Theorem 4.2 is only an identity between two unknown quantities. However, the following result (valid for all reductive groups) determines the left hand side in the theorem for certain special λ.

THEOREM . *([16], [7] Let $\lambda \in C_p$ and suppose Q is a tilting module for G. Then*

$$[Q : T(\lambda)] = \Sigma_{w \in W_p} (-1)^{l(w)} [Q : \Delta(w \cdot \lambda)].$$

Here the left hand side is the multiplicity of $T(\lambda)$ as a direct summand of Q. On the right hand side only those $w \in W_p$ for which $w \cdot \lambda \in X^+$ are relevant. The exponent $l(w)$ is the length of w.

Note that the right hand side is completely determined by $\operatorname{ch} Q$. In the case where $G = GL(V)$ and $Q = V^{\otimes r}$ we do in fact know the character of Q. Hence this theorem combined with Theorem 4.2 determine $\dim D^\lambda(r)$ for all partitions λ of r corresponding to weights in C_p. For details see [24].

4.4 We now turn to our second application which concerns quantum groups at roots of unity as means of producing invariants of 3-manifolds. Reshetikhin and Turaev showed in [27] how one may obtain invariants of a given 3-manifold from socalled modular Hopf algebras. Without repeating here their definition of such algebras we shall just point out that the following result is one of the key properties which makes U_q together with the set $\mathcal{F} = \{L_q(\lambda) \mid \lambda \in C_l\}$ (using the notation from Section 2) into a modular Hopf algebra.

THEOREM . *([3]) Let $Q, Q' \in \mathcal{C}_q$ be tilting modules. Suppose $[Q : T_q(\lambda)] = 0$ for all $\lambda \in C_l$. Then we have also $[Q \otimes_{\mathbb{C}} Q' : T_q(\lambda)] = 0$ for all $\lambda \in C_l$.*

4.5 Recall that $L_q(\lambda) = T_q(\lambda)$ for $\lambda \in C_l$ (this is the quantum analogue of Remark 1.8 iii)). By 2.7 we may write an arbitrary tilting module in $\mathcal{C}_q$ as

$$(\bigoplus_{\lambda \in C_l} L_q(\lambda)) \oplus Q$$

where Q is a tilting module which satisfies the assumption in Theorem 4.4. This allows us, in particular, to define a tensor product $\underline{\otimes}$ on $add(\mathcal{F})$ (the category of U_q-modules consisting of finite direct sums of modules from $\mathcal{F}$) by the following recipe

$$M_1 \otimes_{\mathbb{C}} M_2 = M_1 \underline{\otimes} M_2 \ \oplus Q, \quad M_1, M_2 \in add(\mathcal{F})$$

where Q consists of all summands of the tilting (use Theorem 3.1) module $M_1 \otimes_{\mathbb{C}} M_2$ with $\lambda \in X^+ \setminus C_l$.

Theorem 4.4 shows that $\underline{\otimes}$ is associative.

4.6 For a proof of Theorem 4.4 and more details on 4.5 we refer to [3] and [4]. In [29] one can find the definition of modular Hopf algebras as well as the full details on how to construct 3-manifold invariants from such data.

References

[1] H. H. Andersen, *The strong linkage principle,* J. Reine Ang. Math. **315** (1980), 53–59.

[2] H. H. Andersen, *The Frobenius morphism on the cohomology of homogeneous vector bundles on G/B*, Ann. of Math. **112**, 113–121.

[3] H. H. Andersen, *Tensor products of quantized tilting modules*, Comm. Math. Phys. **149** (1992), 149-159.

[4] H. H. Andersen, *Quantum groups, invariants of* 3*−manifolds and semisimple tensor categories*, Israel Math. Conf. Proc. **7** (1993), 1–12.

[5] H. H. Andersen, *Tilting modules for algebraic groups*, pp. 25–42 in:R. Carter and J. Saxl (ed.) *Algebraic groups and their representations*, Nato ASI Series C Vol. **517**, Kluwer 1998.

[6] H. H. Andersen, J. C. Jantzen, and W. Soergel, *Representations of quantum groups at a p-th root of unity and of semisimple g roups in characteristic p: Independence of p*, Astérisque **220** (1994), 1–320.

[7] H.H. Andersen, J. Paradowski, *Fusion categories arising from semisimple Lie algebras*, Comm. Math. Phys. 169 (1995) 563-588.

[8] Henning Haahr Andersen, Patrick Polo, and Wen Kexin, *Representations of quantum algebras*, Invent. Math. **104** (1991), 1–59.

[9] Henning Haahr Andersen and Wen Kexin, *Representations of quantum algebras. The mixed case,* J. Reine Angew. Math. **427** (1992), 35–50.

[10] R. Carter and G. Lusztig, *On the modular representations of the general linear groups and symmetric groups*, Math. Z. **136** (1974), 193–242.

[11] E. Cline, B. Parshall, L. Scott and W. van der Kallen, *Rational and generic cohomology* Invent. math. **39** (1977), 143–163.

[12] D.H. Collingwood & R.Irving, *A decomposition theorem for certain self-dual modules in the category $\mathcal{O}$*, Duke Math. J. **58**, 89-102, 1989.

[13] S. Donkin, *A filtration for rational modules*, Math. Z. **177** (1981), 1–8.

[14] S. Donkin, *Rational representations of algebraic groups* (Lecture Notes in Mathematics **1140**, Berlin/Heidelberg/New York/Tokyo (1985), Springer.

[15] S. Donkin,*On tilting modules for algebraic groups*, Math. Z. **212** (1993), 39–60.

[16] G. Georgiev, O. Mathieu: Catégorie de fusion pour les groupes de Chevalley. C. R. Acad. Sci. Paris 315 (1992) 659-662.

[17] M. Kaneda, *Based modules and good filtrations in algebraic groups* Hiroshima Math. J. **28** (1998), no. 2, 337–344.

[18] M. Kashiwara and S. Tanisaki, *Characters of the negative level highest weight modules for affine Lie algebras*, Duke Math.J. **77** (1995), 21–62.

[19] M. Kashiwara and T. Tanisaki *Kazhdan-Lusztig conjecture for affine Lie algebras with negative level, II: Non-integral case*, Duke Math.J. **84** (1996), 771–813.

[20] D. Kazhdan and G. Lusztig, *Tensor structures arising from affine Lie algebras*, J. Amer. Math. Soc. **6** (1993), 905–947, 949–1011 and **7** (1994), 335–381, 383–453.

[21] G. Kempf, *Linear systems on homogeneous spaces*, Ann. of Math (2), **103** (1976), 557–591.

[22] G. Lusztig, *Introduction to quantum groups*, Progress in Math., Birkhäuser 1993

[23] O. Mathieu, *Filtration of G-modules for algebraic groups* Ann. Scient. Norm. Sup. II **23** (1990), 6265–644.

[24] O. Mathieu, *On the dimension of some modular irreducible representations of the symmetric group* Lett. Math. Phys. **38** (1996), no. 1, 23–32.

[25] Jan Paradowski, *Filtrations of modules over the quantum algebra,* Proc. Symp. Pure Math. **56** (1994), Part 2, 93-108.

[26] Claus M. Ringel, *The category of modules with good filtrations over a quasi-hereditary algebra has almost split sequences*, Math. Z. **208** (1991), 209-223.

[27] N. Reshetikhin and V. Turaev, *Invariants of* 3*-manifolds via link polynomials and quantum groups*, Invent. Math. **103**, 547–597

[28] Wolfgang Soergel, *Charakterformeln für Kipp-Moduln über Kac-Moody-Algebren* , Represent. Theory **1** (1997), pp. 115-132.

[29] V. Turaev, *Quantum invariants of knots and* 3*-manifolds*, W. de Gruyter, Berlin 1994.

[30] Wang Jian-pan, *Sheaf cohomology and tensor products of Weyl modules* , J. Alg. **77** (1982), 162-185.

CONNECTIONS BETWEEN GROUP REPRESENTATIONS AND COHOMOLOGY

JON F.CARLSON[*] (jfc@sloth.math.uga.edu)
Department of Mathematics,
University of Georgia,
Athens, GA 30602

Abstract. The purpose of these notes is to outline the connection between the homological algebra of group rings, especially group cohomology, and the modular representation theory of finite groups. The aim of the subject is the development of a module theory for finite groups. It is not surprising that the theory of cohomology of groups would play a significant role. After all things like the extensions of modules are measured by cohomological functors. But there are also deeper connections through the notion of support varieties between the structure of finitely generated modules and the spectrum of cohomology rings. Some of these ideas have been extended to infinitely generated modules.

Key words: modules over group algebras, cohomology of finite groups, support varieties, infinitely generated modules

20C20,20J06:

This paper represents the notes of three lectures that were presented at the meeting on Representation Theory in Constanta, Romania, in August of 2000. The major part of the subject is the support varieties for modules as a subset of the maximal ideal spectrum of the cohomology ring of the group. The support varieties are measured at the level of the elementary abelian subgroups and they are directly connected to module structure. These results apply to both finitely generated and infinitely generated modules but in strangely different ways. The results have been extended to rings other than group algebras. See the survey paper (Nakano,01) for

* Partially supported by a grant from NSF

K.W. Roggenkamp and M. Ştefănescu (eds.), Algebra - Representation Theory, 23–46.

more information on such extensions.

The notes are divided into three parts. In the first section we discuss the rank variety of a module over an elementary abelian p-group and show how the rank variety is related to the cohomological support variety in this case. The rank variety is defined purely through the structure of the module, but it measures the homological properties. In section two we will demonstrate a method of measuring properties of modules over an arbitrary finite group by looking at the restrictions to elementary abelian p-subgroups. One of the most striking consequences is the theorem of Chouinard that says that a module is projective if and only if its restriction to every elementary abelian p-subgroup is projective. There are several other connections here. In the third section we show how the notions of the varieties have been extended to infinitely generated modules. Actually the "variety" of an infinitely generated module is not a variety at all, but just a subset of the maximal ideal spectrum of the cohomology ring. However it still has many of the properties of the varieties defined for finitely generated modules.

Throughout the discussion, G denotes a finite group and k a field of characteristic $p > 0$. We assume that k is algebraically closed, even though that will not always be necessary. We let $\mathbf{mod}(kG)$ denote the category of finitely generated kG-modules and kG-module homomorphisms. The category of all kG-modules will be denoted by $\mathbf{Mod}(kG)$.

I owe a great thanks to Marco Soriano for taking notes for the lectures and providing a first draft of the paper. I want also to thank the Humboldt foundation for their support during my visit to Stuttgart when the plans for the lectures were made. Finally I want to thank Klaus Roggenkamp, Mirela Stefanescu and the other organizers for their help and hospitality during the workshop and conference.

1. Varieties for finitely generated modules

In this section we define the rank variety of a module over an elementary abelian p-group. In order to motivate the definition, we start with a couple of examples. Throughout, we denote by I_n the $n \times n$ identity matrix. For $\alpha \in k$ we set $J_n(\alpha)$ to be the $n \times n$

Jordan block with eigenvalue α, i.e.

$$J_n(\alpha) := \begin{pmatrix} \alpha & 1 & & & \\ & \alpha & 1 & & 0 \\ & & \ddots & \ddots & \\ & 0 & & \ddots & 1 \\ & & & & \alpha \end{pmatrix}_{n\times n} .$$

Finally, let C_n denote the cyclic group with n elements.

EXAMPLE 1.1. *Let $p = 2$ and suppose that $G = \langle x, y\rangle \cong C_2 \times C_2$. There is a complete classification of finitely generated indecomposable kG-modules, which we quote here (see (Heller-Reiner,61)). In most cases we present the modules by giving matrices for the action of the group generators x and y on the vector space of the module. Let n be any natural number. The list of pairwise non-isomorphic, indecomposable kG-modules consists of the following.*

1. *The indecomposable projective module kG. Note that kG, like the modular group algebra of any p-group is a local ring and hence projective modules are necessarily free.*
2. $$x \mapsto \left(\begin{array}{c|c} I_n & 0 \\ \hline 0\ldots 0 & \\ I_n & I_{n+1} \end{array}\right), \quad y \mapsto \left(\begin{array}{c|c} I_n & 0 \\ \hline I_n & I_{n+1} \\ 0\ldots 0 & \end{array}\right).$$
3. $$x \mapsto \left(\begin{array}{cc|c} \multicolumn{2}{c|}{I_{n+1}} & 0 \\ \hline & 0 & \\ I_n & \vdots & I_n \\ & 0 & \end{array}\right), \quad y \mapsto \left(\begin{array}{cc|c} \multicolumn{2}{c|}{I_{n+1}} & 0 \\ \hline 0 & & \\ \vdots & I_n & I_n \\ 0 & & \end{array}\right).$$
4. $$x \mapsto \left(\begin{array}{c|c} I_n & 0 \\ \hline I_n & I_n \end{array}\right), \quad y \mapsto \left(\begin{array}{c|c} I_n & 0 \\ \hline J_n(\alpha) & I_n \end{array}\right), \quad \alpha \in k .$$

   There is one more case in this rough category and that is the case in which $\alpha = 0$ and the matrices for x and y are interchanged.
5. *The trivial module k is the only simple kG-module.*

Let $\gamma_1, \gamma_2 \in k$ and define $u_\gamma := 1 + \gamma_1(x-1) + \gamma_2(y-1) \in kG$. Observe that $u_\gamma^2 = 1$, since char$(k) = 2$. Let M be a given finitely generated kG-module. We ask now the question: for what values of γ_1, γ_2 is M not free as a $k\langle u_\gamma\rangle$-module?

Since $u_\gamma^2 = 1$, we have that $(u_\gamma - 1)^2 = 0$ and hence also that

$$k\langle u_\gamma\rangle = \{a + bu_\gamma \mid a, b \in k\} = \{c + d(u_\gamma - 1) \mid c, d \in k\} \cong k[X]/X^2 .$$

With the basis $\{1, u_\gamma - 1\}$, the free module $k\langle u_\gamma\rangle$ as a u_γ-module is given by:

$$u_\gamma = \begin{pmatrix} 1 & 0 \\ 1 & 1 \end{pmatrix} \text{ or rather } u_\gamma - 1 = \begin{pmatrix} 0 & 0 \\ 1 & 0 \end{pmatrix} .$$

From this we easily get that a $k\langle u_\gamma\rangle$-module is free if and only if the rank of the matrix representing $u_\gamma - 1$ equals half of the dimension of the module M. This is the only criterion that is necessary. Now we can analyze the above modules and decide whether the condition is satisfied or not.

- *The restriction of any projective module $(kG)^n$ is projective, thus free.*
- *The dimensions of the modules of types 2) and 3) are always odd, and so their restrictions can never be free. Of course, the same applies to the trivial module k of type 5.*
- *The interesting case is the modules of type 4). Here the representing matrix of $u_\gamma - 1$ equals*

$$\left(\begin{array}{c|c} 0 & 0 \\ \hline J_n(\gamma_1 + \alpha\gamma_2) & 0 \end{array}\right) .$$

  The Jordan block $J_n(\gamma_1 + \alpha\gamma_2)$ fails to have full rank if and only if $\gamma_1 + \alpha\gamma_2 = 0$. In this case, the restriction of the module in question to u_γ will not be free. So u_γ fails to act freely if and only if $\gamma_1/\gamma_2 = \alpha$.

Next we turn to a slightly more complicated example.

EXAMPLE 1.2. *Suppose that $p = 2$ and $G = \langle x, y, z\rangle \cong (C_2)^3$. Choose arbitrary elements $A, B, C \in k$ and define a 4-dimensional* *module* $M = M(A, B, C)$ *by the matrices presentation*

$$x \mapsto \left(\begin{array}{cc|cc} 1 & 0 & 0 & 0 \\ 0 & 1 & 0 & 0 \\ \hline 1 & 0 & 1 & 0 \\ 0 & 1 & 0 & 1 \end{array}\right), \quad y \mapsto \left(\begin{array}{cc|cc} 1 & 0 & 0 & 0 \\ 0 & 1 & 0 & 0 \\ \hline A & 0 & 1 & 0 \\ 0 & B & 0 & 1 \end{array}\right), \quad z \mapsto \left(\begin{array}{cc|cc} 1 & 0 & 0 & 0 \\ 0 & 1 & 0 & 0 \\ \hline 0 & C & 1 & 0 \\ 1 & 0 & 0 & 1 \end{array}\right).$$

Again we ask for what $(\gamma_1, \gamma_2, \gamma_3) \in k^3$ *is* M *not free as a* $k\langle u_\gamma \rangle$*-module, where* $u_\gamma = 1 + \gamma_1(x-1) + \gamma_2(y-1) + \gamma_3(z-1)$*? In this case* $u_\gamma - 1$ *acts by multiplication by the matrix*

$$\left(\begin{array}{cc|cc} 0 & 0 & 0 & 0 \\ 0 & 0 & 0 & 0 \\ \hline \gamma_1 + \gamma_2 A & \gamma_3 C & 0 & 0 \\ \gamma_3 & \gamma_1 + \gamma_2 B & 0 & 0 \end{array}\right).$$

The restriction of M *is free if and only if* $\operatorname{rank}(u_\gamma - 1) = \frac{1}{2}\dim(M) = 2$*. This can be read off from the determinant of the submatrix in the lower left corner. That is, we ask when is*

$$\det \begin{pmatrix} \gamma_1 + \gamma_2 A & \gamma_3 C \\ \gamma_3 & \gamma_1 + \gamma_2 B \end{pmatrix} = (\gamma_1 + \gamma_2 A)(\gamma_1 + \gamma_2 B) - \gamma_3^2 C = 0\,?$$

Hence the answer to the question is that M *is not free as a* $k\langle u_\gamma \rangle$*-module if and only if* $(\gamma_1, \gamma_2, \gamma_3)$ *is a zero of the polynomial* $(\gamma_1 + \gamma_2 A)(\gamma_1 + \gamma_2 B) - \gamma_3^2 C$.

We can generalize these examples by making the following definition.

DEFINITION 1.3. *(Carlson,81, Carlson,83) Let* $G = \langle x_1, \ldots, x_n \rangle \cong (C_p)^n$ *be an elementary abelian p-group of rank n. For* $(\gamma_1, \ldots, \gamma_n) \in k^n$ *define the element* $u_\gamma := 1 + \sum_{i=1}^n \gamma_i(x_i - 1)$*. Observe that* $u_\gamma^p = 1$*. Let* M *be any finitely generated kG-module. We define the* rank variety *of* M *as the set*

$$V_G^r(M) := \{\gamma \in k^n \mid M\downarrow_{\langle u_\gamma \rangle} \text{ is \underline{not} a free } k\langle u_\gamma \rangle\text{-module}\} \cup \{0\}.$$

REMARK 1. *Adding the zero point in the above definition ensures that* $V_G^r(M)$ *is indeed an affine variety. This fact is essentially proved as in the previous examples in that the non-freeness of the* *restricted module is a condition that depends only on the rank of the matrix of the element $u_\gamma - 1$, and it holds if and only if a certain collection of determinants vanish.*

EXAMPLE 1.4. *We turn back to our first example, where $G = \langle x, y\rangle \cong (C_2)^2$. We denote by M_α the module of type 4), given by the representation $x \mapsto \left(\begin{smallmatrix} I_n & 0 \\ I_n & I_n \end{smallmatrix}\right)$ and $y \mapsto \left(\begin{smallmatrix} I_n & 0 \\ J_n(\alpha) & I_n \end{smallmatrix}\right)$. Then we can compute that*

$$V_G^r(M_\alpha) = \{(\alpha\gamma, \gamma) \in k^2 \,|\, \gamma \in k\},$$

which is a line in the 2-dimensional affine space.

Because group algebras are Hopf algebras, we can define a G-module structure on the tensor product (over k) of any two G-modules. In our example the module $M_\alpha \otimes_k M_\beta$, where the G-action is diagonal, as in $g \cdot (m_1 \otimes m_2) := gm_1 \otimes gm_2$. Then we have that $M_\alpha \otimes M_\beta$ is a free module if and only if $\alpha \neq \beta$. Some forty years ago this fact had to be computed essentially by hand, which is a rather involved task. Now it is just an easy corollary to the following two theorems. They demonstrate some of the uses of the rank variety of a module.

THEOREM 1.5. *(Carlson,81) Let M and N be any finitely generated kG-modules. Then we have that*

$$V_G^r(M \otimes N) = V_G^r(M) \cap V_G^r(N).$$

THEOREM 1.6. *(Dade,78) $V_G^r(M) = \{0\}$ if and only if M is a projective kG-module.*

Thus the rank variety is well behaved with respect to taking tensor products of modules. It also recognizes projective modules. These two theorems have had several applications in the module theory, and they were the tests of success in the search for a definition of the variety for an infinitely generated module. However, the rank variety has an even more important use. It measures the homological behavior of modules. To make the point we need to introduce some new notation and recall some ideas from group cohomology. See the books by (Benson,91) and (Evens,91) for details.

The cohomology ring $\mathrm{H}^*(G,k)$ associated to G is defined by the following process. First we define it as a k-vectorspace. Take a projective resolution

$$\ldots \longrightarrow P_2 \longrightarrow P_1 \longrightarrow P_0 \longrightarrow k \longrightarrow 0$$

of the trivial kG-module k. Applying the functor $\mathrm{Hom}_{kG}(-,k)$ to this resolution, we get the complex

$$\ldots \longleftarrow \mathrm{Hom}_{kG}(P_2,k) \longleftarrow \mathrm{Hom}_{kG}(P_1,k) \longleftarrow \mathrm{Hom}_{kG}(P_0,k) \longleftarrow 0\,.$$

The cohomology of the group is defined as the homology of this complex, i.e.

$$\mathrm{H}^n(G,k) := \mathrm{H}^n(\mathrm{Hom}_{kG}(P_*,k))\,.$$

The usual comparison theorem for projective resolutions guarantees that this definition is independent of the choice of resolution. Actually the nth cohomology group of G with coefficients in k is nothing else but the nth group of extensions between the trivial module k with itself, i.e.

$$\mathrm{H}^n(G,k) \cong \mathrm{Ext}^n_{kG}(k,k)\,.$$

This isomorphism gives an easy way of defining a product structure on $\mathrm{H}^*(G,k)$. We can represent an element in $\mathrm{Ext}^n_{kG}(k,k)$ as the equivalence class of an exact sequence of the form

$$0 \longrightarrow k \longrightarrow X_{n-1} \longrightarrow \ldots \longrightarrow X_1 \longrightarrow X_0 \longrightarrow k \longrightarrow 0\,.$$

Two sequences of this form can be spliced together, thus inducing a multiplication

$$\mathrm{H}^n(G,k) \times \mathrm{H}^m(G,k) \longrightarrow \mathrm{H}^{n+m}(G,k)\,.$$

As an example we consider an element $\zeta \in \mathrm{Ext}^2_{kG}(k,k)$ represented by

$$(\zeta) : 0 \longrightarrow k \xrightarrow{\alpha} X_1 \xrightarrow{\beta} X_0 \xrightarrow{\gamma} k \longrightarrow 0\,.$$

Splicing this exact sequence with itself we obtain

$$\begin{array}{ccccccccccccc}
 & & & & 0 & \longrightarrow & k & \xrightarrow{\alpha} & X_1 & \xrightarrow{\beta} & X_0 & \xrightarrow{\gamma} k \longrightarrow 0 \\
 & & & & & & \uparrow{\scriptstyle\gamma} & \nearrow{\scriptstyle\alpha\gamma} & & & & \\
0 & \longrightarrow & k & \xrightarrow{\alpha} & X_1 & \xrightarrow{\beta} & X_0 & & & & &
\end{array}$$

so the element ζ^2 of degree 4 is represented by the equivalence class of the sequence

$$(\zeta^2) : 0 \longrightarrow k \xrightarrow{\alpha} X_1 \xrightarrow{\beta} X_0 \xrightarrow{\alpha\gamma} X_1 \xrightarrow{\beta} X_0 \xrightarrow{\gamma} k \longrightarrow 0\,.$$

There are several other equivalent ways of defining the product structure on $\mathrm{H}^*(G,k)$. One of them is composition of chain maps between (shifted) projective resolutions, which is a most useful method sometimes for doing explicit calculations. We should note that the cohomology ring is a graded commutative ring, so that if we have two homogeneous elements $\gamma_1, \gamma_2 \in \mathrm{H}^*(G,k)$, then

$$\gamma_1\gamma_2 = (-1)^{\deg(\gamma_1)\cdot\deg(\gamma_2)}\gamma_2\gamma_1\,.$$

Thus even degree elements commute with everything, and odd degree elements anticommute. The following examples will be important to us.

EXAMPLE 1.7. *Let $G = \langle x\rangle \cong C_p$. Observing that $(x-1)^p = x^p - 1 = 0$, we get a minimal periodic projective resolution of the trivial module k in the form*

$$\ldots \longrightarrow kG \xrightarrow{(x-1)^{p-1}} kG \xrightarrow{(x-1)} kG \xrightarrow{(x-1)^{p-1}} kG \xrightarrow{(x-1)} kG \longrightarrow k \longrightarrow 0\,.$$

Notice that the resolution has period 2 if $p > 2$, and period 1 if $p = 2$. From the resolution we see immediately that $\mathrm{H}^n(G,k) \cong k$ for any n. As for the product structure, we have that

$$\mathrm{H}^*(G,k) = \begin{cases} k[\zeta] & , \textit{if } p = 2,\ (\deg(\zeta) = 1)\,; \\ k[\zeta,\eta]/(\eta^2 = 0)\,, & \textit{if } p > 2,\ (\deg(\zeta) = 2, \deg(\eta) = 1)\,. \end{cases}$$

The maps ζ and η (of appropriate degree) are induced from the only non-trivial map from kG to k obtained by factoring out the radical of kG.

This example can be easily generalized to cyclic p-groups of any order, and we can also compute the cohomology with coefficients in the universal ring $\mathbb{Z}$. Instead we use the example for sketching a way of computing the cohomology ring of an elementary abelian group.

So let $G = \langle x_1, \ldots, x_n \rangle \cong (C_p)^n$ be an elementary abelian p-group of rank n. The group ring kG is isomorphic to the tensor product $k\langle x_1 \rangle \otimes \cdots \otimes k\langle x_n \rangle$, where each of the $k\langle x_i \rangle$ is isomorphic to the group ring kC_p considered above. The same applies to resolutions in that we obtain a projective resolution of k as kG-module by tensoring the projective resolutions of k as $k\langle x_i \rangle$-module for each i. Using, then, the Künneth Theorem, which relates the cohomology of the tensor product of complexes to the cohomology of each of the factors, we easily obtain that

$$\mathrm{H}^*(G,k) \cong \mathrm{H}^*(\langle x_1 \rangle, k) \otimes \mathrm{H}^*(\langle x_2 \rangle, k) \otimes \cdots \otimes \mathrm{H}^*(\langle x_n \rangle, k)\,.$$

Since we know already the cohomology ring of a cyclic group of order p, we have the following.

THEOREM 1.8. *Let G be an elementary abelian p-group of rank n. Then we have*

$$\mathrm{H}^*(G,k) = \begin{cases} k[\zeta_1, \ldots, \zeta_n] & , \text{ if } p = 2, \ (\deg(\zeta_i) = 1)\,; \\ k[\zeta_1, \ldots, \zeta_n] & \\ \otimes \Lambda(\eta_1, \ldots, \eta_n)\,, & \text{if } p > 2, \ (\deg(\zeta_i) = 2, \deg(\eta_i) = 1)\,. \end{cases}$$

Here $\Lambda(\eta_1, \ldots, \eta_n)$ denotes the exterior algebra on n generators, i.e. we have the relations $\eta_i^2 = 0$ and $\eta_i \cdot \eta_j = -\eta_j \cdot \eta_i$ for $i \neq j$. Because the elements η_i have square zero, they belong to the radical of the cohomology ring. So in any case,

$$\mathrm{H}^*(G,k)\,/\mathrm{Rad}(\mathrm{H}^*(G,k)) \cong k[\zeta_1, \ldots, \zeta_n]\,,$$

is a polynomial ring.

Now we define the cohomological support variety $V_G(k)$ of the trivial module k as the maximal ideal spectrum of the ring $\mathrm{H}^*(G,k)$. Since we are considering an algebraically closed field, the support variety is affine n-dimensional space k^n. That is a maximal ideal $\mathfrak{m}$ is the kernel of a map that is given by evaluation at a point $\alpha = (\alpha_1, \ldots \alpha_n) \in k^n$. As $\mathrm{H}^*(G,k)$, modulo its radical, is a polynomial ring in n variables, we have an exact sequence

$$0 \longrightarrow \mathfrak{m} \longrightarrow \mathrm{H}^*(G,k) \xrightarrow{\mathrm{eval}_\alpha} k \longrightarrow 0$$

given by $f \to f(\alpha_1, \ldots \alpha_n)$ for f a polynomial in $\zeta_1, \ldots, \zeta_n$.

The definition of cohomological support variety can easily be extended to any finite group G and finitely generated module M. First we define $V_G(k)$ as the maximal ideal spectrum of $\mathrm{H}^*(G,k)$, just as in the above example. Then we observe that $\mathrm{Ext}^*_{kG}(M,M)$ is a ring by iself and is also a module over the cohomology ring $\mathrm{H}^*(G,k) = \mathrm{Ext}^*_{kG}(k,k)$. Let $J(M)$ be the annihilator ideal of $\mathrm{Ext}^*_{kG}(M,M)$ inside $\mathrm{H}^*(G,k)$. That is, given an exact sequence that begins and ends with k, we can tensor the sequence with M to get an exact sequence representing an element of $\mathrm{Ext}^*_{kG}(M,M)$. What makes all of this possible is the theorem of Evens and Venkov (see the standard references (Benson,91) or (Evens,91)) that $\mathrm{H}^*(G,k)$ is a finitely generated algebra over k and that for any finitely generated kG-modules M and N, $\mathrm{Ext}^*_{kG}(M,N)$ is a finitely generated module over $\mathrm{H}^*(G,k)$.

DEFINITION 1.9. *The* cohomological support variety $V_G(M) = V_G(J(M))$ *of the module* M *is the set of all maximal ideals in* $\mathrm{H}^*(G,k)$ *that contain* $J(M)$. *That is,*

$$V_G(M) := \{\mathfrak{m} \in V_G(k) \mid J(M) \subseteq \mathfrak{m}\}.$$

REMARK 2. *Note that the set of maximal ideals $V_G(k)$ is topologized by the Zariski topology. That is, a closed set is precisely the set of all maximal ideals containing a given ideal. Thus $V_G(M)$ is a closed set.*

We can think of $J(M)$ in the following way. We represent an element $\zeta \in \mathrm{H}^n(G,k)$ as (the equivalence class of) an exact sequence

$$E_\zeta : 0 \longrightarrow k \longrightarrow B_{n-1} \longrightarrow \dots \longrightarrow B_0 \longrightarrow k \longrightarrow 0.$$

Applying the exact functor $- \otimes_k M$ to the sequence gets us the sequence

$$E_\zeta \otimes_k M : 0 \to M \to B_{n-1} \otimes M \to \dots \to B_0 \otimes M \to M \to 0,$$

which represents $\zeta \otimes 1$. So $J(M)$ consists of all classes of sequences E_ζ, such that $E_\zeta \otimes M$ splits, i.e. corresponds to the zero element of $\mathrm{Ext}^n_{kG}(M,M)$.

The cohomological support variety behaves as desired with respect to the tensor product of modules. That is, we have, for any finitely generated modules M and N, that

$$V_G(M \otimes N) = V_G(M) \cap V_G(N).$$

It is also capable of recognizing projectivity, in the sense that

$V_G(M) = \{0\}$ if and only if M is a projective module.

The problem is that $V_G(M)$ may be extremely hard to compute, in contrast to the rank variety $V_G^r(M)$. However in the case of elementary abelian groups the two varieties are essentially the same.

THEOREM 1.10. *Let G be an elementary abelian group, M a finitely generated kG-module. Then the cohomological support variety $V_G(M)$ and the rank variety $V_G^r(M)$ are isogenic.*

The situation can be briefly described as follows. For purposes of the argument assume that $p > 2$. In the notation of Theorem 1.8 and Definition 1.3, if $f = f(\zeta_1, \ldots, \zeta_n) \in \mathrm{H}^{2m}(G, k)$ and if $\gamma = (\gamma_1, \ldots, \gamma_n) \in k^n$ then

$$\mathrm{res}_{G,\langle u_\gamma \rangle}(f) = f(\gamma_1^p, \ldots, \gamma_n^p) \cdot \zeta_u^m$$

where ζ_u is a generator for $\mathrm{H}^2(\langle u_\gamma \rangle, k)$. So the actual correspondence between varieties involves a shift by the Frobenius map. Because the Frobenius map (raising to p^{th}-powers) has no polynomial inverse, the map on varieties is only an isogeny and not an isomorphism.

2. Induction from elementary abelian subgroups

The reason why we are paying so much attention to elementary abelian groups is that, to some extent, they determine the homological algebra. This had already been observed more than 30 years ago by Atiyah and others who were working on algebraic topology. The main breakthrough came with Quillen's dimension theorem, which applied only to the cohomology ring of the group in its original version. It was later adapted to the varieties of modules. In what follows we attempt to make this general principle more precise.

One of the most powerful and general theorems in this direction, having heavy structural implications, is the following.

THEOREM 2.1. *(Carlson,00) There exist elementary abelian p-subgroups $E_1, \ldots, E_t$ of G with the following property. Let M be any kG-module. There exists a kG-module M' such that the direct sum $M \oplus M'$ has a filtration of the form*

$$\{0\} = L_0 \subseteq L_1 \subseteq \ldots \subseteq L_n \subseteq M \oplus M' ,$$

where for each i, the factor L_i/L_{i-1} is isomorphic to an induced module $W_i \uparrow_{E_i}^G$ for some kE_i-module W_i.

Here $W_i \uparrow_{E_i}^G = kG \otimes_{kE_i} W_i$, by definition.

REMARK 3.

1. *In categorical terms, the theorem states that the thick subcategory of modules generated by modules induced from elementary abelian subgroups is the whole stable module category of kG (definitions of the involved categorical terms appear further on).*
2. *The module M in the theorem could as well be infinitely generated. In case M is finitely generated, then the module M' is also finitely generated.*
3. *The same theorem has been proved for the universal ring $\mathbb{Z}$, i.e. for integral $\mathbb{Z}G$-modules.*

It is sufficient to prove the theorem in the case that $M = k$ is the trivial module. For a sketch of this part, start with a filtration as given by the theorem. That is suppose that we have a filtration

$$\{0\} = L_0 \subseteq L_1 \subseteq \ldots \subseteq L_n \subseteq k \oplus V ,$$

where $L_i/L_{i-1} \cong W_i \uparrow_{E_i}^G$. Tensoring the filtration with the module M, we obtain a new filtration of the form

$$\{0\} = L_0 \otimes_k M \subseteq L_1 \otimes_k M \subseteq \ldots \subseteq L_n \otimes_k M \subseteq M \ \oplus \ (V \otimes_k M) .$$

For each factor, we have a short exact sequence

$$0 \longrightarrow L_{i-1} \longrightarrow L_i \longrightarrow W_i \uparrow^G \longrightarrow 0 .$$

Now, applying the exact functor $- \otimes_k M$ to this sequence, we get a new short exact sequence:

$$0 \longrightarrow L_{i-1} \otimes_k M \longrightarrow L_i \otimes_k M \longrightarrow W_i \uparrow^G \otimes_k M \longrightarrow 0 .$$

From this sequence we can compute the quotient appearing in the filtration. Note that

$$L_i \otimes_k M / L_{i-1} \otimes_k M \cong W_i \uparrow^G \otimes_k M \cong (W \otimes_k M \downarrow_{E_i}) \uparrow^G ,$$

where in the last isomorphism Frobenius reciprocity has been used. But the last module is induced from an elementary abelian p-subgroup, as desired.

Throughout the rest of the section we fix the sequence of elementary abelian p-subgroups $E_1, \ldots, E_t$ and the modules $W_1, \ldots, W_t$ for which we have a filtration

$$\{0\} = L_0 \subseteq L_1 \subseteq \ldots \subseteq L_t \subseteq k \oplus V ,$$

where $L_i/L_{i-1} \cong W_i \uparrow_{E_i}^G$. The theorem has many applications. One of the easiest is the following. We give a short proof.

THEOREM 2.2. *(Chouinard, 76) Suppose the module M has the property that $M \downarrow_{E_i}$ is free for all possible E_i. Then M is a projective module.*

Proof. The tensor product $W_i \otimes_k M \downarrow_{E_i}$ is kE_i-projective. Inducing, we obtain that $(W_i \otimes_k M \downarrow_{E_i}) \uparrow^G$ is a kG-projective module. But projective and injective modules are the same, so now an easy induction over the filtration of $M \oplus M'$ implies that this module is projective. But then M itself is projective, as claimed.

Before stating the next theorem, we introduce some new notation. Let H be any subgroup of G. Then we can restrict the action of G to H, on both modules and homomorphisms so that we have maps

$$\begin{aligned} \mathrm{res}_{G,H} &: \mathrm{H}^*(G,k) \longrightarrow \mathrm{H}^*(H,k) \\ \text{and } \mathrm{res}_{G,H} &: \mathrm{Ext}^*_{kG}(M,N) \longrightarrow \mathrm{Ext}^*_{kH}(M,N) \end{aligned}$$

The maps are in general not surjective. Pulling back maximal ideals, the restriction map $\mathrm{res}_{G,H}$ induces a map between the respective maximal ideal spectra,

$$\mathrm{res}^*_{G,H} : V_H(k) \longrightarrow V_G(k) ,$$

which is always finite to one. We shall be especially interested in the restriction maps $\mathrm{res}_{G,E}$ to elementary abelian p-subgroups E of G.

We denote by $\mathcal{A}_p(G)$ the set of all elementary abelian p-subgroups of G.

THEOREM 2.3. *Let $X_0, X_1, \ldots, X_t$ be kG-modules and suppose that elements $\zeta_i \in \mathrm{Ext}_{kG}^{n_i}(X_{i-1}, X_i)$ are given with $res_{G,E_i}(\zeta_i) = 0$. Then the product $\zeta_t \cdots \zeta_2 \cdot \zeta_1 = 0$.*

As an immediate consequence of this theorem, we obtain a formulation of Quillen's Dimension Theorem (Quillen,71), where one is considering the case $X_0 = X_1 = \ldots = X_t = k$.

COROLLARY 2.4. *(Quillen,71)(See also (Quillen-Venkov,72).) Let*

$$I := \{\zeta \in \mathrm{H}^*(G,k) \mid res_{G,E}(\zeta) = 0 \ \ \textit{for all } E \in \mathcal{A}_p(G)\},$$

which is a 2-sided ideal in $\mathrm{H}^*(G,k)$. Then I is nilpotent with nilpotence degree t, i.e. $I^t = \{0\}$.

Quillen's Dimension Theorem is often stated in the following way:

$$V_G(k) = \bigcup_{E \in \mathcal{A}_p(G)} \mathrm{res}^*_{G,E}(V_E(k))\,.$$

Here, the importance of elementary abelian groups for cohomology becomes evident. In this way, the theorem is describing structural properties of the graded commutative ring $\mathrm{H}^*(G,k)$. That is, *maximal* elementary abelian p-subgroups correspond to the components of $V_G(k)$, which in turn correspond to the minimal primes in $\mathrm{H}^*(G,k)$.

Next we give a sketch of the proof of Theorem 2.3. Define the module M as the direct sum of all modules involved in the given extensions, i.e. $M = X_0 \oplus X_1 \oplus \ldots \oplus X_t$. Then we know by Theorem 2.1 that there exists a module M' and a filtration

$$\{0\} = L_0 \subseteq L_1 \subseteq \ldots \subseteq L_t \subseteq M \oplus M' = N\,,$$

where $L_i/L_{i-1} \cong W_i \uparrow_{E_i}^G$ for some kE_i-modules W_i. Observe that we can consider each of the modules X_i as a direct summand of N, so that we have

$$\zeta_i \in \mathrm{Ext}_{kG}^{n_i}(X_{i-1}, X_i) \subseteq \mathrm{Ext}_{kG}^{n_i}(N,N)\,.$$

The short exact sequence

$$0 \longrightarrow L_{i-1} \longrightarrow L_i \longrightarrow W_i \uparrow_{E_i}^G \longrightarrow 0$$

induces a long exact sequence in cohomology

$$\ldots \to \operatorname{Ext}^r_{kG}(N, L_{i-1}) \to \operatorname{Ext}^r_{kG}(N, L_i) \to \operatorname{Ext}^r_{kG}(N, W_i \uparrow_{E_i}^G) \to \ldots$$

Using the above observation we can multiply each of the terms in this long sequence with ζ_i, inducing a chain map of degree n_i as in the following diagram.

$$\begin{array}{ccccccccc}
\cdots & \longrightarrow & \operatorname{Ext}^r_{kG}(N, L_{i-1}) & \longrightarrow & \operatorname{Ext}^r_{kG}(N, L_i) & \longrightarrow & \operatorname{Ext}^r_{kG}(N, W_i \uparrow_{E_i}^G) & \longrightarrow & \cdots \\
 & & \downarrow \cdot\zeta_i & & \downarrow \cdot\zeta_i & & \downarrow \cdot\zeta_i & & \\
\cdots & \longrightarrow & \operatorname{Ext}^{r+n_i}_{kG}(N, L_{i-1}) & \longrightarrow & \operatorname{Ext}^{r+n_i}_{kG}(N, L_i) & \longrightarrow & \operatorname{Ext}^{r+n_i}_{kG}(N, W_i \uparrow_{E_i}^G) & \longrightarrow & \cdots
\end{array}$$

Now notice that by assumption the rightmost vertical map in the above diagram of the chain map is zero. That is, when $i = 0$ and $L_t = N$ we have that $\mathrm{Id}_N \cdot \zeta_t$ is in the image of $\operatorname{Ext}^{n_t}_{kG}(N, L_{t-1})$. Then using an easy induction argument, we get that

$$\mathrm{Id}_N \cdot \zeta_t \cdot \zeta_{t-1} \cdots \zeta_1 \in \operatorname{image}(\operatorname{Ext}^*_{kG}(N, \underbrace{L_0}_{=0!})) = 0\,.$$

We end this section by giving a rough idea of how to obtain the filtration of Theorem 2.3. First remember that it is enough to find a suitable filtration for the trivial kG-module k. Then the next step is a reduction to the case of p-groups. So let P be a Sylow p-subgroup of G. With k_P we denote the trivial kP-module. It is a fact that k is a direct summand of the induced module $k_P \uparrow^G$. That is $k_P \uparrow^G \cong k \oplus U$ for some U. So if $k_P \oplus V$ has a filtration by kP-modules of the desired form,

$$\{0\} = L_0 \subseteq \ldots \subseteq L_t \subseteq k_P \oplus V\,,$$

then we only have to induce it to G, and we are done.

In the rest of the proof, a deep theorem of Serre (Serre,65) is required. Let G be a p-group, not elementary abelian. The theorem states that there exist maximal subgroups $H_1, \ldots, H_n$ of G and elements $\beta_{H_i} \in \operatorname{Ext}^2_{kG}(k,k)$ such that

$$\beta_{H_1} \cdot \beta_{H_2} \cdots \beta_{H_n} = 0\,.$$

The elements β_{H_i} are called the "Bocksteins" corresponding to the maximal subgroup H_i. That is, they are the images under the Bockstein homomorphism that sends the class of a homomorphism with kernel a given maximal subgroup in $\mathrm{Hom}(G, \mathbb{Z}/p) \cong \mathrm{H}^1(G, \mathbb{F}_p)$ to an element of $\mathrm{H}^2(G, \mathbb{F}_p)$. For our purposes it is only necessary to understand the structure of the representing sequence that we get after extending the coefficients to k. So if $G = \langle H_i, x_i \rangle$ with $G/H_i \cong C_p$, then a sequence representing β_{H_i} has the form

$$\beta_{H_i} \ : \quad 0 \longrightarrow k \longrightarrow k_{H_i} \uparrow^G \ \stackrel{x_i-1}{\longrightarrow} \ k_{H_i} \uparrow^G \longrightarrow k \longrightarrow 0 \, .$$

Now we splice the elements together getting a sequence

$$0 \to k \to k_{H_1} \uparrow^G \to \cdots \to k_{H_{n-1}} \uparrow^G \to k_{H_n} \uparrow^G \to k_{H_n} \uparrow^G \to k \to 0$$

that represents $\beta_{H_1} \beta_{H_2} \cdots \beta_{H_t}$. It has the following two properties.

1. All of the terms except the ends are induced from proper subgroups.
2. It represents 0 in $\mathrm{Ext}^{2n}_{kG}(k, k)$.

Now we truncate both ends of this sequence (the terms that are isomorphic to the trivial module k) and get a complex

$$C_* \, : 0 \longrightarrow C_{2n-1} \longrightarrow C_{2n-2} \longrightarrow \ldots \longrightarrow C_1 \longrightarrow C_0 \longrightarrow 0 \, .$$

Next we push the complex into a single degree. That is, we collapse it to a module by taking a projective resolution, truncating in a degree beyond the degrees of the nonzero homology of the complex and then translating the homology of the truncated complex back to degree zero. The details of the process can be found in (Carlson,00). Thus we get a kG-module X from the complex. Then X has the following properties

1. By (1) above, it is filtered by modules induced from maximal subgroups of G.
2. By (2) above, it is isomorphic to $k \oplus \Omega^{-2n+1}(k)$, where $\Omega^m(k)$ denotes the mth syzygy of a kG-module k.

Now the process needs to be repeated until we get down to elementary abelian p-subgroups.

Quillen's Dimension Theorem has been generalized to arbitrary finitely generated modules.

THEOREM 2.5. *(Alperin-Evens,82) (Avrunin,81) Let G be a finite group and M a finitely generated kG-module. Then we have that*

$$V_G(M) = \bigcup_{E \in \mathcal{A}_p(G)} res^*_{G,E}(V_E(M)) \, .$$

3. Varieties for infinitely generated modules

In this section we generalize this and other quoted theorems to the case of infinitely generated modules as well. But serious problems appear. The celebrated Theorem of Evens and Venkov tells us that the cohomology ring $\mathrm{H}^*(G,k)$ is a finitely generated k-algebra, and that the $\mathrm{H}^*(G,k)$-module $\mathrm{Ext}^*_{kG}(M,M)$ is finitely generated in case M is finitely generated. This is no longer true for M infinitely generated, and it creates severe difficulties. We start with an example showing that the rank "variety" for infinitely generated modules is no longer a variety in the usual sense.

EXAMPLE 3.1. *Again we turn back to the fours group, $G = \langle x, y \rangle \cong C_2 \times C_2$ and $p = 2$. We pick an infinite series of indecomposable kG-modules. Let M_n have a basis $a_1, \ldots, a_n$ and $b_1, \ldots, b_{n+1}$ consisting of $2n+1$ elements, and where the group G acts as follows:*

$$(x-1)a_i = b_i \, , \qquad (y-1)a_i = b_{i+1} \, ,$$
$$(x-1)b_i = (y-1)b_i = 0 \, , \quad \textit{for every } i \, .$$

We can picture this action in the following way. We let a point correspond to a basis element, an arrow down to the left correspond to multiplication by the radical generator $x-1$, and similarly an arrow down to the right correspond to multiplication by $y-1$. For M_n one obtains

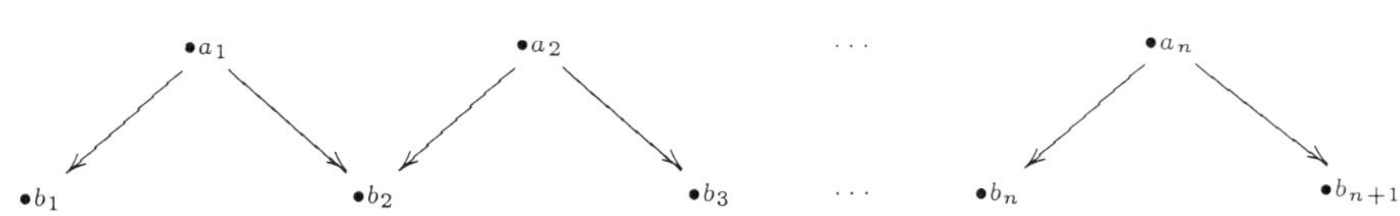

As we already know, in this case we have that $V^r_G(M_n) = k^2$. But we could also continue the picture infinitely to the right, still

giving an indecomposable kG-module M_∞. *This means we have basis elements* a_i *and* b_i *for* $i \in \mathbb{N}$. *Let the G-action be defined by the above equations. Hence* M_∞ *is not finitely generated. Letting* $u_\gamma = 1 + \gamma_1(x-1) + \gamma_2(y-1)$ *for a pair* $(\gamma_1, \gamma_2) \in k^2$, *we ask the same question as in the beginning: for which parameters does the element* u_γ *fail to act freely on* M_∞? *A straightforward analysis shows that only* $x-1$ *acts freely, but that is all. So we have that*

$$V_G^r(M_\infty) = k^2 \backslash \{(a,0) \mid a \in k\} .$$

This is not an affine variety, but rather an affine open set with respect to the Zariski topology on k^2.

The formal definition of the variety of an infinitely generated module goes as follows. Let $G = \langle x_1, \ldots, x_n \rangle \cong (C_p)^n$ be an elementary abelian p-group of rank n. Let $\mathcal{V}_G(k) = \operatorname{proj}(\mathrm{H}^*(G,k))$ denote the projective prime ideal spectrum of the cohomology ring $\mathrm{H}^*(G,k)$, i.e. the set of all homogeneous prime ideals with the usual Zariski topology. Let M be any kG-module. In particular, it could be infinitely generated as in the above example. Now take an algebraically closed extension field K of k of large transcendence degree, at least $n-1$. For an arbitrary n-tuple $(\gamma_1, \ldots, \gamma_n) \in K^n$ we define $u_\gamma := 1 + \sum_{i=1}^n \gamma_i(x_i - 1)$.

DEFINITION 3.2. *(See (Benson-Carlson-Rickard,96) or (Carlson,96)) The rank variety of M is defined as*

$$\mathcal{V}_G^r(M) := \{\gamma \in K^n \mid (K \otimes_k M) \downarrow_{\langle u_\gamma \rangle} \text{ is not projective } \} \cup \{0\} .$$

We would like to identify the rank variety $\mathcal{V}_G^r(M)$ with an appropriate subset of $\mathcal{V}_G(k)$. The next lemma is the key to making such an identification.

LEMMA 3.3. *Let* $\gamma, \delta \in K^n$ *be given, and suppose that both are generic for the same homogeneous prime ideal. Then*

$$\gamma \in \mathcal{V}_G^r(M) \text{ if and only if } \delta \in \mathcal{V}_G^r(M) .$$

Informally, one can identify $\mathcal{V}_G^r(M)$ with a subset of $\mathcal{V}_G(k)$ by taking generic points. We wish to avoid the technicalities here, but an example should suffice to illustrate the point.

EXAMPLE 3.4. *We consider the homogeneous prime ideal* $\mathfrak{p} = \langle x^2 - yz \rangle$ *in the commutative ring* $k[x, y, z]$*. Let* K *be a field extension of* k *with transcendence degree bigger than 2. Choose a pair* δ_1, δ_2 *of algebraically independent elements over* k*. Then the triple* $\gamma = (\delta_1\delta_2, \delta_1^2, \delta_2^2)$ *is a generic point for* $\mathfrak{p}$*. This shows that there are myriads of generic points for* $\mathfrak{p}$*, corresponding to the choice of the pair* (δ_1, δ_2)*.*

With this extension of the notion of rank variety to arbitrary kG-modules, it is possible to prove the following

PROPOSITION 3.5. *(See (Benson-Carlson-Rickard,96))*

1. $\mathcal{V}_G^r(M) = \{0\}$ *if and only if* M *is projective.*
2. *For* kG*-modules* M *and* N *one has:*

$$\mathcal{V}_G^r(M \otimes N) = \mathcal{V}_G^r(M) \cap \mathcal{V}_G^r(N)\,.$$

3. *Given any subset* S *of* $\mathcal{V}_G(k)$ *there exists a* kG*-module* M *with* $\mathcal{V}_G^r(M) = S$*.*

The first two assertions show that the generalized notion of rank variety behaves as desired.

Now what about a cohomological support variety? We need some notation to begin. If M is an arbitrary kG-module M, let $P(M)$ be its projective cover and $I(M)$ its injective hull. The first syzygy module $\Omega(M)$ is defined via the short exact sequence

$$0 \longrightarrow \Omega(M) \longrightarrow P(M) \longrightarrow M \longrightarrow 0\,,$$

and the module $\Omega^{-1}(M)$ through the short exact sequence

$$0 \longrightarrow M \longrightarrow I(M) \longrightarrow \Omega^{-1}(M) \longrightarrow 0\,.$$

Of course, one can iterate this procedure:

$$\Omega^n(M) := \Omega(\Omega^{n-1}(M))\,, \quad \Omega^{-n}(M) := \Omega^{-1}(\Omega^{-n+1}(M))\,, \quad n \geq 2\,.$$

The problem is that taking syzygies does not behave functorially in the module category of kG. If we want Ω to become a functor, we must pass to an appropriate quotient category, namely, the stable module category $\mathbf{stmod}(kG)$. The objects of $\mathbf{stmod}(kG)$ are finitely generated kG-modules. For any two modules M and N we define $\operatorname{PHom}_{kG}(M,N)$ to be the set of all those kG-homomorphisms from M to N that factor through a projective module. The family $\operatorname{PHom}_{kG}(M,N)$ for all possible M and N forms an ideal (in the categorical sense), so that one can consider the quotient $\overline{\operatorname{Hom}}_{kG}(M,N) := \operatorname{Hom}_{kG}(M,N)/\operatorname{PHom}_{kG}(M,N)$ as the morphisms in the category $\mathbf{stmod}(kG)$. An important point is that a projective kG-module P becomes the zero object in $\mathbf{stmod}(kG)$, since the identity map Id_P belongs to $\operatorname{PHom}_{kG}(P,P)$.

What is true is that $\mathbf{stmod}(kG)$ is not an abelian category, but rather it is triangulated. See the books (Happel,88, Weibel,94) for details. Roughly speaking, we trade exact sequences in $\mathbf{mod}(kG)$, for triangles in $\mathbf{stmod}(kG)$. Since this part of the theory is rather technical, we just want to get some glimpse of what a triangle actually is. If we have a short exact sequence of kG-modules

$$0 \longrightarrow A \xrightarrow{\alpha} B \xrightarrow{\beta} C \longrightarrow 0,$$

then we can add to β a map $\iota : B \longrightarrow P$ where P is a projective module such that

$$0 \longrightarrow B \overset{\beta' := (\beta,\iota)}{\longrightarrow} C \oplus P$$

is exact. Note that $P = I(B)$ with the canonical injection would do. Notice that β and β' represent the same morphism in $\mathbf{stmod}(kG)$. Then it is an exercise to show that the cokernel of β' has the form $\Omega^{-1}(A) \oplus Q$, where Q is a projective module (depending on our choice of P). So finally we have an exact sequence

$$0 \longrightarrow B \xrightarrow{\beta'} C \oplus P \xrightarrow{\gamma} \Omega^{-1}(A) \oplus Q \longrightarrow 0.$$

Taking representatives modulo projective homomorphisms, both short exact sequences fit together into one triangle in $\mathbf{stmod}(kG)$ of the form

$$\ldots \longrightarrow A \xrightarrow{\alpha} B \xrightarrow{\beta} C \xrightarrow{\gamma} \Omega^{-1}(A) \overset{\Omega^{-1}(\alpha)}{\longrightarrow} \ldots$$

Indeed, Ω^{-1} is a functor on $\mathbf{stmod}(kG)$, even an autoequivalence. The functor Ω^{-1} is called the translation functor for obvious reasons.

Let $\mathcal{C}$ be a thick subcategory of $\mathbf{stmod}(kG)$. We mean by this that $\mathcal{C}$ is a triangulated subcategory closed under the taking of direct summands. The following theorem was proved by Jeremy Rickard. It is an adaptation to the stable category of Brown's Representability Theorem in homotopy and it has had several surprising applications.

PROPOSITION 3.6. *(Rickard,97) Let $\mathcal{C}$ be a thick subcategory of* $\mathbf{stmod}(kG)$. *For any object M in* $\mathbf{stmod}(kG)$ *there exists a distinguished triangle of the form*

$$\mathcal{E}_{\mathcal{C}}(M) \longrightarrow M \longrightarrow \mathcal{F}_{\mathcal{C}}(M) \longrightarrow \Omega^{-1}(\mathcal{E}_{\mathcal{C}}(M)),$$

where

1. *$\mathcal{E}_{\mathcal{C}}(M)$ is a direct limit of objects in $\mathcal{C}$, and*
2. *$\mathcal{F}_{\mathcal{C}}(M)$ is $\mathcal{C}$-local, i.e. $\underline{\mathrm{Hom}}_{kG}(C, \mathcal{F}_{\mathcal{C}}(M)) = 0$ for all $C \in \mathcal{C}$.*

Moreover the two properties characterize the triangle.

Thick subcategories are closely related to cohomology, as we see in the following example. Let S be any collection of subvarieties of $V_G(k)$, and suppose that S is closed under finite unions and specialization. By closed under specialization we mean that if $U \subseteq V$ are subvarieties of $V_G(k)$ and if $V \in S$, then U also belongs to S. A motivation for making this additional condition is the following.

LEMMA 3.7. *Let*

$$0 \longrightarrow L \longrightarrow M \longrightarrow N \longrightarrow 0$$

be an exact sequence of kG-modules. Then we have for the corresponding support varieties that $V_G(M) \subseteq V_G(L) \cup V_G(N)$.

Now let $\mathcal{C}_S$ be the full subcategory of $\mathbf{stmod}(kG)$ consisting of all modules M such that $V_G(M) \in S$.

THEOREM 3.8. *$\mathcal{C}_S$ is a thick subcategory of* $\mathbf{stmod}(kG)$.

Before giving the rather involved definition of cohomological support variety for an arbitrary module, we illustrate the notion with an easy example. Remember that for the trivial module k there is a distinguished triangle

$$\mathcal{E}_{\mathcal{C}}(k) \longrightarrow k \longrightarrow \mathcal{F}_{\mathcal{C}}(k) \longrightarrow \Omega^{-1}(\mathcal{E}_{\mathcal{C}}(k)).$$

EXAMPLE 3.9. *We return to the fours group presented as $G = \langle x, y \rangle \cong C_2 \times C_2$ at the prime $p = 2$. Let S be the set of all lines in k^2. So S is closed under finite unions and specialization. Let $\mathcal{C} = \mathcal{C}_S$. Then we have that*

$$\mathcal{V}_G(\mathcal{E}_{\mathcal{C}}(k)) = S\,, \qquad \mathcal{V}_G(\mathcal{F}_{\mathcal{C}}(k)) = \mathcal{V}_G(k) \backslash S\,.$$

Note that $\mathcal{V}_G(\mathcal{F}_{\mathcal{C}}(k))$ is far from being empty, since S consists of k-rational lines and the construction is realized over the bigger field K.

More generally, in the case of an elementary abelian group G, $\mathcal{V}_G(\mathcal{E}_{\mathcal{C}}(k)) \subseteq \mathcal{V}_G(k)$ consists of all homogeneous closed irreducible subvarieties of $V_G(k)$ contained in an appropriate S, and $\mathcal{V}_G(\mathcal{F}_{\mathcal{C}}(k))$ consists of the complement $\mathcal{V}_G(k) \backslash \mathcal{V}_G(\mathcal{E}_{\mathcal{C}}(k))$.

Let T be an irreducible closed subvariety of $V_G(k)$. We let S be the set consisting of T and all possible subvarieties. With S' we denote the set of all proper subvarieties of T. Let $\mathcal{C} = \mathcal{C}_S$ and $\mathcal{C}' = \mathcal{C}_{S'}$. Then we define the module

$$\mathcal{K}(T) := \mathcal{E}_{\mathcal{C}}(k) \otimes \mathcal{F}_{\mathcal{C}'}(k)\,.$$

The variety of this module consists basically of one point, meaning $\mathcal{V}_G(\mathcal{K}(T)) = \{T\}$. With the help of this module we can finally define the cohomological variety.

DEFINITION 3.10. *Let M be an arbitrary kG-module. The cohomological support variety $\mathcal{V}_G(M)$ of M consists of the set of all irreducible closed subvarieties of $V_G(k)$ such that the module $\mathcal{K}(T) \otimes M$ is not projective.*

With this definition we again have all of the desired properties of a "support variety". Most essentially we have the following.

- If M is a finitely generated kG-module then $\mathcal{V}_G(M)$ is the set of all $S \in \mathcal{V}_G(k)$ such that as an affine variety S is a closed set in $V_G(M)$.
- $\mathcal{V}_G(M) = \emptyset$ if and only if M is projective.
- $\mathcal{V}_G(M) \cong \mathcal{V}_G^r(M)$ if G is elementary abelian.
- $\mathcal{V}_G(M) = \cup_E \mathrm{res}^*_{G,E}(\mathcal{V}_E(M))$ where the union runs over the collection of elementary abelian p-subgroups of G.
- $\mathcal{V}_G(M \otimes N) = \mathcal{V}_G(M) \cap \mathcal{V}_G(N)$ for arbitrary kG-modules M and N.
- The extension property on short exact sequences is satisfied, as stated in the above lemma 3.7.

There are many important applications of the notions of this section. For example, it is possible to give a complete classification of the thick subcategories of **stmod**(kG) in the case that G is a p-group (Benson-Carlson-Rickard,97). Benson (Benson,95) has given a complete proof of the Nucleus conjecture of (Benson-Carlson-Robinson,90). He shows that if M is a finitely generated module in the principal block of kG and if $\mathrm{H}^*(G, M) = 0$ then $V_G(M)$ is contained in the nucleus, which is $\cup_E \mathrm{res}^*_{G,E}(V_E(M))$ where the union is taken over all elementary abelian subgroups $E \subseteq G$ such that $C_G(E)$ is not p-nilpotent.

References

Alperin-Evens,82. J. L. Alperin and L. Evens, *Varieties and elementary abelian subgroups*, J. Pure Appl. Algebra **26**(1982), 221-227.

Avrunin,81. G. S. Avrunin, *Annihilators of cohomology modules*, J. Algebra **69**(1981), 150-154.

Benson,91. D. J. Benson, *Representations and Cohomology* II*: Cohomology of Groups and Modules*, Cambridge Studies in Advanced Mathematics **31**, Cambridge University Press, Cambridge, 1991.

Benson,95. D. J. Benson, *Cohomology of modules in the principal block of a finite group*, New York J. Math. **1** (1995), 196–205.

Benson-Carlson-Rickard,96. D. J. Benson, J. F. Carlson and J. Rickard, *Complexity and varieties for infinitely generated modules*, II, Math. Proc. Cam. Phil. Soc. **120**(1996), 597-615.

Benson-Carlson-Rickard,97. D. J. Benson, J. F. Carlson and J. Rickard, *Thick subcategories of the stable category*, Fund. Math. **153**(1997), 59-80.

Benson-Carlson-Robinson,90. D. J. Benson, J. F. Carlson and G. R. Robinson, *On the vanishing of group cohomology*, J. of Alg. **131** (1990), 40–73.

Carlson,81. J. F. Carlson, *The complexity and varieties of modules*, Proceedings of Oberwolfach Conference on Integral Representations and Applications, Springer Lecture Notes No. 882, (1981), 415-422.

Carlson,83. J. F. Carlson, *The varieties and the cohomology ring of a module*, J. Algebra **85** (1983), 104-143.

Carlson,96. J. F. Carlson, *Modules and Group Algebras*, Birkhäuser, Basel, 1996.

Carlson,00. J. F. Carlson, *Cohomology and induction from elementary abelian subgroups*, Quarterly J. Math., **51** (2000), 169-181.

Chouinard,76. L. Chouinard, *Projectivity and relative projectivity over group algebras*, J. Pure Appl. Algebra, **7**(1976), 278-302.

Dade,78. E. C. Dade, *Endo-permutation modules over p-groups,* II, Ann. Math. **108** (1978), 317–346.

Evens,91. L. Evens, *The Cohomology of Groups*, Oxford University Press, New York, 1991.

Happel,88. D. Happel, *Triangulated Categories in the Representation Theory of Finite-Dimensional Algebras*, Cambridge University Press, Cambridge, 1988.

Heller-Reiner,61. A. Heller and I. Reiner, *Indecomposable representations*, Ill. J. Math. **5** (1961), 314-323.

Nakano,01. D. K. Nakano, *Complexity and support varieties for finite dimensional algebras*, Cont. Math. (to appear).

Quillen,71. D. Quillen, *The spectrum of an equivariant cohomology ring* I, II , Ann. Math. **94** (1971), 549-602.

Quillen-Venkov,72. D. Quillen and B. B. Venkov, *Cohomology of finite groups and elementary abelian subgroups*, Topology **11**(1972), 552-556.

Rickard,97. J. Rickard, *Idempotent modules in the stable category*, J. London Math Soc. (2) **56**(1997), 149-170.

Serre,65. J.-P. Serre, *Sur la dimension cohomologique des groupes profinis,* Topology **3**(1965), 413-420.

Weibel,94. C. A. Weibel, *An Introduction to Homological Algebra*, Cambridge University Press, Cambridge, 1994.

INVARIANT THEORY OF ALGEBRA REPRESENTATIONS

M. DOMOKOS
Rényi Institute of Mathematics,
Hungarian Academy of Sciences
P. O. Box 127, 1364 Budapest, Hungary

1. Polynomial invariants and semisimple representations

A *quiver* Γ is a finite directed graph consisting of Γ_0, the set of vertices, and Γ_1, the set of arrows. For an arrow $a \in \Gamma_1$ we denote by $a' \in \Gamma_0$ its starting vertex and by $a'' \in \Gamma_0$ its terminating vertex.

We fix an algebraically closed base field K. A *dimension vector* $\alpha = (\alpha_i \mid i \in \Gamma_0)$ assigns a non-negative integer α_i with each vertex i. The *space of representations* of Γ with dimension vector α is

$$X := X(\Gamma, \alpha) = \bigoplus_{a \in \Gamma_1} K^{\alpha_{a'} \times \alpha_{a''}},$$

that is, a point of X is a collection of matrices $(x(a) \mid a \in \Gamma_1)$, where $x(a)$ is a matrix of size $\alpha_{a'} \times \alpha_{a''}$. The vector space X is endowed with a linear action of the group $G := GL(\alpha) = \prod_{i \in \Gamma_0} GL(\alpha_i, K)$, given by the rule $g \cdot (x(a) \mid a \in \Gamma_1) = (g_{a'} x(a) g_{a''}^{-1} \mid a \in \Gamma_1)$, $g \in G$, $x \in X$. Our motivating aim is to classify G-orbits in X. For example, when the quiver has a single vertex and m loops, we obtain the well known problem of classifying m-tuples of $n \times n$ matrices up to the simultaneous conjugation action of $GL(n, K)$.

X is called *representation space* because it parameterizes modules over the *path algebra* $K\Gamma$, which is spanned as a K-vector space by the oriented paths of Γ, and multiplication is given by concatenation of paths. By definition $K\Gamma$ contains a path e_i of length zero at

K.W. Roggenkamp and M. Ştefănescu (eds.), Algebra - Representation Theory, 47–61.

the vertex i for any $i \in \Gamma_0$, hence $K\Gamma$ is unitary with $1 = \sum_{i\in\Gamma_0} e_i$. There is a natural map from X to mod-$K\Gamma$, the category of finite dimensional right $K\Gamma$-modules. This assigns to $x = (x(a) \mid a \in \Gamma_1)$ the module V_x. As a K-vector space, $V_x = \bigoplus_{i\in\Gamma_0} K^{\alpha_i}$, where K^{α_i} is the space of row vectors of dimension α_i. Multiplication by e_i projects V_x to the component K^{α_i} corresponding to $i \in \Gamma_0$, and multiplication by $a \in \Gamma_1$ annihilates $V_x e_i$ if $i \neq a'$, and maps $V_x e_{a'} = K^{\alpha_{a'}}$ to $V_x e_{a''} = K^{\alpha_{a''}}$ as the right multiplication of row vectors by the matrix $x(a)$. The action of G on X was defined so that for $x, y \in X$ we have $V_x \cong V_y$ if and only if $G \cdot x = G \cdot y$. We refer to (ARS) for more background on modules over path algebras.

1.1. POLYNOMIAL INVARIANTS OF QUIVERS

Denote by $K[X]$ the coordinate ring of X, this is a $\sum_{a\in\Gamma_1} \alpha_{a'} \cdot \alpha_{a''}$-variable commutative polynomial algebra. Our first object is

$$K[X]^G = \{f \in K[X] \mid \forall g \in G, \forall x \in X : f(g \cdot x) = f(x)\},$$

the *algebra of polynomial invariants.* Let $C = a_1 \dots a_s$ be an oriented cycle in Γ, i.e. an oriented path starting and ending at the same vertex $a_1' = a_s''$, we put α_C for the value of α at this vertex. For $j = 1, \dots, \alpha_C$ denote by σ_C^j the function mapping $x \in X$ to the jth coefficient of the characteristic polynomial of the matrix $x(a_1) \cdots x(a_s)$. The following theorem is due to Le Bruyn-Procesi (LP) when $\mathrm{char}(K) = 0$ (see (D1) for an alternative approach) and to Donkin (Dn2) when $\mathrm{char}(K) = p > 0$.

THEOREM 1.1.1. *$K[X]^G$ is generated as a K-algebra by the elements σ_C^j, $j = 1, \dots, \alpha_C$, as C ranges over the oriented cycles in Γ.*

Since G is reductive, the algebra $K[X]^G$ is finitely generated, so in the above theorem it is sufficient to run C over a finite set O of oriented cycles. When $\mathrm{char}(K) = 0$, one can take for O the set of cycles whose length is at most $(\sum_{i\in\Gamma_0} \alpha_i)^2$.

The algebra $K[X]^G$ is the coordinate ring of some affine algebraic variety X/G, and the algebra injection $K[X]^G \to K[X]$ corresponds to a morphism of affine varieties

$$\pi = \pi_{X,G} : X \to X/G,$$

which has the following properties (see for example (N)):

- π is constant on G-orbits.
- π is a categorical quotient, that is, any morphism $X \to Y$ of affine varieties which is constant on G-orbits factors through π.
- π is surjective and any fibre of π contains a unique closed orbit.

The morphism π can be given explicitly as $\pi : x \mapsto (\sigma_C^j(x) \mid C \in O, j = 1, \ldots, \alpha_C)$. Furthermore it was observed by Artin (A) that

- $G \cdot x$ is closed in X if and only if V_x is semisimple;
- $\pi(x) = \pi(y)$ if and only if V_x and V_y have the same Jordan-Hölder factors in mod-$K\Gamma$.

Thus X/G parameterizes the isomorphism classes of semisimple modules over $K\Gamma$ of dimension vector α.

1.2. SEMISIMPLE REPRESENTATIONS OF QUIVERS

Recall that the combinatorial data of Γ is encoded in the *Euler bilinear form* on $\mathbb{Z}^{\Gamma_0}$ defined by

$$\langle \beta, \gamma \rangle = \sum_{i \in Q_0} \beta(i)\gamma(i) - \sum_{a \in Q_1} \beta(a')\gamma(a'').$$

It has a homological interpretation, namely if V, W are $K\Gamma$-modules and $[V] = (\dim_K(Ve_i) \mid i \in \Gamma_0)$ denotes the dimension vector of V, then

$$\langle [V], [W] \rangle = \dim \operatorname{Hom}_{K\Gamma}(V, W) - \dim \operatorname{Ext}^1_{K\Gamma}(V, W).$$

Next we define the representation type of $x \in X$. Assume that the Jordan-Hölder factors of V_x are $S_1, \ldots, S_t$ with multiplicity $m_1, \ldots, m_t$. Denote by α^i the dimension vector of S_i. Then we call the *representation type* of x the unordered t-tuple $\tau(x) = (\alpha^1, m_1; \ldots; \alpha^t, m_t)$ of simple dimension vectors with multiplicities. For a fixed representation type τ we set

$$X_\tau = \{x \in X \mid \tau(x) = \tau\} \quad \text{and} \quad (X/G)_\tau = \pi(X_\tau).$$

We shall need the following partial order $\preceq$ on the set of representation types: For any representation type $\tau = (\alpha^1, m_1; \ldots; \alpha^t, m_t)$

writing one of the α^i, say α^1 as a sum $\alpha^1 = \beta + \gamma$ of two simple dimension vectors, we get $(\beta, m_1; \gamma, m_1; \alpha^2, m_2; \dots; \alpha^t, m_t) \prec \tau$. If there are two equal dimension vectors among the α^i, say $\alpha^1 = \alpha^2$, then $(\alpha^1, m_1 + m_2; \alpha^3, m_3; \dots; \alpha^t, m_t) \prec \tau$. Making $\preceq$ transitive we get the desired partial order.

Following (LP) we associate a new quiver $\Gamma(\tau)$ with the representation type τ, and call it the *local quiver*. Its vertex set is $\{1, \dots, t\}$, and for any pair (i, j) of vertices $\Gamma(\tau)$ has $\delta_j^i - \langle \alpha^i, \alpha^j \rangle = \dim \mathrm{Ext}^1_{K\Gamma}(S_i, S_j)$ arrows from i to j (here δ_j^i is 1, if $i = j$, and 0 otherwise). Then $\alpha(\tau) = (m_1, \dots, m_t)$ is a dimension vector for $\Gamma(\tau)$, and take the representation space $X(\tau) = X(\Gamma(\tau), \alpha(\tau))$ of the local quiver $\Gamma(\tau)$ with the action of $G(\tau) = GL(\alpha(\tau))$.

Given a point $x \in X$ such that $G \cdot x$ is closed, and $\tau(x) = \tau$, we identify the group $G(\tau)$ with the stabilizer subgroup G_x in G: Clearly, G_x is isomorphic to the group of units in $\mathrm{End}_{K\Gamma}(V_x)$, and by Schur's Lemma this latter group is isomorphic to $G(\tau)$.

In the next theorem we collect the main facts coming from the representation theoretic interpretation of the Luna stratification of X/G. These results are due to Le Bruyn-Procesi (LP), when the base field is of characteristic zero. Recently it was shown in (DZ2) that the same holds in any characteristic. The theme with various applications is discussed also in the book (L2).

THEOREM 1.2.1. *Assume that $x \in X$ with $G \cdot x$ closed, and $\tau(x) = \tau$. Then*

- *(i) there exists an étale morphism from a neighbourhood of $\bar{0} \in X(\tau)/G(\tau)$ onto a neighbourhood of $\pi(x) \in X/G$.*
- *(ii) $\pi^{-1}(\pi(x)) \cong G \times^{G(\tau)} \pi^{-1}_{X(\tau),G(\tau)}(\pi_{X(\tau),G(\tau)}(0))$ as G-varieties.*

Furthermore,

- *(iii) $\bigcup_\nu (X/G)_\nu$ is a finite stratification of X/G into locally closed, irreducible, smooth subvarieties.*
- *(iv) $(X/G)_\nu$ belongs to the Zariski closure of $(X/G)_\tau$ if and only if $\nu \preceq \tau$.*
- *(v) The dimension of the stratum $(X/G)_\tau$ equals to the number of loops in the quiver $\Gamma(\tau)$.*

Recall that the morphism $\phi : Y \to Z$ is called *étale* if the induced homomorphism $\phi_* : \hat{O}_{\phi(y)}(Z) \to \hat{O}_y(Y)$ between the local ring completions is an isomorphism for any $y \in Y$. In particular, when we work over the field of complex numbers, an étale morphism is a local analytic isomorphism.

In statement (ii) above the fibre of $\bar{0}$ at the morphism $\pi_{X(\tau),G(\tau)}$ is a $G(\tau)$-variety, and $G(\tau)$ is identified with the subgroup G_x of G, so the homogeneous fibre space $G \times^{G(\tau)} \pi^{-1}_{X(\tau),G(\tau)}(\bar{0})$ makes sense. We recall this construction. Let H be a closed subgroup of G, and let Z be an H-variety. Then $G \times H$ acts on $G \times Z$ by $(g,h) \cdot (x,z) = (gxh^{-1}, hz)$. The orbit space $G \times^H Z$ for this action has a natural structure of an algebraic G-variety. If Z is affine and H is reductive, then $G \times^H Z$ is also affine. (See the survey (B) for various applications of homogeneous fibre spaces in the geometric study of modules.)

Modulo some combinatorial and elementary linear algebra arguments the closure relation between the various strata given in Theorem 1.2.1 (iv) implies the following description of simple dimension vectors. This is due to Le Bruyn-Procesi (LP) in characteristic zero; see (DZ2) for an argument which shows clearly that the result is a consequence of Theorem 1.2.1. For a dimension vector α we denote by $\mathrm{supp}(\alpha)$ the *support* of α, ie. the full subquiver of Γ spanned by the vertices where α is non-zero. A quiver is *strongly connected* if any pair of its vertices belongs to an oriented cycle.

THEOREM 1.2.2. *There is a simple $K\Gamma$-module with dimension vector α if and only if one of the following alternatives hold:*

(i) $\mathrm{supp}(\alpha)$ *is a single vertex i, and $\alpha = \varepsilon^i$, where $\varepsilon^i_j = \delta^i_j$.*
(ii) $\mathrm{supp}(\alpha)$ *is an indecomposable cycle, and all the non-zero values of α are equal to* 1.
(iii) $\mathrm{supp}(\alpha)$ *is strongly connected, not an indecomposable cycle or a single vertex, and*

$$\langle \alpha, \varepsilon^i \rangle \leq 0, \quad \langle \varepsilon^i, \alpha \rangle \leq 0 \quad \textit{for all } i \in \Gamma_0. \tag{1}$$

It is well known (and follows from Theorem 1.2.1) that if α is simple, then V_x is simple for x taken from an open subset of X. It follows easily from Theorem 1.1.1 that in this case $\dim(X(\Gamma,\alpha)/GL(\alpha))$ is 0, 1, or ≥ 2 depending on whether α is of type (i), (ii), or (iii).

Since $X = X(\Gamma, \alpha)$ is an irreducible algebraic variety, there exists a unique representation type $\tau_g = \tau_g(\Gamma, \alpha)$, called the *generic representation type*, such that the stratum X_{τ_g} is open (and hence dense) in X, or equivalently, $(X/G)_{\tau_g}$ is open in X/G. By Theorem 1.2.1 (iv) this means that τ_g is the unique maximal element with respect to $\preceq$ of the set of representation types of (Γ, α). The generic representation type is determined in (LP). Here we state the result in a slightly more precise and stronger form, which brings out some extra information on X/G.

We need an easy combinatorial Lemma. Below $\beta \leq \alpha$ means $\beta_i \leq \alpha_i$ for all $i \in \Gamma_0$.

LEMMA 1.2.3. *For any dimension vector α there exists a unique maximal dimension vector $\tilde{\alpha}$ such that $\tilde{\alpha} \leq \alpha$ and $\tilde{\alpha}$ satisfies the inequalities (1). The support of $\alpha - \tilde{\alpha}$ does not contain any oriented cycle. In particular, if* $\text{supp}(\alpha)$ *is strongly connected, then* $\text{supp}(\tilde{\alpha}) = \text{supp}(\alpha)$.

Proof. The first statement follows from the observation that if $\beta, \gamma \leq \alpha$ satisfy (1) then $(\max\{\beta_i, \gamma_i\} \mid i \in \Gamma_0)$ also satisfies (1). Moreover, for an oriented cycle C the dimension vector $(1, \ldots, 1)_C$ taking value 1 at all the vertices belonging to C and 0 everywhere else clearly satisfies (1). If C was contained in $\text{supp}(\alpha - \tilde{\alpha})$, then $\tilde{\alpha} + (1, \ldots, 1)_C \leq \alpha$ would satisfy (1), contradicting the definition of $\tilde{\alpha}$. □

Given α introduce an equivalence relation on $\text{supp}(\alpha)_0$ by saying that the vertices i, j are equivalent if they belong to an oriented cycle in $\text{supp}(\alpha)$. The full subquivers of $\text{supp}(\alpha)$ spanned by the equivalence classes are called the *strongly connected components* of $\text{supp}(\alpha)$. Clearly, α decomposes as $\alpha = \alpha^1 + \cdots + \alpha^k$, where $\text{supp}(\alpha^1), \ldots, \text{supp}(\alpha^k)$ are the strongly connected components of $\text{supp}(\alpha)$. Since by Theorem 1.2.2 the simple dimension vectors have strongly connected support, any representation type τ for (Γ, α) can be decomposed as $\tau = (\tau^1; \ldots; \tau^k)$, where τ^i is a representation type for (Γ, α^i). ¿From the description of the simple dimension vectors and the partial order $\preceq$ one easily obtains the following results:

THEOREM 1.2.4.

(i) The generic representation type for (Γ, α) *is given by the generic representation types of the strongly connected components, namely* $\tau_g(\Gamma, \alpha) = (\tau_g(\Gamma, \alpha^1); \ldots; \tau_g(\Gamma, \alpha^k))$.

(ii) If $\mathrm{supp}(\alpha)$ *is strongly connected and not an indecomposable cycle, then* $\tau_g(\Gamma, \alpha) = (\tilde{\alpha}, 1; \varepsilon^i, \alpha_i - \tilde{\alpha}_i \mid i \in \mathrm{supp}(\alpha - \tilde{\alpha}))$.

(iii) If $\mathrm{supp}(\alpha)$ *is an indecomposable cycle, then* $\tilde{\alpha} = m\gamma$, *where* $\gamma = (1, \ldots, 1)_{\mathrm{supp}(\alpha)}$ *and* m *is the minimal positive value of* α. *In this case*

$$\tau_g(\Gamma, \alpha) = (\gamma, 1; \ldots; \gamma, 1; \varepsilon^i, \alpha_i - \tilde{\alpha}_i \mid i \in \mathrm{supp}(\alpha - \tilde{\alpha})).$$

(iv) If α *is a strongly connected component of the dimension vector* β, *and* $i \in \mathrm{supp}(\alpha - \tilde{\alpha})$ *(in particular, there is no loop at* i*), then any* $K\Gamma$*-module of dimension vector* β *has the unique simple module* E^i *of dimension vector* ε^i *as a Jordan-Hölder factor with multiplicity* $\geq \alpha_i - \tilde{\alpha}_i$.

Proof. (i) follows from the observation that $\tau \preceq \nu$ if and only if $\tau^1 \preceq \nu^1, \ldots, \tau^k \preceq \nu^k$. The types we claim generic in (ii), (iii) exist by Theorems 1.2.2 and 1.1.1, and they are clearly maximal with respect to $\preceq$. For the representation type ν of a module whose dimension vector is β we have $\nu \preceq \tau_g(\Gamma, \beta)$, so (iv) follows from (i),(ii),(iii) and the definition of $\preceq$. □

When $\mathrm{supp}(\alpha)$ is strongly connected one can get $\tilde{\alpha}$ by the following algorithm: if the inequalities (1) hold for all $i \in \mathrm{supp}(\alpha)_0$, then $\tilde{\alpha} = \alpha$. Otherwise take an i such that (1) is violated, replace α by $\alpha - \varepsilon^i$, and repeat the same process. We note that by elementary linear algebra arguments, if $\langle \alpha, \varepsilon^i \rangle > 0$ (resp. $\langle \varepsilon^i, \alpha \rangle > 0$), then E^i is a submodule (resp. factor module) of any module of dimension vector α. Furthermore, any β has such a strongly connected component α that there is no arrow in $\mathrm{supp}(\beta)$ going out from $\mathrm{supp}(\alpha)$. Obviously, any module of dimension vector β has a submodule whose dimension vector is this α. These elementary observations yield a direct proof of Theorem 1.2.4 (iv), which does not refer to Theorem 1.2.1.

Once τ_g is determined, one can express $\dim(X/G) = \dim((X/G)_{\tau_g})$ in terms of the Euler form by Theorem 1.2.1 (v).

COROLLARY 1.2.5.

(i) We have the isomorphism of algebraic varieties

$$X(\Gamma,\alpha)/GL(\alpha) \cong X(\Gamma,\alpha^1)/GL(\alpha^1) \times \cdots \times X(\Gamma,\alpha^k)/GL(\alpha^k),$$

where α^i are the strongly connected components of α.

(ii) If supp(α) *is strongly connected, then we have the isomorphism of algebraic varieties $X(\Gamma,\alpha)/GL(\alpha) \cong X(\Gamma,\tilde{\alpha})/GL(\tilde{\alpha})$.*

Proof. (i) is implicit in (L, Lemma 3.4).

To prove (ii), embed $X(\Gamma,\tilde{\alpha})$ into $X(\Gamma,\alpha)$ by adding $\alpha_{a'} - \tilde{\alpha}_{a'}$ number of zero rows and $\alpha_{a''} - \tilde{\alpha}_{a''}$ number of zero columns to the matrix component $x(a)$ of $x = (x(a) \mid a \in \Gamma_1) \in X(\Gamma,\tilde{\alpha})$. Denote by $\iota(x)$ the image of x under this embedding. Clearly, $V_{\iota(x)} \cong V_x \oplus \sum^{\oplus}_{i\in \mathrm{supp}(\alpha-\tilde{\alpha})}(E^i)^{\alpha_i-\tilde{\alpha}_i}$. The embedding ι induces a homomorphism $\iota^* : K[X(\Gamma,\alpha)] \to K[X(\Gamma,\tilde{\alpha})]$, which by Theorem 1.1.1 restricts to a surjective homomorphism $\varphi : K[X(\Gamma,\alpha)]^{GL(\alpha)} \to K[X(\Gamma,\tilde{\alpha})]^{GL(\tilde{\alpha})}$. By Theorem 1.2.4 (iv) for any $y \in X(\Gamma,\alpha)$ there exist an $x \in X(\Gamma,\tilde{\alpha})$ such that V_y and $V_{\iota(x)}$ have the same Jordan-Hölder factors, hence for any $f \in K[X(\Gamma,\alpha)]^{GL(\alpha)}$ we have $f(y) = f(\iota(x)) = \varphi(f)(x)$. Thus $\varphi(f) = 0$ implies $f = 0$, and φ is an isomorphism of the coordinate rings of $X(\Gamma,\alpha)/GL(\alpha)$ and $X(\Gamma,\tilde{\alpha})/GL(\tilde{\alpha})$. □

When supp(β) is an indecomposable cycle, $X(\Gamma,\beta)/GL(\beta)$ is known to be an affine space of dimension $\min\{\beta_i \mid i \in \mathrm{supp}(\beta)\}$. So investigating the structure of the quotient varieties $X(\Gamma,\alpha)/GL(\alpha)$, in view of Corollary 1.2.5 we can reduce to the case when α is a simple dimension vector for Γ.

1.3. ÉTALE SLICES

The proof of Theorem 1.2.1 is based on the theory of étale slices due to Luna (Lu). If char$(K) = 0$, and G is a reductive group acting linearly on a vector space X, there exists an étale slice at any $x \in X$ with $G \cdot x$ closed by (Lu), and the statements follow by general arguments after interpreting the Luna stratification in terms of representation theory, and taking into account the exact structure of the étale slices (see (LP)). When char$(K) = p > 0$, the existence of étale slices is not automatic. However, there is a necessary and sufficient condition for the existence, due to Bardsley-Richardson (BR):

There exists an étale slice at $x \in X$ *with* $G \cdot x$ *closed if and only if the morphism* $G \to G \cdot x$, $g \mapsto gx$ *is separable, and the tangent space* $T_x(G \cdot x)$ *of the orbit* $G \cdot x$ *has a* G_x*-stable direct complement* N_x *in* $T_x(X)$. *Moreover, in this case we can take for an étale slice a principal affine open neighbourhood* $(N_x)_h = \{y \in N_x \mid h(y) \neq 0\}$ *of* x *in* N_x, *where* $h \in K[N_x]^{G_x}$.

So we have to verify the conditions of the above criterion in our situation $X = X(\Gamma, \alpha)$ and $G = GL(\alpha)$. In (LP) it is observed that $T_xX/T_x(G \cdot x) \cong X(\tau)$ as $G_x \cong G(\tau)$-modules. The authors compute the character of both sides, and conclude that they are the same. This proof does not work in positive characteristic. However, inserting a third term in the middle of this isomorphism we get the following quiver version of Voigt's Lemma, which clearly holds in any characteristic:

LEMMA 1.3.1. *Let* V_x *be a semisimple representation of type* τ. *Then we have the isomorphisms*

$$T_xX/T_x(Gx) \cong \mathrm{Ext}^1_{K\Gamma}(V_x, V_x) \cong X(\tau)$$

of $G_x \cong \mathrm{End}_{K\Gamma}(V_x)^* \cong G(\tau)$*-modules.*

(The first isomorphism is obtained following the usual proof of Voigt's Lemma (V), whereas the second is obtained decomposing V_x into the sum of simple modules, and using the bilinearity of Ext^1.)

Therefore we have the exact sequence

$$0 \to T_x(G \cdot x) \to T_xX \to X(\tau) \to 0$$

of $G(\tau)$-modules. Now our group $G(\tau)$ is of the form $\prod_{i=1}^t GL(V_i)$, and as $\prod_{i=1}^t GL(V_i)$-modules, both $X(\tau)$ and $T_x(G \cdot x)$ are direct sums of modules of the form $\mathrm{Hom}_K(V_i, V_j) \cong V_i^* \otimes V_j$. Such $\prod_{i=1}^t GL(V_i)$-modules have no non-trivial extensions with each other. For example, they are *tilting modules*, and any short exact sequence of tilting modules splits (see the lectures of Andersen (An) in the same volume). Thus the above short exact sequence splits, and the Bardsley-Richardson Criterion is fulfilled (the explicit analysis of $T_x(G \cdot x)$ shows also the separability of the orbit map).

Then almost all statements of Theorem 1.2.1 follow by general arguments in the same way as in characteristic zero (see (DZ2) for the details). There is however a slight difference in the proof of the smoothness of the Luna strata: the usual general argument does not work in characteristic $p > 0$. So we have to look at the concrete situation, and apply Theorem 1.1.1.

2. Semi-invariants of quivers

In this section K is an arbitrary infinite field. The theory described in Section 1 becomes trivial when Γ has no oriented cycles. However, there can be non-trivial rational invariants in this case, and other kind of quotient varieties provided by geometric invariant theory (see (K)). Since any rational invariant is the quotient of relative invariants belonging to the same weight, this motivates the study of the so-called semi-invariants.

Consider the subgroup $SL(\alpha) = \prod_{i\in\Gamma_0} SL(\alpha_i, K)$ of $GL(\alpha)$, then $K[X]^{SL(\alpha)}$ is called the *algebra of semi-invariants* of (Γ,α). It is a $\mathbb{Z}^{\Gamma_0}$-graded algebra, i.e. $K[X]^{SL(\alpha)} = \bigoplus_{\theta\in\mathbb{Z}^{\Gamma_0}} K[X]^{G,\theta}$, where

$$K[X]^{G,\theta} = \{f \in K[X] \mid \forall g \in G, \forall x \in X : \ f(g^{-1}x) = \prod_{i\in\Gamma_0} \det(g_i)^{\theta_i} f(x)\}$$

is the *space of relative invariants of weight* θ.

There are three recent independent descriptions of semi-invariants of quivers in (DW), (DZ1), (SV). We present the approach of (DZ1) here.

2.1. RELATIVE INVARIANTS OF BIPARTITE QUIVERS

First we deal with *bipartite quivers*. In a bipartite quiver the vertices are divided into two disjoint subsets $\Gamma_0 = \Gamma_0^- \cup \Gamma_0^+$, and all arrows start in Γ_0^- and terminate in Γ_0^+. Accordingly, a dimension vector is written as $\alpha = (\alpha^-, \alpha^+)$, where $\alpha^- \in \mathbb{Z}^{\Gamma_0^-}$ and $\alpha^+ \in \mathbb{Z}^{\Gamma_0^+}$. Denote by $\tilde{\alpha}$ the weight $\tilde{\alpha} = (-\alpha^-, \alpha^+)$.

Take a pair of dimension vectors α, β, and consider the representation spaces $X = X(\Gamma,\alpha)$ and $Y = X(\Gamma,\beta)$. For $x \in X$ and

$y \in Y$ build the matrix

$$M^{\alpha,\beta}(x,y) = \sum_{a \in \Gamma_1} x(a) \otimes y(a) \otimes e_{a',a''},$$

where $\otimes$ stands for the Kronecker product of matrices, and $e_{a',a''}$ is the matrix unit of size $|\Gamma_0^-| \times |\Gamma_0^+|$ containing 1 in the (a', a'') position. One can imagine $M^{\alpha,\beta}(x,y)$ as a matrix having a double block structure. The large block rows (resp. columns) are indexed by Γ_0^- (resp. Γ_0^+). A large block in the (i^-, j^+) position is again partitioned into $\beta_{i^-} \times \beta_{j^+}$ number of small blocks; the small block in the (k,l) position is the linear combination of the matrices $\{x(a) \mid a' = i^-, a'' = j^+\}$, where the coefficient of $x(a)$ is the (k,l) entry of $y(a)$.

This is a square matrix if

$$\sum_{i \in \Gamma_0^-} \alpha_i \beta_i = \sum_{j \in \Gamma_0^+} \alpha_j \beta_j. \tag{2}$$

In this case we define the function $\chi^{\alpha,\beta}$ on $X \times Y$ by

$$\chi^{\alpha,\beta}(x,y) = \det(M^{\alpha,\beta}(x,y)).$$

By multiplicativity of the determinant $\chi^{\alpha,\beta} \in K[X \times Y]^{GL(\alpha)\times GL(\beta),(\tilde{\beta},\tilde{\alpha})}$, that is, $\chi^{\alpha,\beta}$ is a relative invariant of weight $(\tilde{\beta}, \tilde{\alpha})$ with respect to the natural action of $GL(\alpha)\times GL(\beta)$ on $X \times Y$.

$\chi^{\alpha,\beta}(x,-) \in K[Y]$ is a polynomial whose variables are the coordinate functions on Y, we call it the *β-characteristic polynomial* of the representation x. Considered as functions on X, the coefficients of the β-characteristic polynomial are relative invariants of weight $\tilde{\beta}$.

THEOREM 2.1.1. *Let Γ be a bipartite quiver. If there is a non-zero relative invariant of weight θ on $X = X(\Gamma, \alpha)$, then $\theta = \tilde{\beta}$ for some dimension vector β satisfying (2). In this case $K[X]^{G,\theta} = \mathrm{Span}_K\{\chi^{\alpha,\beta}(-,y) \mid y \in Y\}$, or in other words, the coefficients of the β-characteristic polynomial span the space $K[X]^{G,\tilde{\beta}}$.*

We note that in (DZ1) there is a stronger formulation involving some necessary arithmetical restrictions on the multi-degrees of the non-zero relative invariants. The proof is basically a translation of the first fundamental theorem of classical invariant theory. To get the result in any characteristic we apply the theory of modules with good filtration.

We identify the weights with linear functions on the dimension vectors, as $\theta(\alpha) = \sum_{i\in\Gamma_0}\theta_i\alpha_i$. We say that V is *θ-stable*, if $\theta([V]) = 0$, and for any $0 < W < V$ we have $\theta([W]) > 0$.

The assumption that a module V_y has a proper submodule means that in the G-orbit of y there is a point whose matrix components have certain simultaneous block lower triangular form. Rearranging rows and columns this yields a lower block triangular form of the matrix $M^{\alpha,\beta}(x, y)$. Hence if the zero block is too large, then its determinant is zero, and if the diagonal blocks are of square format, then its determinant factors as the product of the determinants of the diagonal blocks.

COROLLARY 2.1.2. *Let Γ be a bipartite quiver, $X = X(\Gamma, \alpha)$. Then as a K-algebra $K[X]^{SL(\alpha)}$ is generated by those $\chi^{\alpha,\beta}(-, y)$, for which the module V_y is $\tilde{\alpha}$-stable. In particular, β is a Schur root in this case, i.e. there is a Zariski open subset of $X(\Gamma, \beta)$ such that for any point z from this subset V_z is indecomposable.*

2.2. REDUCTION TO BIPARTITE QUIVERS

Start with an arbitrary quiver Γ, and construct the quiver $\hat{\Gamma}$ as follows: double the vertex set of Γ as

$$\hat{\Gamma}_0 = \Gamma_0^- \bigcup \Gamma_0^+ = \{i^- \mid i \in \Gamma_0\} \bigcup \{i^+ \mid i \in \Gamma_0\},$$

for any $a \in \Gamma_1$ put an arrow $\hat{a} \in \hat{\Gamma}_1$ with $\hat{a}' = (a')^-$, $\hat{a}'' = (a'')^+$, and for any $i \in \Gamma_0$ add an extra arrow a_i with $a_i' = i^-$, $a_i'' = i^+$. Since all arrows of $\hat{\Gamma}$ go from a $-$ vertex to a $+$ vertex, $\hat{\Gamma}$ is bipartite.

If α is a dimension vector for Γ, then $\hat{\alpha} = (\alpha, \alpha)$ is a dimension vector for $\hat{\Gamma}$. Consider $\hat{X} = X(\hat{\Gamma}, \hat{\alpha})$ and its $SL(\hat{\alpha})$-stable Zariski closed subset $Z = \{x \in \hat{X} \mid \forall i \in \Gamma_0 : \det(x(a_i)) = 1\}$. The crucial observation is the following Lemma:

LEMMA 2.2.1. *We have the isomorphism of $SL(\hat{\alpha})$-varieties*

$$Z \cong SL(\hat{\alpha}) \times^{SL(\alpha)} X(\Gamma, \alpha).$$

(To form this homogeneous fibre space we identify $SL(\alpha)$ with the diagonal subgroup of $SL(\alpha) \times SL(\alpha) \cong SL(\hat{\alpha})$.)

By Frobenius reciprocity we get from Lemma 2.2.1 that $K[X]^{SL(\alpha)} \cong K[Z]^{SL(\hat{\alpha})}$, hence we can deduce generators of $K[X]^{SL(\alpha)}$ from the generators of $K[X]^{SL(\hat{\alpha})}$.

COROLLARY 2.2.2. *Specializing $x(a_i)$ to the identity matrix for each $i \in \Gamma_0$ and identifying the coordinates of $x(\hat{a})$ with the coordinates of $x(a)$ we obtain a surjective homomorphism $K[\hat{X}]^{SL(\hat{\alpha})} \to K[X]^{SL(\alpha)}$.*

To conclude the surjectivity of the above homomorphism in positive characteristic it is essential that Z is a good subvariety in X (see (Dn1)).

Example. Let Γ be the quiver consisting of a single vertex 1 and a single loop a, $\alpha = (n)$. Then $\hat{\Gamma}$ is the Kronecker quiver consisting of two arrows $\hat{a}$, a_1 going from 1^- to 1^+, and $\hat{\alpha} = (n, n)$. The only interesting weight is $\tilde{\beta} = (-1, 1)$, where $\beta = (1, 1)$. For $x = (x(\hat{a}), x(a_1)) = (A, B) \in K^{n\times n} \times K^{n\times n}$ and $y = (y(\hat{a}), y(a_1)) = (s, t) \in K^2$ we have

$$\chi^{(n,n),(1,1)}(x, y) = \det(sA + tB) = s^n \det(A) + \cdots + t^n \det(B).$$

Substituting the identity matrix into B we get that the coefficients of the ordinary characteristic polynomial of $x(a)$ generate $K[X(\Gamma, (n))]^{SL(n,K)}$, where $SL(n, K)$ acts by conjugation on $n \times n$ matrices.

Semi-invariants of (extended) Dynkin quivers have an extensive literature (see (DZ1) for references). Some concrete examples relating the generalized Kronecker quiver are studied quantitatively in (D2), (D3). We note also that similar problems have relevance in the more general framework of quivers with relations (see (K)). For example, the so-called canonical algebras are studied in (DL), (SW) from this point of view.

2.3. RELATIVE INVARIANTS AND PERPENDICULAR MODULES

Starting with a projective resolution $0 \to P_1 \xrightarrow{y} P_0 \to V \to 0$ of some $K\Gamma$-module V with $\langle [V], \alpha \rangle = 0$, Schofield (S) constructed a relative invariant $\kappa_y \in K[X(\Gamma, \alpha)]^{GL(\alpha),\theta}$, where $\theta(\beta) = -\langle [V], \beta \rangle$ for any dimension vector β. By construction, for any $x \in X$ we have

$$\kappa_y(x) \neq 0 \Leftrightarrow V_x \in V^{\perp},$$

where $V^{\perp} = \{W \in \text{mod-}K\Gamma \mid \text{Hom}_{K\Gamma}(V, W) = 0 = \text{Ext}^1_{K\Gamma}(V, W)\}$ denotes the *right perpendicular category* of V.

Derksen-Weyman (DW) (assuming that Γ has no oriented cycles) and Schofield-Van den Bergh (SV) (assuming that $\text{char}(K) = 0$) proved that the relative invariants κ_y span the space of relative invariants of weight θ. This result relates semi-invariants and modules directly.

To make the connection between Schofield's relative invariants and our construction in 2.1, we note finally that if Γ is a bipartite quiver, β is a dimension vector with (2), and $y \in X(\Gamma, \beta)$, then one can associate naturally with y an injective homomorphism $P_1 \xrightarrow{y} P_0$ of projective modules, such that κ_y and $\chi^{\alpha,\beta}(-, y)$ coincide.

This research was partially supported by OTKA No. F32325.

References

An. H. Andersen, Tilting modules for algebraic and quantum groups, this volume.

A. M. Artin, On Azumaya algebras and finite dimensional representations of rings, *J. Algebra* **11** (1969), 532–563.

ARS. M. Auslander, I. Reiten and S. Smalø, "Representation Theory of Artin Algebras", Cambridge University Press, 1995.

BR. P. Bardsley, R. W. Richardson, Étale slices for algebraic transformation groups in characteristic p, *Proc. London Math. Soc.*, Ser 3, **51** (1985), 295–317.

B. K. Bongartz, Some geometric aspects of representation theory, Algebras and Modules I, *CMS Conf. Proc.* **23** (1998), 1–27, (I. Reiten, S. O. Smalø, Ø. Solberg, editors).

DW. H. Derksen and J. Weyman, Semi-invariants of quivers and saturation for Littlewood-Richardson coefficients, *J. Amer. Math. Soc.* **13** (2000), 467–479.

D1. M. Domokos, Invariants of quivers and wreath products, *Comm. Alg.* **26** (1998), 2807–2819.

D2. M. Domokos, Relative invariants of 3×3 matrix triples, *Lin. Multilin. Alg.* **47** (2000), 175–190.

D3. M. Domokos: Poincaré series of semi-invariants of 2×2 matrices, *Lin. Alg. Appl.* **310** (2000), 183–194.

DL. M. Domokos and H. Lenzing, Invariant theory of canonical algebras, *Journal of Algebra* **228** (2000), 738–762.

DZ1. M. Domokos and A. N. Zubkov, Semi-invariants of quivers as determinants, *Transformation Groups* **6** (2001), 9–24.

DZ2. M. Domokos and A. N. Zubkov, Semisimple representations of quivers in characteristic p, *Algebras and Representation Theory*, to appear.

Dn1. S. Donkin, The normality of conjugacy classes of matrices, *Inv. Math.* **101** (1990), 717–736.

Dn2. S. Donkin, Polynomial invariants of representations of quivers, *Comment. Math. Helvetici* **69** (1994), 137–141.

K. A. D. King, Moduli of representations of finite dimensional algebras, *Quart. J. Math. Oxford* (2), **45** (1994), 515–530.

L. L. Le Bruyn, Singularities of quantum groups, *Proc. London Math. Soc.* **80** (2000), 304–336.

L2. L. Le Bruyn, Noncommutative Geometry (book in preparation), http://win-www.uia.ac.be/u/lebruyn/Book.html

LP. L. Le Bruyn and C. Procesi, Semisimple representations of quivers, *Trans. Amer. Math. Soc.* **317** (1990), 585–598.

Lu. D. Luna, Slices étales, *Bull. Soc. Math. Mém.* **33** (1973), 81–105.

N. P. E. Newstead, Introduction to Moduli Problems and Orbit Spaces, *Tata Institute Lecture Notes*, Springer-Verlag, 1978.

S. A. Schofield, Semi-invariants of quivers, *J. London Math. Soc. (2)* **43** (1991), 385–395.

SV. A. Schofield and M. Van den Bergh, *Semi-invariants of quivers for arbitrary dimension vectors*, Indag. Math., to appear.

SW. A. Skowroński and J. Weyman, Semi-invariants of canonical algebras, *Manuscripta Math.* **100** (1999), 391–403.

V. D. Voigt, Endliche algebraische Gruppen, *Lecture Notes in Math.* **592**, Springer-Verlag, 1977.

REPRESENTATION THEORY OF ORDERS

OSAMU IYAMA
Department of Mathematics,
Kyoto University,
Kyoto, 606-8502,
Japan

Introduction

Orders over complete discrete valuation rings give fruitful examples of non-commutative noetherian rings of Krull dimension one. They often play important roles in number theory as well. Some results about their representation and relation to finite dimensinal algebras and higher dimensional orders will be explained. Main objects are: Bass orders (§4), AR quivers (§5,§6,§7), finite type (§7) and tame type (§8,§9). Each topics are closely related to one another.

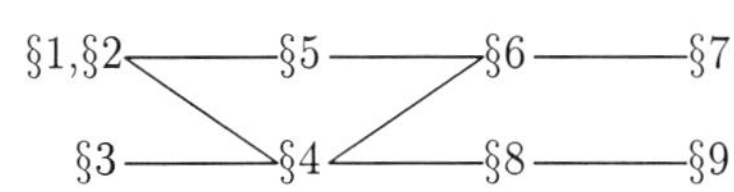

Throughout this paper, let R be a complete regular local ring of dimension $d \geq 0$ and K a quotient field of R. An R-algebra Λ is called an R-**order** if it is finitely generated free as an R-module. For an R-order Λ, a left Λ-module L is called a Λ-**lattice** if it is finitely generated free as an R-module. We denote by $\operatorname{lat}\Lambda$ (resp. $\operatorname{pr}\Lambda$) the category of Λ-lattices (resp. projective Λ-lattices). Then $(\ )^* = \operatorname{Hom}_R(\ , R)$ gives a duality between $\operatorname{lat}\Lambda$ and $\operatorname{lat}\Lambda^{op}$. Let $\operatorname{rin}\Lambda := (\operatorname{pr}\Lambda^{op})^*$ be the category of **relative injective** Λ-lattices. We denote by J_Λ the Jacobson radical of Λ. For $d = 0$ (R is a field), an R-algebra is an R-order iff it is finite dimensional over R. In this case, $\operatorname{lat}\Lambda$ is denoted by $\operatorname{mod}\Lambda$, which is the category of finite dimensional Λ-modules.

K.W. Roggenkamp and M. Ştefănescu (eds.), Algebra - Representation Theory, 63–96.

An additive category $\mathcal{C}$ is called **Krull-Schmidt** if any object is isomorphic to a finite direct sum of objects whose endomorphism rings are local. Then any object can be written uniquely as a finite direct sum of indecomposable objects. We denote by $\mathfrak{I}(\mathcal{C})$ the set of isomorphism classes of indecomposable objects in $\mathcal{C}$. It is a basic result that $\operatorname{lat}\Lambda$ *is a Krull-Schmidt category for an R-order* Λ.

For an R-order Λ and $L \in \operatorname{lat}\Lambda$, we put $\widetilde{L} := L \otimes_R K$. The finite dimensional K-algebra $\widetilde{\Lambda}$ is called the **ambient algebra** of Λ, and Λ is called an order **in** $\widetilde{\Lambda}$. An R-order Λ is called of **finite type** if $\mathfrak{I}(\operatorname{lat}\Lambda)$ is a finite set, and called of **bounded type** if $\sup_{L\in\mathfrak{I}(\operatorname{lat}\Lambda)} l_{\widetilde{\Lambda}}(\widetilde{L}) < \infty$, where $l_A(X)$ is the length of an A-module X.

An R-order Λ is called an **isolated singularity** if $\operatorname{gl.dim}\Lambda_{\wp} = \operatorname{ht}\wp$ holds for any non-maximal prime ideal $\wp$ of R and $\Lambda_{\wp} := \Lambda \otimes_R R_{\wp}$. It is well known that any order of finite type is an isolated singularity [**A2**], and any isolated singularity of bounded type is of finite type (Brauer-Thrall I)[**Y**]. For $d = 1$, Λ is an isolated singularity iff the ambient algebra $\widetilde{\Lambda}$ is semisimple.

In this paper, *assume* $\dim R = 1$ *unless explicitly stated otherwise.*

1 Overorders and Overrings

Let $\dim R$ be arbitrary. An R-order Γ is called an **overring** (resp. **overorder**) of Λ if $\widetilde{\Lambda}/I \supset \Gamma \supseteq (\Lambda + I)/I$ holds for some ideal I of $\widetilde{\Lambda}$ (resp. $\widetilde{\Lambda} \supset \Gamma \supseteq \Lambda$ holds). Then the natural morphism $\Lambda \to \Gamma$ induces a full faithful functor $\operatorname{lat}\Gamma \to \operatorname{lat}\Lambda$. Thus $\operatorname{lat}\Gamma$ (resp. $\mathfrak{I}(\operatorname{lat}\Lambda) - \mathfrak{I}(\operatorname{lat}\Gamma)$) can be regarded as a full subcategory of $\operatorname{lat}\Lambda$ (resp. subset of $\mathfrak{I}(\operatorname{lat}\Lambda)$). We call such a full subcategory of $\operatorname{lat}\Lambda$ (resp. subset of $\mathfrak{I}(\operatorname{lat}\Lambda)$) **rejective** (resp. **rejectable**). Naturally, we have bijections between (overrings of Λ), (rejective subcategory of $\operatorname{lat}\Lambda$) and (rejectable subsets of $\mathfrak{I}(\operatorname{lat}\Lambda)$).

It is a basic method in representation theory of orders to compare $\operatorname{lat}\Lambda$ with $\operatorname{lat}\Gamma$. To study rejective subcategories, first notice that the inclusion functor $\operatorname{lat}\Gamma \to \operatorname{lat}\Lambda$ has a right adjoint $(\ )^- := \operatorname{Hom}_\Lambda(\Gamma,\ )$ and a left adjoint $(\ )^+ := (\Gamma\otimes_\Lambda\ )^{**}$. The following shows that these properties characterize rejective subcategories of $\operatorname{lat}\Lambda$. It will be used to define rejective subcategories for τ-categories in §6.

1.1 (Adjoint functors) *Let* Λ *be an* R*-order and* $\dim R \leq 2$. *A subcategory* $\mathcal{C}'$ *of* $\mathcal{C} := \operatorname{lat}\Lambda$ *is rejective iff the natural inclusion* $\mathcal{C}' \to \mathcal{C}$ *has a right adjoint* $(\ )^- : \mathcal{C} \to \mathcal{C}'$ *with a counit* ϵ^- *such that* ϵ_X^- *is a monomorphism for any* $X \in \mathcal{C}$, *and a left adjoint* $(\ )^+ : \mathcal{C} \to \mathcal{C}'$ *with a unit* ϵ^+ *such that* ϵ_X^+ *is an epimorphism for any* $X \in \mathcal{C}$.

1.2 For rejectability of a singleton set, there is the following criterion due to Drozd-Kirichencko [**DK1**]. In §6, we will study its generalization.

(DK Rejection Lemma) *Let* Λ *be an* R*-order and* $\dim R \leq 1$. *Then a singleton set* $\{X\}$ *is rejectable iff* $X \in \operatorname{pr}\Lambda \cap \operatorname{rin}\Lambda$.

In this case, assume that $\dim R = 1$ *and* Λ *is a non-maximal (1.3) local order. Then* $\Gamma := O_l(J_\Lambda)$ *satisfies* $\mathfrak{I}(\operatorname{lat}\Lambda) - \mathfrak{I}(\operatorname{lat}\Gamma) = \{X\}$, *where we put* $O_l(L) := \{x \in \widetilde{\Lambda} \mid xL \subseteq L\}$ *for* $L \in \operatorname{lat}\Lambda$.

1.3 Maximal orders and hereditary orders Assume $\dim R = 1$. An R-order Λ is called **maximal** if Λ has no proper overorder, and called **hereditary** if $\operatorname{gl.dim}\Lambda = 1$. Then the following results are known [**CR**].

(1) Any order in a semisimple algebra has a maximal overorder since any increasing chain of orders in a semisimple algebra stops.

(2) Any maximal order is hereditary, and the ambient algebra of any hereditary order is semisimple.

(3)**(Structure theorem)** In any division K-algebra D, there exists a unique maximal R-order $\mathcal{O}_D$. An R-order is hereditary (resp. maximal) iff it is Morita equivalent to a direct product of $\mathrm{T}_l(\mathcal{O}_D)$ (resp. $\mathcal{O}_D$), where D is a division K-algebra and $\mathrm{T}_l(\mathcal{O}_D)$ is the subring of $\mathrm{M}_l(\mathcal{O}_D)$ consisting of $(x_{ij})_{i,j}$ such that $x_{ij} \in J_{\mathcal{O}_D}$ for any $i < j$.

2 AR sequences and AR quivers

Let $\dim R$ be arbitrary. We will explain AR sequences whose meaning we can understand essentially by considering functor categories.

2.1 Modules over additive categories Let $\mathcal{C}$ be a Krull-Schmidt category which is skeletally small, namely isomorphism classes of objects form a set. We denote by $\mathcal{C}(X,Y)$ the set of morphisms from X to Y, and by $fg \in \mathcal{C}(X,Z)$ the composition of $f \in \mathcal{C}(X,Y)$ and $g \in \mathcal{C}(Y,Z)$. A **$\mathcal{C}$-module** is a contravariant additive functor from $\mathcal{C}$ to the category $\mathcal{A}b$ of abelian groups. For $\mathcal{C}$-modules M and M', we denote by $\mathrm{Hom}(M,M')$ the set of natural transformations from M to M'. Thus we obtain the abelian category $\mathrm{Mod}\,\mathcal{C}$ of $\mathcal{C}$-modules [**A1**].

Define functors $H^{\mathcal{C}} : \mathcal{C} \to \mathrm{Mod}\,\mathcal{C}$ and $H_{\mathcal{C}} : \mathcal{C} \to \mathrm{Mod}\,\mathcal{C}^{op}$ by $H_X^{\mathcal{C}} := \mathcal{C}(\ ,X)$ and $H_{\mathcal{C}}^X := \mathcal{C}(X,\ )$. We say that a $\mathcal{C}$-module M is **finitely generated** if there exists an exact sequence $H_X^{\mathcal{C}} \to M \to 0$ for some $X \in \mathcal{C}$. Then Yoneda's Lemma shows that the functor $H^{\mathcal{C}}$ (resp. $H_{\mathcal{C}}$) induces an equivalence from $\mathcal{C}$ to the category of finitely generated projective $\mathcal{C}$-modules (resp. $\mathcal{C}^{op}$-modules).

We denote by $\mathcal{J}_{\mathcal{C}}$ the **Jacobson radical** of $\mathcal{C}$, which is an ideal of $\mathcal{C}$ such that $\mathcal{J}_{\mathcal{C}}(\ ,X)$ (resp. $\mathcal{J}_{\mathcal{C}}(X,\ )$) is the **radical** (=intersection of all maximal submodules) of $H_X^{\mathcal{C}}$ (resp. $H_{\mathcal{C}}^X$) for any $X \in \mathcal{C}$. Define functors
$S^{\mathcal{C}} : \mathcal{C} \to \mathrm{Mod}\,\mathcal{C}$ and $S_{\mathcal{C}} : \mathcal{C} \to \mathrm{Mod}\,\mathcal{C}^{op}$ by $S_X^{\mathcal{C}} := H_X^{\mathcal{C}} / \mathcal{J}_{\mathcal{C}}(\ ,X)$ and $S_{\mathcal{C}}^X := H_X^{\mathcal{C}} / \mathcal{J}_{\mathcal{C}}(X,\ )$. Then $S^{\mathcal{C}}$ (resp. $S_{\mathcal{C}}$) induces a bijection from $\mathfrak{I}(\mathcal{C})$ to the set of isomorphism classes of simple $\mathcal{C}$-modules (resp. $\mathcal{C}^{op}$-modules). We simply denote $H_X^{\mathcal{C}}$ (resp. $H_{\mathcal{C}}^X$, $S_X^{\mathcal{C}}$, $S_{\mathcal{C}}^X$) by H_X (resp. H^X, S_X, S^X).

We denote by $\mathrm{pd}\,M$ the projective dimension of $\mathcal{C}$-module M. For $n \geq 0$, we denote by $\mathfrak{I}_n^+(\mathcal{C})$ (resp. $\mathfrak{I}_n^-(\mathcal{C})$) the subset of $\mathfrak{I}(\mathcal{C})$ consisting of X such that $\mathrm{pd}\,S_X \leq n$ (resp. $\mathrm{pd}\,S^X \leq n$).

$\mathcal{C}$ is called **left artinian** (resp. **right artinian**) if H_X (resp. H^X) has a finite length for any $X \in \mathcal{C}$, and called **artinian** if $\mathcal{C}$ is left and right artinian.

2.2 AR sequences It is an impotant result that *an R-order Λ is an isolated singularity iff* $\mathrm{lat}\,\Lambda$ *has AR sequences* [**A2**]. This means the second part of the following theorem.

2.2.1 Theorem *Let Λ be an R-order and $d := \dim R$.*

(1) For any $X \in \mathfrak{I}(\mathrm{pr}\,\Lambda)$ (resp. $X \in \mathfrak{I}(\mathrm{rin}\,\Lambda)$), $\mathrm{pd}\,S_X = d$ (resp. $\mathrm{pd}\,S^X = d$) holds if $d > 0$, and $\mathrm{pd}\,S_X \leq 1$ (resp. $\mathrm{pd}\,S^X \leq 1$) holds if $d = 0$. Moreover, S_X (resp. S^X) has a minimal projective

resolution whose terms are finitely generated.

(2) Λ *is an isolated singularity iff, for any* $X \in \mathfrak{I}(\operatorname{lat}\Lambda) - \mathfrak{I}(\operatorname{pr}\Lambda)$ *(resp.* $X \in \mathfrak{I}(\operatorname{lat}\Lambda) - \mathfrak{I}(\operatorname{rin}\Lambda)$*),* $\operatorname{pd} S_X = 2$ *(resp.* $\operatorname{pd} S^X = 2$*) holds and* S_X *(resp.* S^X*) has a minimal projective resolution whose terms are finitely generated.*

Assume that Λ is an isolated singularity. For any $X \in \operatorname{lat}\Lambda$, we can take a complex $(X] = (\tau^+X \overset{\nu_X^+}{\to} \theta^+X \overset{\mu_X^+}{\to} X)$ (resp. $[X) = (X \overset{\mu_X^-}{\to} \theta^-X \overset{\nu_X^-}{\to} \tau^-X))$ in $\operatorname{lat}\Lambda$ such that $H_{\tau^+X} \to H_{\theta^+X} \to H_X \to S_X \to 0$ (resp. $H^{\tau^-X} \to H^{\theta^-X} \to H^X \to S^X \to 0$) gives the first part of a minimal projective resolution. Then we can show the following fact easily.

2.2.2 Proposition *Let* Λ *be an isolated singularity and* $d := \dim R$.

(1) If $X \in \mathfrak{I}(\operatorname{lat}\Lambda)$ *satisfies* $\operatorname{pd} S_X = 2$ *(resp.* $\operatorname{pd} S^X = 2$*), then* $(X] = [\tau^+X)$ *(resp.* $[X) = (\tau^-X]$*) holds. We call* $(X]$ *(resp.* $[X)$*) an* **AR sequence**.

(2) τ^+ *and* τ^- *give mutually inverse bijections between* $\mathfrak{I}(\operatorname{lat}\Lambda) - \mathfrak{I}(\operatorname{pr}\Lambda)$ *and* $\mathfrak{I}(\operatorname{lat}\Lambda) - \mathfrak{I}(\operatorname{rin}\Lambda)$*. Moreover, if* $d = 2$*, then* τ^+ *and* τ^- *give mutually inverse bijections between* $\mathfrak{I}(\operatorname{pr}\Lambda)$ *and* $\mathfrak{I}(\operatorname{rin}\Lambda)$.

2.3 AR quivers (1) $Q = (Q, Q^p, Q^i, \tau^+, d, d')$ is called a **translation quiver** if Q is a set, Q^p and Q^i are subsets of Q, τ^+ is a bijection $Q - Q^p \to Q - Q^i$, and d and d' are maps $Q \times Q \to \mathbb{N}_{\geq 0}$ such that $d(Y, X) = d'(\tau^+X, Y)$ holds for any $X \in Q - Q^p$ and $Y \in Q$.

Usually, we draw Q as a directed graph: Q is the set of vertices, and we draw valued arrows $X \xrightarrow{(d(X,Y),d'(X,Y))} Y$ for any $X, Y \in Q$ such that $d(X, Y) \neq 0$, and dotted arrows from X to τ^+X for any $X \in Q - Q^p$. We usually omit a valuation $(1, 1)$.

Moreover, a translation quiver Q is called **admissible** if there exists a map $c : Q \to \mathbb{N}_{>0}$ such that $c(X)d(X, Y) = d'(X, Y)c(Y)$ holds for any X, $Y \in Q$.

(2) For a set Q, we denote by $\mathbb{Z}Q$ (resp. $\mathbb{N}Q$) the free $\mathbb{Z}$-module (resp. free abelian monoid) generated by Q. We introduce an inner

product $\langle \ , \ \rangle$ on $\mathbb{Z}Q$ by taking Q as an orthonormal base. For $X \in \mathbb{Z}Q$, put $\operatorname{supp} X := \{Y \in Q \mid \langle X, Y\rangle \neq 0\}$, and write $X = X_+ - X_-$ where $X_+, X_- \in \mathbb{N}Q$ and $\operatorname{supp} X_+ \cap \operatorname{supp} X_- = \emptyset$.

For example, the set of isomorphism classes of objects in a Krull-Schmidt category $\mathcal{C}$ is given by $\mathbb{N}\mathfrak{I}(\mathcal{C})$.

(3) Let Λ be an isolated singularity, $\dim R \leq 2$ and $\mathcal{C} := \operatorname{lat}\Lambda$. Then a translation quiver $\mathbb{A}(\mathcal{C}) = (Q, Q^p, Q^i, \tau^+, d, d')$ called the **AR quiver** of Λ is defined by $Q := \mathfrak{I}(\mathcal{C})$, $Q^p := \mathfrak{I}_1^+(\mathcal{C})$, $Q^i := \mathfrak{I}_1^-(\mathcal{C})$, $d(X,Y) := \langle\theta^+ Y, X\rangle$ and $d'(X,Y) := \langle\theta^- X, Y\rangle$. Then $\mathbb{A}(\mathcal{C})$ is admissible by $k := R/J_R$ and $c(X) := \dim_k \operatorname{End}_\Lambda(X)/J_{\operatorname{End}_\Lambda(X)}$.

$\mathbb{A}(\operatorname{lat}\Lambda)$ displays terms of each $(X]$ and $[X)$ diagrammatically. We will give a characterization of $\mathbb{A}(\operatorname{lat}\Lambda)$ for orders of finite type in §7.

(4) For a translation quiver $Q = (Q, Q^p, Q^i, \tau^+, d, d')$, define elements θ^+, θ^-, τ^+ and τ^- of $\operatorname{End}_{\mathbb{Z}}(\mathbb{Z}Q)$ as follows. Put $\theta^+ X := \sum_{Y\in Q} d(Y,X)Y$ and $\theta^- X := \sum_{Y\in Q} d'(X,Y)Y$ for $X \in Q$. Put $\tau^+ X := 0$ for $X \in Q^p$, $\tau^- X := (\tau^+)^{-1}(X)$ for $X \in Q - Q^i$ and $\tau^- X := 0$ for $X \in Q^i$.

For $f \in \operatorname{End}_{\mathbb{Z}}(\mathbb{Z}Q)$ and a subset S of Q, define $f_{Q/S} \in \operatorname{End}_{\mathbb{Z}}(\mathbb{Z}(Q - S))$ by $f_{Q/S}(X) := f(X)|_{Q-S}$ where $|_{Q-S} : \mathbb{Z}Q \to \mathbb{Z}(Q - S)$ is the natural projection.

2.3.1 Example Let Λ be an isolated singularity, $d := \dim R \leq 2$ and $Q := \mathbb{A}(\operatorname{lat}\Lambda)$.

(1) If $d \leq 1$, then $Q^p = \mathfrak{I}(\operatorname{pr}\Lambda)$ and $Q^i = \mathfrak{I}(\operatorname{rin}\Lambda)$. If $d = 2$, then $Q^p = Q^i = \emptyset$.

(2) For $\Lambda := \mathrm{T}_l(\mathcal{O}_D)$ in 1.3(3), Q is a cycle 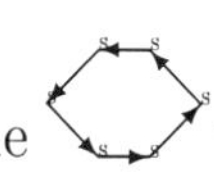 with l-vertices such that $Q = Q^p = Q^i$. Hence the AR quiver of a hereditary order is a disjoint union of such cycles, and the AR quiver of a maximal order is a disjoint union of loops $\circlearrowright$.

3 Quadratic extensions

As we will see in §4 and §8, some kind of quadratic extension often appears in representation theory of orders [**I4**]. In this section, any modules and algebras are assumed to be finitely generated R-modules.

3.1 Definition (1) An (A,B)-bimodule M is called a **Nakayama bimodule** if the set of sub A-modules of Mf is totally ordered by inclusion for any primitive idempotent f of B, and the set of sub B^{op}-modules of eM is totally ordered by inclusion for any primitive idempotent e of A. An algebra A is called a **Nakayama algebra** if A is a Nakayama (A,A)-bimodule.

(2) Let $\mathcal{C}_l$ and $\mathcal{C}_r$ be Krull-Schmidt categories. A $(\mathcal{C}_l,\mathcal{C}_r)$**-bimodule** is a covariant additive functor $M : \mathcal{C}_r \to \mathrm{Mod}\,\mathcal{C}_l$. In other word, a multiliner map $\mathcal{C}_l(W,X)\times M(X,Y)\times\mathcal{C}_r(Y,Z) \to M(W,Z)$ is given for any $W,X \in \mathcal{C}_l$ and $Y,Z \in \mathcal{C}_r$ such that the associativity holds.

M is called a **Nakayama bimodule** if $M(X,Y)$ is a Nakayama $(\mathcal{C}_l(X,X),\mathcal{C}_r(Y,Y))$-bimodule for any $X \in \mathcal{C}_l$ and $Y \in \mathcal{C}_r$.

3.2 Definition Let $f : B \to A$ be an algebra monomorphism.

(1) We call f a **minimal quadratic extension** if f is Morita equivalent to one of the following (O)–(V), where D is a division algebra and D' is a sub-division algebra of D of index 2.

(O): Identity map $D \to D$.

(I·II): Regular representation $D \to \mathrm{M}_2(D')$.

(III): Diagonal embedding $D \to D \times D$.

(IV): Injection $D' \to D$.

(V): $D \times D \to \mathrm{M}_2(D)$, $(x,y) \mapsto \left(\begin{smallmatrix} x & 0 \\ 0 & y \end{smallmatrix}\right)$.

Assume that R is a valuation field. A morhism of type(I·II) is called of **type(I)** (resp. **type(II)**) if it is unramified (resp. ramified). A morhism of type(IV) is called of **type(IVa)** (resp. **type(IVb)**) if it is unramified (resp. ramified).

(2) Assume that A is semisimple. Then we call f a **normal quadratic extension** (resp. **triangular extension**) if f is Morita equivalent to a direct product of minimal quadratic extensions (resp. injections $\mathrm{T}_l(D) \to \mathrm{M}_l(D)$, where D is a division algebra and $\mathrm{T}_l(D)$ is the upper triangular matrix ring). We call f a **quadratic extension** if f is a composition $B \xrightarrow{g} C \xrightarrow{h} A$ of a normal quadratic extension g and a triangular extension h.

For $k := R/J_R$, we can show that f *is a quadratic extension iff* $2\dim_k eBe \geq \dim_k eAe$ *holds for any idempotent* e *of* B.

In general, we call f a **quadratic extension** (resp. **normal quadratic extension**, **triangular extension**) if A is a Nakayama algebra, $J_A \subset B$ and $B/J_A \to A/J_A$ is a quadratic extension (resp. normal quadratic extension, triangular extension) above.

(3) An additive functor $\mathbb{I} : \mathcal{C} \to \mathcal{D}$ of Krull-Schmidt categories is called a **quadratic extension** (resp. **normal quadratic extension**, **triangular extension**) if $\mathbb{I}_{X,X} : \mathcal{C}(X,X) \to \mathcal{D}(\mathbb{I}(X), \mathbb{I}(X))$ is a quadratic extension (resp. normal quadratic extension, triangular extension) for any $X \in \mathcal{C}$. Then any quadratic extension $\mathbb{I}$ is uniquely written as a composition $\mathcal{C} \overset{\mathbb{J}}{\to} \mathcal{E} \overset{\mathbb{K}}{\to} \mathcal{D}$ of a normal quadratic extension $\mathbb{J}$ and a triangular extension $\mathbb{K}$. This decomposition is called a **normalization** of $\mathbb{I}$.

(4) For a normal quadratic extension $\mathbb{I} : \mathcal{C} \to \mathcal{D}$, define a graph $G(\mathbb{I})$ as follows. The set of vertices is $\mathfrak{I}(\mathcal{D})$. For any $X \in \mathfrak{I}(\mathcal{D})$, put $Y := \bigoplus_{Z \in \mathfrak{I}(\mathcal{C}), X \le \mathbb{I}(Z)} Z \in \mathcal{C}$. Then the induced morphism $\mathbb{I}_{Y,Y} : \mathcal{C} / \mathcal{J}_{\mathcal{C}}(Y,Y) \to \mathcal{D} / \mathcal{J}_{\mathcal{D}}(\mathbb{I}(Y), \mathbb{I}(Y))$ is a minimal quadratic extension. If it is of type(I·II)(IV) or (V), then we draw an edge X—X. If it is of type(III), then $\mathbb{I}(Y) = X \oplus X'$ holds for some $X \neq X' \in \mathfrak{I}(\mathcal{D})$, and we draw an edge X—X'.

3.3 Example Let Λ be an R-order and $\dim R = 1$.

Then Λ is a hereditary order iff Λ is a Nakayama algebra iff the injection $\Lambda \to \Gamma$ is a triangular extension for some (any) overorder Γ [**HN2**]. On the other hand, we call Λ a **quadratic order** if the injection $\Lambda \to \Gamma$ is a quadratic extension for some overorder Γ.

Recall that Λ is called **Bäckström** if there exists a hereditary overorder Γ such that $\Lambda \supset J_\Lambda = J_\Gamma$ [**RR**]. Then any quadratic order is Bäckström.

4 Bass orders

In this section, assume $\dim R = 1$. An R-order Λ is called **Gorenstein** if $\Lambda^* = \mathrm{Hom}_R(\Lambda, R) \in \mathrm{pr}\,\Lambda$, and called **Bass** if any overorder of Λ is Gorenstein. It is easily shown that any hereditary order is Bass.

4.1 For simplicity, let $\Lambda^{(0)} := \Lambda$ be a local Bass order. If Λ is not hereditary, then there exists an overorder $\Lambda^{(1)}$ such that $\mathfrak{I}(\mathrm{lat}\,\Lambda^{(0)}) - \mathfrak{I}(\mathrm{lat}\,\Lambda^{(1)}) = \{\Lambda^{(0)}\}$ by DK Rejection Lemma (1.2). Again $\Lambda^{(1)}$ is Bass by definition, and we can show that $\Lambda^{(1)}$ is local if $\Lambda^{(1)}$ is not hereditary. Thus we obtain a unique (finite or infinite) sequence $\Lambda^{(0)} \subset \Lambda^{(1)} \subset \Lambda^{(2)} \subset \cdots$ satisfying $\mathfrak{I}(\operatorname{lat}\Lambda^{(i)}) - \mathfrak{I}(\operatorname{lat}\Lambda^{(i+1)}) = \{\Lambda^{(i)}\}$, where the last term $\Lambda^{(s)}$ is hereditary if the sequence is finite, and other terms are non-hereditary local Bass orders.

This sequence is fundumental for our study of Bass orders. As an application, we can obtain the following result [**DKR**][**Ro**][**HN2**], where we call an algebra **purely non-semisimple** if it does not have a simple ring direct summand.

Theorem *(1) A Gorenstein order Λ is Bass iff any minimal overorder of Λ is Gorenstein.*

(2) Any Bass order is Morita equivalent to a direct product of the following orders.

(i) A hereditary order.

(ii) $\begin{pmatrix} \Omega & \Omega \\ J_\Omega^{s+1} & \Omega \end{pmatrix}$ *where Ω is a local maximal order and $s > 0$.*

(iii) A local Bass order in a semisimple algebra.

(iv) A Bass order in a purely non-semisimple algebra.

In 1.3, we already have detailed description for (i) and (ii). In 4.2 (resp. 4.3), we will review some results for (iii) (resp. (iv)).

4.2 Local Bass orders in semisimple algebras A chain $\{\Lambda_i\}_{0\le i\le s}$ $(0 \le s \le \infty)$ is called a **primary Bass chain** of length s if Λ_i is a unique minimal overorder of a local Bass order Λ_{i+1} for any i $(0 \le i < s)$ and Λ_0 is hereditary. Then a pair $(\Lambda_0, J_{\Lambda_1})$ is called the **dominating pair**. We can show that it is one of the following (I)–(IVb) where Ω and Ω' are local maximal orders.

(I) $(\mathrm{M}_2(\Omega), \mathrm{M}_2(J_\Omega))$. (II) $\left(\begin{pmatrix} \Omega & \Omega \\ J_\Omega & \Omega \end{pmatrix}, \begin{pmatrix} J_\Omega & \Omega \\ J_\Omega & J_\Omega \end{pmatrix}\right)$.

(III) $(\Omega \times \Omega', J_\Omega \times J_{\Omega'})$. (IVa) (Ω, J_Ω). (IVb) (Ω, J_Ω^2).

The argument in 4.1 and 1.3(1) show that, *for any local Bass order Λ in a semisimple algebra, there exists a unique primary Bass chain $\{\Lambda_i\}_{0\le i\le s}$ of length $s(<\infty)$ such that $\Lambda = \Lambda_s$.* In this case, s is called the **rank** of Λ, and Λ is called of **type(T)**$_s$ where $T \in \{\text{I,II,III,IVa,IVb}\}$ is determined by the dominating pair. Moreover, an order in 4.1(2)(ii) is called of **type(V)**$_s$.

Describing AR quivers, we obtain the following good correspondence between types of Bass orders and classical Dynkin diagrams with an involution, where τ^+ is the identity map except $\tau^+ X = Y$ and $\tau^+ Y = X$ in type(III)$_s$.

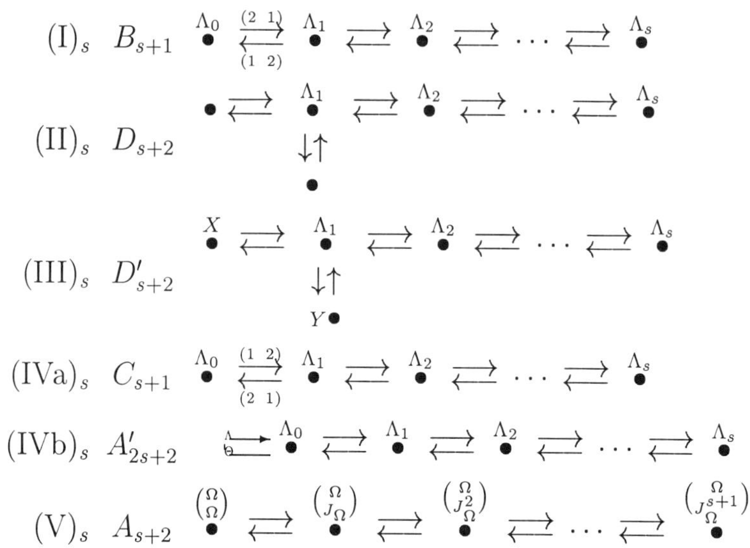

In general, it is a very difficult problem to classify local Bass orders completely since it depends heavily on the arithmetic of a basic ring R. But, under weak assumptions on R and A, a classification is given in [**HN1**] which is rather number theoretic. On the other hand, there is the following very clear description of primary Bass chains of infinite length [**HN1**].

4.2.1 (primary Bass chain of infinite length) *Let A be a semisimple K-algebra. Then there exists a bijection from (the set of primary Bass chain $\{\Lambda_i\}_{0\le i}$ in A of infinite length) to (the set of division subalgebra B of A of index 2), which is given by $\{\Lambda_i\}_{0\le i} \mapsto B := K(\bigcap_{0\le i} \Lambda_i)$. Moreover, the type of a primary Bass chain of infinite length defined above coincides with the type of the corresponding injection $B \to A$ defined in 3.2(1).*

4.3 Bass orders in purely non-semisimple algebras In general, for an order Λ, it is effective to consider the ambient algebra $\tilde{\Lambda}$ and an order $(\Lambda + J_A)/J_A$ in a semisimple algebra A/J_A. It is quite remarkable that we obtain the following clear description of Bass orders in purely non-semisimple algebras [**HN2**].

Theorem *Let Λ be an R-order in a purely non-semisimple algebra A. Then Λ is Bass iff ($l_A(Ae) = l_A(eA) = 2$ for any primitive idempotent e of A, and $(\Lambda + J_A)/J_A$ is a maximal order in A/J_A).*

Notice that, for a purely non-semisimple algebra A, $l_A(Ae) = l_A(eA) = 2$ for any primitive idempotent e of A iff A is a quasi-Frobenius algebra satisfying $J_A^2 = 0$.

4.4 Almost Bass orders An R-order Λ is called **almost Bass** if Λ and $O_l(J_\Lambda)$ are Gorenstein. By 1.2 and 4.1(1), *a local order is almost Bass iff it is Bass.*

Let $\Lambda^{(0)} := \Lambda$ be a ring-indecomposable almost Bass order. Using DK Rejection repeatedly, we obtain a unique (finite or infinite) sequence $\Lambda^{(0)} \subset \Lambda^{(1)} \subset \Lambda^{(2)} \subset \cdots$ satisfying $\mathfrak{I}(\operatorname{lat}\Lambda^{(i)}) - \mathfrak{I}(\operatorname{lat}\Lambda^{(i+1)}) = \mathfrak{I}(\operatorname{pr}\Lambda^{(i)})$, where the last term $\Lambda^{(s)}$ is hereditary if the sequence is finite, and other terms are ring-indecomposable non-hereditary almost Bass orders (cf. 4.1).

If $\widetilde{\Lambda}$ is seimsimple, then this sequence is finite by 1.3(1), and s is called the **rank** of Λ. We can obtain the following structure theorem of almost Bass orders [**H**].

4.4.1 Theorem *Let Λ be a ring-indecomposable order in a semisimple algebra.*

(1) Λ is an almost Bass order of rank 1 iff there exists a hereditary overorder Γ of Λ such that $J_\Lambda = J_\Gamma$ and the injection $\Lambda/J_\Lambda \to \Gamma/J_\Gamma$ is Morita equivalent to a direct product of morphisms of type (I·II), (III), (IV) and (V) (in 3.2(1)).

(2) Λ is an almost Bass order of rank $s \geq 2$ iff it is Morita equivalent to one of the following orders.

$$\begin{pmatrix} \Delta_i & \Delta_i & \cdots & \Delta_i & \Delta_{i-1} & \cdots & \Delta_{i-1} \\ J_{\Delta_i} & \ddots & \ddots & & \ddots & \ddots & \vdots \\ \vdots & \ddots & & & & \ddots & \Delta_{i-1} \\ J_{\Delta_i} & & & & & & \Delta_i \\ J_{\Delta_{i+1}} & \ddots & & & & \ddots & \vdots \\ \vdots & \ddots & \ddots & & \ddots & \ddots & \Delta_i \\ J_{\Delta_{i+1}} & \cdots & J_{\Delta_{i+1}} & J_{\Delta_i} & \cdots & J_{\Delta_i} & \Delta_i \end{pmatrix} \quad \begin{pmatrix} \Delta_{i,1} & \Delta_{i-1} & \Delta_{i-1} & \cdots & \Delta_{i-1} \\ J & \Delta_{i,2} & \ddots & \ddots & \vdots \\ J & \ddots & \ddots & \ddots & \Delta_{i-1} \\ \vdots & \ddots & \ddots & \Delta_{i,n-1} & \Delta_{i-1} \\ J & \cdots & J & J & \Delta_{i,n} \end{pmatrix}$$

Here, for the left order, (number of rows of Δ_i)+(number of rows of J_{Δ_i})=(size of the matrix) holds, and either $\{\Delta_l\}_l$ is a primary Bass chain, $\Delta_l = \begin{pmatrix} \Omega & J_\Omega^l \\ J_\Omega^l & \Omega \end{pmatrix}$ or $\Delta_l = \begin{pmatrix} \Omega & J_\Omega^l \\ J_\Omega^{l+1} & \Omega \end{pmatrix}$ (Ω is a local maximal order). For the right order, $\{\Delta_{l,j}\}_l$ is a primary Bass chain such that $J = J_{\Delta_{i,j}}$ and $\Delta_{i-1} = \Delta_{i-1,j}$ for any j.

The AR quiver of an order Λ in (2) above is determined by the rank s, the size n of the matrix and the type of the corresponding Bass orders $\{\Delta_l\}_l$. For example, $\mathbb{A}(\operatorname{lat}\Lambda)$ is the following if $n = 3$ and the type of Δ_l is either (I)(II) or (V), where the bottom vertices are identified with the top vertices and τ^+ is the downward shift.

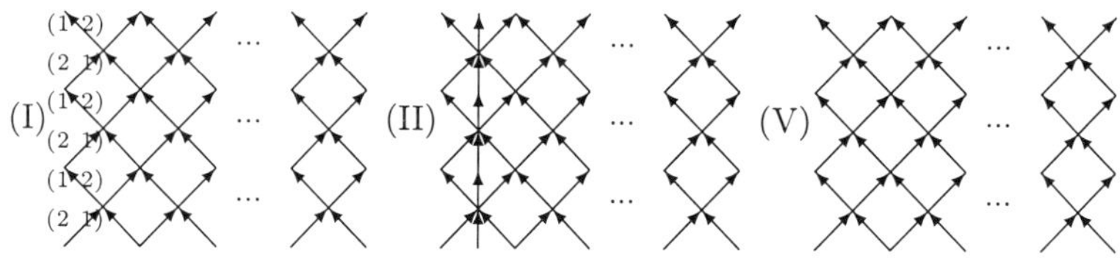

4.4.2 As we have seen in 4.2.1, Bass orders have strong relationship to quadratic extensions. The following result shows that almost Bass orders have another relationship to quadratic extensions [**I4**]. It will be used in §9.

Theorem *Let Λ be an almost Bass order and $\mathbb{G} := (\Lambda/J_\Lambda)\otimes_\Lambda : \operatorname{lat}\Lambda \to \operatorname{mod}\Lambda/J_\Lambda$. Then the induced functor $\operatorname{lat}\Lambda/\ker\mathbb{G} \to \operatorname{mod}\Lambda/J_\Lambda$ is a quadratic extension (in 3.2(3)).*

4.5 Remark (1) Using Bass orders, we can obtain a classification of local orders of finite type [**DK2**][**HN3**]. Describing AR quivers, we obtain a good correspondense between "types" of local orders of finite type and Dynkin diagrams.

(2) Let Λ be a ring-indecomposable finite dimensional algebra. It is well known that Λ is local Nakayama (resp. quasi-Frobenius Nakayama) iff Λ/I is quasi-Frobenius for any ideal I of Λ (resp. Λ and Λ/J_Λ are quasi-Frobenius). Thus Bass orders (resp. almost Bass orders) can be regarded as one-dimensional analogy of local Nakayama algebras (resp. quasi-Frobenius Nakayama algebras).

5 τ-categories

Although our first aim is to study lat Λ, it is often much better to study a wider class of additive categories called **τ-categories** [**I3**]. Results in this section will be used in §7, where our general treatment works essentially. It is also interesting that τ-categories are closely related to Auslander-Gorenstein rings (5.5, 7.4).

5.1 Definition A Krull-Schmidt category $\mathcal{C}$ is called a **τ-category** if, for any $X \in \mathcal{C}$, there exist complexes $(X] = (\tau^+X \overset{\nu_X^+}{\to} \theta^+X \overset{\mu_X^+}{\to} X)$ and $[X) = (X \overset{\mu_X^-}{\to} \theta^-X \overset{\nu_X^-}{\to} \tau^-X)$ such that $H_{\tau^+X} \to H_{\theta^+X} \to H_X \to S_X \to 0$ and $H^{\tau^-X} \to H^{\theta^-X} \to H^X \to S^X \to 0$ are minimal projective resolutions, and $H^{\theta^+X} \to H^{\tau^+X} \to S^{\tau^+X} \to 0$ and $H_{\theta^-X} \to H_{\tau^-X} \to S_{\tau^-X} \to 0$ are exact. We sometimes denote $(X]$ (resp. $[X)$, $\theta^\pm$, $\tau^\pm$) by $(X]_\mathcal{C}$ (resp. $[X)_\mathcal{C}$, $\theta_\mathcal{C}^\pm$, $\tau_\mathcal{C}^\pm$).

We can show that $(X] = [\tau^+X)$ (resp. $[X) = (\tau^-X]$) holds for any $X \in \mathfrak{I}(\mathcal{C}) - \mathfrak{I}_1^+(\mathcal{C})$ (resp. $X \in \mathfrak{I}(\mathcal{C}) - \mathfrak{I}_1^-(\mathcal{C})$). Hence τ^+ and τ^- give mutually inverse bijections between $\mathfrak{I}(\mathcal{C}) - \mathfrak{I}_1^+(\mathcal{C})$ and $\mathfrak{I}(\mathcal{C}) - \mathfrak{I}_1^-(\mathcal{C})$. Thus we can define the **AR quiver** $\mathbb{A}(\mathcal{C})$ of $\mathcal{C}$ by 2.3(3). A τ-category $\mathcal{C}$ is called **strict** if $\operatorname{pd} S_X \le 2$ and $\operatorname{pd} S^X \le 2$ hold for any $X \in \mathcal{C}$.

For example, if Λ is an isolated singularity and $\dim R \le 2$, then lat Λ is a strict τ-category by 2.2. But our introduction of τ-categories is motivated by another important example more strongly.

Igusa-Todorov introduced **τ-species** (=modulated translation quiver in [**IT2**]), which are "algebraic realization" of translation quivers (2.3(1)) by skew fields and their bimodules. For each τ-species $\mathcal{Q}$, we can define a τ-category $\widehat{\mathbb{M}}(\mathcal{Q})$ called the **mesh category** of $\mathcal{Q}$, which is familiar in representation theory [**R**][**BG**][**IT2**]. (In 5.6, we will give a quick review of τ-species and their mesh categories.) Conversely, for each τ-category $\mathcal{C}$, we can define the **AR species** $\widehat{\mathbb{A}}(\mathcal{C})$, which is an algebraic realization of the AR quiver $\mathbb{A}(\mathcal{C})$. Then the following result 5.2 shows that τ-category is a natural domain to study mesh categories [**I3**].

5.2 Let $\mathcal{C}$ be a Krull-Schmidt category. Define the **associated completely graded category** $\widehat{\mathbb{G}}(\mathcal{C})$ of $\mathcal{C}$ by $\widehat{\mathbb{G}}(\mathcal{C}) := \prod_{i\geq 0} \mathcal{J}_{\mathcal{C}}^{i} / \mathcal{J}_{\mathcal{C}}^{i+1}$. Then $\mathcal{C}$ is called **completely graded** if $\mathcal{C}$ is equivalent to $\widehat{\mathbb{G}}(\mathcal{C})$.

Theorem *(1) $\widehat{\mathbb{G}}(\mathcal{C})$ is a τ-category for any τ-category $\mathcal{C}$.*

(2) $\widehat{\mathbb{A}}(\widehat{\mathbb{M}}(\mathcal{Q})) = \mathcal{Q}$ holds for any τ-species $\mathcal{Q}$, and $\widehat{\mathbb{M}}(\widehat{\mathbb{A}}(\mathcal{C})) = \widehat{\mathbb{G}}(\mathcal{C})$ holds for any τ-category $\mathcal{C}$. Thus $\widehat{\mathbb{A}}$ and $\widehat{\mathbb{M}}$ give mutually inverse bijections between the set of τ-species and the set of completely graded τ-categories.

5.3 Ladders A morphism $a \in \mathcal{C}(X,Y)$ is called **special** if $a+f$ is isomorphic to a as a complex for any $f \in \mathcal{J}_{\mathcal{C}}^{2}(X,Y)$. The following theorem is fundamental for τ-categories [**I3**][**IT1**]. In particular, it will be used to translate categorical conditions for $\mathcal{C}$ to combinatorial conditions for $\mathbb{A}(\mathcal{C})$.

Theorem *Let $\mathcal{C}$ be a τ-category, $a \in \mathcal{C}(X_0, Y_0)$ a special morphism and $L := \operatorname{Cok} H_a$.*

(1) There exist the following commutative diagram (called the **ladder** *of a), $U_i \in \mathcal{C}$ and $h_i \in \mathcal{C}(U_i, Z_{i-1})$ such that $a = \binom{b_0}{0} \in \mathcal{C}(Z_0 \oplus U_0, Y_0)$ and*

$$(Y_{i-1}] = (Z_i \oplus U_i \xrightarrow{\left(\begin{smallmatrix} b_i & -g_i \\ 0 & h_i \end{smallmatrix}\right)} Y_i \oplus Z_{i-1} \xrightarrow{\left(\begin{smallmatrix} f_i \\ b_{i-1} \end{smallmatrix}\right)} Y_{i-1})$$

for any i.

$$\begin{array}{ccccccccc}
Y_0 & \xleftarrow{f_1} & Y_1 & \xleftarrow{f_2} & Y_2 & \xleftarrow{f_3} & Y_3 & \xleftarrow{f_4} & \cdots \\
\uparrow{\scriptstyle b_0} & & \uparrow{\scriptstyle b_1} & & \uparrow{\scriptstyle b_2} & & \uparrow{\scriptstyle b_3} & & \cdots \\
Z_0 & \xleftarrow{g_1} & Z_1 & \xleftarrow{g_2} & Z_2 & \xleftarrow{g_3} & Z_3 & \xleftarrow{g_4} & \cdots
\end{array}$$

(2) The diagram in (1) induces the following commutative diagram, where each vertical complexes are minimal projective resolutions and $\mathcal{J}_{\mathcal{C}}^{n-1} L \to \mathcal{J}_{\mathcal{C}}^{n} L$ is the natural inclusion for any $n > 0$.

$$\begin{array}{ccccccccc}
L & \longleftarrow & \mathcal{J}_{\mathcal{C}} L & \longleftarrow & \mathcal{J}_{\mathcal{C}}^{2} L & \longleftarrow & \mathcal{J}_{\mathcal{C}}^{3} L & \longleftarrow & \cdots \\
\uparrow & & \uparrow & & \uparrow & & \uparrow & & \cdots \\
H_{Y_0} & \xleftarrow{H_{f_1}} & H_{Y_1} & \xleftarrow{H_{f_2}} & H_{Y_2} & \xleftarrow{H_{f_3}} & H_{Y_3} & \xleftarrow{H_{f_4}} & \cdots \\
\uparrow{\scriptstyle H_{b_0}} & & \uparrow{\scriptstyle H_{b_1}} & & \uparrow{\scriptstyle H_{b_2}} & & \uparrow{\scriptstyle H_{b_3}} & & \cdots \\
H_{Z_0} & \xleftarrow{H_{g_1}} & H_{Z_1} & \xleftarrow{H_{g_2}} & H_{Z_2} & \xleftarrow{H_{g_3}} & H_{Z_3} & \xleftarrow{H_{g_4}} & \cdots
\end{array}$$

(3) Assume that L is semisimple. Then we can obtain terms Y_i and Z_i by $Z_0 = X_0 - (\theta^+ Y_0 - X_0)_-$ and the following **recursion fomula**, *where we use notations in 2.3(2). In particular, we can compute them from (X_0, Y_0) and $\mathbb{A}(\mathcal{C})$.*

$$Y_i = \theta^+ Y_{i-1} - Z_{i-1}, \quad Z_i = \tau^+ Y_{i-1} - (\theta^+ Y_i - \tau^+ Y_{i-1})_- \quad (i > 0)$$

As a conclusion, we obtain the following corollaries.

(4) $\mathcal{C}$ is left artinian iff the following condition holds.

For any $X \in \mathfrak{I}(\mathcal{C})$, put $Y_0 := 0$, $Y_1 := X$ and $Y_i := (\theta^+ Y_{i-1} - \tau^+ Y_{i-2})_+$ for $i \geq 2$. Then $Y_i = 0$ holds for sufficiently large i.

(5) Assume $\bigcap_{i \geq 0} \mathcal{J}_{\mathcal{C}}^i = 0$. Then $\mathcal{C}$ is a strict τ-category iff the following condition holds.

For any $X \in \mathfrak{I}(\mathcal{C})$, put $Y_0 := 0$, $Y_1 := X$ and $Y_i := \theta^+ Y_{i-1} - \tau^+ Y_{i-2}$ for $i \geq 2$. Then $Y_i \geq 0$ holds for any $i \geq 0$.

5.4 Quotient categories In the rest of this paper, any subcategory $\mathcal{C}'$ of a Krull-Schmidt category $\mathcal{C}$ is assumed to be *full, closed under isomorphisms, direct sums and direct summands.* Then the correspondence $\mathcal{C}' \mapsto \mathfrak{I}(\mathcal{C}')$ gives a bijection between subcategories of $\mathcal{C}$ and subsets of $\mathfrak{I}(\mathcal{C})$. We denote by $S \mapsto \operatorname{add} S$ the inverse of this correspondence.

In this case, we denote by $I_{\mathcal{C}'}$ the ideal of $\mathcal{C}$ consisting of morphisms which factor through some object in $\mathcal{C}'$. Define the quotient category $\overline{\mathcal{C}} := \mathcal{C} / \mathcal{C}'$ by $Ob(\overline{\mathcal{C}}) := Ob(\mathcal{C})$ and $\overline{\mathcal{C}}(X, Y) := \mathcal{C}(X, Y) / I_{\mathcal{C}'}(X, Y)$ for $X, Y \in \mathcal{C}$.

We call $\mathcal{C}'$ **coartinian** if $\overline{\mathcal{C}} := \mathcal{C} / \mathcal{C}'$ is artinian, namely $H_X^{\overline{\mathcal{C}}}$ and $H_{\overline{\mathcal{C}}}^X$ have finite length for any $X \in \mathcal{C}$. We call a subset S of $\mathfrak{I}(\mathcal{C})$ **artinian** if $\mathcal{C}' := \operatorname{add}(\mathfrak{I}(\mathcal{C}) - S)$ is a coartinian subcategory of $\mathcal{C}$.

Proposition *Let $\mathcal{C}'$ be a subcategory of a τ-category $\mathcal{C}$ and $\overline{\mathcal{C}} := \mathcal{C}/\mathcal{C}'$. Then $\overline{\mathcal{C}}$ is also a τ-category. For any $X \in \mathfrak{I}(\mathcal{C}) - \mathfrak{I}(\mathcal{C}')$, $\overline{(X]_{\mathcal{C}}} = \overline{(X]_{\mathcal{C}}}$ holds if $\overline{\theta^+ X} \neq 0$, and $(X]_{\overline{\mathcal{C}}} = (0 \to 0 \to X)$ holds if $\overline{\theta^+ X} = 0$.*

5.5 τ-categories with an order structure and Auslander orders For an additive category $\mathcal{C}$, we say that $\mathcal{C}$ has an **order structure** if there exists a complete discrete valuation ring R and an R-order Λ such that $\mathcal{C}$ is equivalent to $\operatorname{pr}\Lambda$.

As we have seen in 5.1, $\operatorname{lat}\Delta$ is a τ-category for any R-order Δ. Conversely, the following theorem 5.5.2 shows that a τ-category $\mathcal{C}$ is equivalent to $\operatorname{lat}\Delta$ for an order Δ if $\mathcal{C}$ has an order structure and the ambient category is semisimple [**I3**]. This result is closely related to **Auslander orders** and **Auslander-Gorenstein rings** [**ARo**][**FGR**].

5.5.1 Definition (1) Assume $\dim R = 1$. An R-order Λ is called an **Auslander order** if $\operatorname{gl.dim}\Lambda \leq 2$ and a minimal projective resolution $0 \to P_1 \to P_0 \to \Lambda^* \to 0$ of Λ^* satisfies $P_0 \in \operatorname{rin}\Lambda$. This definition is slightly different from that of [**ARo**] since we do not assume that $\Lambda \otimes_R K$ is semisimple.

(2) Let Λ be a noetherian ring and $0 \to \Lambda \to I^0 \to I^1 \to \cdots$ a minimal injective resolution of a Λ-module Λ. We say that Λ satisfies the (l,n)-**condition** $(l, n \geq 0)$ if $\operatorname{fd} I^i < l$ holds for any i $(0 \leq i < n)$. Moreover, Λ is called **Auslander regular** (resp. **Auslander-Gorenstein**) if $\operatorname{gl.dim}\Lambda < \infty$ (resp. $\operatorname{id}{}_\Lambda\Lambda < \infty$ and $\operatorname{id}\Lambda_\Lambda < \infty$) and Λ satisfies the (l,l)-condition for any $l > 0$.

5.5.2 Theorem *Let Λ be an R-order and $\dim R = 1$. Then the following conditions (1)–(3) are equivalent. Moreover, if $\widetilde{\Lambda}$ is seimsimple, then (4) is also equivalent.*

(1) $\operatorname{pr}\Lambda$ *is a τ-category.*

(2) Λ is an Auslander order.

(3) $\operatorname{gl.dim}\Lambda \leq 2$ *and Λ is an Auslander regular ring.*

(4) There exists an R-order Δ of finite type such that $\operatorname{pr}\Lambda$ *is equivalent to* $\operatorname{lat}\Delta$.

5.6 Appendix: τ-species and mesh categories In this subsection, we will give a quick review of τ-species and their mesh categories.

5.6.1 $\mathcal{Q} = (Q, D_X, {}_XM_Y)$ is called a **species** if Q is a set, D_X is a skew field for any $X \in Q$ and ${}_XM_Y$ is a (D_X, D_Y)-bimodule such that $\sum_{X\in Q} \dim_{D_X} {}_XM_Y < \infty$ and $\sum_{Y\in Q} \dim_{D_Y} {}_XM_Y < \infty$ for any $X, Y \in Q$.

For $X, Y \in Q$, put $P_0(X,Y) := 0$ if $X \neq Y$, and $P_0(X,X) := D_X$. Put

$$P_n(X,Y) := \bigoplus_{Z_1,\cdots,Z_{n-1}\in Q} {}_X M_{Z_1} \otimes_{D_{Z_1}} \cdots \otimes_{D_{Z_{n-1}}} {}_{Z_{n-1}}M_Y \text{ for } n > 0,$$
$$P_n(A,B) := \prod_{X,Y\in Q} \mathrm{Mat}_{\langle A,X\rangle,\langle B,Y\rangle}(P_n(X,Y))$$
$$\text{for } A, B \in \mathbb{N}Q \text{ and } n \geq 0.$$

We have a natural map $P_n(X,Y) \times P_m(Y,Z) \to P_{n+m}(X,Z)$, $(f,g) \mapsto fg := f \otimes g$ for any $X, Y, Z \in Q$. Using a matrix multiplication, we have a natural map $P_n(A,B) \times P_m(B,C) \to P_{n+m}(A,C)$ for any $A, B, C \in \mathbb{N}Q$.

Define additive categories $\widehat{\mathbb{P}}(\mathcal{Q})$ and $\mathbb{P}(\mathcal{Q})$ by $Ob(\widehat{\mathbb{P}}(\mathcal{Q})) = Ob(\mathbb{P}(\mathcal{Q})) := \mathbb{N}Q$, and $\widehat{\mathbb{P}}(\mathcal{Q})(A,B) := \prod_{n\geq 0} P_n(A,B)$ and $\mathbb{P}(\mathcal{Q})(A,B) := \bigoplus_{n\geq 0} P_n(A,B)$ for $A, B \in \mathbb{N}Q$, where the composition is given by $(f_n)_{n\geq 0} \cdot (g_n)_{n\geq 0} := (\sum_{i=0}^{n} f_i g_{n-i})_{n\geq 0}$.

5.6.2 Proposition *Let $\mathcal{Q}$ be a species. Then $\widehat{\mathbb{P}}(\mathcal{Q})$ is a completely graded Krull-Schmidt category called the* **tensor category** *of $\mathcal{Q}$.*

5.6.3 $\mathcal{Q} = (Q, Q^p, Q^i, D_X, {}_XM_Y, \tau^+, a, b)$ is called a **τ-species** if $(Q, D_X, {}_XM_Y)$ is a species, Q^p and Q^i are subsets of Q, τ^+ is a bijection $Q - Q^p \to Q - Q^i$, $a_X : D_X \to D_{\tau^+X}$ is a ring isomorphism and $b_{X,Y} : \mathrm{Hom}_{D_Y}({}_{\tau^+X}M_Y, D_Y) \to {}_YM_X$ is a (D_Y, D_X)-isomorphism for any $X \in Q - Q^p$ and $Y \in Q$, where ${}_{\tau^+X}M_Y$ is regarded as a D_X-module through a_X.

Putting $d(X,Y) := \dim_{D_X} {}_XM_Y$ and $d'(X,Y) := \dim_{D_Y} {}_XM_Y$, we obtain a translation quiver $|\,\mathcal{Q}\,| := (Q, Q^p, Q^i, \tau^+, d, d')$ called the **underlying quiver** of $\mathcal{Q}$.

Since $\mathrm{Hom}_{D_Y}(\mathrm{Hom}_{D_Y}({}_{\tau^+X}M_Y, D_Y), {}_YM_X) = {}_{\tau^+X}M_Y \otimes_{D_Y} {}_YM_X$, we can regard $b_{X,Y}$ as an element of ${}_{\tau^+X}M_Y \otimes_{D_Y} {}_YM_X$. Put $\gamma(X) := \sum_{Y\in Q} b_{X,Y} \in P_2(\tau^+X, X)$. Then $\gamma(X)f = a_X(f)\gamma(X)$ holds for any $f \in D_X$ since $b_{X,Y}$ is a D_X^{op}-morphism. Let I be the ideal of $\mathbb{P}(\mathcal{Q})$ generated by $\{\gamma(X) | X \in Q\}$. Then we can write $I = \bigoplus_{n\geq 0} I_n$ $(I_n \subseteq P_n)$. Put $\widehat{\mathbb{M}}(\mathcal{Q}) := \widehat{\mathbb{P}}(\mathcal{Q})/\widehat{I}$, where $\widehat{I} := \prod_{n\geq 0} I_n$ is an ideal of $\widehat{\mathbb{P}}(\mathcal{Q})$.

5.6.4 Proposition *Let $\mathcal{Q}$ be a τ-species. Then $\widehat{\mathbb{M}}(\mathcal{Q})$ is a completely graded τ-category called the* **mesh category** *of $\mathcal{Q}$.*

6 Generalized Rejection Lemma

In this section, assume that Λ is an isolated singularity and $\dim R \leq 1$. As we have seen in §4, our theory of Bass orders are constructed on DK Rejection. In this section, we will study its generalization. In 6.1, we will characterize artinian (5.4) rejectable subsets of $\mathfrak{I}(\mathrm{lat}\,\Lambda)$ by a language of $\mathbb{A}(\mathrm{lat}\,\Lambda)$ combinatorially [**I3**].

In general, for treating AR quivers, it is important to consider relationship between (1) representation theoretic conditions for Λ, (2) categorical conditions for $\mathrm{lat}\,\Lambda$, and (3) combinatorial conditions for $\mathbb{A}(\mathrm{lat}\,\Lambda)$. We often use 5.3 and 5.4 to study relationship between (2) and (3).

We will treat not only categories of lattices but also τ-categories, and such a general treatment will be necessary in §7. Here, we call a subcategory (5.4) $\mathcal{C}'$ of a τ-category $\mathcal{C}$ **rejective** if the natural inclusion $\mathcal{C}' \to \mathcal{C}$ has a right adjoint $(\ )^- : \mathcal{C} \to \mathcal{C}'$ with a counit ϵ^- such that ϵ_X^- is a monomorphism for any $X \in \mathcal{C}$, and a left adjoint $(\ )^+ : \mathcal{C} \to \mathcal{C}'$ with a unit ϵ^+ such that ϵ_X^+ is an epimorphism for any $X \in \mathcal{C}$. Then we call a subset $\mathfrak{I}(\mathcal{C}) - \mathfrak{I}(\mathcal{C}')$ of $\mathfrak{I}(\mathcal{C})$ **rejectable**. By 1.1, this definition is compatible with that in §1.

It is quite remarkable that *rejectability of a subset S of $\mathfrak{I}(\mathcal{C})$ depends only on (i) the translation quiver obtained by the restriction of $\mathbb{A}(\mathcal{C})$ to S, and (ii) the subsets $S \cap \mathfrak{I}_1^+(\mathcal{C})$ and $S \cap \mathfrak{I}_1^-(\mathcal{C})$ of S.*

6.1 Theorem *Let $\mathcal{C}$ be a τ-category, $\mathcal{C}'$ a coartinian subcategory of $\mathcal{C}$ and $\overline{\mathcal{C}} := \mathcal{C} / \mathcal{C}'$. Then the following conditions are equivalent.*

(1)(representation theoretic condition) $\mathcal{C}'$ is a rejective subcategory of $\mathcal{C}$.

(2)(categorical condition) μ_X^- is a monomorphism in $\overline{\mathcal{C}}$ for any $X \in \mathfrak{I}(\mathcal{C}) - \mathfrak{I}_1^-(\mathcal{C})$ and μ_X^+ is an epimorphism in $\overline{\mathcal{C}}$ for any $X \in \mathfrak{I}(\mathcal{C}) - \mathfrak{I}_1^+(\mathcal{C})$.

(3)(combinatorial condition) The following (i) and (ii) hold.

(i) For any $X \in \mathfrak{I}(\mathcal{C}) - \mathfrak{I}_1^-(\mathcal{C})$, put $Y_0 := \theta_{\overline{\mathcal{C}}}^- X$, $Y_1 := \theta_{\overline{\mathcal{C}}}^+ \theta_{\overline{\mathcal{C}}}^- X - X$ and $Y_i := \theta_{\overline{\mathcal{C}}}^+ Y_{i-1} - \tau_{\overline{\mathcal{C}}}^+ Y_{i-2}$ for $i \geq 2$. Then $Y_i \geq 0$ holds for any $i \geq 0$.

(ii) For any $X \in \mathfrak{I}(\mathcal{C}) - \mathfrak{I}_1^+(\mathcal{C})$, put $Y_0 := \theta_{\overline{\mathcal{C}}}^+ X$,

$Y_1 := \theta^-_{\overline{\mathcal{C}}}\theta^+_{\overline{\mathcal{C}}}X - X$ *and* $Y_i := \theta^-_{\overline{\mathcal{C}}}Y_{i-1} - \tau^-_{\overline{\mathcal{C}}}Y_{i-2}$ *for* $i \geq 2$. *Then* $Y_i \geq 0$ *holds for any* $i \geq 0$.

6.2 It is often convenient to use the concept of **trivial** subcategories. A subcategory $\mathcal{C}'$ of $\mathcal{C}$ is called trivial if any proper subcategory $\mathcal{C}''$ of $\mathcal{C}$ containing $\mathcal{C}'$ is not a rejective subcategory of $\mathcal{C}$. We have the following result [**I1**][**I3**].

Theorem *Let* $\mathcal{C}$ *be a* τ*-category,* $\mathcal{C}'$ *a coartinian subcategory of* $\mathcal{C}$ *and* $\overline{\mathcal{C}} := \mathcal{C}\,/\,\mathcal{C}'$. *Then the following conditions are equivalent.*

(1)(representation theoretic condition) $\mathcal{C}'$ *is a trivial subcategory of* $\mathcal{C}$.

(2)(categorical condition) $\overline{\mathcal{C}}(P, I) = 0$ *for any* $P \in \mathfrak{I}^+_1(\mathcal{C})$ *and* $I \in \mathfrak{I}^-_1(\mathcal{C})$.

(3)(combinatorial condition) For any $X \in \mathfrak{I}^-_1(\mathcal{C})$, *put* $Y_0 := X$, $Y_1 := \theta^+_{\overline{\mathcal{C}}}X$ *and* $Y_i := (\theta^+_{\overline{\mathcal{C}}}Y_{i-1} - \tau^+_{\overline{\mathcal{C}}}Y_{i-2})_+$ *for* $i \geq 2$. *Then* $Y_i|_{\mathfrak{I}^+_1(\mathcal{C})} = 0$ *holds for any* $i \geq 0$.

6.3 Example Let Λ be an R-order, $\dim R = 1$ and S a subset of $\mathfrak{I}(\operatorname{lat}\Lambda)$ such that $\#S \leq 4$. Then S is a minimal artinian rejectable subset iff S has one of the following forms, where $S \cap \mathfrak{I}(\operatorname{pr}\Lambda) = \{P\}$ and $S \cap \mathfrak{I}(\operatorname{rin}\Lambda) = \{I\}$.

(1) $\overset{P=I}{\bullet}$ (DK Rejection)

(2) $\overset{P}{\bullet}\longrightarrow\overset{I}{\bullet}$

(3) $\overset{P}{\bullet}\xrightarrow{(a\ b)}\bullet\longrightarrow\overset{I}{\bullet}$ $ab \leq 2$

(4) $\overset{P}{\bullet}\longrightarrow\bullet\xrightarrow{(a\ b)}\overset{I}{\bullet}$ $ab \leq 2$

(5) $\overset{P}{\bullet}\xrightarrow{(a\ b)}\bullet\xrightarrow{(b\ a)}\overset{I}{\bullet}$ $P = \tau^+ I, ab \leq 3$

(6) $\overset{P}{\bullet}\xrightarrow{(a\ b)}\bullet\longrightarrow\bullet\longrightarrow\overset{I}{\bullet}$ $ab \leq 2$

(7) $\overset{P}{\bullet}\longrightarrow\bullet\xrightarrow{(a\ b)}\bullet\longrightarrow\overset{I}{\bullet}$ $ab \leq 2$

(8) $\overset{P}{\bullet}\longrightarrow\bullet\longrightarrow\bullet\xrightarrow{(a\ b)}\overset{I}{\bullet}$ $ab \leq 2$

(9) $\overset{P}{\bullet}\xrightarrow{(a\ b)}\bullet\xrightarrow{(b\ a)}\overset{X}{\bullet}\longrightarrow\overset{I}{\bullet}$ $P = \tau^+ X, ab \leq 3$

(10) 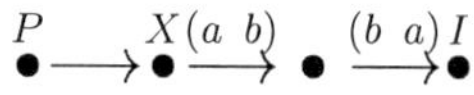 $X = \tau^+ I, ab \leq 3$

(11) 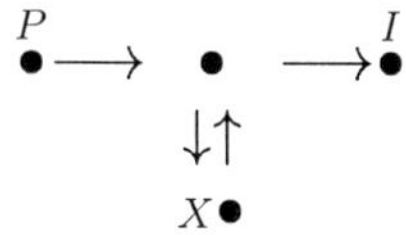 $P = \tau^+ I, X = \tau^+ X$

(12) 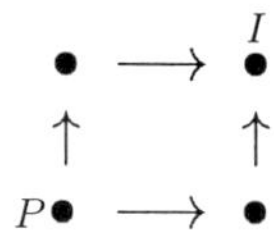 $P = \tau^+ I$

6.4 Remark Assume $\dim R = 1$. A rejectable subset $S = \mathfrak{I}(\operatorname{lat} \Lambda) - \mathfrak{I}(\operatorname{lat} \Gamma)$ is artinian iff $\# S < \infty$ and Γ is an overorder of Λ. More generally than 6.1, we can obtain a characterization of finite rejectable subsets by summing up with a result in [**I2**], where we gave a characterization of finite rejectable subsets with a form $\mathfrak{I}(\operatorname{lat} \Lambda) - \mathfrak{I}(\operatorname{lat} e\Lambda)$ (e is a central idempotent e of $\widetilde{\Lambda}$) by using additive functions of translation quivers.

7 Orders of finite type

There are many study about orders of finite type, and classifications were given for some classes, for example, local orders (4.5(1)), Bäckström orders [**RR**], tiled orders [**ZK**] and its generalization [**S**]. But, it seems to be very difficult to give a classification without any restriction. In this section, we will give an approach to orders of finite type by using AR quivers, whose viewpoint is slightly different from classification theory. In 7.2, we will give an answer to the following problem (P_1) [**I3**] by progressing philosophy of [**ARo**] and [**A1**] which developped a connection between homological properties and representation theoretic properties.

(P_d) Give a combinatorial characterization of finite translation quivers which are realized as an AR quiver $\mathbb{A}(\operatorname{lat} \Lambda)$ of an order Λ over a complete regular local ring of dimension d.

Although answers to (P_0) and (P_2) are given by [**IT2**] and [**RV**] and partial results for (P_1) are given in [**W**], one shall meet a sheer difficulty to get a general solution for (P_1), which was not existent in other cases. We will overcome the difficulty by applying results for τ-categories in §5 and §6.

7.1 Orderlike τ-categories

7.1.1 Definition Let $\mathcal{C}$ be a τ-category such that $\bigcap_{i\geq 0}\mathcal{J}_{\mathcal{C}}^i = 0$ and $\mathfrak{I}(\mathcal{C}) < \infty$.

(1) For A, $B \in \mathfrak{I}(\mathcal{C})$ and $n \geq 0$, we say that (A, B) is a **Nakayama pair** of distance n if there exists the following commutative diagram which satisfies $a_0 = \mu_A^-$, $a_n = \mu_B^+$, and $(Y_{i-1}] = [X_i) = (X_i \xrightarrow{(-a_i\ g_i)} Y_i \oplus X_{i-1} \xrightarrow{\binom{f_i}{a_{i-1}}} Y_{i-1})$ for any i $(1 \leq i \leq n)$.

$$\begin{array}{ccccccccc}
Y_0 & \xleftarrow{f_1} & Y_1 & \xleftarrow{f_2} \cdots \xleftarrow{f_{n-1}} & Y_{n-1} & \xleftarrow{f_n} & B = Y_n \\
\uparrow{\scriptstyle a_0} & & \uparrow{\scriptstyle a_1} & \cdots & \uparrow{\scriptstyle a_{n-1}} & & \uparrow{\scriptstyle a_n} \\
A = X_0 & \xleftarrow{g_1} & X_1 & \xleftarrow{g_2} \cdots \xleftarrow{g_{n-1}} & X_{n-1} & \xleftarrow{g_n} & X_n
\end{array}$$

In this case, we write $B = n^-(A)$ and $A = n^+(B)$. Put $s^-(A) := \bigcup_{i=0}^n \operatorname{supp} X_i$ and $s^+(B) := \bigcup_{i=0}^n \operatorname{supp} Y_i$.

(2) We call $\mathcal{C}$ **orderlike** if n^- induces a bijection $\mathfrak{I}_1^-(\mathcal{C}) \to \mathfrak{I}_1^+(\mathcal{C})$ and $\mathfrak{I}(\mathcal{C}) = \bigcup_{A\in\mathfrak{I}_1^-(\mathcal{C})} s^-(A)$ holds.

The following gives a reason why we call such (A, B) Nakayama pair, and such $\mathcal{C}$ orderlike.

7.1.2 Example Let Λ be an R-order of finite type, $\dim R = 1$ and $\mathcal{C} := \operatorname{lat}\Lambda$. Then $\mathcal{C}$ is an orderlike τ-category such that $n^+ : \mathfrak{I}_1^+(\mathcal{C}) \to \mathfrak{I}_1^-(\mathcal{C})$ is given by the Nakayama functor $\operatorname{Hom}_\Lambda(\ ,\Lambda)^*$. Take $A \in \mathfrak{I}_1^-(\mathcal{C})$ (resp. $B \in \mathfrak{I}_1^+(\mathcal{C})$) and $X \in \mathfrak{I}(\mathcal{C})$. Then $X \in s^-(A)$ (resp. $X \in s^+(B)$) holds iff an injective hull $X \to I$ (resp. projective cover $P \to X$) of X satisfies $A \leq I$ (resp. $B \leq P$). Moreover, $\mathcal{C}\,/\,\mathfrak{I}_1^-(\mathcal{C})$ and $\mathcal{C}\,/\,\mathfrak{I}_1^+(\mathcal{C})$ are artinian.

7.2 Theorem *Let Q be an admissible translation quiver with a finite number of vertices. Then the following conditions are equivalent.*

(1)(representation theoretic condition) There are a complete discrete valuation ring R and an R-order Λ such that Q is the AR quiver of Λ.

(2)(categorical condition) If a τ-category $\mathcal{C}$ satisfies $\mathbb{A}(\mathcal{C}) = Q$, then $\mathcal{C}$ is orderlike and $\mathcal{C}\,/\,\mathfrak{I}_1^+(\mathcal{C})$ is artinian.

(3)(combinatorial condition) Q is an admissible translation quiver satisfying the following conditions, where we use notations

in 2.3(2)(4).

(i) For any $X \in Q^i$*, put* $Y_0 := \theta^- X$*,* $Y_1 := \theta^+\theta^- X - X$ *and* $Y_i := \theta^+ Y_{i-1} - \tau^+ Y_{i-2}$ *for* $i \geq 2$*, and* $t(X) := \bigcup_{i\geq 0} \operatorname{supp} Y_i$*. Then there exists* $n \geq 0$ *such that* $Y_i \in \mathbb{N}(Q - Q^p)$ *for any* i $(0 \leq i < n)$*,* $Y_n \in Q^p$ *and* $Y_{n+1} = 0$*. Moreover,* $Q = \bigcup_{X \in Q^i} t(X)$ *holds.*

(ii) For any $X \in Q^i$*, put* $Y_0 := 0$*,* $Y_1 := X$ *and* $Y_i := (\theta^+_{Q/Q^p} Y_{i-1} - \tau^+_{Q/Q^p} Y_{i-2})_+$ *for* $i \geq 2$*. Then* $Y_i = 0$ *holds for sufficiently large* i*.*

(2)⇔(3) follows from 5.3, and (1)⇒(2) follows from 7.1.2. Thus we only have to show (2)⇒(1). We can take a τ-species $\mathcal{Q}$ such that $|\mathcal{Q}| = Q$ since Q is admissible. We only have to show that $\widehat{\mathbb{M}}(\mathcal{Q})$ is equivalent to $\operatorname{lat}\Lambda$ for an order Λ. By 5.5.2, we only have to show that $\widehat{\mathbb{M}}(\mathcal{Q})$ has an order structure. This is the most difficult part, which is essential and peculiar to the case $\dim R = 1$. We can overcome it by the following lemma, which is proved by using a reduction to rejective subcategories as an application of results in §6.

7.2.1 Lemma *Let* $\mathcal{C}$ *be a completely graded orderlike* τ*-category,* k *a field and* $k[[x]]$ *a formal power series ring. Assume that* $\mathcal{C}$ *is a* k*-category,* $\mathcal{C} / \mathcal{J}_{\mathcal{C}}(X, X)$ *is a finite field extension of* k *for any* $X \in \mathfrak{I}(\mathcal{C})$ *and* $\mathcal{C} / \mathfrak{I}_1^+(\mathcal{C})$ *is artinian. Then* $\mathcal{C}$ *has a* $k[[x]]$*-order structure.*

7.3 Remark As an application of our characterization, we can give a complete classification of Gorenstein orders of finite type whose tree types are classical Dynkin diagrams. It is given by combining with generalized Rejection in §6, and a remarkable fact is that such an order has an almost Bass overorder canonically.

In general, it is an interesting problem to give an algorithm to recover Λ from $\mathbb{A}(\operatorname{lat}\Lambda)$ using generalized Rejection since mesh categories are not crucial when $\dim R = 1$. Notice that, for $\dim R = 1$, Gabriel quivers (=presentation of rings by quivers with relations) do not seem to play as crucial role as the case in [**BGRS**] ($\dim R = 0$ and R is an algebraically closed field) since one of the most important orders in applications are orders over $R = \mathbb{Z}_p$ and they are not easily presented by quivers.

7.4 Appendix: Artinian τ-categories In this subsection, let Γ be a finite dimensional algebra over a field R and $\mathcal{C} := \mathrm{pr}\,\Gamma$. Assume that $\mathcal{C}$ is a strict τ-category and we will study relationship between (0) homological conditions for Γ, (1) representation theoretic conditions for $\mathcal{C}$, and (2) categorical conditions for $\mathcal{C}$. Using 5.3, we can easily translate (2) into a combinatorial conditions for $\mathbb{A}(\mathcal{C})$. In particular, the equivalence of (1-4) and (2-4) gives an answer to (P_0) in [**IT2**].

Theorem *Let Γ be a finite dimensional algebra over a field, $0 \to \Gamma \to I^0 \to I^1 \to \cdots$ a minimal injective resolution of a Γ-module Γ, and $\mathcal{C} := \mathrm{pr}\,\Gamma$. Then the following conditions (0-i), (1-i) and (2-i) are equivalent for each i ($1 \leq i \leq 4$).*

(0-i)(homological conditions) $\mathrm{gl.dim}\,\Gamma \leq 2$*, and*

(0-1) Γ and Γ^{op} satisfy the $(2,2)$-condition (in 5.5.1).

(0-2) Γ is Auslander regular.

(0-3) Γ is Auslander regular and I^1 does not have a non-zero projective direct summand.

(0-4) Γ satisfies the $(1,2)$-condition (We call such Γ an **Auslander algebra***).*

(1-i)(representation theoretic conditions)

(1-1): There exists a finite dimensional algebra Λ and a torsion theory $(\mathcal{T},\mathcal{F})$ on $\mathrm{mod}\,\Lambda$ *such that $\mathcal{C}$ is equivalent to $\mathcal{F}$.*

(1-2): (1-1) and $(\mathcal{T},\mathcal{F})$ is a hereditary torsion theory.

(1-3): (1-1) and $\mathcal{F}$ is the category of Λ-modules whose socles are projective.

(1-4): (1-1) and $\mathcal{F} = \mathrm{mod}\,\Lambda$.

(2-i)(categorical conditions)

(2-1): $\mathcal{C}$ is a strict τ-category.

(2-2): (2-1) and n^- gives a map $\mathfrak{I}_1^-(\mathcal{C}) - \mathfrak{I}_0^-(\mathcal{C}) \to \mathfrak{I}(\mathcal{C})$.

(2-3): (2-1) and n^- gives a map $\mathfrak{I}_1^-(\mathcal{C}) - \mathfrak{I}_0^-(\mathcal{C}) \to \mathfrak{I}_1^+(\mathcal{C})$.

(2-4): (2-1) and n^- gives a map $\mathfrak{I}_1^-(\mathcal{C}) - \mathfrak{I}_0^-(\mathcal{C}) \to \mathfrak{I}(\mathcal{C}) - \mathfrak{I}_1^+(\mathcal{C})$.

Another answer to (P_0) was given in [**B**] which uses additive functions on translation quivers. It is closely related to non-artinian rejection in 6.4 and results in [**RV**][**RVo**]. It is an open problem to give an answer to (P_1) using additive functions.

8 Orders of tame type In this section, assume $\dim R \leq 1$ unless explicitly stated otherwise. We say that (R, Λ) is in **classical situation** if ($\dim R = 0$ and R is an algebraically closed field) or ($\dim R = 1$, R is the formal power series ring $k[[x]]$ over an algebraically closed field k and the ambient algebra $\widetilde{\Lambda}$ is semisimple). For classical situation, there are well known definitions of **tame type** and **wild type**, and we have **Dichotomy Theorem**, which asserts that *any order is either tame or wild, and not the both* [**DG**][**D**].

Unfortunately, for general situation, no reasonable definitions of tame type and wild type which satisfy Dichotomy Theorem seem to be established. It will be difficult since assumptions for classical situation are essentially used in the proof of Dichotomy Theorem.

In this section, we will not study dichotomy problem. In 8.2.1, we will show that some kinds of orders have a good representation theoretic property, which seems to give a definition of tame type for general situation. We call such orders of **pseudo tame type**, which are defined by using finite dimensional hereditary algebras.

8.1 Definition (1) Let H be a finite dimensional hereditary algebra over a field k [**DR**]. Then

$$q_H(X) := \dim_k \mathrm{End}_H(X) - \dim_k \mathrm{Ext}^1_H(X, X)$$

gives a quadratic form q_H over $\mathbb{Q} \otimes_{\mathbb{Z}} K_0(H)$, where $K_0(H)$ is a Grothendieck group of H. A hereditary algebra H is called **tame** (resp. **wild**) if q_H is positive semidefinite (resp. not positive semidefinite).

(2) Let R be a complete discrete valuation ring, $k := R/J_R$ and Λ an R-order. We say that Λ is of **pseudo tame type** if there exist a countable number of full subcategories $\mathcal{C}_i$ of $\mathrm{lat}\,\Lambda$, tame hereditary k-algebras H_i and full functors $\mathbb{F}_i : \mathcal{C}_i \to \mathrm{mod}\, H_i$ such that $\mathfrak{I}(\mathrm{lat}\,\Lambda) = \bigcup_i \mathfrak{I}(\mathcal{C}_i)$ and ($\mathbb{F}_i(X) = 0$ iff $X = 0$). We say that Λ is of **pseudo wild type** if there exist a full subcategory $\mathcal{C}'$ of $\mathrm{lat}\,\Lambda$, a wild hereditary k-algebra H and a full dense functor $\mathbb{F} : \mathcal{C}' \to \mathrm{mod}\, H$.

Unfortunately, their dichotomy is an open problem.

8.2 Orders of pseudo tame type and almost Bass orders

8.2.1 Theorem *Let* Λ *be an* R*-order,* $\dim R = 1$ *and* Γ *an almost Bass overorder of* Λ*. Assume that* $J_\Lambda = J_\Gamma$ *and the injection* $\overline{\Lambda} := \Lambda/J_\Lambda \to \overline{\Gamma} := \Gamma/J_\Gamma$ *is a normal quadratic extension (in 3.2(2)). Then* Λ *is of pseudo tame type.*

8.2.2 Definition (1) Let Λ and Γ be R-orders in 8.2.1. Define a graph $G(\Lambda)$ with thin edges and thick edges as follows.

Since $\mathbb{F} := \overline{\Gamma}\otimes_{\overline{\Lambda}} : \operatorname{mod}\overline{\Lambda} \to \operatorname{mod}\overline{\Gamma}$ is a normal quadratic extension, we have a graph $G(\mathbb{F})$ by 3.2(4). For $\mathbb{G} := \overline{\Gamma}\otimes_\Gamma : \operatorname{lat}\Gamma \to \operatorname{mod}\overline{\Gamma}$, the induced functor $\operatorname{lat}\Gamma/\ker\mathbb{G} \to \operatorname{mod}\overline{\Gamma}$ is a quadratic extension by 4.4.2. Taking a normalization $\operatorname{lat}\Gamma/\ker\mathbb{G} \stackrel{\mathbb{J}}{\to} \mathcal{E} \stackrel{\mathbb{K}}{\to} \operatorname{mod}\overline{\Gamma}$, we obtain a graph $G(\mathbb{J})$ by 3.2(4).

We obtain $G(\Lambda)$ by replacing each edges in the disjoint union of $G(\mathbb{F})$ and $G(\mathbb{J})$ by thick edges, and drawing a thin edge between each vertex $X \in \mathfrak{I}(\mathcal{E})$ (in $G(\mathbb{J})$) and $\mathbb{K}(X) \in \mathfrak{I}(\operatorname{mod}\overline{\Gamma})$ (in $G(\mathbb{F})$).

(2) A **cycle** on $G(\Lambda)$ is an infinite sequence $(\cdots, X_{-1}, X_0, X_1, X_2, \cdots)$ of vertices in $G(\Lambda)$ such that and X_{2i-1} and X_{2i} are connected by a thin edge for any i, X_{2i} and X_{2i+1} are connected by a thick edge for any i, and there exists an even $p > 0$ such that $X_{i+p} = X_p$ for any i.

A **walk** on $G(\Lambda)$ is a sequence $(X_1, X_2, \cdots, X_{2n})$ $(n > 0)$ of vertices in $G(\Lambda)$ such that X_{2i-1} and X_{2i} are connected by a thin edge for any i $(0 < i \leq n)$, and X_{2i} and X_{2i+1} are connected by a thick edge for any i $(0 < i < n)$. Let V be the subset of vertices in $G(\Lambda)$ consisting of $X \in G(\mathbb{J})$ such that there is a thick edge starting from X. A walk is called **good** if neither X_1 nor X_{2n} are not in V.

8.2.3 Theorem *Let* Λ *be an* R*-order in 8.2.1. For any cycle and good walk* $\mathbf{W}$ *on* $G(\Lambda)$*, we can define a full subcategory* $\mathcal{N}^{\mathbf{W}}$ *of* $\operatorname{lat}\Lambda$ *which satisfies the following (1)–(3).*

(1) $\mathfrak{I}(\operatorname{lat}\Lambda)$ *is a disjoint union of* $\mathfrak{I}(\mathcal{N}^{\mathbf{W}})$ *for any cycle and good walk* $\mathbf{W}$*.*

(2) $\#\mathfrak{I}(\mathcal{N}^{\mathbf{W}}) = 1$ *or* 2 *for any good walk* $\mathbf{W}$*.*

(3) For any cycle $\mathbf{W}$*, there exist a tame hereditary algebra* $H^{\mathbf{W}}$ *and a full functor* $\mathbb{F}^{\mathbf{W}} : \mathcal{N}^{\mathbf{W}} \to \operatorname{mod} H^{\mathbf{W}}$ *such that (*$\mathbb{F}^{\mathbf{W}}(X) = 0$ *iff* $X = 0$*).*

Notice that 8.2.1 follows from 8.2.3, and we can describe the image of $\mathbb{F}^{\mathbf{W}}$ in 8.2.3(3). As a conclusion, we obtain the following corollary since $\#\mathcal{N}^{\mathbf{W}} = \infty$ holds for any cycle $\mathbf{W}$.

8.2.4 Corollary *Λ in 8.2.1 is of bounded type iff $G(\Lambda)$ does not have a cycle.*

8.2.5 Example

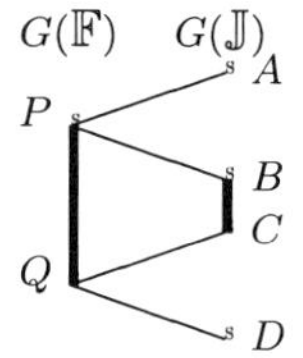

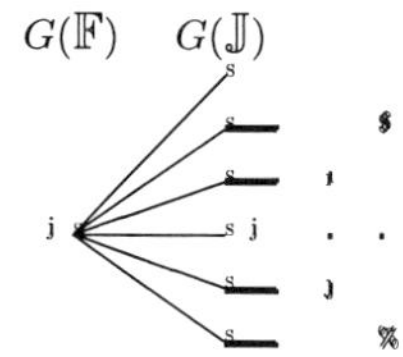

(1) Let Ω be a local maximal order, $\Gamma := \begin{pmatrix} \Omega & J_\Omega \\ J_\Omega^s & \Omega \end{pmatrix}$ and $\Lambda := \{\begin{pmatrix} w & x \\ y & z \end{pmatrix} \in \Gamma \mid w - z \in J_\Omega\}$. Then Λ and Γ satisfy the condition in 8.2.1.

Put $s = 1$ for example. Then $G(\Lambda)$ is the left diagram above, and the following list shows all good walks $\mathbf{W}$ and the corresponding element of $\mathfrak{I}(\mathcal{N}^{\mathbf{W}})$, where we put

$$(\Omega{=}\Omega) := \{(x, y) \in \Omega \times \Omega \mid x - y \in J_\Omega\}.$$

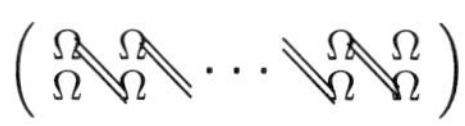 $(Q{-}C{=}B{-}P{=}\cdots{=}Q{-}C{=}B{-}P)$

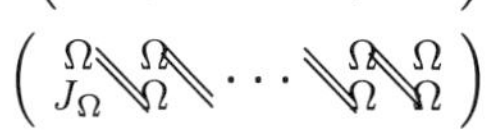 $(A{-}P{=}Q{-}C{=}B{-}P{=}\cdots{=}Q{-}C{=}B{-}P)$

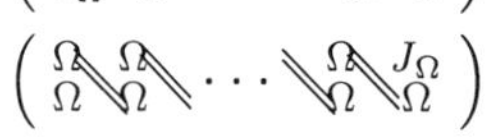 $(Q{-}C{=}B{-}P{=}\cdots{=}Q{-}C{=}B{-}P{=}Q{-}D)$

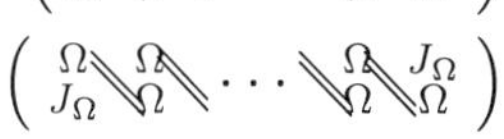 $(A{-}P{=}Q{-}C{=}B{-}P{=}\cdots{=}Q{-}D)$

There is a unique cycle

$$\mathbf{W} = (\cdots B{-}P{=}Q{-}C{=}B{-}P{=}Q{-}C\cdots).$$

For simplicity, assume that $k := \Omega/J_\Omega$ is an algebraically closed field. Then $\mathfrak{I}(\mathcal{N}^{\mathbf{W}}) = \{L_{n,a} \mid n > 0,\ a \in k - \{0\}\}$, where $L_{n,a} := p^{-1}(V_{n,a})$, $p : \mathrm{M}_{2,n}(\Omega) \to \mathrm{M}_{2,n}(k)$ is a natural surjection and

$$V_{n,a} := \{\begin{pmatrix} x_1 & x_2 & x_3 & \cdots & x_n \\ ax_1 & x_1+ax_2 & x_2+ax_3 & \cdots & x_{n-1}+ax_n \end{pmatrix} \in \mathrm{M}_{2,n}(k) \mid x_i \in k\}.$$

(2) Let Γ be a local Bass order of rank s. Assume that Λ is an order such that $J_\Gamma = J_\Lambda \subset \Lambda \subseteq \Gamma$ and $\Lambda/J_\Lambda \to \Gamma/J_\Gamma$ is of type(I·II), (III), (IV) or (V) (in 3.2(1)). Then Λ and Γ satisfy the condition in 8.2.1. For example, if $s = 3$, then $G(\Lambda)$ is the right diagram above.

8.3 Orders in purely non-semisimple algebras Recall that we call Λ in 8.2.1 a **quadratic order** if Γ is hereditary (3.3). For representation type of orders in purely non-semisimple algebras, we obtain the following simple criterion [**I4**]. Notice that (1) follows from 8.2.1, and the condition in (1) is a generalization of that in 4.3,

8.3.1 Theorem *Let Λ be an R-order in a purely non-semisimple algebra A and* $\dim R = 1$.

(1) If $l_A(Ae) \leq 2$ and $l_A(eA) \leq 2$ for any primitive idempotent e of A and $(\Lambda + J_A)/J_A$ is a quadratic order in A/J_A, then Λ is of pseudo tame type.

(2) If Λ is not of pseudo wild type, then Λ satisfies the conditions in (1).

(3) In particular, Λ is of bounded type iff Λ satisfies the condition in (1) and $G(\Lambda)$ (in 8.2.2) does not have a cycle.

8.4 Some results for $\dim R = 0$ **or** 2 In this section, we will give results for $\dim R = 0$ or 2, which is closely related to 8.2.1. As we will see in 8.5, we can prove them by reducing to same kind of problem though each reductions are rather different. It is remarkable that quadratic orders appear in each results.

8.4.1 Theorem *Let Λ be a quadratic R-order,* $\dim R = 1$ *and I a 2-sided ideal of Λ such that $\widetilde{\Lambda} = \widetilde{I}$. Then the artin algebra Λ/I is of pseudo tame type.*

Notice that algebras above give a generalized class of special biserial algebras and clannish algebras which are shown to be of tame type in [**WW**][**CB2**].

8.4.2 Theorem *Let Λ be an R-order,* $\dim R = 2$, *Γ an overorder of Λ such that* $\mathrm{gl.dim}\,\Gamma = 2$. *Assume that I is a principal ideal of Γ contained in Λ and $S := R/R \cap I$ is a complete discrete valuation ring with a quotient field L. If Γ/I is a quadraic S-order and*

$\Lambda/I \otimes_S L \to \Gamma/I \otimes_S L$ is a quadratic extension (in 3.2(2)), then Λ is of pseudo tame type.

8.5 Sketch of proof In this subsection, we will explain proof of theorems above. First, we reduce each lat Λ to some Drozd's bimodule problem $\mathrm{Mat}(\mathcal{C}, M)$ [**D**][**CB1**].

8.5.1 Definition Let $\mathcal{C}$ be an additive category $\mathcal{C}$. For a $(\mathcal{C},\mathcal{C})$-bimodule M (3.1(2)), define an additive category $\mathrm{Mat}(\mathcal{C}, M)$ as follows. An objects of $\mathrm{Mat}(\mathcal{C}, M)$ is a pair Xm consisting of an object $X \in \mathcal{C}$ and an element $m \in M(X,X)$. Put

$$\mathrm{Hom}(Xm, X'm') := \{f \in \mathcal{C}(X,X') \mid mf = fm'\},$$

and the composition is given by that of $\mathcal{C}$.

Using 8.4.1 and 8.2.1, we will explain how to reduce.

For 8.4.1, put $\overline{\Lambda} := \Lambda/I$. Then we have a functor $\mathrm{Mat}(\mathrm{pr}\,\overline{\Lambda} \times \mathrm{pr}\,\overline{\Lambda}, \mathcal{J}_{\mathrm{pr}\,\overline{\Lambda}}) \to \mathrm{mod}\,\overline{\Lambda}$, $(P_1,P_0)f \mapsto \mathrm{Cok}\,f$. This functor is full dense since any $L \in \mathrm{mod}\,\overline{\Lambda}$ has a minimal projective resolution $P_1 \xrightarrow{f} P_0 \to L \to 0$.

For 8.2.1, we use the notation in 8.2.2. By considering a canonical injection $(\overline{\Lambda} \otimes_\Lambda L \to \overline{\Gamma} \otimes_\Gamma (\Gamma L))$ for any $L \in \mathrm{lat}\,\Lambda$, we obtain a full functor $\mathrm{lat}\,\Lambda \to \mathrm{Mat}(\mathrm{mod}\,\overline{\Lambda} \times \mathrm{lat}\,\Gamma/\ker\mathbb{G}, \mathrm{mod}\,\overline{\Gamma})$.

For each cases, using 4.4.2 and so on, we can show that thus constructed M is a **quadratic** $(\mathcal{C},\mathcal{C})$-bimodule which we will explain in §9. We will see that $\mathrm{Mat}(\mathcal{C}, M)$ is of pseudo tame type in 9.3. Thus we finish to prove theorems in this section.

9 Quadratic bimodules

In this section, fix a commutative local ring R and put $k := R/J_R$. Any modules and algebras are assumed to be finitely generated R-modules.

9.1 Definition Let $\mathbb{I} = (\mathbb{I}_l\ \mathbb{I}_r) : \mathcal{C} \to \mathcal{C}_l \times \mathcal{C}_r$ be a quadratic extension (resp. normal quadratic extension) (3.2(3)) and M a Nakayama $(\mathcal{C}_l,\mathcal{C}_r)$-bimodule (3.1(2)). Then we call $(M;\mathcal{C}_l,\mathcal{C}_r)$ (or simply M) a **quadratic** (resp. **normal quadratic**) $(\mathcal{C},\mathcal{C})$-bimodule.

For any quadratic $(\mathcal{C},\mathcal{C})$-bimodule M, we can take $\mathcal{C}_l \times \mathcal{C}_r$ such that $(M;\mathcal{C}_l,\mathcal{C}_r)$ is a normal quadratic $(\mathcal{C},\mathcal{C})$-bimodule by taking the normalization (3.2(3)).

9.2 Definition Let $(M; \mathcal{C}_l, \mathcal{C}_r)$ be a normal quadratic $(\mathcal{C}, \mathcal{C})$-bimodule.

(1) Define a graph $G(\mathcal{C}, M)$ with thin edges and thick edges as follows. Let $G(\mathbb{I})$ be the graph of $\mathbb{I} : \mathcal{C} \to \mathcal{C}_l \times \mathcal{C}_r$ defined by 3.2(4). We obtain $G(\mathcal{C}, M)$ by replacing edges in $G(\mathbb{I})$ by thick edges, and drawing $l_{\mathcal{C}_l(X,X)}(M(X,Y))$ thin edges between each $X \in \mathfrak{I}(\mathcal{C}_l)$ and $Y \in \mathfrak{I}(\mathcal{C}_r)$.

(2) We define a **cycle** and a **walk** on $G(\mathcal{C}, M)$ by a same way as in 8.2.2(2). Notice that walks and cycles can be regarded as a generalization of Green walks [**G**] and strings [**WW**].

9.3 Theorem *Let $(M; \mathcal{C}_l, \mathcal{C}_r)$ be a normalized quadratic $(\mathcal{C}, \mathcal{C})$-bimodule. For any cycle and walk $\mathbf{W}$ on $G(\mathcal{C}, M)$, we can define a full subcategory $\mathcal{N}^{\mathbf{W}}$ of $\mathrm{Mat}(\mathcal{C}, M)$ which satisfies the following (1)–(3).*

(1) $\mathfrak{I}(\mathrm{Mat}(\mathcal{C}, M)) - \mathfrak{I}(\mathrm{Mat}(\mathcal{C}, 0))$ is a disjoint union of $\mathfrak{I}(\mathcal{N}^{\mathbf{W}})$ for any cycle and walk $\mathbf{W}$.

(2) $\#\mathfrak{I}(\mathcal{N}^{\mathbf{W}}) = 1$ or 2 for any walk $\mathbf{W}$.

(3) For any cycle $\mathbf{W}$, there exist a tame hereditary algebra $H^{\mathbf{W}}$ and a full functor $\mathbb{F}^{\mathbf{W}} : \mathcal{N}^{\mathbf{W}} \to \mathrm{mod}\, H^{\mathbf{W}}$ such that ($\mathbb{F}^{\mathbf{W}}(X) = 0$ iff $X = 0$).

9.4 Example Let R be a complete discrete valuation ring and Γ a local maximal R-order.

(1) Put $\Lambda := \{(x, y) \in \Gamma \times \Gamma \mid x - y \in J_\Gamma\}$. Then $(\mathrm{pr}\,\Gamma; \mathrm{pr}\,\Gamma, \mathrm{pr}\,\Gamma)$ is a quadratic $(\mathrm{pr}\,\Lambda, \mathrm{pr}\,\Lambda)$-bimodule with the left graph above. The set of isomorphism classes of objects in $\mathrm{Mat}(\mathrm{pr}\,\Lambda, \mathrm{pr}\,\Gamma)$ is given by $\bigcup_{n \geq 0} \mathrm{Gl}_n(\Lambda) \backslash \mathrm{M}_n(\Gamma)$, where $\mathrm{Gl}_n(\Lambda)$ acts $\mathrm{M}_n(\Gamma)$ by $(x, y) \cdot m := xmy^{-1}$.

For simplicity, assume that $\Gamma = R = k[[x]]$ and k is an algebraically closed field. Then $\mathfrak{I}(\mathcal{N}^{\mathbf{W}})$ consists of the following left matrix for any walk $\mathbf{W} = (P \overset{l_1}{-} Q \textbf{—} P \overset{l_2}{-} Q \textbf{—} \cdots \textbf{—} P \overset{l_{n-1}}{-} Q)$, and $\mathfrak{I}(\mathcal{N}^{\mathbf{W}})$ consists of the following right matrix for any cycle

$\mathbf{W} = (\cdots P \overset{l_1}{-} Q\textbf{—}P \overset{l_2}{-} Q\textbf{—}\cdots\textbf{—}P \overset{l_n}{-} Q\textbf{—}P \overset{l_1}{-} Q\cdots)$, where $T \in \mathrm{M}_m(k) \subseteq \mathrm{M}_m(\Gamma)$ is a Jordan block of non-zero eigenvalue.

$$\begin{pmatrix} 0 & x^{l_1} & 0 & \cdots & 0 & 0 & 0 \\ 0 & 0 & x^{l_2} & \ddots & 0 & 0 & 0 \\ 0 & 0 & 0 & \ddots & 0 & 0 & 0 \\ \vdots & \vdots & \vdots & \ddots & \ddots & \ddots & \vdots \\ 0 & 0 & 0 & \cdots & 0 & x^{l_{n-2}} & 0 \\ 0 & 0 & 0 & \cdots & 0 & 0 & x^{l_{n-1}} \\ 0 & 0 & 0 & \cdots & 0 & 0 & 0 \end{pmatrix} \begin{pmatrix} O & x^{l_1}I & O & \cdots & O & O & O \\ O & O & x^{l_2}I & \ddots & O & O & O \\ O & O & O & \ddots & O & O & O \\ \vdots & \vdots & \vdots & \ddots & \ddots & \ddots & \vdots \\ O & O & O & \cdots & O & x^{l_{n-2}}I & O \\ O & O & O & \cdots & O & O & x^{l_{n-1}}I \\ x^{l_n}T & O & O & \cdots & O & O & O \end{pmatrix}$$

(2) Let $\Gamma \supset \Lambda_i \supset J_\Gamma$ be a local Bass order of rank 1 for $i = 1, 2$ and put $\mathcal{C} := \mathrm{pr}\,\Lambda_1 \times \mathrm{pr}\,\Lambda_2$. Then $(\mathrm{pr}\,\Gamma; \mathrm{pr}\,\Gamma, \mathrm{pr}\,\Gamma)$ is a quadratic $(\mathcal{C}, \mathcal{C})$-bimodule with the right graph above. The set of isomorphism classes of objects in $\mathrm{Mat}(\mathcal{C}, \mathrm{pr}\,\Gamma)$ is given by $\bigcup_{n \geq 0} \mathrm{Gl}_n(\Lambda_1) \backslash \mathrm{M}_n(\Gamma) / \mathrm{Gl}_n(\Lambda_2)$.

9.5 Reduction theorem In the rest, put

$$\underline{\mathrm{Mat}}(\mathcal{C}, M) := \mathrm{Mat}(\mathcal{C}, M) / \mathcal{J}_{\mathcal{C}}$$

for a $(\mathcal{C}, \mathcal{C})$-bimodule M. For an algebra H, we denote by $\mathrm{mod}_0\, H$ the category of finitely generated H-modules without simple direct summand. The following reduction is fundamental for quadratic bimodules.

9.5.1 (Quadratic reduction) *Let $(M; \mathcal{C}_l, \mathcal{C}_r)$ be a normal quadratic $(\mathcal{C}, \mathcal{C})$-bimodule and M' a maximal sub $(\mathcal{C}_l, \mathcal{C}_r)$-bimodule of M. Then the following (1) and (2) hold.*

(1) There exist an additive category $\mathcal{C}'$, a quadratic extension $\mathbb{I}' : \mathcal{C}' \to \mathcal{C}_l \times \mathcal{C}_r$ and a full dense functor (called **reduction functor***) $\mathbb{R} : \underline{\mathrm{Mat}}(\mathcal{C}, M) \to \underline{\mathrm{Mat}}(\mathcal{C}', M')$.*

(2) Define a full subcategory $\mathcal{M}$ of $\mathrm{Mat}(\mathcal{C}, M)$ by $\Im(\mathcal{M}) = \{X \in \Im(\mathrm{Mat}(\mathcal{C}, M)) - \Im(\mathrm{Mat}(\mathcal{C}, 0)) \mid \mathbb{R}(X) \in \Im(\mathrm{Mat}(\mathcal{C}', 0))\}$. Then there exist a tame hereditary algebra H and a full dense functor $\mathbb{R}' : \mathcal{M} \to \mathrm{mod}_0\, H$ such that ($\mathbb{R}'(X) = 0$ iff $X = 0$).

As a conclusion, we can reduce a quadratic $(\mathcal{C}, \mathcal{C})$-bimodule M to a quadratic $(\mathcal{C}', \mathcal{C}')$-bimodule M'. For its proof, we use a computation of certain functors for tame hereditary algebras.

9.5.2 Example Using 9.4(1), we will explain the reduction functor. For simplicity, assume that $\Gamma = R = k[[x]]$ and k is an algebraically closed field.

Take $m \in \mathrm{M}_n(\Gamma)$ and we will explain how to define $\mathbb{R}(m)$. Considering the Jordan normal form of $\overline{m} \in \mathrm{M}_n(k)$, we obtain $x \in \mathrm{Gl}_n(\Lambda)$ such that $xm = \left(\begin{smallmatrix} m_1 & m_2 \\ m_3 & m_4 \end{smallmatrix}\right)$ satisfies that $\overline{m_2} = 0 = \overline{m_3}$, $\overline{m_1}$ is a totality of Jordan blocks of eigenvalue 0, and $\overline{m_4}$ is a totality of Jordan blocks of non-zero eigenvalue. It is easily shown that we may assume $m_2 = 0 = m_3$ by taking proper x.

For simplicity, assume $\overline{m_1} = \begin{pmatrix} 0&0&0&0&0&0 \\ 0&0&1&0&0&0 \\ 0&0&0&0&0&0 \\ 0&0&0&0&1&0 \\ 0&0&0&0&0&1 \\ 0&0&0&0&0&0 \end{pmatrix}$. Then it is easily shown that there exists $x' \in \mathrm{Gl}_n(\Lambda)$ such that $x'm_1 = \begin{pmatrix} a_{11}&a_{12}&0&a_{13}&0&0 \\ 0&0&1&0&0&0 \\ a_{21}&a_{22}&0&a_{23}&0&0 \\ 0&0&0&0&1&0 \\ 0&0&0&0&0&1 \\ a_{31}&a_{32}&0&a_{33}&0&0 \end{pmatrix}$, where each a_{ij} is an element of J_Γ. Then we put $\mathbb{R}(m) := (a_{ij})_{1\le i,j\le 3} \in \mathrm{M}_3(J_\Gamma)$.

Take another $l \in \mathrm{M}_n(\Gamma)$. It can be shown that $\mathrm{Gl}_n(\Lambda)m = \mathrm{Gl}_n(\Lambda)l$ holds if and only if $\overline{m_1}$ is conjugate to $\overline{l_1}$ and $(\Lambda')^\times\mathbb{R}(m) = (\Lambda')^\times\mathbb{R}(l)$ holds, where

$$\Lambda' := \{(x,y) \in \left(\begin{smallmatrix} \Gamma & \Gamma & \Gamma \\ J_\Gamma & \Gamma & \Gamma \\ J_\Gamma & J_\Gamma & \Gamma \end{smallmatrix}\right) \times \left(\begin{smallmatrix} \Gamma & J_\Gamma & J_\Gamma \\ \Gamma & \Gamma & J_\Gamma \\ \Gamma & \Gamma & \Gamma \end{smallmatrix}\right) \mid x - y \in \left(\begin{smallmatrix} J_\Gamma & \Gamma & \Gamma \\ \Gamma & J_\Gamma & \Gamma \\ \Gamma & \Gamma & J_\Gamma \end{smallmatrix}\right)\}.$$

We can apply the argument above for other m with different $\overline{m_1}$. Thus $\mathbb{R}$ gives a bijection from $\mathfrak{I}(\bigcup_{n\ge0} \mathrm{Gl}_n(\Lambda)\backslash \mathrm{M}_n(\Gamma))$ to a disjoint union of (Jordan blocks of non-zero eigenvalue) and $\mathfrak{I}((\Lambda')^\times\backslash \mathrm{M}_3(\Gamma))$ etc.

9.5.3 Remark In general, for a $(\mathcal{C},\mathcal{C})$-bimodule M and a derivation $d : \mathcal{C} \to M$, an additive category $\mathrm{Mat}(\mathcal{C}, M; d)$ is defined [**CB1**]. For them, the following reduction is known, where $d = 0$ does not imply $d' = 0$. Thus, in study of a general $(\mathcal{C},\mathcal{C})$-bimodule M, it is more natural to consider a derivation $d : \mathcal{C} \to M$ since such derivation will appear in a process of reduction. Although 9.5.1 is in the same philosophy, it is quite peculiar phenomenon that we can exclude derivations for quadratic reduction.

Lemma Let M be a $(\mathcal{C},\mathcal{C})$-bimodule, M' a sub $(\mathcal{C},\mathcal{C})$-bimodule and $d : \mathcal{C} \to M$ a derivation. Putting $\mathcal{C}' := \mathrm{Mat}(\mathcal{C}, M/M'; \overline{d})$, we obtain a derivation $d' : \mathcal{C}' \to M'$ and an equivalence $\underline{\mathrm{Mat}}(\mathcal{C}, M; d) \to \underline{\mathrm{Mat}}(\mathcal{C}', M'; d')$.

9.6 Sketch of proof of 9.3 For any cycle and walk $\mathbf{W}$, we can obtain a **reduction sequence**, which is a "successive sequence $\Phi = (\mathbb{R}^{(0)}, \mathbb{R}^{(1)}, \cdots, \mathbb{R}^{(n-1)})$ of reduction functors (in 9.5.1)". We denote by $\mathrm{Red}(\mathcal{C}, M)$ the set of reduction sequences. Then, for any $\Phi \in \mathrm{Red}(\mathcal{C}, M)$, we have a quadratic $(\mathcal{C}^{(j)}, \mathcal{C}^{(j)})$-bimodule $M^{(j)}$ and the following diagram.

$$\begin{array}{ccccccc}
\underline{\mathrm{Mat}}(\mathcal{C}, M) & & \underline{\mathrm{Mat}}(\mathcal{C}^{(1)}, M^{(1)}) & \cdots & \underline{\mathrm{Mat}}(\mathcal{C}^{(n-1)}, M^{(n-1)}) & & \underline{\mathrm{Mat}}(\mathcal{C}^{(n)}, M^{(n)}) \\
\cup & \nearrow^{\mathbb{R}^{(0)}} & \cup & \cdots & \cup & \nearrow^{\mathbb{R}^{(n-1)}} & \cup \\
\underline{\mathrm{Mat}}(\mathcal{C}^{(0)}, N^{(1)}) & & \underline{\mathrm{Mat}}(\mathcal{C}^{(1)}, N^{(2)}) & \cdots & \underline{\mathrm{Mat}}(\mathcal{C}^{(n-1)}, N^{(n)}) & & \underline{\mathrm{Mat}}(\mathcal{C}^{(n)}, 0)
\end{array}$$

Let $\mathcal{M}^{\Phi}$ be the full subcategory of $\underline{\mathrm{Mat}}(\mathcal{C}, M)$ such that $\Im(\mathcal{M}^{\Phi})$ consists of $X \in \Im(\underline{\mathrm{Mat}}(\mathcal{C}, M))$ which satisfies

$$\mathbb{R}^{(n-1)} \circ \cdots \circ \mathbb{R}^{(0)}(X) \in \underline{\mathrm{Mat}}(\mathcal{C}^{(n)}, 0)$$

and

$$\mathbb{R}^{(n-2)} \circ \cdots \circ \mathbb{R}^{(0)}(X) \notin \underline{\mathrm{Mat}}(\mathcal{C}^{(n-1)}, 0).$$

By 9.5.1, we obtain the following lemma, which is crucial to prove 9.3.

Lemma *(1) For any $X \in \Im(\underline{\mathrm{Mat}}(\mathcal{C}, M))$, there exists $\Phi \in \mathrm{Red}(\mathcal{C}, M)$ such that $X \in \Im(\mathcal{M}^{\Phi})$.*

(2) For any $\Phi \in \mathrm{Red}(\mathcal{C}, M)$, there exist a tame hereditary algebra H^{Φ} and a full dense functor $\mathbb{R}^{\Phi} : \mathcal{M}^{\Phi} \to \mathrm{mod}_0\, H^{\Phi}$ such that ($\mathbb{R}^{\Phi}(X) = 0$ iff $X = 0$).

References

[**A1**] M. Auslander: Representation theory of Artin algebras. I, II. Comm. Algebra 1 (1974), 177–268; ibid. 1 (1974), 269–310.

[**A2**] M. Auslander: Isolated singularities and existence of almost split sequences. Representation theory, II (Ottawa, Ont., 1984), 194–242, Lecture Notes in Math., 1178, Springer, Berlin-New York, 1986.

[**ARo**] M. Auslander, K. W. Roggenkamp: A characterization of orders of finite lattice type. Invent. Math. 17 (1972), 79–84.

[**B**] S. Brenner: A combinatorial characterization of finite Auslander-Reiten quivers, (Ottawa, Ont., 1984), 13–49, Lecture Notes in Math., 1177, Springer, Berlin-New York, 1986.

[**BG**] K. Bongartz, P. Gabriel: Covering spaces in representation-theory. Invent. Math. 65 (1981/82), no. 3, 331–378.

[**BGRS**] R. Bautista, P. Gabriel, A. V. Roĭter, L. Salmerón, L: Representation-finite algebras and multiplicative bases. Invent. Math. 81 (1985), no. 2, 217–285.

[**CB1**] W. W. Crawley-Boevey: Matrix problems and Drozd's theorem. Topics in algebra, Part 1 (Warsaw, 1988), 199–222, Banach Center Publ., 26, Part 1, PWN, Warsaw, 1990.

[**CB2**] W. W. Crawley-Boevey: Functorial filtrations. II. Clans and the Gelfand problem. J. London Math. Soc. (2) 40 (1989), no. 1, 9–30.

[**CR**] C. W. Curtis, I. Reiner: Methods of representation theory. Vol. I. With applications to finite groups and orders. Reprint of the 1981 original. Wiley Classics Library. A Wiley-Interscience Publication. John Wiley & Sons, Inc., New York, 1990.

[**D**] Y. A. Drozd: Tame and wild matrix problems. (Russian) Representations and quadratic forms (Russian), pp. 39–74, 154, Akad. Nauk Ukrain. SSR, Inst. Mat., Kiev, 1979.

[**DK1**] Y. A. Drozd, V. V. Kiričenko: The quasi-Bass orders. (Russian) Izv. Akad. Nauk SSSR Ser. Mat. 36 (1972), 328–370.

[**DK2**] Y. A. Drozd, V. V. Kiričenko: Primary orders with a finite number of indecomposable representations. (Russian) Izv. Akad. Nauk SSSR Ser. Mat. 37 (1973), 715–736.

[**DKR**] Y. A. Drozd, V. V. Kiričenko, A. V. Roĭter: Hereditary and Bass orders. (Russian) Izv. Akad. Nauk SSSR Ser. Mat. 31 1967 1415–1436.

[**DG**] Y. A. Drozd, G. M. Greuel: Tame-wild dichotomy for Cohen-Macaulay modules. Math. Ann. 294 (1992), no. 3, 387–394.

[**DR**] V. Dlab, C. M. Ringel: Representation of Graphs and Algebras, Mem. Amer. Math. Soc. **173** (1976).

[**FGR**] R. M. Fossum, P. Griffith, I. Reiten: Trivial extensions of abelian categories. Homological algebra of trivial extensions of abelian categories with applications to ring theory. Lecture Notes in Mathematics, Vol. 456. Springer-Verlag, Berlin-New York, 1975.

[**G**] J. A. Green: Walking around the Brauer Tree. Collection of articles dedicated to the memory of Hanna Neumann, VI. J. Austral. Math. Soc. 17 (1974), 197–213.

[**H**] H. Hijikata: Bass orders in non semisimple algebras, Report of the 4-th Symposium on Representation of Algebras, at Shizuoka, 95-118 (1993) (in Japanese).

[**HN1**] H. Hijikata, K. Nishida: Classification of Bass orders. J. Reine Angew. Math. 431 (1992), 191–220.

[**HN2**] H. Hijikata, K. Nishida: Bass orders in nonsemisimple algebras. J. Math. Kyoto Univ. 34 (1994), no. 4, 797–837.

[**HN3**] H. Hijikata, K. Nishida: Primary orders of finite representation type. J. Algebra 192 (1997), no. 2, 592–640.

[**I1**] O. Iyama: A generalization of rejection lemma of Drozd-Kirichenko. J. Math. Soc. Japan 50 (1998), no. 3, 697–718.

[**I2**] O. Iyama: Some categories of lattices associated to a central idempotent. J. Math. Kyoto Univ. 38 (1998), no. 3, 487–501.

[**I3**] O. Iyama: τ-categories I-III, preprints.

[**I4**] O. Iyama: Quadratic orders I-III, preprints.

[**IT1**] K. Igusa, G. Todorov: Radical layers of representable functors. J. Algebra 89 (1984), no. 1, 105–147.

[**IT2**] K. Igusa, G. Todorov: A characterization of finite Auslander-Reiten quivers. J. Algebra 89 (1984), no. 1, 148–177.

[**R**] C. Riedtmann: Algebren, Darstellungsköcher, Überlagerungen und zurück. (German) Comment. Math. Helv. 55 (1980), no. 2, 199–224.

[**RV**] I. Reiten, M. Van den Bergh: Two-dimensional tame and maximal orders of finite representation type. Mem. Amer. Math. Soc. 80 (1989).

[**RVo**] C. M. Ringel, D. Vossieck, Hammocks, Proc. London Math. Soc. (3) 54 (1987), 216–246.

[**RR**] C. M. Ringel, K. W. Roggenkamp: Diagrammatic methods in the representation theory of orders. J. Algebra 60 (1979), no. 1, 11–42.

[**Ro**] K. W. Roggenkamp: Lattices over orders. II. Lecture Notes in Mathematics, Vol. 142 Springer-Verlag, Berlin-New York 1970.

[**S**] D. Simson: Representation types, Tits reduced quadratic forms and orbit problems for lattices over orders. Trends in the representation theory of finite-dimensional algebras (Seattle, WA, 1997), 307–342, Contemp. Math., 229, Amer. Math. Soc., Providence, RI, 1998.

[**W**] A. Wiedemann: Die Auslander-Reiten Köcher der gitterendlichen Gorensteinordnungen. (German) Bayreuth. Math. Schr. No. 23, (1987), 1–134.

[**WW**] B. Wald, J. Waschbüsch: Tame biserial algebras. J. Algebra 95 (1985), no. 2, 480–500.

[**Y**] Y. Yoshino: Brauer-Thrall type theorem for maximal Cohen-Macaulay modules. J. Math. Soc. Japan 39 (1987), no. 4, 719–739.

[**ZK**] A. G. Zavadskiĭ, V. V. Kiričenko: Semimaximal rings of finite type. (Russian) Mat. Sb. (N.S.) 103(145) (1977), no. 3, 323–345.

ON SOME PROBLEMS IN PI-THEORY IN CHARACTERSITIC P

ALEXANDER KEMER
Ulyanovsk University
Ulyanovsk,432700,Russia

1. Identities and trace identities.

In this section we give the basic definitions and recall some old results.

Consider the free associative algebra $F\langle X\rangle$ over an associative and commutative ring F generated by a countable set X. The elements of the set X are said to be variables and the elements of the algebra $F\langle X\rangle$ are called the (non-commutative) polynomials (depending on the variables from X).

Let A be an arbitrary associative F-algebra, $f = f(x_1, \ldots, x_n) \in F\langle X\rangle$. We say the algebra A satisfies (polynomial) identity $f = 0$ if for every elements $a_i \in A$ the equality

$$f(a_1, \ldots, a_n) = 0$$

holds in A. Denote by $T[A]$ the set of all polynomials $f \in F\langle X\rangle$ such that $f = 0$ is an identity of the algebra A. Evidently the set $T[A]$ forms the ideal of the free algebra $F\langle X\rangle$. This ideal is called the ideal of identities of the algebra A. An arbitrary ideal Γ of the algebra $F\langle X\rangle$ is said to be a T-ideal if Γ is the ideal of identities of some algebra. The class of alebras satisfying the given set of identities is called the variety. Denote by $T[V]$ the set of all polynomials $f \in F\langle X\rangle$ such that $f = 0$ is an identity of every algebra from the variety V. The set $T[V]$ also forms a T-ideal. This ideal is called the ideal of identities of the variety V. We say that the algebra A generates the variety V if $T[A] = T[V]$.

K.W. Roggenkamp and M. Ştefănescu (eds.), Algebra - Representation Theory, 97–111.

A minimal set of identities of the algebra A which implies every identity of this algebra is said to be the base of identities of A. The Finite Bases Problem can be formulated as follows: Is it true that every algebra has a finite base of identities? In the case of the algebras over a field of characteristic 0 the Finite Bases Problem was set up by W. Specht in 1950 (14). The Finite Bases Problem can be formulated for every type of the algebraic systems and can be considered as the problem of classification of the varieties. Indeed if this problem has a positive solution then an arbitrary variety is defined by some finite number of parameters (discrete and continious). Of cource this is not classification in common sense, but usualy the positive solution of the Finite Base Problem implies the theorem which looks like the theorem of classification.

In the case of the associative algebras over a field of characteristic 0 the Finite Bases Problem has a positive solution (2) and the following theorem was proved

Theorem 1. (2). A proper variety of the associative algebras over a field of characteristic 0 is generated by the Grassmann hull of some finitely dimensional superalgebra.

This theorem reduced the classification of the varieties to the classification of the finitely dimensional superalgebras in the language of identities.

In the case of the algebras over an infinite field of characteristic p the Finite Bases Problem was solved negatively by A. Belov (1) in 1999. Little later V. Schigolev (13) has obtained quite general method of the construction of the counterexamples. The negative solution means that it is impossible to classify the varieties but it does not mean that they can not be classified at the multilinear level because all the counterexamples are non-multilinear. Since the multilinear level is the most important and interesting part of PI-theory the Finite Bases Problem at the multilinear level has not lost its urgency.

1.1. TRACE IDENTITIES.

At first we recall the concepts of the algebras with trace and trace identities.

Let A be any associative algebra with unit over F, R be an associative and commutative F-algebra with unit, $C(A)$ be the center of A , $\pi : R \to C(A)$ be a homomorphism of F-algebras. Put $ar = a\pi(r)$ for $a \in A, r \in R$. This turns the algebra A into an R-algebra. Let $Tr : A \to R$ be an arbitrary R-linear mapping, satisfying the property: $Tr(ab) = Tr(ba)$ for every $a, b \in A$. We call the set (A, R, π, Tr) the algebra with trace. Sometimes we say simply A is the algebra with trace, having in mind that the algebra R and the mappings π, Tr are given.

Examples. 1. The full matrix algebra $M_n(F)$ with trace $Tr : M_n(F) \to F$ defined in the usual way, is the algebra with trace.

2. Let G be the Grassmann algebra with unit generated by the countable set, G_0 be the subspace generated by the products of even number of generators ($1 \in G_0$), G_1 be the subspace generated by the products of odd number of generators. Denote by $M_{n,k}$ the subalgebra of the algebra $M_{n+k}(G)$ consisting of all matrix of the form

$$\begin{pmatrix} A & B \\ C & D \end{pmatrix},$$

where A and D are the submatrices of the sizes $n \times n$ and $k \times k$ respectively with the elements from G_0; B and C are the submatrices of the sizes $n \times k$ and $k \times n$ respectively, with the elements from G_1. Define the trace $Tr : M_{n,k} \to G_0$ putting

$$Tr \begin{pmatrix} A & B \\ C & D \end{pmatrix} = Tr(A) - Tr(D),$$

where $Tr(A)$ and $Tr(D)$ are the usual traces of the matrices over commutative algebra G_0.

Let X be a countable set and $F^\sharp\langle X\rangle$ be the free associative algebra with unit generated by the set X. Define an equivalence on the semigroup with unit $\langle X\rangle$ generated by the set X, putting $u_1 \sim u_2$ if and only if there exist the elements $v, w \in \langle X\rangle$ such that $u_1 = vw, u_2 = wv$. If $u \in \langle X\rangle$, then we put $\overline{u} = \{v \in \langle X\rangle | v \sim u\}$. Denote by $T\langle X\rangle$ the free associative and commutative algebra with unit generated by all elements $Tr(\overline{u})$, where $u \in \langle X\rangle$.

The algebra $\widetilde{F}\langle X\rangle = F^\sharp\langle X\rangle \otimes T\langle X\rangle$ is called the free algebra with trace generated by the set X. If we identify the algebras $F^\sharp\langle X\rangle \otimes 1$ and $F^\sharp\langle X\rangle$, then we have the inclusions:

$$X \subseteq F\langle X\rangle \subseteq F^\sharp\langle X\rangle \subseteq \widetilde{F}\langle X\rangle.$$

We identify also the algebras $T\langle X\rangle$ and $1\otimes T\langle X\rangle$. Then an arbitrary element $f\in\widetilde{F}\langle X\rangle$ can be written as an F-linear combination of the elements $u_0Tr(u_1)\cdots Tr(u_n)$, where $u_i\in\langle X\rangle, n\geq 0$. We call the elements of this form trace monomials and the elements of the algebra $\widetilde{F}\langle X\rangle$ the trace polynomials in the variables from X. We also call the polynomials from $F\langle X\rangle$ ordinary polynomials.

Let A be an algebra with trace , $f=f(x_1,\dots,x_n)\in\widetilde{F}\langle X\rangle$. We say the algebra A satisfies a trace identity $f=0$ if for arbitrary $a_1,\dots,a_n\in A$ the equality $f(a_1,\dots,a_n)=0$ is satisfied in A. The ideal

$$\widetilde{T}[A]=\{f\in\widetilde{F}\langle X\rangle | f=0 \text{ is an idenity of } A\}$$

is called the ideal of the trace identities of the algebra A. An arbitrary ideal of the algebra $\widetilde{F}\langle X\rangle$ which is an ideal of the trace identities of some algebra, is called a $\widetilde{T}$-ideal.

Obviously the ideal of trace identities of an arbitrary algebra contains the ideal of the ordinary identities of the same algebra.

A class of all algebras with trace satisfying a given set of trace identities is called a variety of algebras with trace. If $\widetilde{V}$ is a variety of the algebras with trace, then we denote by $\widetilde{T}[\widetilde{V}]$ the ideal of trace identities of $\widetilde{V}$.

Example. Let A be a square matrix of order n, whose elements will be considered as the variables. It is well-known that if char $F=0$ then the coefficients of the characteristic polynomial $X_n(t)$ of the matrix A are the F - linear combinations of the expressions of the form

$$Tr(A^i)\cdots Tr(A^k).$$

It follows from this and the Cayley - Hamilton theorem that the matrix algebra $M_n(F)$ satisfies the trace identity $X_n(x)=0$. The identity $Tr(1)=n$ is also the trace identity (of degree 0) of the algebra $M_n(F)$.

Denote by P a set of all ordinary multilinear polynomials of $F\langle X\rangle$. Let Γ be a T-ideal of $F\langle X\rangle$. The intersection $P\cap\Gamma$ is called the multilinear component of Γ. The multilinear component of the ideal of identities of the variety is said to be a multilinear component of the variety.

Let $\widetilde{P}_n$ be the set of all multilinear polynomials with trace of degree n depending on the variables $x_1, \ldots, x_n$. Put also

$$\widetilde{P} = \bigcup_{n=1}^{\infty} \widetilde{P}_n.$$

It follows from the definition of the free algebra with trace that every polynomial $f \in \widetilde{P}_n$ can be written in the unique way as an F-linear combination of the monomials

$$u_0(Tr(1))^l Tr(u_1) \cdots Tr(u_n), \;\; u_i \in \langle X \rangle, \;\; n, l \geq 0,$$

which belong to $\widetilde{P}_n$ and satisfy the properties:

1. $u_i \neq 1$ for every $i > 0$;
2. For every $i > 0$ the least number j, such that x_j occurs in u_{i+1}, is greater than the least number k, such that x_k occurs in u_i.

Denote by K the subalgebra with unit of the algebra $\widetilde{F}\langle X \rangle$ generated by the element $Tr(1)$. Let KS_{n+1} be the group algebra (over K) of the symmetric group of permutations S_{n+1} acting on the set $\{0, 1, \ldots, n\}$. We define K-linear mapping $\lambda_n : \widetilde{P}_n \to KS_{n+1}$, putting

$$\lambda_n(x_{i_1} \cdots x_{i_s} Tr(x_{j_1} \cdots x_{j_t}) Tr(x_{k_1} \cdots x_{k_l}) \cdots) = \sigma \in S_{n+1},$$

where σ be a permutation whose decomposition into the cycles is the following

$$\sigma = (0, i_1, \ldots, i_s)(j_1, \ldots, j_t)(k_1, \ldots, k_l) \ldots.$$

We can see the symbol 0 plays a role of label which indicates the non-trace part of the monomial. It follows from the definition of the free algebra with trace that the mapping λ_n is the isomorphism of K - modules. If $f \in \widetilde{P}_n, a \in KS_{n+1}$, then we put $fa = \lambda_n^{-1}(\lambda_n(f)a)$, $af = \lambda_n^{-1}(a\lambda_n(f))$. It turns the space $\widetilde{P}_n$ into KS_{n+1}-module.

Example. Put

$$X_n(x_1, \ldots x_n) = \lambda_n^{-1}(\sum_{\sigma \in S_{n+1}} (-1)^\sigma \sigma).$$

It can be proven that the polynomial $X_n(x_1, \ldots, x_n)$ is the full linearization of the Cayley - Hamilton polynomial $X_n(x)$.

The description of the base of trace identetities of the algebra $M_n(F)$ in the case of characterstic 0 was obtained by Yu. P. Razmyslov in 1974 (9). He proved the remarkable theorem.

Theorem 2. (9). If char $F = 0$ then every trace identity of the algebra $M_n(F)$ follows from the Cayley - Hamilton identity $X_n(x) = 0$ and identity $Tr(1) = n$.

In 1995 the author has generalized this theorem

Theorem 3. (3). Let F be a field of an arbitrary characteristic. Every multilinear trace identity of the algebra $M_n(F)$ follows from the identity $X_n(x_1, \ldots, x_n) = 0$ and identity $Tr(1) = n$.

Little later A. Zubkov (15) using the other techniques has also obtained the same result.

In 1985 Yu. P. Razmyslov (10) has described the trace identities of the superalgebras $M_{n,k}$ over a field of characteristic 0. We recall the description.

Consider a rectangle Yuong diagram containing $n+1$ rows of length $k+1$. Put $m = (n+1)(k+1)$. Fill this diagram by the numbers $0, 1, \ldots, m$ in the standard order: From the top to the bottom, from the left to the write. Denote by $D_{n,k}$ the Young tableau we obtained. Let R be the subgroup of the group S_{m+1} consisting of all permutations σ such that for every i the numbers i and $\sigma(i)$ belongs to the same row of $D_{n,k}$, C be the subgroup of all permutations σ such that for every i the numbers i and $\sigma(i)$ belongs to the same column. Put

$$X_{n,k} = X_{n,k}(x_1, \ldots, x_m) = \lambda^{-1}\Big(\sum_{p \in C, q \in R} (-1)^q pq\Big).$$

Theorem 4. (10). If char $F = 0$ then every trace identity of the algebra $M_{n,k}(F)$ follows from the identities $\sigma X_{n,k} = 0, \sigma \in S_{m+1}$ and identity $Tr(1) = n - k$.

We remark that in the case $k = 0$ we obtain the theorem 2.

Recently L. Samoilov (11) has obtained a new much shorter proof of the theorem 4.

2. Prime varieties of associative algebras.

A T-ideal Γ of the algebra $F\langle X\rangle$ is called a verbally prime iff for every T- ideals Γ_1, Γ_2 an inclusion $\Gamma_1\Gamma_2 \subseteq \Gamma$ implies one of the inclusions $\Gamma_1 \subseteq \Gamma$ or $\Gamma_2 \subseteq \Gamma$. A T-ideal Γ is called a verbally semiprime iff there are no non-trival nilpotent modulo Γ T-ideals. A variety of the algebras is called a prime (semiprime) if the ideal of identities of this variety is verbally prime (semiprime). In the same manner we define the prime and semiprime varieties of the algebras with trace.

In the case of characteristic zero the prime varieties were described by the author (2). The following structure theorem is valid.

Theorem 5. (2). Let char $F = 0$. 1. For every non-zero T-ideal Γ of the algebra $F\langle X\rangle$ there exist nilpotent modulo Γ verbally semiprime T-ideal containing Γ. 2. A T-ideal Γ is verbally semiprime iff Γ is the intersection of finite number of the verbally prime T-ideals. 3. A Variety is prime iff it is generated by the algebra $M_{n,k}$ or $M_n(G)$, where G be the Grassmann algebra of infinite rank.

The problem of classification of prime varieties in the case of characteristic p is open. At first we remark that this problem scarcely can be solved in full sense because the Finite Bases Problem has a negative solution. Perhaps it can be solved at the multilinear level.

In this section we talk about some results concerning the multilinear components of the prime varieties.

2.1. CLASSICAL AND NON-CLASSICAL VARIETIES.

Take an arbitrary $\gamma \in F$. We call a $\widetilde{T}$-ideal $\widetilde{\Gamma}$ a γ-classical if

1. $Tr(1) - \gamma \in \widetilde{\Gamma}$;

2. For every n a set $\lambda_n(\widetilde{\Gamma} \cap \widetilde{P}_n)$ is a two-sided ideal of the group algebra KS_{n+1} (the definition of the mapping λ_n was given in the previous chapter).

A variety $\widetilde{V}$ of the algebras with trace is called a γ-classical if the ideal of trace identities of $\widetilde{V}$ is γ-classical. A variety of ordinary algebras V is called a γ-classical if

$$T[V] = \widetilde{T}[\widetilde{V}] \cap F\langle X\rangle$$

for some γ-classical variety of the algebras with trace $\widetilde{V}$.

Using Razmyslov's results (10) it is easy to describe the γ-classical varieties of the algebras with trace in the case of characteristic zero.

Theorem 6. Let char $F = 0, \gamma \in F$. A proper variety of the algebras with trace $\widetilde{V}$ is a γ-classical iff 1. The number γ is integer; 2. The variety $\widetilde{V}$ is generated by the algebra $(M_{n,k})$, where $n-k = \gamma$.

It is also can be proven in the case of characteristic zero the varieties $\mathrm{Var}(M_n(G))$ are non-classical, i.e. they are not γ-classical for every $\gamma \in F$.

An ordinary (non-trace) multilinear polynomial $h = h(x_1, \ldots, x_m)$ is said to be a trace-killer of the matrix algebra $M_k(F)$ if the algebra $M_k(F)$ satisfies the trace identity of the form

$$h(x_1, \ldots, x_m)Tr(x_{m+1}) = g(x_1, \ldots, x_m, x_{m+1})$$

for some ordinary multilinear polynomial g.

Example. It is easy to verify the algebra $M_2(F)$ satisfies the following trace identity

$$[x, y, z]Tr(t) = [[x, y] \circ t, z]$$

$(a \circ b = ab + ba)$. Consequently the polynomial $[x, y, z]$ is a trace-killer of the algebra $M_2(F)$.

Consider a proper T-ideal Γ of the free algebra $F\langle X\rangle$ over a field of characteristic p. It was proved in (3) that $T[M_k(F)] \cap P \subseteq \Gamma$ for some k (we recall P be the set of all multilinear ordinary polynomials). A minimal number k with this property is called the matrix type of Γ. The matrix type of the ideal of identites of the variety is called the matrix type of variety.

Let k be the matrix type of a T-ideal Γ. We call a T-ideal Γ a regular iff Γ does not contain some trace-killer of the algebra $M_k(F)$. A variety is called a regular iff its ideal of identities is regular.

Theorem 7. (4). Every prime variety of the matrix type k is a k-classical.

If the prime variety V of the matrix type k is non-classical then by theorem 7 this variety is irregular. It means that the algebras from V satisfy every identity $h = 0$, where the polynomial h is a trace killer of the algebra $M_k(F)$. This condition is very strong. For example if $k = 2$ then the algebras from V must satisfy the identity $[x, y, z] = 0$. It follows from this immediately that in case $p \neq 2$ there are no such variety V.

In the case $k > 2$ the non-classical prime varieties are not investigated. In conclusion we mention two conjectures about the non-classical prime varieties.

Yu. P. Razmyslov (10) has formulated the conjecture about the identities of the algebras $M_n(G)$.

Conjecture 1. If char $F = 0$ then

$$T[M_n(G)] = (\widetilde{T}[M_{n,n}] + \{Tr(x)\}^{\widetilde{T}}) \cap F\langle X\rangle$$

($\{g\}^{\widetilde{T}}$ be a $\widetilde{T}$-ideal generated by g).

The following conjecture about the multilinear components of the non-classical prime varieties in arbitrary characteristic generalizes the conjecture 1 and also looks quite probable.

Conjecture 2. If V be a non-classical prime variety then

$$P \cap T[V] = P \cap (\widetilde{T}[\widetilde{V}] + \{Tr(x)\}^{\widetilde{T}})$$

for some 0-classical prime variety $\widetilde{V}$.

2.2. CLASSICAL VARIETIES AND THE IDEALS OF THE ALGEBRA FS_∞.

Denote by $\widetilde{P}'_n$ the subspace of $\widetilde{P}_n$ generated by all multilinear trace monomials which do not contain $Tr(1)$ (the definition of $\widetilde{P}_n$ was given in the previous chapter).

Put

$$\widetilde{P}' = \bigcup_{n=1}^{\infty} \widetilde{P}'_n.$$

Let S_∞ be the group of the finitary permutations acting on the set $\{0, 1, 2, \dots\}$, S_{n+1} the subgroup of all permutations $\sigma \in S_\infty$ such that $\sigma(i) = i$ for all $i > n$. Denote by FS_∞, FS_{n+1} respectively the group algebras of these groups.

We define a mapping $\lambda : \widetilde{P} \to FS_\infty$ putting $\lambda(f) = \lambda_n(f)$, if $f \in \widetilde{P}_n$. If $\widetilde{V}$ be a variety of the algebras with trace then we put

$$I_{\widetilde{V}} = \lambda(\widetilde{P} \cap \widetilde{T}[\widetilde{V}]).$$

Let $\gamma \in F$. We define an F-linear mapping $T_\gamma : FS_\infty \to FS_\infty$ as follows: If $\sigma \in S_\infty$, then

$$T_\gamma(\sigma) = \gamma\sigma, \text{ if } \sigma(0) = 0, T_\gamma(\sigma) = \tau_{\sigma(0)}\sigma, \text{ if } \sigma(0) \neq 0,$$

where τ_j be the cycle $(0, j)$. We call an ideal I of the algebra FS_∞ a γ- classical iff $T_\gamma(I) \subseteq I$.

Theorem 8. (4). A variety $\widetilde{V}$ is γ-classical iff the set $I_{\widetilde{V}}$ is a γ-classical ideal of the algebra FS_∞. A γ-classical variety $\widetilde{V}$ is prime (semiprime) iff the ideal $I_{\widetilde{V}}$ is prime (semiprime).

This theorem reduces the description of the multilinear components of the γ-classical varieties to the description of the γ-classical ideals of the algebra FS_∞.

2.3. MULTILINEAR GENERATORS OF THE γ-CLASSICAL $\widetilde{T}$-IDEALS.

The traditional natural way of description of the multilinear components of the γ-classical varieties is the following. Let $\widetilde{\Gamma}$ be a γ-classical $\widetilde{T}$-ideal. At first we choose some non-zero polynomial $f_1 \in \widetilde{\Gamma}$ of minimal degree and generate by this polynomial a γ-classical $\widetilde{T}$-ideal $\widetilde{\Gamma}_1$. Then (if it is possible) we choose a polynomial of minimal degree f_2 belonging to the set

$$\widetilde{\Gamma} \setminus \widetilde{\Gamma}_1$$

and generate by f_1 and f_2 a γ-classical $\widetilde{T}$-ideal $\widetilde{\Gamma}_2$ etc. Obviously we can assume that the polynomials f_i satisfy the additional properties: 1. $f_i \in \widetilde{P}_{n_i}$ for some n_i; 2. The left and right FS_{n_i+1} - modules generated by the polynomial f_i are irreducible modulo FS_{n_i+1} - bimodule $N_i = \widetilde{P}_{n_i} \cap \widetilde{\Gamma}_{i-1}$ ($\widetilde{\Gamma}_0 = 0$). Put

$$M_i = FS_{n_i+1}f_i + N_i/N_i.$$

Theorem 9. (4). The FS_{n_i+1}-module M_i is an irreducible as the FS_{n_i}-module.

It is well-known that in the case of characteristic zero an FS_{n+1}-module M is irreducible as an FS_n-module iff M corresponds to the rectangular Young diagram. In the case of characteristic p the description of such modules was obtained by A. Kleshchev (6).

Theorem 10. (6). An FS_{n+1} - module M is irreducible as FS_n - module iff M corresponds to a p-regular partition $\lambda = (l_1^{(a_1)}, \ldots, l_k^{(a_k)})$ satisfying the property: For every $j \leq k$ and for every $i < j$ the numbers

$$B(i,j) = l_i - l_j + \sum_{k=i}^{j} a_k$$

are divisible by p.

This theorem shows that the multilinear generators of the classical $\widetilde{T}$-ideals satisfies some nice module properties, but nevertheless there are at least two reasons which make the problem of description of the multilinear componenets of prime varieties in the terms of the multilinear generators quite wild.

The first reason. In the case of characteristic zero every γ-classical $\widetilde{T}$-ideal is verbally prime (see theorems 5 and 6). In the case of characteristic p this is not so. Moreover recently the author has proved the following theorem (unpublished)

Theorem 11. If char$F \neq 0$ then for every element γ from the simple subfield there exists a non-semiprime γ-classical variety.

This theorem and theorem 8 immediately imply

Corallory. The algebra FS_∞ has non-semiprime ideals. In particular the algebra FS_∞ has non-semiprimitive ideals.

If we want to describe the multilinear componenets of prime γ-classical varieties in the terms of the multilinear generators then we should be able to answer the question: How do we recognize that the γ-classical $\widetilde{T}$-ideal, given by its multilinear generators, is verbally prime? This problem looks wild.

The second reason. In the next subsection we'll see that the multilinear generators can not be calculated effectively even in the simple cases.

2.4. PRIME SUBVARIETIES OF VAR$M_2(F)$.

In this subsection we give the description of the multilinear components of the prime subvarieties of the variety Var$M_2(F)$ generated by the matrix algebra of order 2 over a field F of characteristic $p > 0$ (5).

Let $\widetilde{Z}\langle X\rangle$, $\widetilde{Q}\langle X\rangle$ are the free algebras with trace over the ring of integers Z and over the field of the rational numbers Q respectively. To formulate the description we need the following lemma.

Lemma. Let $n = p^k - 2$, l be a maximal number with property: p^l devides $(n+1)!$. There exists a uniqely defined modulo $\widetilde{T}[M_2(Q)]$ polynomial $g_k \in \widetilde{Z}\langle X\rangle$ such that

$$X_{0,n} = p^l g_k$$

modulo $\widetilde{T}[M_2(Q)]$.

We recall the formula

$$\lambda_n(X_{0,n}) = \sum_{\sigma \in S_{n+1}} \sigma.$$

Let ϕ be the natural homomorphism $\widetilde{Z}\langle X\rangle \to \widetilde{F}\langle X\rangle$. Put $f_k = \phi(g_k)$. Denote by $\widetilde{\Gamma}_k$ the $\widetilde{T}$-ideal of the algebra $\widetilde{F}\langle X\rangle$ generated by the polynomial f_k, the Cayley - Hamilton polynomial X_2 and $Tr(1) - 2$. Let V_k be the variety corresponding to the T-ideal

$$\Gamma_k = F\langle X\rangle \cap \widetilde{\Gamma}_k.$$

We call a subvariety V of Var$M_2(F)$ a trivial, if either V is a subvariety of the variety of commutative algebras or the multilinear components of the varieties V and Var$M_2(F)$ are equal.

Theorem 12. (5). 1. If V be a non-trivial prime subvariety of Var$M_2(F)$ then for some k the multilinear component of the varieties V and V_k are equal; 2. For every k there exists a prime subvariety V whose multilinear component equals the multilinear component of V_k; 3. If $k < s$ then

$$P \cap \widetilde{\Gamma}_k \supset P \cap \widetilde{\Gamma}_s.$$

It is easy to calculate the polynomial f_1 in every characteristic:

$$f_1 = -X_{p-2}.$$

The prime variety V_1 is known and was found by Yu. P. Razmyslov (9). The relatively free algebra of countable rank of this variety is very interesting. This algebra satisfies Engel identity of degree $p-1$ but is not Lie nilpotent.

It is also easy to calculate the polynomial f_2 for $p = 2$.

$$f_2 = x_1 \circ x_2 + Tr(x_1)Tr(x_2).$$

The relatively free algebra of countable rank satisfying identities $f_2 = 0$ and $Tr(1) = 0$ is also very interesting and can be considered as the Grassmann algebra in characteristic 2.

Recently T. Antipova, A. Antipov and the author with a help of computer have calculated the polynomials f_3 for $p = 2$ and f_2 for $p = 3$ (the paper is in preparation). The first of theese polynomials is a sum of 189 trace monomials, the second one contains more then 1000 summands. We are sure that the other polynomials f_k (may be except f_4 in characteristic 2) can not be calculated by any computer.

The $\widetilde{T}$-ideals $\widetilde{\Gamma}_k$ can be described in the other terms. Let $A \subseteq \widetilde{Z}\langle X\rangle$. We denote by $A_Z^{\widetilde{T}}$ the $\widetilde{T}$-ideal of the algebra $\widetilde{Z}\langle X\rangle$ generated by the set A.

Theorem 13.

$$\widetilde{\Gamma}_k = F\phi(\widetilde{Z}\langle X\rangle \cap \frac{1}{p^l}\{X_2, X_{n,0}, Tr(1)-2\}_Z^{\widetilde{T}})$$

(the numbers n and l are the same as in lemma).

Now we give another approach which looks quite promising.

Let n, k, γ, m are the integers satisfying the properties: 1. n, k, m are non-negative; 2. $n - k = \gamma$ modulo p. Put

$$\widetilde{\Gamma}_{n,k}(\gamma, m) = F\phi(\widetilde{Z}\langle X\rangle \cap \frac{1}{p^m}\{\sigma X_{n,k}, Tr(1)-\gamma | \sigma \in S_{(n+1)(k+1)}\}_Z^{\widetilde{T}}).$$

(the polynomial $X_{n,k}$ was defined ealier).

Theorem 14. $\widetilde{\Gamma}_k = \widetilde{\Gamma}_{2,p^k}(2+p^{l+1}, k+l)$.

This theorem shows that the variety V_k is slightly transformed variety $\mathrm{Var}M_{2,p^k}$. In conclusion we formulate the first conjecture about the multilinear components of the classical prime varieties.

Conjecture 3. If V be a classical prime variety then for some n, k, γ, l the multilinear component of the variety V is equal to the multilinear component of the $\widetilde{T}$-ideal $\widetilde{\Gamma}_{n,k}(\gamma, m)$.

3. Matrix type of Grassman algebra, irregular varieties and conjecture of C.Procesi.

In this short chapter we talk about some recent results which are close to the results of the last part of the previous chapter. The first of them is

Theorem 15. The matrix type of the Grassmann algebra of infinite rank over the field of characteristic p is equal to p.

This theorem can be formulated in the form which reminds the lemma from the subsection 2.3. Let G be the Grassmann algebra of infinite rank over the field of rational numbers Q. The polynomial

$$S_n(x_1, \ldots, x_n) = \sum_{\sigma \in S_{n+1}} (-1)^\sigma x_{\sigma(1)} \cdots x_{\sigma(n)}$$

is called the standard polynomial of degree n.

Theorem 16. For sufficiently large n there exists the multilinear polynomial $g \in Z\langle X \rangle$ such that

$$S_n(x_1, \ldots, x_n) = p^m g$$

modulo $T[G] \cap T[M_{p-1}(Q)$, where p^m be the maximal power of p deviding $n!$.

We have calculated the matrix type of the Grassmann algebra having in mind a more general problem which is very important in PI-theory.

Problem 1. To calculate the matrix type of the algebras $M_n(G)$.

The next problem is easier then Problem 1 but is not solved either

Problem 2. To calculate the minimal degree of the standard identity satisfied by the algebra $M_n(G)$.

Let R_Z, R_{F_p} are the algebras of generic matrices of order n over the ring of integers Z and over a simple field F_p respectively. The conjecture of C. Procesi is well-known: A kernel of the canonical epimorphism $R_Z \to R_{F_p}$ equals pR_Z.

Let p, n are given, char $F = p$. Denote by ψ the natural homomorphism $Z\langle X \rangle \to F\langle X \rangle$. The conjecture of C. Procesi can be also formulated in the same form as theorem 16: If the algebra $M_n(Q)$ satisfies identity $\psi(f) = 0$ then $f = pg$ modulo $T[M_n(Q)]$ for some polynomial $g \in Z\langle X \rangle$.

At first we remark that this conjecture is true at the multilinear level for the trace polynomials. This follows from the theorems 2 and 3.

In (8) C. Procesi has proved his conjecture for $n = 2, p > 2$. In 1985 W. Schelter (12) gave a negative answer to this conjecture for $p = 2, n = 2$.

Now we give the negative answer in every characteric.

Theorem 17. For every prime p there exists $n \leq p$ such that the conjecture of Procesi is not true for pair p, n.

The conjecture of Procesi is close to the problem of computing of the minimal matrix type of the irregular prime varieties. Let n

be a minimal number for which the conjecture of Procesi is not true (for given p); m be a minimal number such that there exists an irregular prime variety of matrix type m.

Theorem 18. $p \geq m \geq n$.

References

1. Belov A. (1999) On non-spechtian varieties, *Fund. and appl. math. (russian)*, **Vol 5, no 1**, pp 47-66.
2. Kemer A.R. (1991) Ideals of identities of associative algebras, *Amer.Math. Soc. Translations of Math. Monographs*, **Vol 87, Providence, R.I.**
3. Kemer A. (1995) Multilinear identities of the algebras over a field of characteristic p, *Intern. J. of Algebra and Computations*, **no 2**, pp 1-9.
4. Kemer A. (1998) Multilinear components of the regular prime varieties, *Lecture notes in pure and applied mathematics*, **Vol 198**, pp 171-183.
5. Kemer A. (2001) Multilinear components of the prime subvarieties of the variety Var$(M_2(F)$, *Algebras and Representations (submited)*.
6. Kleshchev A (1994) On restrictions of irreducible modular representations of semisimple algebraic groups and symmetric groups to some natural subgroups, I, *Proc. Lond. Math. Soc.*, **Vol 69, no 3**, pp 515-540.
7. Procesi C. (1976) The invariant theory of $n \times n$ matrices, *Adv. in Math.*, **Vol 19**, pp 306-381.
8. Procesi C. (1984) Computing with 2×2 matrices, *J. Algebra*, **Vol. 87, no 2**, pp 342-359.
9. Razmyslov Yu. P. (1974) Trace identities of the matrix algebras over field of characteristic zero, *Izv. AN SSSR (russian)*, **Vol. 38, no 4**, pp 723-756.
10. Razmyslov Yu. P. (1985) Trace identities and central polynomials of the matrix superalgebras $M_{n,k}$, *Mat. sbornik (russian)*, **Vol. 128, no 2**, pp 194-215.
11. Samoilov L.M. (2000) New proof of the theorem of Yu. P Razmyslov on the identities of the matrix superalgebra, *Fund. and appl. math. (russian) (submitted)*.
12. Schelter W. F. (1985) On question concerning generic matrices over the integers, *J. Algebra*, **Vol. 96**, pp 48-53.
13. Schigolev V. (1999) Examples of the infinitely based T-ideals, *Fund. and appl. math. (Russian)*, **Vol 5, no 1**, pp 307-313.
14. Specht W. (1950) Gesetze in Ringen.1, *Math. Z.*, **Vol. 52, no 5**, pp 557-589.
15. Zubkov A. N. (1996) On the generalization of the theorem of Procesi - Razmyslov, *Algebra i Logika*, **Vol. 35, no 4**, pp. 433-457.

BLOCKS OF CATEGORY O
DOUBLE CENTRALIZER PROPERTIES,
AND ENRIGHT'S COMPLETIONS

STEFFEN KOENIG*
Department of Mathematics and Computer Science
University of Leicester
University Road
Leicester, LE1 7RH
UK,
email: sck5@mcs.le.ac.uk
www: http://www.mcs.le.ac.uk/~skoenig

Category $\mathcal{O}$ is a fundamental object of study in representation theory of semisimple complex Lie algebras. Moreover it serves as a model for various other categories studied in representation theory of Lie algebras and of algebraic or quantum groups.

This survey article starts with recalling the definition and some basic properties of $\mathcal{O}$ in section one. More information can be found in the books by Dixmier (26) and Jantzen (44). At the end of the first section we list references to more recent results on the structure of blocks of $\mathcal{O}$. Most of these results are, however, too sophisticated to be formulated here. Sections 2 and 3, which form the main part of this article, then survey and discuss recent results of the author, obtained in joint work with Volodymyr Mazorchuk, and with Inger Heidi Slungård and Changchang Xi. These results aim at exhibiting abstract structure for blocks of $\mathcal{O}$ or certain subcategories, and thus to explain (and reprove) in a more general context results like Soergel's double centralizer property or Enright's conjecture. The main tool is the notion of 'dominant dimension', or more generally, Auslander's homological framework for representation

* Supported by DFG. Not supported by NATO.

K.W. Roggenkamp and M. Ştefănescu (eds.), Algebra - Representation Theory, 113–134.

theory of finite dimensional associative algebras. The results reported in sections 2 and 3 show that this point of view can clarify and simplify Lie theoretic considerations.

1. Definition and properties of category $\mathcal{O}$

Let $\mathfrak{g}$ be a finite dimensional semisimple complex Lie algebra; fix a Cartan subalgebra $\mathfrak{h}$ and a triangular decomposition $\mathfrak{g} = \mathfrak{n}_- \oplus \mathfrak{h} \oplus \mathfrak{n}_+$; then $\mathfrak{b}_+ = \mathfrak{n}_+ \oplus \mathfrak{h}$ and $\mathfrak{b}_- = \mathfrak{n}_- \oplus \mathfrak{h}$ are Borel subalgebras. The corresponding universal enveloping algebras are denoted by $\mathcal{U}(\mathfrak{g})$, $\mathcal{U}(\mathfrak{b}_+)$, and so on.

The objects of the **BGG–category $\mathcal{O}$** — as defined by J.Bernstein, I.M.Gelfand, and S.I.Gelfand in (11) — are left $\mathfrak{g}$–modules M with the following properties:

M is finitely generated as $\mathcal{U}(\mathfrak{g})$–module,

M is $\mathfrak{h}$–diagonalizable (that is: $M = \oplus_{\mu \in \mathfrak{h}^*} M_\mu$, where M_μ is the μ–**weight space** $\{m \in M : h \cdot m = \mu(h)m \text{ for all } h \in \mathfrak{h}\}$)

and M is **locally $\mathfrak{n}_+$–finite** (that is: for each $m \in M$ the space $\mathcal{U}(\mathfrak{n}_+) \cdot m$ is finite dimensional over $\mathbb{C}$).

The morphisms in $\mathcal{O}$ are arbitrary $\mathfrak{g}$–module homomorphisms.

(Note that this definition — and all what follows — depends on the choice of a triangular decomposition of $\mathfrak{g}$.)

Category $\mathcal{O}$ is not extension closed within the category of all $\mathfrak{g}$–representations. Choosing this definition thus implies a choice of a cohomology theory for the objects of $\mathcal{O}$.

Important objects of category $\mathcal{O}$ are the highest weight modules, which are defined as follow. Let M be a $\mathfrak{g}$–module. An element $m \in M_\lambda$ (for some λ) which is annihilated by $\mathfrak{n}_+$ is called a **highest weight vector** of weight λ. If m generates M, then M is called a **highest weight module** of highest weight λ. All finite dimensional, and many other, simple $\mathfrak{g}$–modules (in particular, all simple objects in $\mathcal{O}$) are highest weight modules.

Category $\mathcal{O}$ contains all the highest weight modules, and in particular the **Verma modules**, which are defined as follows: Fix a weight $\lambda \in \mathfrak{h}^*$. The Verma module associated with λ (i.e. having highest weight λ) is defined as

$$\Delta(\lambda) := \mathcal{U}(\mathfrak{g}) \otimes_{\mathcal{U}(\mathfrak{b}_+)} \mathbb{C}_\lambda,$$

where $\mathbb{C}_\lambda$ is the one–dimensional representation on which each $h \in \mathfrak{h} \subset \mathfrak{b}_+$ acts by multiplication with $\lambda(h)$ and $\mathfrak{b}_+$ acts via the quotient map $\mathfrak{b}_+ \to \mathfrak{h}$.

Equivalently, $\Delta(\lambda)$ can be defined as follows: Let $I(\lambda)$ be the left ideal in $\mathcal{U}(\mathfrak{g})$ which is generated by $\mathfrak{n}_+$ together with all elements $h - \lambda(h)$ for $h \in \mathfrak{h}$; then $\Delta(\lambda)$ equals the quotient $\mathcal{U}(\mathfrak{g})/I(\lambda)$.

It follows from either definition that the Verma module $\Delta(\lambda)$ is the universal highest weight module to the weight λ; it maps onto any other highest weight module having the same highest weight λ.

Viewing Verma modules as left $\mathcal{U}(\mathfrak{n}_-)$–modules, there is an isomorphism $\mathcal{U}(\mathfrak{n}_-) \simeq \Delta(\lambda)$, sending u to $u \otimes 1$. This is a consequence of the Poincaré–Birkhoff–Witt theorem, which implies that $\mathcal{U}(\mathfrak{g})$ is free over $\mathcal{U}(\mathfrak{b}_+)$ (on each side). Hence $\Delta(\lambda)$ is a free $\mathcal{U}(\mathfrak{n}_-)$–module of rank one.

In (10) (see also (26)) the following result on **composition factors of Verma modules** has been shown. Part (b) will be quoted as the **BGG–theorem**.

Let λ and μ be elements of $\mathfrak{h}^\star$.

(a) Then the $\mathbb{C}$–dimension of $Hom_{\mathcal{O}}(\Delta(\lambda), \Delta(\mu))$ is either 0 or 1. All non–zero homomorphisms are injective.

(b) The following assertions are equivalent:

(I) There is an inclusion $\Delta(\lambda) \subset \Delta(\mu)$;

(II) the simple module $L(\lambda)$ (which is a quotient of $\Delta(\lambda)$) is a composition factor of $\Delta(\mu)$;

(III) there are positive roots $\gamma_1, \ldots, \gamma_n$ such that there is a chain of inequalities $\mu \geq s_{\gamma_1}(\mu) \geq \cdots \geq s_{\gamma_n} \cdots s_{\gamma_1}(\mu) = \lambda$ (where s_γ denotes the reflection associated with γ).

The theorem does not say anything about composition multiplicities of simple modules in Verma modules.

In (11) the following categorical **properties** of category $\mathcal{O}$ have been shown: Category $\mathcal{O}$ is an abelian category with finite direct sums. Homomorphism spaces are finite dimensional.

Category $\mathcal{O}$ decomposes into a direct sum of categories $\mathcal{O}_\theta$ (indexed by central characters θ of $\mathcal{U}(\mathfrak{g})$) as follows: Fix a central character θ, that is a homomorphism from the center $Z(\mathfrak{g})$ of $\mathcal{U}(\mathfrak{g})$ to $\mathbb{C}$, and a module M in $\mathcal{O}$. Then the subset M_θ of M containing all m which are annihilated by a power (which may depend on m) of

$z - \theta(z)$ is the direct summand of M which lies in $\mathcal{O}_\theta$. The summand $\mathcal{O}_\theta$ may decompose further. Its indecomposable direct summands are called **blocks** of $\mathcal{O}$.

Enough projective and injective objects exist and there is a duality which fixes simple objects and sends projective objects to injective ones and vice versa. Every object is a quotient of a projective object. There is a bijection between isomorphism classes of indecomposable projective objects and isomorphism classes of simple objects. The isomorphism classes of simple objects are indexed by the set of **weights** $\lambda \in \mathfrak{h}^*$.

A result of Harish–Chandra (26) says that two simple modules belong to the same block only if their highest weights λ and μ satisfy $W \cdot \lambda = W \cdot \mu$ (where W is the Weyl group). In particular, for all objects in the fixed block, all occuring weights can be written as linear combinations involving finitely many fixed weights (the highest weights of the simple modules of the block) with coefficients zero or one and arbitrary positive roots, all of them with negative integer coefficients. This means that for a fixed weight κ large enough powers of $\mathfrak{n}_+$ annihilate all weight vectors of weight κ of all objects in the block.

The **construction of projective objects** in $\mathcal{O}$ goes as follows. Fix a central character θ and a weight $\lambda \in \mathfrak{h}^\star$. Then Bernstein, Gelfand and Gelfand constructed (11) an object Q in $\mathcal{O}$ (depending on λ and θ) which has a generator q and the following property; for each M in $\mathcal{O}_\theta$, sending a map φ to its evalation $\varphi(q)$ gives an isomorphism

$$Hom_{\mathcal{O}}(Q, M) \simeq M_\lambda$$

Since sending M to a weight space M_λ is an exact functor, the functor $Hom_{\mathcal{O}}(Q, -)$ also is exact and Q is a projective module. So is the maximal direct summand of Q which lies in $\mathcal{O}_\theta$. Thus it remains to construct Q.

Let I be the left ideal in $\mathcal{U}(\mathfrak{g})$ which is generated by all elements $h - \lambda(h)$ (for $h \in \mathfrak{h}$) and by the Nth powers of $\mathfrak{n}_+$, where $N \in \mathbb{N}$ is chosen large enough such that $(\mathfrak{n}_+)^N$ annihilates all vectors of weight λ of objects in the block $\mathcal{O}_\theta$. Now define Q as the quotient of $\mathcal{U}(\mathfrak{g})$ modulo the left ideal I and q as the image of 1. Then q is a generator of weight λ, hence the evaluation map

$Hom_{\mathcal{O}}(Q, M) \to M_\lambda$ is well–defined and injective. Conversely, choose a vector v in M_λ. Then there is a map $\mathcal{U}(\mathfrak{g}) \to M$ sending 1 to v. This map factors via Q by the choice of N.

The assertion in (III) of the BGG–theorem defines a **partial order** on the set of weights. It turns out that this ordering corresponds to the **Bruhat order** (see (40)) on the Weyl group in the following way. If κ is a regular dominant integral weight and $\lambda = w_1 \cdot \kappa$ and $\mu = w_2 \cdot \kappa$ are in the orbit of κ under the dot action of the Weyl group, then $\lambda \leq \mu$ in the above ordering if and only if $w_1 \geq w_2$ in the Bruhat ordering. Let W' be the stabilizer of κ under the dot action. Sending w to $w \cdot \kappa$ defines a natural bijection between the cosets W/W' and the set of isomorphism classes of simple objects in the given block. This bijection identifies the partial order in the BGG–theorem with the Bruhat order on W/W'. A block with W' being the group with one element is called **regular**, any other block is called **singular**. One of the regular blocks is the **principal block**, which contains the trivial one–dimensional module $\mathbb{C}$.

Let us now see how **finite dimensional associative algebras** arise in this context: From the properties of $\mathcal{O}$ it follows that a block is equivalent to the category of finitely generated right modules over a finite dimensional $\mathbb{C}$–algebra A_θ (which is unique up to isomorphism and Morita equivalence) in the following way: Fix a finitely generated projective object P in the block such that each indecomposable projective object in this block is isomorphic to a direct summand of P. Let A be the endomorphism ring $End_{\mathcal{O}}(P)$. Then the functor $Hom_{\mathcal{O}}(P, -)$ defines an equivalence between this block and $mod - A$. We call this algebra A 'the' algebra associated with the given block. From now on, these algebras will be considered as 'the blocks' of $\mathcal{O}$.

In (11) some other properties of the blocks of category $\mathcal{O}$ have been shown, which will be discussed now. Technically, the main property is that **each projective module has a finite filtration the subquotients of which are Verma modules**, that is, the algebras A are quasi–hereditary in the sense of (16). This is obtained as follows: Consider the universal enveloping algebra $\mathcal{U}(\mathfrak{n}_+)$ and factor out the left ideal generated by $(\mathfrak{n}_+)^N$ (where N is the above integer). Choose a basis in this vector space and preimages $x_1, \ldots, x_s$ of this basis such that x_i has weight λ_i and $\lambda_i < \lambda_j$ implies $i > j$. Then the submodule of Q generated by $x_1 \cdot q$ (where q is the generator occuring in the above construction) is isomorphic to the Verma module $\Delta(\lambda_1)$. In fact, $x_1 \cdot q$ has maximal weight in Q. Hence it is annihilated by $\mathfrak{n}_+$, thus a highest weight vector. It follows that $\Delta(\lambda_1)$ maps into Q with image the submodule of Q generated by $x_1 \cdot q$. Now the PBW theorem implies that this map is injective. Moreover the quotient of Q modulo this submodule contains a submodule (generated by the image of $x_2 \cdot q$) which is isomorphic to $\Delta(\lambda_2)$ and so on. This gives the desired filtration, and also an ordering condition on the subquotients; the first subquotient has highest weight λ_1 being maximal among the λ_i, and so on. Direct summands inherit such filtrations.

The multiplicity of a Verma module in such a filtration of this special module Q (which is denoted $(Q : \Delta)$) does not depend on the choice of the filtration and can be computed by using PBW (and Kostant's partition function). The dimension of $Hom_{\mathcal{O}}(Q, \Delta(\mu))$ can also be computed, since (as we have seen before) it equals the dimension of the weight space $\Delta(\mu)_\lambda$. The result is an equality which by passing to indecomposable direct summands proves the following **BGG reciprocity formula**:

$$(P(\lambda) : \Delta(\mu)) = [\Delta(\mu) : L(\lambda)]$$

Note that the reciprocity formula and the arguments leading to it do not determine the multiplicities occuring in this formula. In fact, the module Q we have constructed (and whose filtration by Verma modules we know to some extent) usually is not indecomposable, and it is difficult to determine the multiplicities of its indecomposable direct summands.

From the filtration of projective modules by Verma modules Bernstein, Gelfand and Gelfand obtained several consequences on cohomology: In particular, they showed that the global dimension of the category $\mathcal{O}$ is bounded by $2 \cdot (l(w_0) - 2)$ where $l(w_0)$ is the length of the longest element in the Weyl group.

We close this section with a list of numerical and homological properties of blocks of $\mathcal{O}$. No details are given; a reader not yet familiar with some of these properties may view this list as a suggestion fur further reading.

The BGG–theorem mentioned above yields the isomorphism classes of composition factors of Verma modules, but not their multiplicities. To determine those (or the multiplicities of Verma modules in filtrations of indecomposable projective modules, which by the reciprocity formula is an equivalent problem) is much more difficult. Kazhdan and Lusztig (47) proposed a solution in combinatorial terms (using certain bases of Hecke algebras). This formula, known as **Kazhdan–Lusztig conjecture**, is a theorem independently due to Beilinson and Bernstein (4) and to Brylinski and Kashiwara (15). See also (60, 6) for background information and (40) for an introduction into the combinatorics of Kazhdan–Lusztig polynomials. The proofs given in (4, 15) rely on a sequence of reformulations (based on equivalences of certain categories). First the problem is translated into one on modules over rings of differential operators. Viewing such $\mathcal{D}$–modules (or presentations of them) as differential equations one can use the Riemann–Hilbert correspondence to move on to the solutions of these differential equations. These solutions are (perverse) sheaves on the flag manifold G/B (where G is the Lie group associated with the Lie algebra $\mathfrak{g}$, and B is a Borel subgroup). Using Gabber's decomposition theorem (see (6)), which asserts the semisimplicity of perverse sheaves satisfying certain assumptions, one can in fact compute the multiplicities in question.

No purely algebraic proof of Kazhdan–Lusztig conjecture is known at present. To have such a proof available could be very useful for potential generalizations and analogous problems, in particular on representations of algebraic groups in prime characteristic.

Vogan's conjecture (63) is an equivalent version of Kazhdan–Lusztig conjecture in homological terms. Variations on this conjecture (see (17)) assert the vanishing or non–vanishing of certain cohomology groups (giving extensions between simple modules or between simples and Verma modules). In particular, this information fully determines the **quivers** of the blocks of $\mathcal{O}$. The relations of these finite dimensional algebras are not known in general.

A stronger homological property of the algebras A is that they are **Koszul** ((56, 8), see also (17, 18)). The principal block (containing the trivial one–dimensional module) is even isomorphic to its own cohomological dual $A \simeq Ext^*(L, L)$ (where L is a transversal of simple modules and the right hand side is an algebra by the Yoneda product). This result has been conjectured by Beilinson and Ginzburg (in an unpublished manuscript) and proven by Soergel (56). All blocks of $\mathcal{O}$ are **quadratic** algebras.

Irving (41, 42) has studied ring theoretic properties of blocks of $\mathcal{O}$. In particular, he has shown that Verma modules are **rigid**, that is, their socle and radical filtrations coincide ((42), see (8) for another proof). Moreover, he proved that Kazhdan–Lusztig conjecture is equivalent to the validity of certain bounds on the Loewy lengths of Verma modules.

By a result of Soergel (59), category $\mathcal{O}$ is its own **Ringel dual**. That means, the endomorphism ring of a full tilting module of a block of $\mathcal{O}$ is Morita equivalent to a block of $\mathcal{O}$.

The algebras A have **triangular decompositions** analogous to those of the Lie algebra $\mathfrak{g}$ (49, 50).

2. Double centralizer property

A fundamental property of $\mathcal{O}$ is the following result, which is due to Soergel.

THEOREM 2.1 (Soergel (56), 'Struktursatz 9'). *Let A be a block of category $\mathcal{O}$. Then there is an indecomposable projective–injective A–module Ae with local endomorphism ring eAe such that there is the following double centralizer property:*

$$A \simeq End(Ae_{eAe})$$

It is well–known that the indecomposable projective–injective module Ae is unique (up to isomorphism) in this block. It can be constructed as a direct summand of a tensor product of a Verma module, which is projective and simple (in a block different from that belonging to A), and a finite dimensional module (see (38)). From the construction it follows immediately that in the Verma module filtration of Ae each Verma module occurs precisely once.

The local algebra eAe has a combinatorial description (see (56), 'Endomorphismensatz 7') in terms of invariant theory. In the most important case, that of the principal block, eAe actually equals the **coinvariant algebra** $S(\mathfrak{h})/S(\mathfrak{h})_+^W$, which is commutative and isomorphic to the **cohomology algebra** $H^*(G/B)$ of the flag manifold G/B. In general, eAe is the invariant subalgebra of the coinvariant algebra under the action of a certain subgroup (the 'stabilizer' subgroup of the block A) of the Weyl group W. The structure of Ae as an eAe–module has been described in combinatorial terms by Soergel (56). Therefore, the theorem states that the whole algebra A can be recovered from combinatorial data.

One important consequence of this result is that for given $\mathfrak{g}$, there are only finitely many Morita equivalence classes of blocks of $\mathcal{O}$. Moreover, the description also is valid for non–integral weights. This implies the validity of Kazhdan–Lusztig conjecture in the non–integral situation (whereas the proofs of this conjecture mentioned earlier were restricted to blocks having integral weights).

The coinvariant algebra is a geometric object, and Soergel's result allows to pass directly from the algebraic objects to their geometric counterparts (perverse sheaves on the flag manifold) without the traditional intermediate use of differential equations ($\mathcal{D}$-modules).

This combinatorial description of blocks of $\mathcal{O}$ can be used to compute the algebra A by quiver and relations, see (61) for examples.

Soergel's result is somehow similar to the classical Schur–Weyl duality and its modern extensions:

THEOREM 2.2. **('Schur–Weyl duality')** *[Schur, de Concini–Procesi, Green, Jimbo, Dipper–James, Du–Parshall–Scott, e.a.] Let n and r be two natural numbers and k an infinite field of any characteristic. Let the general linear group $GL_n(k)$ act diagonally from the left on $E = (k^n)^{\otimes r}$ (with natural action on k^n) and let the symmetric group Σ_r act from the right by place permuta-*

tions. Denote by $S_k(n,r)$ the algebra generated by the image of the GL_n–action (the 'Schur algebra').
(a) Suppose $n \geq r$. Then there is a double centralizer property

$$S_k(n,r) \simeq End_{k\Sigma_r}(E)$$

$$k\Sigma_r \simeq End_{S_k(n,r)}(E)$$

(b) Suppose $n < r$. Denote by B the quotient of $k\Sigma_r$ modulo the kernel of the action of $k\Sigma_r$ on E. Then there is a double centralizer property

$$S_k(n,r) \simeq End_B(E)$$

$$B \simeq End_{S_k(n,r)}(E)$$

(c) Parts (a) and (b) remain true if one replaces the Schur algebra $S_k(n,r)$ by the quantized Schur algebra $S_q(n,r)$ and the group algebra $k\Sigma_r$ of the symmetric group by the Hecke algebra $\mathcal{H}_q(r)$ $(q \neq 0)$ of type A.

Schur proved this result in characteristic zero in order to relate the representation theories of the general linear and the symmetric groups. There are various proofs available for the general assertion; in particular, it can be seen as a reformulation of the first and the second fundamental theorem of invariant theory for general linear groups (see (22) for a characteristic free approach). The quantized version is due in the 'generic case' independently to Jimbo (45) and Dipper and James (25) and in full generality to Du, Parshall and Scott (29) (and at roots of unity to Du (28)).

The main observation of the article (52) (which is joint work with I.H.Slungård and C.C.Xi) is that both the blocks of category $\mathcal{O}$ appearing in Soergel's result and the (classical or quantized) Schur algebras with $n \geq r$ have **dominant dimension at least two**. This means that such an algebra A is the first term of an exact sequence $0 \to A \to T_1 \to T_2$ where both T_1 and T_2 are projective–injective A–modules. In the case of category $\mathcal{O}$ this follows from well–known elementary properties; in the other case, this observation is almost trivial.

For rings of dominant dimension at least two, Tachikawa and others developped about 25 years ago a general theory focussing on examples rather far away from algebraic Lie theory. However, the technology of their theory fits quite well to the examples we are interested in. One gets in a short and elementary way a criterion (see theorem 2.4 below) for double centralizer properties which implies Soergel's theorem and the $n \geq r$–part of Schur–Weyl–duality.

In the case $n < r$ this approach cannot work; in fact, $S_k(n,r)$ need not have a faithful projective–injective module. Therefore, we have to generalize our technology appropriately to handle this case as well. It turns out that the right way to do that is as follows. Instead of projective–injective modules we are using (partial) tilting modules, that is, self–dual modules which are filtered by Weyl modules. Again we use an exact sequence $0 \to A \xrightarrow{\delta} T_1 \to T_2$, but now T_1 and T_2 are tilting modules. We have to require that the map δ is an $add(T_1)$–approximation in the sense of Auslander and Reiten (2). This condition is automatic in the above case of projective–injective modules, whereas it is a (necessary) restriction in the more general situation. We note that this condition always can be satisfied by replacing T by T^l, for a large enough integer l. In this situation we say that A has T–dominant dimension at least two.

This more general situation has not been studied by Tachikawa; but it is covered by Auslander and Solberg's relative homological algebra (3). Their theory also implies that the conditions we give below for a double centralizer property to hold are both necessary and sufficient.

Altogether, we have the following result:

THEOREM 2.3. *(a) A block of category $\mathcal{O}$ has dominant dimension at least two.*
(b) A (classical or quantized) Schur algebra $S(n,r)$ with $n \geq r$ has dominant dimension at least two.
(c) For a (classical or quantized) Schur algebra $S(n,r)$ with $n < r$ there is a tilting module T such that $S(n,r)$ has T–dominant dimension at least two.

In each of the situations of the theorem, a ring theoretic criterion then implies the validity of a double centralizer property. This criterion combines a resolution from the definition of dominant dimension with an approximation property 2.4 and a description of the double centralizer as a left module which is given in another purely ring theoretic result 2.5.

THEOREM 2.4. *Let A be an algebra and T an A–module. Suppose there exists an injective left add(T)–approximation* $0 \rightarrow A \xrightarrow{\delta} T$ *which can be continued to an exact sequence* $0 \rightarrow A \xrightarrow{\delta} T \xrightarrow{\epsilon} T^n$ *for some $n \in \mathbb{N}$ (this implies that A has T–dominant dimension at least two). Denote by B the centralizer algebra $B := End({}_AT)$ and by C the double centralizer $C := End(T_B)$. Then C equals A.*

THEOREM 2.5. *Let A be an algebra and T an A–module. Suppose there exists an injective left add(T)–approximation* $0 \rightarrow A \xrightarrow{\delta} T$. *Denote by B the centralizer algebra $B := End({}_AT)$ and by C the double centralizer $C := End(T_B)$. Then C can be identified (as an A–module) with a subspace of T as follows:*

$$C \simeq Q := \bigcap_{f \in B, f(A)=0} ker(f)$$

It is very easy to derive Theorem 2.5 from Theorem 2.4. Just write ϵ as $(\epsilon_1, \ldots, \epsilon_n)$ and check that all the ϵ_i qualify as morphisms f in the formula for C.

In the case of Schur–Weyl duality we get a stronger result than just the existence of a double centralizer property (with a Schur algebra on the left hand side and an endomorphism ring on the right hand side). In fact, the right hand side can be identified with a quotient of the group algebra of a symmetric group. This algebra obviously is contained in the endomorphism ring, and the problem is to prove that the endomorphism ring cannot be larger. Our results give an 'upper bound' for the endomorphism ring in question. In the case $n < r$ we have to use additional structure (of T being a tilting module) to make sure that the endomorphism ring on the right hand side is a quotient of the group algebra of the symmetric group. In the case of category $\mathcal{O}$, the dimension of the endomorphism ring is clear (from the fact that every Verma module occurs exactly once in the filtration of Ae and each Verma module has simple socle). There our methods unfortunately do not yield any additional information about the endomorphism ring appearing on the right hand side of the double centralizer property.

Our proof of the double centralizer property for category $\mathcal{O}$ immediately carries over to another situation studied in (31, 32, 33). There a generalized category $\mathcal{O}$ has been investigated where generalized Verma modules are produced by inducing up infinite dimensional modules over a parabolic subalgebra of the given Lie algebra. The parabolic category $\mathcal{O}(\mathfrak{P}, \mathfrak{K})$, to be mentioned in the next section, is an example of such a generalized category $\mathcal{O}$. Our method reproves Theorem B in (32) and one assertion in the more general Main Theorem in (33) (both stating the validity of a double centralizer property). The finite dimensional associative algebras arising in this situation are not quasi–hereditary any more, but our method still works. In fact, the only place where we have used that Schur algebras are quasi–hereditary was in the case $n < r$, when we had to rely on T being a tilting module.

3. Enright's completion and subalgebras of blocks of $\mathcal{O}$

In his study of fundamental series representations in (30), Enright associated with each simple root of $\mathfrak{g}$ an endofunctor on the category of $\mathfrak{g}$-modules, which sends a module M into what is called a **relative completion** of M. Relative completions preserve $\mathcal{O}$, and on a certain subcategory of $\mathcal{O}$ they can be written as the composition of a localization functor with restriction and with taking a certain locally nilpotent part (23, 54) as follows.

Denote by Π the set of simple roots. For $\alpha \in \Pi$ let U_α denote the Ore localization of $U(\mathfrak{g})$ with respect to $\{X_{-\alpha}^m | m \in \mathbb{N}\}$.

For $\alpha \in \Pi$ we denote by r_α the composition of the following functors:

$$U(\mathfrak{g}) - mod \overset{U_\alpha \otimes_{U(\mathfrak{g})} -}{\longrightarrow} U_\alpha - mod \overset{res}{\longrightarrow} U(\mathfrak{g}) - mod \overset{locfin}{\longrightarrow} U(\mathfrak{g}) - mod,$$

where $locfin$ is the functor of taking locally X_α-finite part. This is a (relative) completion functor (with respect to α).

Since the aim is to start with a small module and to produce a larger one from it (thus in particular something non–zero), it makes sense to restrict attention to the objects of a subcategory, here denoted by $\mathcal{O}'$, which contains all those representations on which all the $X_{-\alpha}$ act injectively.

Enright defined the relative completion functor in a different way. He first defined it for representations of $\mathfrak{sl}(2)$ and then used the fact that in general a semisimple Lie algebra $\mathfrak{g}$ is generated by certain subalgebras which all are isomorphic to $\mathfrak{sl}(2)$. He managed to show that one can perform relative completions on the restrictions of a $\mathfrak{g}$–module to such subalgebras and get another $\mathfrak{g}$–module (not just an $\mathfrak{sl}(2)$–module) as output.

Let us look at the principal block of $\mathfrak{sl}(2)$. It has five indecomposable modules which are the vertices in the following Auslander Reiten quiver.

Γ_A :

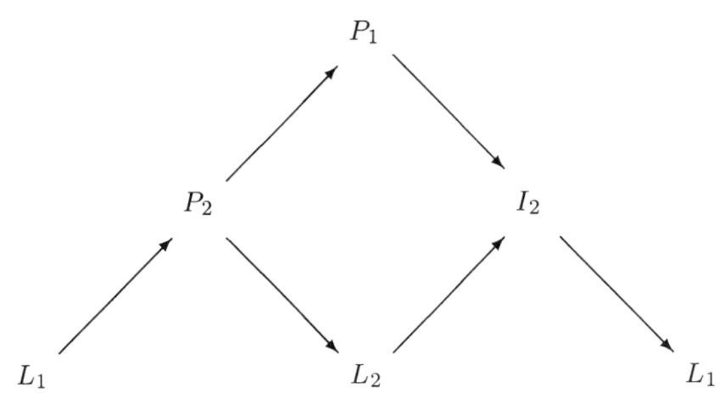

Going through the picture from left to right we find the following modules. There is a simple Verma module L_1 which is the socle of the projective Verma module P_2 (which is uniserial of length two). This projective module P_2 maps onto the other simple module L_2 (which is the trivial one–dimensional module). Moreover, P_2 is a submodule of the projective–injective module P_1 (which is the Ae in Soergel's theorem). P_1 is both the projective cover and the injective envelope of L_1. P_1 is uniserial of length three. It maps onto I_2 which has length two (it is dual to P_2) and which is the injective envelope of L_2. I_2 maps onto L_1 which is the same as the L_1 we started with.

Now the completion r_α with respect to the unique simple root α is given by mapping the modules in the previous picture to those in the following one.

$r_\alpha(\Gamma_A)$:

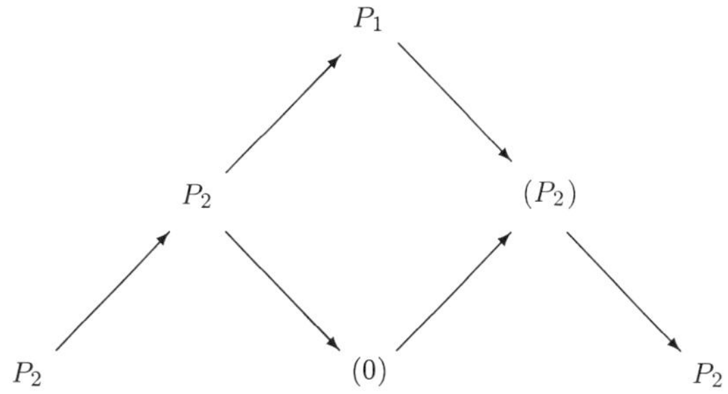

The parentheses indicate that L_2 and I_2 are not in the subcategory $\mathcal{O}'$; in fact, L_2 is sent to zero and I_2 has the same image as its quotient modulo L_2. The two projective modules P_1 and P_2 are complete – they are fixed points under the completion. The simple module L_1 is sent to P_2 which contains L_1.

As this pictures show, a relative completion is an idempotent functor. Thus its image consists of (relatively) complete modules in an obvious sense. At this point it is, however, not clear whether there are absolutely complete modules which remain unchanged under each relative completion functor. Having defined relative completions in (30), Enright conjectured that (on a certain subcategory) these relative completion functors satisfy the braid relations which are valid in the Weyl group W of $\mathfrak{g}$. **Enright's conjecture** guarantees the existence of absolutely complete modules. In fact, applying a sequence of relative completions given by a reduced expression for the longest word w_0 in the Weyl group to any module one arrives at a complete module.

Enright's conjecture has been proven independently by Deodhar ((23)) and by Bouaziz ((13)). Later on Joseph ((46)) extended the result to the whole category $\mathcal{O}$. Since each braid relation involves only two simple roots, it is enough to check the rank two cases. However, the categories occuring there are wild, thus one cannot just go through all cases.

One of the main results of (53) is a reformulation of Enright's completions in an abstract setup, and this reformulation enables us to reprove the Bouaziz-Deodhar result in an easy way.

More precisely, in (53) it is shown that the subcategory $\mathcal{O}'$ equals the category of injectively cogenerated modules. Moreover, both the relative completions and Enright's absolute completions (from now on just called completions), which in this way are shown to be compositions of certain relative completions, have very natural interpretations in terms of the finite dimensional algebras associated with blocks of $\mathcal{O}$. In fact, up to equivalence the (relatively or absolutely) **complete modules** (i.e. the images of completion functors) are precisely the injectively copresented modules for certain choices of injective modules. Thus Enright's completion fits into the general framework established by Auslander (1).

Moreover, all the (relative or absolute) completion functors can be viewed as approximation functors in the sense of Auslander and Reiten. At this point we can revisit Theorem 2.3 from the previous section. In the case of a block A of $\mathcal{O}$ this Theorem asserts the existence of a short exact sequence

$$0 \to A \to (Ae)^m \to (Ae)^l$$

where (Ae) is projective–injective. In the notation of the present section, this is precisely the statement that **projective A–modules are complete**. More generally, a module M is complete if and only if it is the first term of some short exact sequence

$$0 \to M \to (Ae)^m \to (Ae)^l$$

for certain integers m and l. There is a similar description for relatively complete modules, where Ae has to be replaced by a certain sum of injective modules. Thus, again dominant dimension is the key concept to understand the situation.

Putting completions into the framework set up by Auslander in (1) yields in particular that each block of the above categories of (relatively or absolutely) complete modules is equivalent to the module category of an algebra eAe where A is a block of $\mathcal{O}$ and e is an idempotent, the primitive summands of which are naturally indexed by cosets of the Weyl group. Thus, these subcategories carry abelian structures, which are, in fact, not obtained by restriction from the abelian structure of $\mathcal{O}$.

In the case of absolutely complete modules we get:

THEOREM 3.1. *Let $M \in \mathcal{O}$ be a module with integral support. Then M is complete if and only if it is copresented by projective-injective modules.*

Fix a block, its algebra A and the indecomposable projective–injective A–module Ae. Then an A–module is complete if and only if it is copresented by sums of copies of Ae. The category of complete A–modules is equivalent to the category of eAe–modules.

Combining this with Soergel's result on eAe one gets that the category of complete modules in the principal block is equivalent to the category of modules over the coinvariant algebra.

From the abstract framework it is clear that the category of injectively copresented modules (for some choice of injectives) is equivalent to the category of projectively presented modules (for the corresponding choice of projectives). Explicit versions of the last categories occur in (9, II 5.9) in the context of projective functors. There it is shown that certain translation functors are equivalences between categories of Harish-Chandra modules and categories of projectively presented modules in $\mathcal{O}$.

There is yet another equivalent version of these categories, namely, the parabolic generalization of the category $\mathcal{O}$ mentioned above, which has been introduced and studied in (31, 32, 33) by Lie theoretic methods. For example, the abelian structure of such categories was discovered in (33) as a quite exotic looking property, which gets a natural explanation in the present framework.

Altogether we get the following picture - it contains five categories, corresponding blocks of which are all equivalent (for suitable choices of the defining parameters):

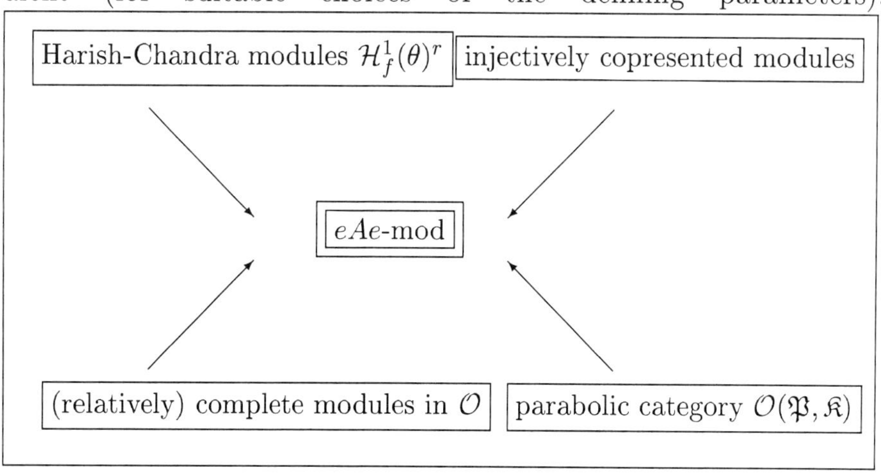

Here, the idempotent e has to be chosen according to the simple roots defining the parabolic subalgebra. This generalizes the situation of absolutely complete modules described above. In the relative situation, e is not primitive any more.

These algebras eAe have interesting structures (in particular, they are standardly stratified); see (53) for details.

Another byproduct of our proof of Enright's conjecture is the fact that complete modules have a Verma flag. This yields a lower bound for the representation type of the category of modules having a Verma flag (and, of course, also for all of $\mathcal{O}$).

The **representation types** (finite / tame / wild) are known for the blocks of $\mathcal{O}$ ((34), see also (14)) and for their subcategories $\mathfrak{F}(\Delta)$ of modules having Verma flags (14). The classification is in terms of Weyl groups and stabilizer subgroups. The paper (14) uses the fact that complete modules have a Verma flag.

THEOREM 3.2. *(14) Fix a block and its finite dimensional algebra A. Then the representation type of the category $\mathcal{F}\Delta$ of modules having a Verma flag is given by the following table. All cases not listed are wild.*

representation type	Φ	Φ_0
finite	*any* Φ	$\Phi_0 = \Phi$
	A_1	$\emptyset$
	A_2	A_1
	A_3	A_2
	B_2	A_1
	A_4	A_3
tame	$A_1 \times A_1$	$\emptyset$
	A_5	A_4
	B_3	B_2

Of course, an equality $\Phi = \Phi_0$ occurs precisely for A a simple algebra.

As a consequence one obtains:

THEOREM 3.3. *(34, 14) Fix a block and its finite dimensional algebra A. Then the representation type of the category A–mod is given by the following table. All cases not listed are wild.*

representation type	Φ	Φ_0
finite	*any* Φ	$\Phi_0 = \Phi$
	A_1	$\emptyset$
	A_2	A_1
tame	A_3	A_2
	B_2	A_1

The proof given earlier by Futorny, Nakano and Pollack in (34) uses different methods, similar to techniques used by these authors to deal with analogous problems for various kinds of Schur algebras. To find the analogue of Theorem 3.2 for Schur algebras is still an open problem.

References

1. M.Auslander, Representation theory of Artin algebras I. Commun. in Alg. 1 (1974), 177-268.
2. M.Auslander and I.Reiten, Applications of contravariantly finite subcategories. Adv. Math. 86, 111–152 (1991).
3. M.Auslander and O.Solberg, Relative homology and representation theory. I. Relative homology and homologically finite subcategories. Comm. Algebra 21, 2995–3031 (1993). II. Relative cotilting theory. Comm. Algebra 21, 3033–3079 (1993). III. Cotilting modules and Wedderburn correspondence. Comm. Algebra 21, 3081–3097 (1993).
4. A.Beilinson and J.Bernstein, Localisation de $\mathfrak{g}$–modules. C.R. Acad. Sci. Paris 292, 15–18 (1981).
5. A.Beilinson and J.Bernstein, A proof of Jantzen conjectures. In: S.Gelfand and S.Gindikin (Eds.), I.M.Gelfand Seminar, Advances in Soviet Mathematics, Vol. 16, Part 1, 1–50 (1993).
6. A.Beilinson, J.Bernstein and P.Deligne, Faisceaux pervers. Astérisque 100 (1983).
7. A.Beilinson and V.Ginzburg, Wall-crossing functors and $\mathcal{D}$-modules. Represent. Theory 3, 1–31 (1999).

8. A.BEILINSON, V.GINZBURG AND W.SOERGEL, Koszul duality patterns in representation theory. J. Amer. Math. Soc. 9, 473–527 (1996).
9. I.N.BERNSTEIN AND S.I.GELFAND, Tensor products of finite- and infinite-dimensional representations of semisimple Lie algebras. Compositio Math. 41 (1980), 245-285.
10. J.BERNSTEIN, I.M.GELFAND AND S.I.GELFAND, Structure of representations generated by vectors of highest weight. Funct. Anal. and Appl. 5 , 1–8 (1971).
11. J.BERNSTEIN, I.M.GELFAND AND S.I.GELFAND, A category of $\mathfrak{g}$–modules. Funct. Anal. and Appl. 10, 87–92 (1976).
12. J.BERNSTEIN, I.M.GELFAND AND S.I.GELFAND, Differential operators on the base affine space and a study of $\mathfrak{g}$–modules. In: I.M. GELFAND (Ed.), Lie groups and their representations, 21–64. A. Hilger, London (1975).
13. A.BOUAZIZ, Sur les représentations des algèbres de Lie semi-simples construites par T.Enright. In: Non-commutative harmonic analysis and Lie groups, Springer LN Mathematics 880 (1981), 57–68.
14. TH.BRÜSTLE, S.KÖNIG AND V.MAZORCHUK, The coinvariant algebra and representation types of blocks of category $\mathcal{O}$. Preprint (2000). Available at " http://www.mcs.le.ac.uk/~skoenig/".
15. J.BRYLINSKI AND M.KASHIWARA, Kazhdan–Lusztig conjecture and holonomic systems. Invent.math. 64, 387–410 (1981).
16. E.CLINE, B.PARSHALL AND L.SCOTT, Finite dimensional algebras and highest weight categories. J. Reine Angew. Math. 391, 85–99 (1988).
17. E.CLINE, B.PARSHALL AND L.SCOTT, Abstract Kazhdan–Lusztig theories. Tohoku Math. J. 45, 511–534 (1993).
18. E.CLINE, B.PARSHALL AND L.SCOTT, The homological dual of a highest weight category. Proc. London Math. Soc. 68, 294–316 (1994).
19. E.CLINE, B.PARSHALL AND L.SCOTT, Simulating perverse sheaves in modular representation theory. In: W.Haboush and B.Parshall (Eds.): Algebraic groups and their generalizations, Vol.1: Classical methods. Proc.Symp. Pure Math. 56, 63–104 (1994).
20. E.CLINE, B.PARSHALL AND L.SCOTT, Infinitesimal Kazhdan–Lusztig theories. In: V.Deodhar (Ed.), Kazhdan–Lusztig theory and related topics. Contemp.Math. 139, 43–74 (1992).
21. D.COLLINGWOOD AND R.IRVING, A decomposition theorem for certain self–dual modules in the category $\mathcal{O}$. Duke Math.J. 58, 89–102 (1989).
22. C.DE CONCINI AND C.PROCESI, A characteristic free approach to invariant theory. Advances in Math. 21, 330–354 (1976).
23. V.V.DEODHAR, On a construction of representations and a problem of Enright, Invent. Math. 57 (1980), 101–118.
24. V.DEODHAR, A brief survey of Kazhdan–Lusztig theory and related topics. In: W.Haboush and B.Parshall (Eds.): Algebraic groups and their generalizations, Vol.1: Classical methods. Proc.Symp. Pure Math. 56 (1994).

25. R.Dipper and G.James, The q-Schur algebra. Proc. London Math. Soc. 59, 23–50 (1989).
26. J.Dixmier, Algèbres enveloppantes. Gauthiers–Villars, Paris–Bruxelles–Montréal 1974.
27. V.Dlab and C.M.Ringel, Quasi–hereditary algebras. Illinois J. Math. 33, 280–291 (1989).
28. J.Du, A note on quantized Weyl reciprocity at roots of unity. Algebra Colloq. 2, 363–372 (1995).
29. J.Du, B.Parshall and L.Scott, Quantum Weyl reciprocity and tilting modules. Comm. Math. Phys. 195, 321–352 (1998).
30. T.J.Enright, On the fundamental series of a real semisimple Lie algebra: their irreducibility, resolutions and multiplicity formulae, Annals Math., 110 (1979), 1–82.
31. V.Futorny, S.König and V.Mazorchuk, Categories of induced modules and standardly stratified algebras. To appear in Algebras and Representation Theory.
32. V.Futorny, S.König and V.Mazorchuk, A combinatorial description of blocks in $\mathcal{O}$ associated with $sl(2)$ induction, Journal of Algebra 231, 86–103 (2000).
33. V.Futorny, S.König and V.Mazorchuk, $\mathcal{S}$-subcategories in $\mathcal{O}$, Manuscripta mathematica 102, 487–503 (2000).
34. V. Futorny, D. Nakano and D. Pollack, Representation type of the blocks of category $\mathcal{O}$. Preprint (1999).
35. O.Gabber and A.Joseph, On the Bernstein–Gelfand–Gelfand resolution and the Duflo sum formula. Compos.Math. 43, 107–131 (1981).
36. O.Gabber and A.Joseph, Towards the Kazhdan–Lusztig conjecture. Ann. Sci. Ecole Norm. Sup. (4) 14, 261–302 (1983).
37. J.Humphreys, Introduction to Lie algebras and representation theory. Springer (1972).
38. J.Humphreys, A construction of projective modules in the category $\mathcal{O}$ of Bernstein-Gelfand-Gelfand. Nederl. Akad. Wetensch. Proc. Ser. A 80(=Indag. Math. 39), 301–303 (1977).
39. J.Humphreys, Highest weight modules for semisimple Lie algebras. In: V.Dlab, P.Gabriel (Ed.): Representation theory I. Springer LN 831 (1980).
40. J.Humphreys, Reflection groups and Coxeter groups. Cambridge University Press, 1990. xii+204 pp.
41. R.Irving, Projective modules in the category $\mathcal{O}_S$: Loewy series. Trans.A.M.S. 291, 733–754 (1985).
42. R.Irving, The socle filtration of a Verma module. Ann.Sci.Ec.Norm.Sup. 21, 47–65 (1988).
43. R.Irving, BGG–algebras and the BGG–reciprocity principle. J.Alg. 135, 363–380 (1990).
44. J.C.Jantzen, Einhüllende Algebren halbeinfacher Lie–Algebren. Springer (1983).

45. M.Jimbo, A q-analogue of $U(\mathfrak{gl}(N+1))$, Hecke algebra, and the Yang Baxter equation. Lett. Math. Phys. 11, 247–252 (1986).
46. A.Joseph, The Enright functor on the Bernstein-Gelfand-Gelfand category $\mathcal{O}$. Invent. Math. 67 (1982), 423–445.
47. D.Kazhdan und G.Lusztig, Representations of Coxeter groups and Hecke algebras. Invent.math. 83, 165–184 (1979).
48. D.Kazhdan and G.Lusztig, Schubert varieties and Poincaré duality. Proc.Symp. Pure Math. 36, 185–203 (1980).
49. S.König, Exact Borel subalgebras of quasi–hereditary algebras, I. With an appendix by L.Scott. Math. Z. 220, 399–426 (1995).
50. S.König, Strong exact Borel subalgebras of quasi–hereditary algebras and abstract Kazhdan–Lusztig theory. Adv. Math. 147, 110–137 (1999).
51. S.König, Cartan decompositions and BGG–resolutions. Manuscripta Math. 86, 103–111 (1995).
52. S.König, I.H.Slungård and C.C.Xi, Double centralizer properties, dominant dimension and tilting modules. To appear in J.Alg.
53. S.König and V.Mazorchuk, Enright's completions and injectively copresented modules. Preprint (1999). Available at "http://www.mcs.le.ac.uk/~skoenig/".
54. O.Mathieu, Classification of irreducible weight modules. To appear in Ann. Inst. Fourier.
55. C.M.Ringel, The category of modules with good filtrations over a quasi–hereditary algebra has almost split sequences. Math.Z. 208, 209–223 (1991).
56. W.Soergel, Kategorie $\mathcal{O}$, perverse Garben und Moduln über den Koinvarianten zur Weylgruppe. J. Amer. Math. Soc. 3, 421–445 (1990).
57. W.Soergel, The combinatorics of Harish–Chandra bimodules. J. reine angew. Math. 429, 49–74 (1992).
58. W.Soergel, Combinatorics of Harish-Chandra modules. Representation theories and algebraic geometry (Montreal, PQ, 1997), 401–412, NATO Adv. Sci. Inst. Ser. C Math. Phys. Sci., 514, Kluwer Acad. Publ., Dordrecht (1998).
59. W.Soergel, Character formulas for tilting modules over Kac-Moody algebras. Represent. Theory 2, 432–448 (1998).
60. T.Springer, Quelques applications de la cohomologie d'intersection. Bourbaki Seminar, Vol. 1981/1982, pp. 249–273, Astérisque, 92–93, Soc. Math. France, Paris, 1982.
61. C.Stroppel, Quivers of category $\mathcal{O}$. Preprint (2000).
62. D.N.Verma, Structure of certain induced representations of complex semisimple Lie algebras. Bull.A.M.S. 74, 160–166 (1968).
63. D.Vogan, Irreducible representations of semisimple Lie groups, I and II. Duke Math.J. 46, 61–108 and 805–859 (1979).

THE GELFAND-KIRILLOV DIMENSIONS OF ALGEBRAS ARISING FROM REPRESENTATION THEORY

ZONGZHU LIN *
Department of Mathematics
Kansas State University
Manhattan, KS 66506
zlin@math.ksu.edu

Abstract. This note is to study a variety of graded algebras that arise from the induced representations for semisimple algebraic groups and quantum groups. These algebras will play an important role in a study of the cohomology groups of line bundles over the flag varieties. This short note concentrates on the calculation of the Gelfand-Kirillov dimensions of these algebras.

1. Introduction

In this paper, we calculate the Gelfand-Kirillov dimensions of certain graded algebras arising from the representation theory of semisimple algebraic groups and quantum groups.

For a given semisimple algebraic group G and a Borel subgroup B over a field k. Each rational B-module M defines a vector bundle over the flag variety G/B and the space of global sections of the vector bundle is a subspace of $k[G] \otimes M$. In particular when M is 1-dimensional, the space of global sections can be viewed as a subspace of the coordinate algebra $k[G]$. If M is a commutative algebra on which B acts as algebra automorphisms, then the space of global sections is also an algebra on which the algebraic group G acts as algebra automorphisms. The aim of the paper is to study

* Research was supported in part by NSF grant DMS-9970603

K.W. Roggenkamp and M. Ştefănescu (eds.), Algebra - Representation Theory, 135–157.

the properties of these algebras as far as the GK-dimension is concerned. For algebraic groups, the geometry of these algebras can be described. But for their quantum parts, the geometric interpretation still needs to be worked out. So far the evidence shows that in the quantum case, the Gk-dimension is closely related to the space classifying the flag modules of these algebras as studied by Ohn.

For a given line bundle $\mathcal{L}_\lambda$ associated to an integral weight λ of the root system R. $A(\lambda) = \oplus_{n=0}^{\infty} H^0(G/B, \mathcal{L}_\lambda^{\otimes n})$ is naturally a graded associative algebra with multiplication being the tensor product map

$$H^0(G/B, \mathcal{L}_\lambda^{\otimes n}) \otimes H^0(G/B, \mathcal{L}_\lambda^{\otimes m}) \to H^0(G/B, \mathcal{L}_\lambda^{\otimes (n+m)}).$$

$A(\lambda)$ is a rational G-module on which G acts as automorphism of graded algebra. If λ is not dominant, the algebra $A(\lambda)$ is just the field k. For dominant λ, the algebra $A(\lambda)$ is finitely generated (using the fact that tensor product of two induced modules have a good filtration). The algebra $A(\lambda)$ is actually a subalgebra of the coordinate algebra and thus defines an irreducible projective variety of dimension GK-dim$(A(\lambda))-1$. This variety actually is the G-orbit of the highest weight space in the projective space $\mathbb{P}(V(\lambda))$ of the Weyl module $V(\lambda)$ of highest weight λ. The geometric property of this algebra is of special interests, in particular, in classifying the point modules of the graded algebra. For any submonoid Λ of the dominant weight monoid X_+, $M(\Lambda) = \oplus_{\lambda \in \Lambda} k_\lambda$ is a B-algebra and $H^0(G/B, \mathcal{L}(M(\Lambda)))$ is a G-algebra and its GK-dimension will also be computed. However, their geometric interpretation is to be worked out.

One can also construct analogous graded algebras $A_q(\lambda)$ for quantum groups. Although the algebras $A_q(\lambda)$ have the same GK-dimensions as their counter parts for G, they fail to be commutative. In this case an interesting question is to classify the point modules and to relate the point modules with the G-orbit of the highest weight space in the projective space of the corresponding Weyl modules for the corresponding algebraic group G. When the parameter is not a root of 1 in the field of complex numbers, its known that the quantum coordinate algebra possesses infinite dimensional irreducible modules. But it is not clear whether these subalgebras possesses infinite dimensional irreducible modules. The computation of the GK-dimensions of these algebras works for all characteristic and any non-zero parameter q.

This paper is organized as follows. In Section 2, we set more general induction functor for general Hopf algebra (even bialgebras) and prove that the induced module for a Hopf algebra from a Hopf module algebra of a Hopf subalgebra remains a Hopf module algebra. As a special example, a skew polynomial extension can be viewed as an induced module over polynomial algebras. Zhang (Zh2) proved skew polynomial extension will increase the GK-dimension by at most 1. The results in later sections seem to suggest a more general phenomena. This setting of induced representation will cover both algebraic group and quantum group situations. Section 3 is devoted to computing the GK-dimensions of the induced algebras for both quantum groups and algebraic groups. This dimension seems to appear naturally from the geometry of generalized flag varieties. Section 4 contains a few general remarks on interesting appearance of GK-dimensions. In particular the GK-dimension of the extension algebra of a module for a finite dimensional algebra is closely related to the complexity of the modules. For finite groups, it is proved by Carlson (C) that the Krull dimension (which equals the GK-dimension in this case) of the extension algebra of a finite dimensional module is the same as the complexity of the module. For cocommutative Hopf algebras, the results of Friedlander and Suslin (FS) can be used to prove that the same to be true for all finite dimensional modules.

2. Induced Representations

2.1 Let k be a field and H a Hopf algebra over k. A full subcategory $\mathcal{C}_H$ of the H-module category is called admissible if all modules in $\mathcal{C}$ are locally finite, $\mathcal{C}$ includes the trivial modules and is closed under direct limit, tensor product, and finite dimensional dual. As studied in (Lin1, Lin2), such category is uniquely determined by a class of cofinite ideals

$$\mathcal{I} = \{\mathrm{Ann}_H(M) \mid M \in \mathrm{Ob}(\mathcal{C}_H) \text{ and } \dim M < \infty\}.$$

For any Hopf subalgebra D of H, the collection of ideals $\mathcal{I}_D = \{D \cap I \mid I \in \mathcal{I}\}$ defines an admissible category $\mathcal{C}_D$ for the

Hopf algebra D and the restriction functor from the category of all H-modules to the category of all D-modules sends all modules in $\mathcal{C}_H$ to $\mathcal{C}_D$. The restriction functor $\mathrm{Res}_D^H : \mathcal{C}_H \to \mathcal{C}_D$ has a right adjoint functor $\mathrm{Ind}_D^H : \mathcal{C}_D \to \mathcal{C}_H$ which is called the induction functor (see (Lin1, Lin2)).

For any D-module M, $\mathrm{Hom}_D(H, M)$ is an H-module with H-action defined by $(hf)(x) = f(xh)$ for all $x, h \in M$ and $f \in \mathrm{Hom}_D(H, M)$. If M is in $\mathcal{C}_D$, then $\mathrm{Ind}_D^H(M)$ is the unique largest H-submodule of $\mathrm{Hom}_D(H, M)$ which is in $\mathcal{C}_H$.

EXAMPLE 1. *Let* $H = \mathrm{Dist}(G)$ *be the distribution algebra of an algebraic group scheme* G *and* $\mathcal{C}_H$ *be the category of all* G*-rational modules (possibly infinite dimensional). For any closed subgroup scheme* N *of* G, $\mathrm{Dist}(N)$ *is a Hopf subalgebra of* H *and* $\mathcal{C}_D$ *is exactly the category of all rational* N*-modules and* Ind_D^H *is isomorphic to the induction functor* Ind_N^G *as defined in (Jan) for algebraic group schemes.*

EXAMPLE 2. *Let* $H = U_q(\mathfrak{g})$ *be the quantum enveloping algebra, as defined in (L3), of a complex semisimple Lie algebra* $\mathfrak{g}$ *over* $\mathbb{C}(q)$ *with* q *either an indeterminate or* $0 \neq q \in \mathbb{C}$. *Let* $\mathcal{C}_H$ *be the category of all integrable modules of type* $\mathbf{1}$. *If* D *is any parabolic subalgebra* $U(I, J)$ *as defined in (APW), then the category* $\mathcal{C}_D$ *defined here and the category defined in (APW) for* $U(I, J)$ *are the same and the induction functor* Ind_D^H *is also the same as the one defined in (APW, 2.1).*

PROPOSITION 2.2. *Let* M *be a* k*-algebra. If* M *is further a* D*-module algebra (Sw) such that* M *is in* $\mathcal{C}_D$ *as* D*-module, then* $\mathrm{Ind}_D^H M$ *is an* H*-module algebra.*

Proof. Since the multiplication map $M \otimes M \to M$ is a homomorphism of D-modules, we have the induced homomorphism $\mathrm{Ind}_D^H M \otimes \mathrm{Ind}_D^H M \to M \otimes M \to M$ of D-modules, which is the composition of the multiplication map with tensor product of the evaluation map $\mathrm{Ind}_D^H M \to M$ (Lin1). Applying the Frobenius reciprocity (Lin1), we get a homomorphism $\mathrm{Ind}_D^H M \otimes \mathrm{Ind}_D^H M \to \mathrm{Ind}_D^H M$. One can verify, using the associativity of the multiplication of M, that the above defined H-module homomorphism is an associative multiplication on $\mathrm{Ind}_D^H M$. The identity of the multiplication is the counit $\varepsilon : H \to k1 \subseteq M$. (By the definition of D-module algebra, $k1$ is a trivial D-submodule of M.)

Note that $\mathrm{Ind}_D^H(M)$ is a subalgebra of $\mathrm{Hom}_D(H, M)$ which is also a subalgebra of $\mathrm{Hom}_k(H, M)$ with the convolution multiplication defined by

$$(fg)(x) = \sum_{(x)} f(x_{(1)})g(x_{(2)})$$

for all $f, g \in \mathrm{Hom}_k(H, M)$ and $x \in H$ using Sweedler's notation for comultiplication (Sw). In particular, when $M = k$ and $D = k$, $k[H] = \mathrm{Ind}_k^H k$ is an H-module algebra (as a subalgebra of the algebra $\mathrm{Hom}_k(H, k)$). If all modules in $\mathcal{C}_H$ are locally finite dimensional, then $k[H]$ has a Hopf algebra structure, which we will not use in this paper. (see (Lin1) for more details.)

COROLLARY 2.3. *If $M = \oplus_{\lambda\in\Lambda} k_\lambda$ as D-module algebra with k_λ being 1-dimensional D-module and algebra structure is given by $k_\lambda \otimes k_\mu \to k_{\lambda\mu}$ for a set Λ of characters of D including ϵ, where $(\lambda\mu)(d) = \sum_{(d)} \lambda(d_{(1)})\mu(d_{(2)})$ for all $d \in D$, then $\mathrm{Ind}_D^H(M) = \oplus_{\lambda\in\Lambda} \mathrm{Ind}_D^H k_\lambda$ is isomorphic to an D-invariant subalgebra of $k[H]$.*

Proof. Using the H-module algebra structure of $k[H]$, there is another way to see the H-module algebra structure on $\mathrm{Ind}_D^H M$ as follows. First we can embed $M \otimes_k k[H]$ in $\mathrm{Hom}_k(H, M)$ by

$$(a \otimes f)(x) = f(x)a, \quad (\forall x, \in H, f \in k[H], a \in M).$$

$k[H]$ has a $(H \otimes H)$-bimodule structure with actions given by $((h \otimes d)f)(x) = f(\gamma(d)xh)$ for all $x, h, d \in H$ and $f \in k[H]$, where $\gamma : H \to H$ is the antipode. This makes $k[H]$ an $H \otimes D$-module with $H \otimes D$ being the tensor product Hopf algebra. Note that $k[H]$ is an $H \otimes k = H$-module algebra. By letting H act on M trivially, we can make M an $H \otimes D$-module algebra. But with the $D \cong k \otimes D$-action, $k[H]$ is not a D-module algebra in general. However we have the following

$$(1 \otimes d)(fg) = \sum_d ((1 \otimes d_{(2)})f)((1 \otimes d_{(1)})g)$$

for all $d \in D$ and $f, g \in k[H]$. Therefore $k[H] \otimes M$ with the tensor product algebra structure is an $H \otimes D$-module. Note that $D \cong k \otimes D$

is a normal Hopf subalgebra of $H \otimes D$ with quotient Hopf algebra being H (Lin1). Thus the D-fixed point space

$$(M \otimes k[H])^D = \{x \in M \otimes k[H] \mid d(x) = \varepsilon(d)x \ \forall d \in D\}$$

is closed under multiplication and is an H-module subalgebra. In fact, for any $f_i, g_j \in k[H]$ and $a_i, b_j \in M$, such that $\sum_i f_i \otimes a_i$, $\sum_j g_j \otimes b_j \in (M \otimes k[H])^D$, then

$$\begin{aligned}(1 \otimes d)(\sum_{i,j} a_i b_j \otimes f_i g_j) &= \sum_{(d)} \sum_{i,j} d_{(1)}(a_i b_j) \otimes (1 \otimes d_{(2)})(f_i g_j) \\ &= \sum_{(d)} \sum_{i,j} (d_{(1)} a_i)(d_{(2)} b_j) \otimes (d_{(4)} f_i)(d_{(3)} g_j).\end{aligned}$$

Now since M is a direct sum of the 1-dimensional D-modules, we can assume a_i, b_j are in the 1-dimensional submodules and $(d_{(1)} a_i)$ and $(d_{(2)} b_j)$ are scalars. Apply this to the above one shows that $(M \otimes k[H])^D$ is closed under multiplication.

Note that by (Lin1) $\operatorname{Ind}_D^H M \cong (M \otimes k[H])^D$ as H-module. Note that $\operatorname{Ind}_D^H k_\lambda = k_\lambda \otimes k[H]^\lambda$ where

$$k[H]^\lambda = \{f \in k[H] \mid f(dx) = \lambda(d) f(x), \forall d \in D\}.$$

We get a natural isomorphism of vector spaces $\oplus_{\lambda \in \Lambda} \operatorname{Ind}_D^H k_\lambda \rightarrow \oplus_{\lambda \in \Lambda} k[H]^\lambda$. Note that D acts on $k[H]^\lambda$ by the character $\lambda^* = \lambda \circ \gamma$. By the definition of the product $\operatorname{Ind}_D^H k_\lambda \otimes \operatorname{Ind}_D^H k_\mu \rightarrow \operatorname{Ind}_D^H(k_{\lambda\mu})$, it is easy to check that under the above isomorphism, multiplication is preserved and the multiplication is an isomorphism of H-module algebras.

2.4 For each D-module M in $\mathcal{C}_D$, the tensor algebra $T(M)$ is a (graded) D-module algebra. Thus $\operatorname{Ind}_D^H T(M) = \oplus_{n=0}^\infty \operatorname{Ind}_D^H(M^{\otimes n})$ (Ind_D^H is additive and commutes with direct limits). Furthermore, if D is cocommutative, then the symmetric algebra $S(M)$ is also a grade D-module algebra. Thus we can consider the induced algebras from $T(M)$ or from $S(M)$ if D is commutative. An interesting question is to compare that GK-dimension of $\operatorname{Ind}_D^H(M)$ with the GK-dimension of M for a D-module algebra M.

2.5 The definition of admissible categories for bialgebras (without the requirement of antipode) and its subalgebra can also be defined by simply removing the condition about dual modules. Then we can define the induced modules for bialgebras in a similar way.

Let A be a k-algebra and σ an algebra endomorphism of A. The polynomial extension of A with respect to σ is the algebra $A[x, \sigma]$ generated by A and an indeterminate x subject to the commuting condition $xa = \sigma(a)x$. We can make the polynomial algebra $k[y]$ a bialgebra by setting $\Delta(y) = y \otimes y$ and $\varepsilon(y) = 1$. Define a new bialgebra $k[y, x]$ with $\Delta(x) = 1 \otimes x + x \otimes y$. Then $k[y]$ is a sub-bialgebra of $k[y, x]$. $k[y]$ acts on A by letting $y \to \sigma$. Then the multiplication of A is a homomorphism of $k[y]$-modules. We can define the admissible category $\mathcal{C}$ for $k[x, y]$ to be the full subcategory consisting of all $k[x, y]$-modules on which x acts nilpotently. Now the induced algebra $\mathrm{Ind}_{k[y]}^{k[y,x]} A$ is isomorphic to $A[x, \sigma]$. In (Zh2), Zhang proved that GK-dim$(A[x, \sigma]) \leq$ GK-dim(A). Let F_y be the quotient field of the polynomial ring $k[y]$. Note that the GK-dimension of the F_y-algebra $k[x, y] \otimes_{k[y]} F_y$ has GK-dimension 1. The results in the next section suggests that something more general along this line should hold. The result should be closely related certain geometry.

3. Algebras from representation theory

3.1 In this section we set H and $\mathcal{C}_H$ as in Examples 1-2. Let us fix a root system R with $\Delta \subseteq R^+$ being the sets of simple roots and positive roots. In the case of Example 1, we assume that G is a simply connected semisimple algebraic group and N is a split Borel subgroup B with a maximal torus T. Thus D is the distribution algebra of the Borel subgroup B. For any D-module M in $\mathcal{C}_D$ (thus a rational B-module), $\mathrm{Ind}_D^H M$ is the vector space of global sections of the vector bundle associated to M over the flag variety G/B.

In the case of Example 2, we fix D to be the subalgebra $U(\emptyset, \Delta)$ of $U_q(\mathfrak{g})$ generated by $U^0, F_i^{(m)}$'s. (See (APW) for more details on the notations.)

Let $X(T)$ be the character group of T, which is isomorphic to the integral weight lattice of the root system R (Hu3). Then all irreducible D-modules (up to isomorphism) in $\mathcal{C}_D$ in both cases are in one-to-one correspondence to the set $X(T)$, while the irreducible H-modules in $\mathcal{C}_H$ are in one-to-one correspondence to the set X_+ of all dominant weights in $X(T)$. There is a character of U^0 associated to each $\lambda \in X(T)$, which is an algebra homomorphism $\lambda : U^0 \to \mathbb{C}(q)$ such that $\lambda(K_i) = q^{d_i\langle\lambda,\alpha\rangle}$. For each H or D-module M and $\lambda \in X(T)$, we denote by M_λ the T (or U^0)-weight space of weight λ in the sense of (APW). In particular each λ defines a character for the algebra D. Although the Hopf algebra D is not cocommutative, the convolution product of characters is commutative as they factors through the cocommutative Hopf algebra U^0. Thus we can also regard $X(T)$ as the character group of D in the quantum case and write it additively.

3.2 For H and D as mentioned above, we can see that $k[H]$ has no zero divisors. For $\lambda \in X(T)$, $\operatorname{Ind}_D^H k_\lambda = (k_\lambda \otimes k[H])^D = k[H]^\lambda$ which is the largest D-submodule of $k[H]$ on which D acts via the character $\lambda^* = \lambda \circ \gamma$. In fact, for any $f \in k[H]$, we have

$$d(1 \otimes f) = \sum (d_{(1)}1 \otimes d_{(2)}f) = \sum 1 \otimes \lambda(d_{(1)})d_{(2)}f.$$

Thus $1 \otimes f \in \operatorname{Ind}_D^H k_\lambda$ if and only if $d(1 \otimes f) = \varepsilon(d)(1 \otimes f)$ for all $d \in D$ and

$$\operatorname{Ind}_D^H k_\lambda = (k_\lambda \otimes k[H])^D = \{f \in k[H] \mid d(1 \otimes f) = \varepsilon(d)(1 \otimes f)\}.$$

In the case of algebraic groups, the above condition is equivalent to $b(1 \otimes f) = (1 \otimes f)$ for all $b \in B$, i.e., $1 \otimes f = \lambda(b)1 \otimes bf$ and $bf = \lambda(b^{-1})f$. In the case of quantum groups, the condition is equivalent to $K_i(1 \otimes f) = 1 \otimes \lambda(K_i)K_i f$ and

$$0 = F_i(1 \otimes f) = F_i 1 \otimes K_i^{-1} f + 1 \otimes F_i f = 1 \otimes F_i f,$$

i.e., $K_i f = \lambda(\gamma(K_i))f$ and $F_i f = 0$ for all i. In the case that q is a root of 1, there are a few more conditions need to be checked, which we omit here.

Let Λ be an additive subset of $X(T)$ containing zero, i.e., Λ is a submonoid of $X(T)$. We define a D-module $M(\Lambda) = \oplus_{\lambda\in\Lambda} k_\lambda$. The tensor product $k_\lambda \otimes k_\mu \to k_{\lambda+\mu}$ defines a k-algebra structure on $M(\Lambda)$. We will discuss the GK–dimension of the algebra $\operatorname{Ind}_D^H M(\Lambda)$. Since $\operatorname{Ind}_D^H k_\lambda = 0$ unless $\lambda \in X_+$, we have $\operatorname{Ind}_D^H M(\Lambda) = \operatorname{Ind}_D^H M(\Lambda \cap X_+))$. Thus we can assume that $\Lambda \subseteq X_+$.

Note that, by Corollary 2.3, $\operatorname{Ind}_D^H M(\Lambda)$ is always a subalgebra of the coordinate algebra $k[H]$.

THEOREM 3.3. *For $\lambda, \mu \in X_+$, the multiplication map $\operatorname{Ind}_D^H k_\lambda \otimes \operatorname{Ind}_D^H k_\mu \to \operatorname{Ind}_D^H k_{\lambda+\mu}$ is onto. Thus $\operatorname{Ind}_D^H M(\Lambda)$ is finitely generated if Λ is finitely generated as submonoid of X_+. In particular the coordinate algebra $k[G/U] = \sum_{\lambda \in X_+} \operatorname{Ind}_B^G(k_\lambda)$ is finitely generated for both the algebraic group and its quantum analog.*

Proof. We will prove the first part of the theorem only. The second part is a consequence of the first part. Let $\lambda, \mu \in X_+$. $\operatorname{Ind}_D^H k_\lambda \otimes \operatorname{Ind}_D^H k_\mu$ has a filtration with consecutive quotients isomorphic to $\operatorname{Ind}_D^H k_\nu$ with $\nu \in X_+$. For semisimple algebraic groups this is well-known (M, Don, W) (for the characteristic 0 situation, this is trivial due to the semi-simplicity of finite dimensional modules and the characterization of the simple modules by Borel and Weil). For quantum groups when the parameter q is not a root of unit, this is similar to the case of characteristic 0 reductive groups with detailed work in (APW). When q is a p^r-th root of 1, Andersen, Polo, and Wen have shown the result in (APW). The general situation is proved by Paradowski (P) using Lusztig's canonical basis.

To show that $\operatorname{Ind}_D^H k_\lambda \otimes \operatorname{Ind}_D^H k_\mu \to \operatorname{Ind}_D^H k_{\lambda+\mu}$ is onto, one first notes that the dual module of $\operatorname{Ind}_D^H k_\lambda$ is a universal highest weight module, which is called a Weyl module (or the quantum Weyl module). The dual module of $\operatorname{Ind}_D^H k_\lambda \otimes \operatorname{Ind}_D^H k_\mu$ admits a filtration with consecutive quotient being Weyl modules. Now using almost identical argument as in the proof of (W, (3.1)) one can show that the dual module of $\operatorname{Ind}_D^H k_{\lambda+\mu}$ is a submodule of the dual module of $\operatorname{Ind}_D^H k_\lambda \otimes \operatorname{Ind}_D^H k_\mu$. This shows that $\operatorname{Ind}_D^H k_{\lambda+\mu}$ is a quotient of $\operatorname{Ind}_D^H k_\lambda \otimes \operatorname{Ind}_D^H k_\mu$. Now we conclude that the multiplication map is onto since $\operatorname{Hom}_H(\operatorname{Ind}_D^H k_\lambda \otimes \operatorname{Ind}_D^H k_\mu, \operatorname{Ind}_D^H k_{\lambda+\mu})$ is 1-dimensional.

Note that, when $\Lambda = X(T)_+$, $\operatorname{Ind}_B^G(M(\Lambda)$ is the shape algebra of the flag variety G/B. In the quantum case, $\operatorname{Ind}_D^H(M(\Lambda))$ is the shape algebra of the quantum flag variety. In particular the graded algebra $\operatorname{Ind}_D^H(M(\Lambda))$ is generated by the degree 1 space, where the gradation is given by the heights of the highest weights. Ohn (O) was interested in studying the algebra $\operatorname{Ind}_D^H M(\Lambda)$, especially, its geometric properties.

3.4 Let $f : \mathbb{N} \to \mathbb{N}$ be a function. The polynomial growth of f is defined to be the real number

$$\nu(f) = \inf\{\nu \in \mathbb{R} \mid f(n) \leq n^{\nu} \text{ for all } n \gg 0\}.$$

Let A be a k-algebra. Suppose A is finitely generated. Let $V \subseteq A$ be a subspace containing a set of generators of A. Set $A_n = \sum_{i=1}^{n} V^i \subseteq A$ with $V^0 = k1$ and $f(n) = \dim(A_n)$. The GK-dimension (Gelfand-Kirillov dimension) of A is defined to be $\nu(f)$. The GK-dimension of A is independent of the choice of the subspace V and denoted by GK-dim(A). For more general k-algebras A, the GK-dimension of A is defined to be

$$\sup\{\text{GK-dim}(A') \mid A' \text{ is a finitely generated subalgebra of } A\}.$$

In particular, if $f(n) = \dim(A_n)$ is polynomial of n, then GK-dim$(A) = \deg(f(n))$. See (MR) and (KL) for more details about GK-dimensions of algebras and modules.

3.5 To calculate the GK-dimension of the algebra $\operatorname{Ind}_D^H M(\Lambda)$ for any finitely generated submonoid $\Lambda \subseteq X_+$. We consider the special case $\Lambda = \{n\lambda \mid n \in \mathbb{N}\}$ for fixed $\lambda \in X_+$. The algebra $A = \operatorname{Ind}_D^H M(\Lambda)$ is naturally an $\mathbb{N}$-graded algebra. Theorem 3.3 shows that the subspace $\operatorname{Ind}_D^H k_\lambda$ generates the algebra $\operatorname{Ind}_D^H M(\Lambda)$. By Weyl's character formula (Hu3) we have

$$\dim \operatorname{Ind}_D^H k_\lambda = \frac{\prod_{\alpha \in R^+} \langle \lambda + \rho, \alpha \rangle}{\prod_{\alpha \in R^+} \langle \rho, \alpha \rangle} = \prod_{\alpha \in R^+} \left(\frac{\langle \lambda, \alpha \rangle}{\langle \rho, \alpha \rangle} + 1\right). \tag{3.1}$$

Here ρ is the half sum of all positive roots, and $\langle \lambda, \alpha \rangle = \frac{2(\lambda,\alpha)}{(\alpha,\alpha)}$ with $(\ ,\ )$ being the inner product on the Euclidean space containing the root system R.

For the convenience of later calculation, we set $\dim(\lambda) = \dim \operatorname{Ind}_D^H k_\lambda$ and take $V = \operatorname{Ind}_D^H k_\lambda$. Theorem 3.3 shows that $V^n = \operatorname{Ind}_D^H k_{n\lambda}$. Thus $\dim V^n = \dim(n\lambda)$ for all n. By the definition of the GK-dimension, (MR, 8.1.9) we need to calculate the dimension of $A_n = \sum_{i=0}^{n} V^i$. Note that

$$\dim A_n = \sum_{i=0}^{n} \dim V^i = \sum_{i=0}^{n} \dim(i\lambda) = \sum_{i=0}^{n} \prod_{\alpha \in R^+} \left(\frac{\langle \lambda, \alpha \rangle}{\langle \rho, \alpha \rangle} i + 1\right). \tag{3.2}$$

3.6 For fixed λ, $\dim(i\lambda)$ is a polynomial of i, i.e., $\dim(i\lambda) = \sum_{s=0}^{N} a_s i^s$ with $a_s \in \mathbb{Q}$ depending on λ only. In particular $a_N = \prod_{\alpha \in R^+} \frac{\langle \lambda, \alpha \rangle}{\langle \rho, \alpha \rangle}$. Here $N = |R^+|$ is the number of positive roots.

Define $f(n) = \dim(A_n)$. We regard $f(n)$ as a function of n. Then

$$f(n) = \sum_{i=0}^{n} \sum_{s=0}^{N} a_s i^s = \sum_{s=0}^{N} a_s (\sum_{i=0}^{n} i^s). \tag{3.3}$$

The following lemma is well-known, but we can not locate a reference. Since we are going to use it repeatedly, we give a proof here for the completeness.

LEMMA 3.7. *For any nonnegative integer* s, $\sum_{i=0}^{n} i^s = \frac{1}{s+1} n^{s+1} + t_n$. *Here* t_n *is a polynomial of* n *of degree at most* s.

Proof. When $s = 0, 1$, this is trivial. We now use induction on s. Denote $g_s(n) = \sum_{i=1}^{n} i^s$. Assume that $g_s(n) = a_{s+1} n^{s+1} + a_s n^s + \cdots a_0$ is a polynomial of n with coefficients $a_0, a_1, \ldots a_{s+1}$ independent of n and $a_{s+1} = \frac{1}{s+1}$. By using the Abel partial summation formula (or integration by parts) we one gets

$$\begin{aligned} g_{s+1}(n) &= g_s(n)(n+1) - \sum_{i=1}^{n} g_s(i) \\ &= (n+1) \sum_{j=0}^{s+1} a_j n^j - \sum_{j=0}^{s+1} a_j (\sum_{i=1}^{n} i^j) \\ &= (n+1) \sum_{j=0}^{s+1} a_j n^j - \frac{1}{s+1} g_{s+1}(n) - \sum_{j=0}^{s} a_j g_j(n) \end{aligned} \tag{3.4}$$

Thus, solving out $g_{s+1}(n)$ we get

$$g_{s+1}(n) = \frac{1}{s+2} n^{s+2} + \text{ terms of } n \text{ with degrees at most } s+1. \tag{3.5}$$

Hence the induction proves the lemma.

Note that more generally we have $\nu(g_s(n)) = s+1$ for any $s > 0$. In fact we can write $g_s(n) = n^{s+1} \cdot \frac{1}{n} \sum_{i=1}^{n} (\frac{i}{n})^s$ and $g_s(n)/n^{s+1} \to \int_0^1 x^s dx = 1/(s+1)$.

Let us go back to (3.3). $f(n) = \sum_{s=0}^{N} a_s g_s(n)$ is a polynomial of degree $N+1$ by the above lemma if $\langle \lambda, \alpha \rangle > 0$ for all simple roots α.

For more general weight λ, we consider the polynomial

$$d_\lambda(x) = \prod_{\alpha \in R^+} (\frac{\langle \lambda, \alpha \rangle}{\langle \rho, \alpha \rangle} x + 1). \tag{3.6}$$

Let R_λ^+ be the set of all positive roots α such that $\langle \lambda, \alpha \rangle = 0$. Then $d_\lambda(x)$ has degree $N - |R_\lambda^+|$. Thus $f(n)$ has degree $N - |R_\lambda^+| + 1$.

THEOREM 3.8. *The graded algebra* $A(\lambda) = \oplus_{n=0}^{\infty} \operatorname{Ind}_D^H k_{n\lambda}$ *has GK-dimension* $N + 1 - |R_\lambda^+|$.

Let $I_\lambda = \{\alpha \in \Delta \mid \langle \lambda, \alpha^\vee \rangle = 0\}$ and P_λ be the parabolic subgroup containing B and positive simple roots $\alpha \in I_\lambda$. Then the one dimensional B-modules $k_{n\lambda}$ extends to P_λ-module structures. Thus $\operatorname{Ind}_B^G(k_{n\lambda}) = \operatorname{Ind}_{P_\lambda}^G k_{n\lambda}$, which is the space of global sections of the line bundle $k_{n\lambda}$ on G/P_λ. Note that $\dim G/P_\lambda = N - |R_\lambda^+|$. The theorem shows that the GK-dimension of the algebra $\operatorname{Ind}_{P_\lambda}^G \oplus_{n=0}^{\infty} k_{n\lambda}$ is $\dim(G/P_\lambda) + 1$. In the quantum version, we don't have a precise geometric definition of generalized flag manifold G/P_λ. However the GK-dimension of the quantum analog of the graded algebra $\operatorname{Ind}_{P_\lambda}^G (\oplus_{n=0}^{\infty} k_{n\lambda})$ is independent of the choice of the weight λ. It depends on the set R_λ^+ only. This dimension seams to be the dimension of the algebraic variety that parameterizes the isomorphism classes of the point modules of the graded algebras. Later we will see that the variety is independent of the choice of the weight λ.

3.9 As mentioned in the introduction, the graded algebra $A(\lambda)$ defines an irreducible projective variety Y_λ in the commutative case. In fact, using the Serre duality, $\operatorname{Ind}_D^H(k_\lambda)$ is the space of linear functions on the Weyl module $V(\lambda)$. Theorem 3.3 shows that induced algebra homomorphism $k[V(\lambda)] \to A(\lambda)$ is homogeneous and surjective. Thus we can embed Y_λ in $\mathbb{P}(V(\lambda))$ as G-invariant closed subvariety. In fact, Y_λ the homogeneous subvariety generated by highest weight space. Since P_λ stabilizes the highest weight space, then we have a natural morphism $G/P_\lambda \to Y_\lambda$. By comparison of the dimension and the fact that both varieties are complete, therefore, the above map is finite. In the quantum case, it is interesting to classify the point modules of the algebra $A(\lambda)$ and relate the variety of all point modules in the sense of Artin-Tate-Van der Bergh (ATV) to the variety Y_λ or G/P_λ.

3.10 Next we consider the case when $\Lambda = \sum_{i=1}^{r} \mathbb{N}\lambda_i \subseteq X_+$ is a free commutative submonoid of rank r with basis $\lambda_1, \lambda_2, \ldots, \lambda_r$. For example $\Lambda = X_+$ is free commutative monoid of rank l with basis $\omega_1, \ldots, \omega_l$ being the set of fundamental weights. Here l is the rank of the root system. By Theorem 3.3, then algebra $A = \mathrm{Ind}_D^H M(\Lambda)$ is generated by the subspace $V = \oplus_{i=1}^{r} \mathrm{Ind}_D^H k_{\lambda_i}$.

$$A_n = \sum_{i=0}^{n} V^i = \bigoplus_{\substack{\lambda \in X_+ \\ \lambda = \sum n_i\lambda_i, \sum n_i \le n}} \mathrm{Ind}_D^H k_\lambda.$$

Then $\dim A_n = \sum \dim(\sum_{i=1}^{r} n_i\lambda_i)$ with the summation taken over all $(n_1, \ldots, n_r) \in \mathbb{N}^r$ such that $\sum n_i \le n$. Using Weyl's dimension formula (3.1) we have

$$\dim(\sum n_i\lambda_i) = \prod_{\alpha\in R^+} \left(1 + \sum_{i=1}^{r} n_i a_i(\alpha)\right) \tag{3.7}$$

where $a_i(\alpha) = \frac{\langle\lambda_i,\alpha\rangle}{\langle\rho,\alpha\rangle}$ (for $i = 1, \cdots, r$) are non-negative rational numbers.

More generally, let us fix an $l \in \mathbb{N}$. For each $n \in \mathbb{N}$, set $C(l,n) = \{(n_1, \ldots, n_l) \in \mathbb{N}^l \mid \sum_{i=1}^{l} n_i = n\}$. For each $t = (t_1, \ldots, t_l) \in \mathbb{N}^l$, we define

$$d_{l,t}(n) = \sum_{(n_1,\ldots,n_l)\in C(l,n)} n_1^{t_1} \cdots n_l^{t_l}.$$

Note that $d_{l,t}(n)$ generalizes the function $g_s(n)$. In fact $g_s(n) = d_{2,(s,0)}(n)$. The following lemma generalizes Lemma 3.7.

LEMMA 3.11. *For each* $t = (t_1, \ldots t_l) \in \mathbb{N}^l$, $d_{l,t}(n)$ *is a polynomial of* n *with degree* $l - 1 + t_1 + \cdots + t_l$ *and the leading coefficient* $\frac{t_1!\cdots t_l!}{(l-1+t_1+\cdots+t_l)!}$.

Proof. We use induction on l. It is trivial when $l = 1$. Assume $l > 1$. We write

$$\begin{aligned} d_{l,t}(n) &= \sum_{n_1=0}^{n} n_1^{t_1} \left(\sum_{(n_2,\ldots,n_l)\in C(l-1,n-n_1)} n_2^{t_2} \cdots n_l^{t_l} \right) \\ &= \sum_{n_1=0}^{n} n_1^{t_1} d_{l-1,t'}(n-n_1) = \sum_{i=0}^{n} (n-i)^{t_1} d_{l-1,t'}(i). \end{aligned} \tag{3.8}$$

Here $t' = (t_2, \dots, t_l) \in \mathbb{N}^{l-1}$. By the induction assumption, we can write $d_{l-1,t'}(i) = \sum_{j=0}^{r} c_j i^j$ with $r = l-2+t_2+\dots+t_l$, $c_0, c_1, \dots, c_r$ being independent of i, and $c_r = \frac{t_2!\cdots t_l!}{(l-2+t_2+\cdots+t_l)!}$. Then, by using Lemma 3.7, we have

$$\begin{aligned}\sum_{i=0}^{n}(n-i)^{t_1}d_{l-1,t'}(i) &= \sum_{i=0}^{n}\sum_{s=0}^{t_1}\binom{t_1}{s}(-1)^s n^{t_1-s}i^s\sum_{j=0}^{r}c_j i^j \\ &= \sum_{j=0}^{r}c_j\sum_{s=0}^{t_1}\binom{t_1}{s}(-1)^s n^{t_1-s}\sum_{i=0}^{n}i^{s+j} \\ &= \sum_{j=0}^{r}c_j\sum_{s=0}^{t_1}\binom{t_1}{s}(-1)^s n^{t_1-s}\Big(\frac{1}{s+j+1}n^{s+j_1}+\cdots\Big) \\ &= \sum_{j=0}^{r}c_j\Big[\Big(\sum_{s=0}^{t_1}\binom{t_1}{s}(-1)^s\frac{1}{s+j+1}\Big)n^{t_1+j+1}+\cdots\Big].\end{aligned} \tag{3.9}$$

The leading term occurs when $j = r$ with leading coefficient being

$$\sum_{s=0}^{t_1}\binom{t_1}{s}\frac{(-1)^s}{s+r+1}.$$

Since $x^r(1-x)^{t_1} = \sum_{s=0}^{t_1}\binom{t_1}{s}(-1)^s x^{r+s}$, we note that

$$B(r+1,t_1+1) = \int_0^1 x^r(1-x)^{t_1}dx = \sum_{s=0}^{t_1}\binom{t_1}{s}\frac{(-1)^s}{s+r+1}$$

is the classical beta-function and $B(r+1,t_1+1) = \frac{r!t_1!}{(r+t_1+1)!} > 0$ since r and t_1 are non-negative integers ($l \geq 2$). Thus the leading coefficient is

$$\begin{aligned}c_r\frac{r!t_1!}{(r+t_1+1)!} &= \frac{t_2!\cdots t_l!}{(l-2+t_2+\cdots+t_l)!}\frac{r!t_1!}{(r+t_1+1)!} \\ &= \frac{t_1!\cdots t_l!}{(l-1+t_1+\cdots+t_l)!}.\end{aligned}$$

PROPOSITION 3.12. *For positive integers r, N, and an $r\times N$ matrix $(a_{ij}$ with non-negative real entries, set $J = \{j \mid \sum_{i=1}^{r} a_{ij} = 0\}$.*

Then

$$d(n) = \sum_{(n_1,\cdots,n_r)\in C(r,n)} \prod_{j=1}^{N} \left(1 + \sum_{i=1}^{r} n_i a_{ij}\right) \tag{3.10}$$

is a polynomial of n *with degree* $N + r - 1 - |J|$.

Proof. Let us consider the polynomial

$$f(x_1, \cdots, x_r) = \prod_{j=1}^{N} \left(\sum_{i=1}^{r} (x_i a_{ij} + 1)\right)$$

which has total degree $N - |J|$ since each factor has degree 1 unless $j \in J$. Without lose of generality, we can assume that $|J| = 0$. Let us write

$f(x_1, \cdots, x_r) = \sum c_{t_1,\cdots t_r} x_1^{t_1} \cdots x_r^{t_r}$. Then

$$\begin{aligned} d(n) &= \sum_{(n_1,\cdots,n_r)\in C(r,n)} f(n_1, \cdots, n_r) \\ &= \sum_{(t_1,\cdots,t_r)} c_{t_1,\cdots,t_r} \left(\sum_{(n_1,\cdots,n_r)\in C(r,n)} n_1^{t_1} \cdots n_r^{t_r} \right) = \sum_{(t_1,\cdots,t_r)} c_{t_1,\cdots,t_r} d_{r,t}(n). \end{aligned} \tag{3.11}$$

By the lemma above, $d_{r,t}(n)$ is a polynomial of n with degree $t_1 + \cdots + t_r + r - 1$ while the total degree of $f(x_1, \cdots, x_r)$ is N. Note that all $c_{t_1,\cdots,t_r}$ are sums of products of a_{ij}'s. Thus monomials of highest total degree N in $f(x_1, \cdots, x_r)$ must have nonnegative coefficients with at least one being positive. On the other hand, the leading coefficients of all $d_{r,t}(n)$'s are positive. Therefore the coefficient of n^{N+r-1} (the leading term) in $d(n)$ must be positive. Hence $d(n)$ has degree $N + r - 1$ since $N = t_1 + \cdots + t_r$ for any monomial $x_1^{t_1} \cdots x_r^{t_r}$ with total degree N.

Let us write $d(n) = \sum_{j=0}^{N+r-1} d_j n^j$ for all $n \in \mathbb{N}$. Here d_j's are independent of n and $d_{N+r-1} > 0$.

THEOREM 3.13. *When* $\Lambda \subseteq X_+$ *is a free commutative submonoid of rank* r *and* $I_\Lambda = \{\alpha \in \Delta \mid \langle \lambda, \alpha^\vee \rangle = 0, \forall \lambda \in \Lambda\}$, *then the GK-dimension of* $\operatorname{Ind}_D^H M(\Lambda)$ *is* $\dim G/P_\Lambda + r$, *where* P_Λ *is the parabolic subgroup containing* B *determined the simple roots in* I_Λ.

Proof. Let $J_\Lambda = \{\alpha \in R^+ \mid \langle \lambda, \alpha^\vee \rangle = 0, \forall \lambda \in \Lambda\}$. Note that

$$\begin{aligned}\dim A_n &= \sum_{i=0}^{n} d(i) = \sum_{i=0}^{n} \left(\sum_{j=0}^{N+r-1-|J|} d_j i^j \right) \\ &= \sum_{j=0}^{N+r-1-|J|} d_j \left(\sum_{i=0}^{n} i^j \right) \\ &= \sum_{j=0}^{N+r-1-|J|} d_j \left(\frac{1}{j+1} n^{j+1} + \cdots \right).\end{aligned} \tag{3.12}$$

Thus $\dim A_n$ is a polynomial of n with the leading term occurs when $j = N+r-1-|J|$ and the leading coefficient being $\frac{d_{N+r-1-|J|}}{N+r-|J|}$. Note that $\dim(G/P_\Lambda) = N - |J|$. The theorem follows.

3.14 A rather more general question is to compute the GK-dimension of $\mathrm{Ind}_D^H(M(\Lambda))$ for more general submonoid Λ. Note that not all submonoids of X_+ are finitely generated. Thus the algebra $\mathrm{Ind}_D^H(M(\Lambda))$ needs not be finitely generated. However, let $L = \mathbb{Z}\Lambda$ be the $\mathbb{Z}$-submodule of $X(T)$ generated by Λ. Then L is finitely generated and free as $\mathbb{Z}$-module. We need the following lemma.

LEMMA 3.15. *If L is a $\mathbb{Z}$-submodule of $\mathbb{Z}^n$ generated by a subset $Y \subseteq \mathbb{N}^n$, then L has basis $\{\lambda_1, \ldots, \lambda_r\}$ such that $Y \subseteq \sum_{i=1}^{r} \mathbb{N}\lambda_i$. In general, $\lambda_1, \ldots, \lambda_r$ not necessarily in $\mathbb{N}^n$.*

Proof. We use induction on n. When $n = 1$, the lemma is trivial. For $n > 1$, let $\{e_1, \ldots, e_n\}$ be the standard basis of $\mathbb{Z}^n$. We consider the projection $\phi : \mathbb{Z}^n \to \mathbb{Z}^{n-1}$ to the first $n-1$ components. Let $\bar{L} = \phi(L)$ and $\bar{Y} = \phi(Y)$. Then $\bar{L}$ in $\mathbb{Z}^{n-1}$ is generated by $\bar{Y}$ as $\mathbb{Z}$-module. By the induction hypothesis, $\bar{L}$ has a $\mathbb{Z}$-basis $\{\bar{\lambda}_1, \ldots, \bar{\lambda}_{r'}\}$ such that $\bar{Y} \subseteq \sum_{i=1}^{r'} \mathbb{N}\bar{\lambda}_i$. If $\phi : L \to \bar{L}$ is not an isomorphism, then the kernel is a rank one free $\mathbb{Z}$-submodule with a basis ae_n for some positive integer a. In this case, $\lambda_i = (\bar{\lambda}_i, 0) \in L$ for $i = 1, \ldots, r'$ and $\{\lambda_1, \ldots, \lambda_{r'}, ae_n\}$ is the required basis.

If $\phi : L \to \bar{L}$ is an isomorphism of $\mathbb{Z}$-modules, let $\lambda_i \in L$ such that $\phi(\lambda_i) = \bar{\lambda}_i$ for $i = 1, \ldots, r'$. Then $\lambda_i = (\bar{\lambda}_i, a_i)$ for uniquely determined $a_i \in \mathbb{Z}$. Each element of Y is of the form $y = (\bar{y}, a)$ with $\bar{y} \in \bar{Y}$ and $a \in \mathbb{N}$. Then $\bar{y} = \sum_{i=1}^{r'} x_i \bar{\lambda}_i$ for some $x_i \in \mathbb{N}$. Then $y = \sum_{i=1}^{r'} x_i \lambda_i$ since ϕ is an isomorphism between L and $\bar{L}$.

We remark that in general one cannot take the basis in $\mathbb{N}^n$, even having $\mathbb{Z}$ replaced by the fields $\mathbb{Q}$ and $\mathbb{R}$. Nevertheless, for $n = 1, 2, 3$, the above inductive argument can be used to prove that if L is a $\mathbb{Q}$-subspace generated by vectors in $(\mathbb{Q}^+)^n$, then L has a $\mathbb{Q}$-basis $\{\lambda_1, \ldots, \lambda_n\}$ such that $L \cap (\mathbb{Q}^+)^n = \sum \mathbb{Q}^+ \lambda_i$. For $\dim L = 1$ or n, the statement is trivial. Thus we only need to consider $\dim L = 2$ for $n = 3$. If the projection $\phi : L \to \bar{L} \subseteq (\mathbb{Q}^+)^2$ is not an isomorphism, then $(0, 0, 1)L$ for some positive $a \in \mathbb{Q}^+$. The induction above will finish the proof. We can assume that projection $\phi : L \to \bar{L}$ is an isomorphism and $a_1 > 0$ and $a_2 < 0$ We can assume that $a_1 = 1$ and $a_2 = -1$. Then $\{\lambda_1, \lambda_2 + \lambda_1\}$ is a $\mathbb{Q}$-basis of L contain in $(\mathbb{Q}^+)^3$. For $\lambda \in L \cap (\mathbb{Q}^+)^3$, $\phi(\lambda) = x_1\bar{\lambda}_1 + x_2\bar{\lambda}_2$ for some $x_1, x_2 \in \mathbb{Q}^+$. Since ϕ is an isomorphism, it is necessary that $x_1 \geq x_2$ and $\lambda = (x_1 - x_2)\lambda_1 + x_2(\lambda_1 + \lambda_2)$.

For $n > 3$, one can take L to be the solution space of $x_1 + x_2 = x_3 + x_4$ in $\mathbb{R}^4$. L is generated by

$$(1, 0, 1, 0), (1, 0, 0, 1), (0, 1, 1, 0), (0, 1, 0, 1).$$

But L does not have a $\mathbb{R}$-basis in $(\mathbb{R}^+)^4$ such that all four generators are non-negative linear combinations of the basis vectors.

3.16 We are now ready to compute the GK-dimension of the induced algebra $\mathrm{Ind}_D^H(M(\Lambda))$ for any submonoid Λ of X_+. Let $L = \mathbb{Z}\Lambda$ be the free $\mathbb{Z}$-submodule of rank r. The subsets $I_L \subseteq \Delta$ and $J_L \subseteq R^+$ are defined similarly. Then we have $I_\Lambda = I_L$ and $J_\Lambda = J_L$. Let P_Λ be the parabolic subgroup of G containing B with I_Λ being the simple roots of the Levi factor.

THEOREM 3.17. *For any submonoid* $\Lambda \subseteq X_+$*, the GK-dimension of the algebra* $\mathrm{Ind}_D^H M(\Lambda)) = \dim(G/P_\Lambda) + 1$.

Proof. First of all $\mathbb{Q}\Lambda$ has a $\mathbb{Q}$-basis $\{\lambda_1, \ldots, \lambda_r\} \subseteq \Lambda$. Let $\Lambda' = \sum_{i=1}^r \mathbb{N}\lambda_i \subseteq \Lambda$. Note that every element of Λ is a $\mathbb{Q}$-linear combination of $\lambda_1, \lambda_2, \ldots, \lambda_r$. Thus we have $I_\Lambda = I_{\Lambda'}$ and $J_\Lambda = J_{\Lambda'}$. Therefore, $P_\Lambda = P_{\Lambda'}$. Since $\mathrm{Ind}_D^H M(\Lambda')$ is a subalgebra of $\mathrm{Ind}_D^H M(\Lambda)$, by Theorem 3.13, the GK-dimension of $\mathrm{Ind}_D^H M(\Lambda)$ is at least $\dim G/P + 1$.

We now need to prove that the GK-dimension of $\mathrm{Ind}_D^H M(\Lambda)$ cannot be larger than $\dim G/P_\Lambda + 1$. Let $L = \mathbb{Z}\Lambda$. By the lemma above, L has a $\mathbb{Z}$-basis $\{\lambda_1, \ldots, \lambda_r\}$ such that $\Lambda \subseteq \sum_{i=1}^r \mathbb{N}\lambda_i$. For each n, we define

$$\Lambda_n = \{\lambda \in \Lambda \mid \lambda = \sum_{i=1}^{r} n_i\lambda_i, n_i \in \mathbb{N}, \sum_{i=1}^{r} n_i = n\}.$$

Note that all $n_i \in \mathbb{N}$ and $\Lambda_n + \Lambda_m \subseteq \Lambda_{n+m}$. This defines a graded algebra structure on $\operatorname{Ind}_D^H(M(\Lambda)) = \oplus_{n=0}^{\infty} \operatorname{Ind}_D^H(M(\Lambda_n))$.

Using the Weyl dimension formula, we have

$$\begin{aligned}
\dim \operatorname{Ind}_D^H M(\Lambda_n) &= \sum_{\lambda\in\Lambda_n} \dim(\lambda) \\
&\leq \sum_{(n_1,\cdots,n_r)\in C(r,n)} \prod_{j=1}^{N} |1 + \sum_{i=1}^{r} n_i a_{ij}| \\
&\leq \sum_{(n_1,\cdots,n_r)\in C(r,n)} \prod_{j=1}^{N} \left(1 + \sum_{i=1}^{r} n_i |a_{ij}|\right) = d(n),
\end{aligned} \tag{3.13}$$

where $a_{ij} = \langle \lambda_i, \alpha_j^\vee \rangle$. By Proposition 3.11, $d(n)$ is a polynomial of degree $N + r - 1 - |J_L|$. Then the sequence

$$f(n) = \dim(\oplus_{m=0}^{n} \operatorname{Ind}_D^H M(\Lambda_m))$$

has growth rate not larger than $N + r - |J_L|$.

Let A be a finitely generated subalgebra of $\operatorname{Ind}_D^H(M(\Lambda))$. There exist n_0 such that the generators are in $V = \oplus_{m=0}^{n_0} \operatorname{Ind}_D^H M(\Lambda_m)$. We can assume that A is generated by V, then $V^n \subseteq \oplus_{m=0}^{nn_0} \operatorname{Ind}_D^H M(\Lambda_m)$. Thus $\dim V^n \leq f(nn_0)$. This shows that the GK-dimension of A is at most $N+r-|J_\Lambda|$. Therefore, by the definition, the GK-dimension of $\operatorname{Ind}_D^H M(\Lambda)$ is at most $N + r - |J_\Lambda|$.

Since $N - |J_\Lambda| = \dim G/P_\Lambda$, we thus have

$$\text{GK-dim}(\operatorname{Ind}_D^H M(\Lambda)) = \dim G/P_\Lambda + \operatorname{rank} \mathbb{Z}\Lambda.$$

REMARK 3.17.1. *If $H = \operatorname{Dist} G$ and $D = \operatorname{Dist}(B)$ for a reductive group G with a Borel subgroup B, then the GK-dimension of $\operatorname{Ind}_D^H M(\Lambda)$ is the same as the Krull dimension and the dimension of the quotient variety G/U with U being the unipotent radical of B. However for the quantum groups, this algebras are not commutative.*

3.18 Note that the algebra $M(\Lambda)$ is graded when a basis as in the above proof for $L = \mathbb{Z}\Lambda$ is given. $\dim M(\Lambda_n) = \binom{r+n-1}{r-1}$ which is a polynomial of n of degree $r-1$. The above proof also shows that GK-dim$(M(\Lambda)) = \text{rank}(\mathbb{Z}\Lambda)$ for any submonoid Λ of X_+.

Thus the above results suggest that

$$\text{GK-dim}\,\text{Ind}_D^H(M) \leq \dim G/P + \text{GK-dim}(M)$$

for any D-module algebra M, where P is parabolic subgroups of G such that M extends to the a P-module structure (or its quantum analog).

4. Final Remarks

4.1 Let H be the quantum enveloping algebra over $k = \mathbb{C}(q)$. The canonical basis defined by Lusztig defines a canonical basis for each $\text{Ind}_D^H k_\lambda$ through the dual basis of the canonical basis of the quantum Weyl module for all $\lambda \in X_+$. The $\mathbb{Z}[q,q^{-1}]$-lattice $A_{\mathbb{Z}[q,q^{-1}]}(\lambda)$ generated by this basis in $\text{Ind}_D^H k_\lambda$ respects the tensor product in the sense that the multiplication $\text{Ind}_D^H k_\lambda \otimes \text{Ind}_D^H k_\mu \to \text{Ind}_D^H k_{\mu+\lambda}$ sends $A_{\mathbb{Z}[q,q^{-1}]}(\lambda) \otimes_{\mathbb{Z}[q,q^{-1}]} A_{\mathbb{Z}[q,q^{-1}]}(\mu)$ to $A_{\mathbb{Z}[q,q^{-1}]}(\lambda+\mu)$ (see the (L1, 25.1.2)). One can also define the induced modules by taking $\mathbb{Z}[q,q^{-1}]$ as the ground ring instead of the field k. Let $H_{\mathbb{Z}[q,q^{-1}]}$ and $D_{\mathbb{Z}[q,q^{-1}]}$ be the $\mathbb{Z}[q,q^{-1}]$-form in H and D respectively, then $A_{\mathbb{Z}[q,q^{-1}]}(\lambda) = \text{Ind}_{D_{\mathbb{Z}[q,q^{-1}]}}^{H_{\mathbb{Z}[q,q^{-1}]}} \mathbb{Z}[q,q^{-1}]_\lambda$. In this way we can get a $\mathbb{Z}[q,q^{-1}]$-form $A_{\mathbb{Z}[q,q^{-1}]}(\Lambda) = \oplus_{\lambda\in\Lambda} A_{\mathbb{Z}[q,q^{-1}]}(\lambda)$ in $\text{Ind}_D^H M(\Lambda)$ for each submonoid Λ see (APW) and (Lin3).

4.2 For any field k and any ring homomorphism $\mathbb{Z}[q,q^{-1}] \to k$, we can form the k-Hopf algebras $H_k = k \otimes_{\mathbb{Z}[q,q^{-1}]} H_{\mathbb{Z}[q,q^{-1}]}$ and $D_k = k \otimes_{\mathbb{Z}[q,q^{-1}]} D_{\mathbb{Z}[q,q^{-1}]}$. Then define $\text{Ind}_{D_k}^{H_k} M_k(\Lambda)$. An easy computation will show that $\text{Ind}_{D_k}^{H_k} M_k(\Lambda) \cong k \otimes_{\mathbb{Z}[q,q^{-1}]} A_{\mathbb{Z}[q,q^{-1}]}(\Lambda)$. In particular, if $\mathbb{Z}[q,q^{-1}] \to k$ sends q to 1, then $\text{Ind}_{D_k}^{H_k} M_k(\Lambda)$ is the algebra we constructed for the corresponding semisimple algebraic group G. Thus $\text{Ind}_D^H M(\Lambda)$ is a quantum deformation of the corresponding algebra of the semisimple algebraic group.

4.3 It is also interesting to note that for the distribution algebra of semisimple algebraic groups in positive characteristic p, the GK-dimension is 0 since the algebra itself is a direct union of finite dimensional hyperalgebras. This shows that different $\mathbb{Z}$-forms in the enveloping algebra of the complex semisimple Lie algebra could give rise to k-algebras with different GK-dimensions. However, for the quantum enveloping algebras defined by Lusztig using "divided" powers, the GK-dimension does not change when the base field is of characteristic 0 even in the root of unity cases.

4.4 For fixed and algebraic group G and Borel subgroup B, the algebra $\mathrm{Ind}_B^G M(\Lambda)$ is a subalgebra of the coordinate algebra $k[G]$. In fact, $\mathrm{Ind}_{D_k}^{H_k} M_k(\Lambda) = \oplus_{\mu \in -\Lambda} k[G]_\mu$ is the subalgebra of B-semi-invariants in $k[G]$. Here G acts on $k[G]$ from left by $(bf)(x) = f(b^{-1}x)$ for all $x \in G$ and all $b \in G$. Since G is connected, this algebra is an integral domain. There is a natural gradation defined by the height of weights in X_+ in the sense of $\mathrm{ht}(\lambda) = \sum_{i=1}^{l} n_i$ if $\lambda = \sum_{i=1}^{l} n_i \omega_i$, with $\omega_1, \cdots, \omega_l$ being the fundamental weights. This graded algebra defines an irreducible projective scheme $X(\Lambda)$, on which G acts. The flag variety G/B is the special case. If $\Lambda = X_+$, then $A(\Lambda)$ contains the homogeneous coordinate algebra of G/B as a subalgebra. This defines a surjective G-equi-variant map $X(\Lambda) \to G/B$. An interesting question is to describe the geometry of the projective scheme $X(\Lambda)$, and their relation with the representations of the algebraic groups, in particular, the cohomology groups of line bundles of over the flag varieties.

4.5 For a finite dimensional algebra A and a finite dimensional A-module M, let $\cdots \to P_1 \to P_0 \to M \to 0$ be the minimal projective resolution. Define $f(n) = \dim P_n$ for all $n \in \mathbb{N}$. The complexity of M, $c(M)$ is defined to be $\nu(f) + 1$.

On the other hand, $\mathrm{Ext}_A^*(M, M) = \oplus_{n=0}^{\infty} \mathrm{Ext}_A^n(M, M)$ is a graded k-algebra with the multiplication being the Yoneda product (see (Ben1). Let $g(n) = \dim \mathrm{Ext}_A^n(M, M)$. It is easy to see that $\text{GK-dim}(\mathrm{Ext}_A^*(M, M)) \leq \nu(g) + 1$. On the other hand, following an argument as in (Ben2, 5.3) we always have $\nu(g) \leq c(M) - 1$. If $\mathrm{Ext}_A^*(M, M)$ is finitely generated as a k-algebra, then

GK-dim($\mathrm{Ext}_A^*(M, M)$) $= \nu(g) + 1$. Furthermore, if $\mathrm{Ext}_A^*(M, S)$ is a finitely generated $\mathrm{Ext}_A^*(M, M)$-module for each finite dimensional A-module S, then we have GK-dim($\mathrm{Ext}_A^*(M, M)$) $= c(M)$. Note that the complexity $c(M)$ was defined as an integer in (Ben2).

Friedlander and Suslin (FS) have proved that for a finite algebraic group scheme G and any finite dimensional rational G-module M, $\mathrm{Ext}_A^*(M, M)$ is finitely generated as a k-algebra and $\mathrm{Ext}_A^*(M, N)$ is a finitely generated as $\mathrm{Ext}_A^*(M, M)$-module for any finite dimensional G-module N. Note that in this case the Yoneda product coincides with the cup product, and $\mathrm{Ext}_A^*(M, M)$ is graded commutative. Thus the Hilbert series of the graded vector space $\mathrm{Ext}_A^*(M, N)$ has the form $p(t)/\prod(1 - t_i^n)$ with $p(t)$ being a polynomial. Therefore $\nu(\dim \mathrm{Ext}_A^n(M, N))$ is an integer.

Now for any finite dimensional Hopf algebra H, the category of all H-modules is isomorphic to the category of all H°-comodules. Here H° is a finite dimensional Hopf algebra. (See (Lin1)). Thus we have

THEOREM 4.6. *If A is a finite dimensional cocommutative Hopf algebra, then for any finite dimensional H-module M, we have* GK-dim$(M) = c(M)$ *which is always an integer.*

References

APW. H.H. Andersen, P. Polo, K. Wen, Representations of quantum algebras, *Invent. Math.* **104** (1991), 1–59.

ATV. M. Artin, J. Tate and M. Van der Bergh, Modules over regular algebras of dimension 3, *Invent. Math.* **106** (1991), 335–388.

Ben1. D.J. Benson, *Representations and cohomology I: Basic representations theory of finite groups and associative algebras*, Cambridge University Press, 1991.

Ben2. D.J. Benson, *Representations and cohomology II: Cohomology of groups and modules*, Cambridge University Press, 1991.

C. J. Carlson, Complexity and Krull dimension. in *Representations of algebras* (Puebla, 1980), pp. 62–67, Lecture Notes in Math., 903, Springer, Berlin/New York, 1981.

Don. S. Donkin, *Rational representations of algebraic groups: tensor products and filtrations*, (Lecture Notes in Mathematics **1140**), Springer-Verlag, Berlin/Heidelberg/New York/Tokyo 1985.

FS. E. Friedlander, A. Suslin, Cohomology of finite group schemes over a field. *Invent. Math.* **127** (1997), no. 2, 209–270.

Ha. R. Hartshorne, *Algebraic Geometry*, Springer-Verlag, New York, 1977.

Hu3. J. E. Humphreys, *Introduction to Lie Algebras and Representation Theory*, Gradate Text Book in Mathematics 9, Springer-Verlag, New York/Heidelberg/Berlin 6th printing, 1994.

Jan. J.C. Jantzen, *Representations of Algebraic Groups*, Academic Press, 1987.

KL. G. Krause and T. H. Lenagan, *Growth of algebras and Gelfand-Kirillov dimension*, Research Notes in Mathematics, Pitman Adv. Publ. Program, vol 116 (1985).

Lin1. Z. Lin, Induced representations of Hopf algebras: applications to quantum groups at roots of 1, *J. Algebra* **154** (1993), 152–187.

Lin2. Z. Lin, A Mackey decomposition theorem and cohomology for the quantum groups at roots of 1, *J. Algebra* **166** (1) (1994), 100-129.

Lin3. Z. Lin, Freeness of quantum coordinate algebra, (unpublished).

LN1. Z. Lin, D.K. Nakano, Algebraic group actions in the representation and cohomology theory of Lie algebras of Cartan type, *to appear in J. Algebra.*

LMO. A. Leroy, J. Matczuk and J. Okninski, On the Gelfand-Kirillov dimension of normal localizations and twisted polynomial rings, *Perspectives in Ring Theory*, F. van Oystaeyen and L. Le Bruyn, Editors, Kluwer Academic Publishers (1988), 205-214.

L1. G. Lusztig, *Introduction to Quantum Groups*, Birkhäuser, Boston/Basel/Berlin 1994

L2. G. Lusztig, Canonical basis arising from quantized enveloping algebras II, (pp. 175–201) in *Common Trends in Mathematics and Quantum Field Theories*, (T. Eguchi et. al., eds.) Progr. Theor. Phys. Supll., Vol. 102, 1990

L3. G. Lusztig, Quantum groups at roots of 1, *Geom. Dedicata* **35** (1990), 89–113.

M. O. Mathieu, Filtrations of G-modules, *Ann. Sci. École Norm. Sup.* **23** (1990), 625–644.

MR. J. C. McConnell and J. C. Robson, *Non-Commutative Noetherian Rings*, Wiley-interscience, Chichester, 1987.

Mu. I. Musson, Gelfand-Kirillov dimension of twisted Laurent Extensions, *Comm. Alg.* 17 (11), 1989, 2853-2856.

O. C. Ohn, "Classical" flag varieties for quantum groups: the standard quantum $SL(n, \mathbb{C})$, Preprint (http://arXiv.org/abs/math.AG/0007005).

P. J. Paradowski, Filtrations of modules over the quantum algebra, *Proceedings of Symp. Pure Math.* Ed. Haboush and Parshall **56** Part 2 (1994), 93–108.

Sw. M. Sweedler, *Hopf Algebras*, Benjamin, New York, 1974.

W. Jian-pan Wang, Sheaf cohomology on G/B and tensor products of Weyl modules, *J. Algebra* **77** (1982), 162–185

Zh1. J. Zhang, On Gelfand-Kirillov transcendence degree, *Trans. Amer. Math. Soc.* **348** (1996), no. 7, 2867–2899.

Zh2. J. Zhang, A note on GK-dimension of skew polynomial extensions, *Proc. Amer. Math. Soc.* **125** (1997), no. 2, 363–373.

THE NORMALIZER OF A FINITE GROUP IN ITS INTEGRAL GROUP RING AND ČECH COHOMOLOGY *

Z. S. MARCINIAK (zbimar@mimuw.edu.pl)
Institute of Mathematics,
Warsaw University,
ul.Banacha 2,
02 - 99 Warsaw, Poland

K. W. ROGGENKAMP
(roggenkamp@mathematik.uni-stuttgart.de)
Mathematisches Institut B,
Universität Stuttgart,
Pfaffenwaldring 57,
D-70550 Stuttgart, Germany

Primary: 16G30

Introduction

Recently M. Hertweck [(He,97)] has constructed two non isomorphic finite groups G and H which have isomorphic integral group rings $\mathbb{Z}G \simeq \mathbb{Z}H$. His construction goes back to the work of A. Zimmermann and the second author (RoZi,95.2), who have constructed examples of non isomorphic finite-by-(infinite cyclic) groups $\mathcal{G}$ and $\mathcal{H}$ with $R\mathcal{G} \simeq R\mathcal{H}$ where R is the ring of integers in a suitably chosen algebraic number field. This latter construction is based on an investigation of the kernel of the natural homomorphism

* Both authors were partially supported by the Deutsche Forschungsgemeinschaft. The first author was also supported by Polish Scientific Grant No. 2P03A01113.

K.W. Roggenkamp and M. Ştefănescu (eds.), Algebra - Representation Theory, 159–188.

$Aut(G) \longrightarrow Aut(\mathbb{Z}G)$, which induces in turn a homomorphism

$$\Phi\colon Out(G) \longrightarrow Out(\mathbb{Z}G).$$

The group $Ker(\Phi)$ has been studied by D. Coleman (Co,64), J. Krempa (Kr,87) and S. Jackowski with the first author (JaMa,87). The crucial observation on which the construction of the above examples is based is due to M. Mazur (Ma,95). He showed that for two automorphisms α and β of the group G, the semi-direct products $G \rtimes_\alpha C_\infty$ and $G \rtimes_\beta C_\infty$, where a fixed generator of the infinite cyclic group C_∞ acts on G via α and β resp., are isomorphic if and only if $\alpha \circ \beta^{-1}$ is inner on G. On the other hand, the group rings $\mathbb{Z}(G \rtimes_\alpha C_\infty)$ and $\mathbb{Z}(G \rtimes_\beta C_\infty)$ are isomorphic provided $\alpha \circ \beta^{-1} \in Ker(\Phi)$.

This observation of Mazur cannot be immediately applied if C_∞ is replaced by a finite cyclic group. Hertweck though pointed out an instance, when the arguments can be modified to apply to finite groups.

One part of the present paper is to elaborate on the concept underlying Hertweck's construction (cf. Section 3). Before we can state the results, we have to introduce some notation.

For an element a in a ring Λ and an automorphism $\beta \in Aut(\Lambda)$ we write ${}^{\beta}a$ for the element $\beta(a)$ and we put

$${}^{N_n(\beta)}a = a \cdot {}^{\beta}a \cdot {}^{\beta^2}a \cdot \cdots \cdot {}^{\beta^{n-1}}a .$$

For an automorphism $\beta \in Aut(G)$ of a group G we put

$$G_\beta(n) = G \rtimes_\alpha C_n =_{\text{as set}} G \times C_n ,$$

where $C_n = \langle x \rangle$ is the cyclic group of order n. Here we assume that the automorphism β has finite order $o(\beta)$ and n is a multiple of this order.

For g, $g' \in G$ and x^i, $x^j \in C_n$ we have

$$(g, x^i) \cdot (g', x^j) = (g \cdot {}^{\beta^i}(g'), x^{i+j}) .$$

Lemma 3.6. Let α, $\beta \in Aut(G)$ such that $\alpha \circ \beta^{-1}$ is inner, say a conjugation with $g_0 \in G$. The map $\phi_n : G_\alpha(n) \longrightarrow G_\beta(n)$ induced by

$$(g, x) \longmapsto (g, 1) \cdot (g_0, 1) \cdot (1, x)$$

is a homomorphism if and only if ${}^{N_n(\beta)}g_0 = 1$. As similar statement holds if $\alpha \circ \beta^{-1} \in Ker(\Phi)$.
For groups G, H with $|H| < \infty$ and given maps $\alpha, \beta \colon H \longrightarrow Aut(G)$ which give rise to the groups $G_\alpha = G \rtimes_\alpha H$ and $G_\beta := G \rtimes_\beta H$ we have the following result on an isomorphism between G_α and G_β.

Proposition 3.10. Suppose we have a ring isomorphism

$$\Psi : \mathbb{Z}(G_\alpha) \longrightarrow \mathbb{Z}(G_\beta)$$

such that $\Psi_{|G} = id$ and $\Psi(h)h^{-1} \in \mathbb{Z}G$ for all $h \in H$. If the order of H is odd then $G_\alpha \simeq G_\beta$.
The rest of the paper is devoted to a further analysis of $Ker(\Phi)$.

Proposition 12.3. Let G be a metabelian group. If a Sylow 2-subgroup of G is abelian, then $Ker(\Phi) = 1$.
To prove our theorems we develop a Čech style cohomology which gives for a finite group G, which is a projective limit, $G = lim.proj.(G_i)$, an obstruction theory of $Ker(\Phi_G)$, provided $Ker(\Phi_{G_i}) = 1$ for every i. (cf. Section 5).
Such an obstruction theory was first applied by L. L. Scott and the second author (RoSc,85) to construct a counterexample to the Zassenhaus conjecture. A theory for it was developed for the Isomorphism Problem and the Zassenhaus conjecture by W. Kimmerle and the second author (KiRo,93).
The positive results we obtain are based on a result of F. Gross (Gr,82), who showed:

Theorem 7.1. Let G be a finite group and p a prime number. If $C_G(O_p(G)) \leq O_p(G)$ then every automorphism of G which is the identity on a Sylow p-subgroup is inner on G.
We give a short proof of this result. The main application is as follows. We know that every finite solvable group G can be written as the projective limit $G = lim.proj.(G_i = G/O_{p_i'}(G))$. By Gross' result we have $Ker(\Phi_{G_i}) = 1$, so we may apply out obstruction theory (cf. Section 8). For example, we can show

Proposition 8.1. Let $G = lim.proj.(G_i)$ with $G_i = G/O_{p_i'}(G)$ and assume that $\rho \in Ker(\Phi)$. Then ρ can be modified by an inner automorphism so that $\gamma_{ij}(\rho) \in Z(G_{ij})$ are involutions.
In particular, if G_{ij} is odd for all $i \neq j$, then Φ is injective. (This extends the result of (JaMa,87).)

For any subgroup $H < G$ we define its WEIL GROUP as $N_G(H)/C_G(H)$. Notice that this group acts naturally on H. The following statement shows a way of making groups with $Ker(\Phi) = 1$ bigger.

Proposition 11.1 . Let $1 \longrightarrow H \longrightarrow G \longrightarrow \overline{G} \longrightarrow 1$ be an exact sequence of finite groups. If $Ker(\Phi(\overline{G})) = 1$ and the groups H and $W_G(H)$ have odd orders then $Ker(\Phi(G)) = 1$.

The final result, Proposition 9.1 gives an upper and lower bound for $Ker(\Phi)$. We do not formulate it here, since it needs a fair amount of notation.

1. Notation

We apply the standard conventions of group theory. Thus, for a finite group G we denote by $\pi(G)$ the set of rational prime divisors of $|G|$. For any prime $p \in \pi(G)$ we denote by p' the set $\pi(G) \setminus \{p\}$. For a subset $\omega \subset \pi(G)$ we denote by $O_\omega(G)$ the maximal normal subgroup H in G for which $\pi(H) \subset \omega$.

For a ring or a group X we denote by $Z(X)$ the CENTER of X. Further, we denote by $Aut(X)$ the group of all automorphisms of X and by $Inn(X)$ its subgroup consisting of inner automorphisms, i.e. conjugations $conj(u)$ with invertible elements $u \in X$. We also write $Out(X) = Aut(X)/Inn(X)$.

A group automorphism will be called CENTRAL if it preserves the conjugacy classes of G. We denote by $Aut_c(G)$ the group of all central automorphisms of G.

For a group G and an integral domain R we denote by RG the GROUP RING of G over R. The AUGMENTATION of RG is the ring homomorphism $\epsilon_{RG}: RG \longrightarrow R$ induced by sending every $g \in G$ to 1. The group $U(RG)$ denotes the units in RG and $V(RG)$ consists of all units of augmentation one. By $(\ )^*: RG \longrightarrow RG$ we denote the natural anti-involution induced by sending $g \in G$ to g^{-1}. We denote by $Aut_{gr}(RG)$ the group of ring automorphisms of RG which leave the subset $G \subset RG$ invariant.

At various places we will make use of the following well-known fact:

PROPOSITION 1.1. *Let N be a normal subgroup of a group G, let*

$$A = \{\alpha \in Aut(G) \,:\, \alpha(n) = n \textit{ and } \alpha(g)\cdot g^{-1} \in N \textit{ for } n \in N\,,\ g \in G\}\,,$$

and put $I = A \cap Inn(G)$. Then there is a canonical isomorphism $A/I \xrightarrow{\simeq} H^1(G/N, Z(N))$. □

The main objective of this paper is the study of the natural homomorphism

$$\Phi_R(G)\colon Out(G) \longrightarrow Out(RG).$$

If there is no confusion we just write Φ.

2. Known facts about $\Phi_R(G)$

We start from recording a few basic results which will be used later.

NOTE 2.1.

- Coleman's result (Co,64) (see also (RoSc,85)) states that if G is a finite group then for any p-subgroup $P \leq G$ holds

$$N_{V(RG)}(P) = N_G(P)\cdot C_{V(RG)}(P) \text{ provided } p\cdot R \neq R\,.$$

- Krempa (Kr,87) has shown that for any group G, finite or infinite, $Ker(\Phi_{\mathbb{Z}}(G))$ is an elementary abelian 2-group.
- Jackowski and the first author (JaMa,87) have shown that $\Phi_{\mathbb{Z}}(G)$ is injective if G is finite and its Sylow 2-subgroup is normal.
- It is easily seen that every automorphism of G, which is inner in RG, is necessarily central.

However, until 1994 not a single example of a finite group G was known for which $Ker(\Phi_R(G)) \neq 1$ when $R = \mathbb{Z}$ or – more generally – when R is the ring of algebraic integers in some global field K.

THEOREM 2.2 (Roggenkamp-Zimmermann (RoZi,95.1)). *There is a finite solvable group G and a non inner automorphism α of G which becomes inner in $\mathbb{Z}_\omega G$ where Z_ω is the semi-localization of $\mathbb{Z}$ at a finite set of primes ω containing $\pi(G)$.* □

3. Our conceptual point of view at Hertweck's construction

M. Mazur, who was aware of the automorphism problem posed by Jackowski and the first author (JaMa,87), elaborated on the following construction relating the homomorphism $\Phi_R(G)$ to semi-direct products.
Recall that for any group homomorphism $\kappa\colon H \longrightarrow Aut(G)$ we can construct a new group $G \rtimes_\kappa H$ as follows. As a set, $G \rtimes_\kappa H$ is the Cartesian product $G \times H$. The multiplication is defined by

$$(g, h) \cdot (g', h') = (g \cdot {}^{\kappa(h)}g', h \cdot h').$$

If we are given just a single automorphism $\alpha \in Aut(G)$ then it gives rise to a homomorphism $\kappa\colon C_\infty = \langle x \rangle \longrightarrow Aut(G)$ which sends the generator x to α. We will denote the group which arises from this particular κ in the above construction simply by $G \rtimes_\alpha C_\infty$. Mazur proved the following

LEMMA 3.1 (Mazur (Ma,95)). *Let $\alpha, \beta \in Aut(G)$. Then*

(i) $G \rtimes_\alpha C_\infty \simeq G \rtimes_\beta C_\infty$ if and only if $\alpha \circ \beta^{-1}$ is inner on G.
(ii) Moreover, we have an isomorphism of group rings $R(G \rtimes_\alpha C_\infty) \simeq R(G \rtimes_\beta C_\infty)$ provided $\alpha \circ \beta^{-1}$ is inner on RG.

□

The first statement is well known. The second statement is proved by repeating the proof of the first one on the ring level. Namely, if $\alpha \circ \beta^{-1} = conj(u)|_G$ for a unit $u \in RG$ then the required group ring isomorphism is given by

$$\phi\colon g \cdot x^i \mapsto g \cdot (ux)^i \in RG, \text{ where } g \in G \text{ and } x^i \in C_\infty. \quad (1)$$

Now, associating Mazur's observation (with $\beta = id$) with Theorem 2.2 we conclude:

REMARK 3.2. There exists a group automorphism α of a finite group G and an algebraic number ring R such that:

$$G \rtimes_\alpha C_\infty \not\simeq G \times C_\infty \text{ and}$$
$$R(G \rtimes_\alpha C_\infty) \simeq R(G \times C_\infty).$$

One might be tempted to apply this construction to finite groups, i. e., to replace C_∞ with a finite cyclic group $C_n = \langle x \mid x^n = 1\rangle$. However, it will not work even if we were able to replace R with $\mathbb{Z}$. Namely, from the "normal subgroup correspondence" (RoTa,92), Theorem IV.1, it follows that for finite groups F_1 and F_2 an isomorphism $\mathbb{Z}F_1 \simeq \mathbb{Z}F_2$ implies that F_1 is a direct product of two groups if and only if F_2 is a direct product of two groups of the same orders.

Let us then see what actually goes wrong with the construction. First of all, the automorphism α is of finite order, say n_0. Thus, for any positive integer $n = \nu \cdot n_0$ we have the "structure map" $\kappa\colon C_n \longrightarrow Aut(G)$, sending x to α. Consequently, we can create the group $G \rtimes_\alpha C_n$.

Now suppose that $\alpha = conj(u)$ for a unit $u \in \mathbb{Z}G$ and set, as before, $\beta = id$. Then the isomorphism ϕ given by Formula 1 maps $1 = x^n$ to $1 = (ux)^n = u^n \cdot x^n = u^n$ in $\mathbb{Z}(G \times C_n)$ and hence the unit u is of finite order. However, this can happen only when α is inner as we have

LEMMA 3.3. *Let $R = \mathbb{Z}$ and let G be any, possibly infinite, group. Then a unit $u \in N_{U(\mathbb{Z}G)}(G)$ is of finite order if and only if $u = \pm g$ for some group element $g \in G$ of finite order.*

Proof. It was shown in (JaMa,87) that $u = \sum_{g\in G} u_g \cdot g \in N_{U(\mathbb{Z}G)}(G)$ if and only if $v := u^* \cdot u$ is central. It follows that u commutes with u^*. Hence if u has finite order then the same holds for u^* and also for v. But a central unit in $\mathbb{Z}G$ which is of finite order and of augmentation one is a group element. As $v_1 = \sum u_g^2 \neq 0$ we see that $v = 1$. In particular $v_1 = 1$ and hence $u = \pm g$ for some $g \in G$ of finite order.

REMARK 3.4.

- Notice that the unit u in the above considerations is not determined uniquely by the automorphism – it is unique only modulo the center of RG. However, all possible units inducing a non inner automorphism are of infinite order.
- This means that just by using a *single automorphism* there is no hope to construct a counterexample to the Isomorphism Problem for finite groups.

It was M. Hertweck's idea to turn then to a *pair of automorphisms.* Take $\alpha, \beta \in Aut(G)$ such that $\alpha \circ \beta^{-1} \in Inn(RG)$ and let n be a common multiple of their orders. If we look again at the homomorphism $\phi\colon R(G \rtimes_\alpha C_n) \longrightarrow R(G \rtimes_\beta C_n)$ given by Formula 1 we see that the situation is completely different this time: The generating element x of C_n is sent to $u \cdot_\beta x \in G \rtimes_\beta C_n$. Though u has infinite order, the element $u \cdot_\beta x$ may very well have finite order. Let us assume, for example, that $x^2 = 1$. Then $\beta^2 = 1$ and we have

$$1 = (u \cdot_\beta x)^2 = u \cdot x \cdot u \cdot x = u \cdot {}^{\beta}u \cdot x^2 = u \cdot {}^{\beta}u\,.$$

If now u were an element inverted by the action of β then $u \cdot_\beta x$ would be an involution and the group-rings $R(G \rtimes_\alpha C_2)$ and $R(G \rtimes_\beta C_2)$ were isomorphic.
Let us consider the case of an arbitrary n.

DEFINITION 3.5.

- For an element a in the ring Λ and an automorphism $\beta \in Aut(\Lambda)$ we put for $n \in \mathbb{N}$

$${}^{N_n(\beta)}a = a \cdot {}^{\beta}a \cdot {}^{\beta^2}a \cdots {}^{\beta^{n-1}}a\,.$$

- For an automorphism $\beta \in Aut(G)$ of a group G we put

$$G_\beta(n) = G \rtimes_\beta C_n$$

where n is a multiple of the order of β.

Recall that for $g, g' \in G$, $x^i, x^j \in C_n$ we have

$$(g \cdot x^i) \cdot (g' \cdot x^j) = g\ {}^{\beta^i}(g') \cdot x^{i+j}.$$

LEMMA 3.6. *Let α, $\beta \in Aut(G)$ be such that $\alpha \circ \beta^{-1}$ is inner, say the conjugation with $g_0 \in G$. The map*

$$\phi_n\colon G_\alpha(n) \longrightarrow G_\beta(n) \text{ induced by } g \cdot x \longmapsto g g_0 \cdot x$$

is a group isomorphism if and only if ${}^{N_n(\beta)}g_0 = 1$.
Proof. The map ϕ_n is clearly a bijection. It is easy to check that ϕ_n is multiplicative if and only if $\phi_n(x^n) = 1$. But $\phi_n(x^n) = {}^{N_n(\beta)}g_0$ whence the lemma.

NOTE 3.7.

- If $\alpha \circ \beta^{-1}$ is conjugation with g_0 then we always can find an $n \in \mathbb{N}$ such that ${}^{N_n(\beta)}g_0 = 1$. It is enough to choose for n a multiple of $n_0 = o(\beta) \cdot o({}^{N_{n_0}(\beta)}g_0)$, as follows from the formula

$$ {}^{N_{\nu n_0}(\beta)}g_0 = ({}^{N_{n_0}(\beta)}g_0)^{\nu}. $$

- We point out that g_0 is only unique up to the center of G.
- If could conceivably happen that ${}^{N_n(\beta)}g_0 \neq 1$ and

$$ {}^{N_n(\beta)}(g_0 \cdot z) = {}^{N_n(\beta)}(g_0) \cdot {}^{N_n(\beta)}(z) \neq 1 $$

for every central element $z \in Z(G)$, so none of the maps ϕ_n can be an isomorphism. This does not mean though that $G_\alpha(n) \not\simeq G_\beta(n)$.
- If in the above observations g_0 is replaced by $g_0 \cdot u$ for some unit $u \in Z(U(\mathbb{Z}G))$ then it might happen that

$$ {}^{N_n(\beta)}(g_0 \cdot u) = {}^{N_n(\beta)}g_0 \cdot {}^{N_n(\beta)}u = 1 . $$

For this we need ${}^{N_n(\beta)}u \in G$ and hence ${}^{N_n(\beta)}u \in Z(G)$. This then implies that $\mathbb{Z}G_\alpha(n) \simeq \mathbb{Z}G_\beta(n)$.
- So it seems possible to construct a counterexample to the Isomorphism Problem in this way.

The above idea has been used by M. Hertweck (He,97) in the case $n = 2$ to prove:

THEOREM 3.8 (Hertweck(He,97)). *There is a finite group G and automorphisms α and β of G such that:*
(i) The automorphism $\alpha \circ \beta^{-1}$ becomes inner in $\mathbb{Z}G$;
(ii) There is an integer $n \in \mathbb{N}$ such that
- *The finite groups $G \rtimes_\alpha C_n$ and $G \rtimes_\beta C_n$ are not isomorphic.*
- *The group rings $\mathbb{Z}(G \rtimes_\alpha C_n)$ and $\mathbb{Z}(G \rtimes_\beta C_n)$ are isomorphic.*

NOTE 3.9.

- The groups Hertweck constructs have order $2^{28} \cdot 97^{21}$ and so checking for non-isomorphism is definitely not so easy. In fact, Hertweck uses a theoretical criterion coming from Čech cohomology (KiRo,93).

– The isomorphism between $\mathbb{Z}(G \rtimes_\alpha C_n)$ and $\mathbb{Z}(G \rtimes_\alpha C_n)$ is such that it is the identity on $\mathbb{Z}G$ and the identity on $\mathbb{Z}((G \rtimes_\alpha C_n)/G)$.

Analyzing Hertweck's construction one can observe that the number n in the above theorem is even. This has a good, much more general, reason. We prove for groups G, H with $|H| < \infty$ and given homomorphisms α, $\beta \colon H \longrightarrow Aut(G)$ which give rise to the groups $G_\alpha = G \rtimes_\alpha H$ and $G_\beta = G \rtimes_\beta H$ the following result.

PROPOSITION 3.10. *Suppose we have a ring isomorphism*

$$\Psi \colon \mathbb{Z}G_\alpha \longrightarrow \mathbb{Z}G_\beta$$

such that $\Psi|_G = id$ *and* $\Psi(h) \cdot h^{-1} \in \mathbb{Z}G$ *for all* $h \in H$. *If the order of* H *is odd then* $G_\alpha \simeq G_\beta$.

Proof. Consider the map $\delta : H \longrightarrow \mathbb{Z}G$, $\delta(h) = \Psi(h) \cdot h^{-1}$. It is easy to check that δ is a derivation. Moreover, the values of δ lie in $N_{U(\mathbb{Z}G)}(G)$, as for $x \in G$ holds:

$$\delta(hk) = \Psi(h)\Psi(k)k^{-1}h^{-1} = \delta(h)h\delta(k)h^{-1} = \delta(h) \cdot {}^{\beta(h)}\delta(k).$$

Clearly, $\delta(h)$ are units in $\mathbb{Z}(G)$. In fact, $\delta(h) \in N_U(G) := N_{U(\mathbb{Z}G)}(G)$ as for $x \in G$ holds

$$\begin{aligned} {}^{\alpha(h)}x = \Psi({}^{\alpha(h)}x) = \Psi(h \cdot x \cdot h^{-1}) = \Psi(h) \cdot x \cdot \Psi(h)^{-1} = \\ \delta(h) \cdot h \cdot x \cdot h^{-1} \cdot \delta(h)^{-1} = {}^{\delta(h)\beta(h)}x. \end{aligned}$$

In particular, $conj(\delta(h))|_G = \alpha(h)\beta(h)^{-1}$.

Now, the β-action of H on G extends by linearity to an action on $\mathbb{Z}G$ which preserves the multiplicative subgroups $G \leq N_U(G) < U(\mathbb{Z}G)$. Hence the quotient group $E = N_U(G)/G$, which is an elementary abelian 2-group by Note 2.1.2, becomes a $\mathbb{Z}H$-module.

Let $\pi : N_u(G) \longrightarrow E$ be the natural map. Then $\overline{\delta} = \pi \circ \delta : H \longrightarrow E$ is a 1-cocycle. As $|H|$ is odd we have $H^1(H; E) = 0$, i.e. $\overline{\delta}$ is also a coboundary: there exists $\overline{v} \in E$ such that $\overline{\delta}(h) = {}^{\beta(h)}\overline{v} - \overline{v}$ for all $h \in H$.

Pick any $v \in N_U(G)$ with $\pi(v) = \overline{v}$. Then $\delta(h) = \gamma(h) \cdot v^{-1} \cdot {}^{\beta(h)}v$ for a suitable function $\gamma : H \longrightarrow G$. Let $\sigma = conj(v)|_G \in Aut(G)$. We define a map $\phi \colon G_\alpha \longrightarrow G_\beta$ by the formula

$$\phi(x \cdot h) = \sigma(x \cdot \gamma(h)) \cdot h \text{ for } x \in G\,,\; h \in H\,.$$

A staightforward calculation shows that ϕ is a group homomorphism. From the commutative diagram

$$\begin{array}{ccccccccc} 0 & \longrightarrow & G & \longrightarrow & G_\alpha & \longrightarrow & H & \longrightarrow & 0 \\ & & \downarrow{\scriptstyle\sigma} & & \downarrow{\scriptstyle\phi} & & \downarrow{\scriptstyle id_H} & & \\ 0 & \longrightarrow & G & \longrightarrow & G_\beta & \longrightarrow & H & \longrightarrow & 0 \end{array}$$

it follows that ϕ is an isomorphism.

4. Čech Cohomology

The interpretation of a finite solvable group G as a projective limit

$$G = lim.proj.(G/O_{p_i'}(G))$$

has been very successful in the study of the Zassenhaus Conjecture and the Isomorphism Problem (cf. (KiRo,93)); it was also used in Hertweck's proof that his two groups are not isomorphic. We shall next develop a similar theory as in (KiRo,93) for $Ker(\Phi_R(G))$. Let us first recall the basic concepts. We shall set up the Čech cohomology in a general framework. Let $\mathcal{K}$ be either the category of groups or of rings. For an object X we denote by $Z(X)$ the center of X.

DEFINITION 4.1. Let us be given objects of $\mathcal{K}$ and epimorphisms

$$\{X_i, X_{ij} = X_{ji}\,,\, X_{ii} = X_i \mid \phi_{ij} : X_i \longrightarrow X_{ij},\, 1 \le i, j \le n\}\,.$$

Then we can form the PROJECTIVE LIMIT

$$X := lim.proj.(X_i) = \{(x_i) \mid \phi_{ij}(x_i) = \phi_{ji}(x_j)\} \subset \prod_{i=1}^{n} X_i\,.$$

We shall use $\underline{X}$ to stress that we are working with a projective limit.

– The CENTRAL COCYCLES are defined as:

$$Z(\underline{X}, \mathcal{Z}) = \{(\gamma) = (\gamma_{ij}) : \gamma_{ij} \in Z(X_{ij}) \mid \gamma_{ij} = \gamma_{ji}^{-1}, \gamma_{ii} = 1\}.$$

– The CENTRAL COBOUNDARIES are defined as

$$B_c(\underline{X}, \mathcal{Z}) = \{(\gamma_{ij}) \in Z(\underline{X}, \mathcal{Z}) \mid \gamma_{ij} = \gamma_i \cdot \gamma_j^{-1} \text{ for } \gamma_i \in Z(X_i)\}. \tag{2}$$

– The COBOUNDARIES are defined as

$$B(\underline{X}, \mathcal{Z}) = \{(\gamma_{ij}) \in Z(\underline{X}, \mathcal{Z}) \mid \gamma_{ij} = \gamma_i \cdot \gamma_j^{-1} \text{ for } \gamma_i \in X_i\}. \tag{3}$$

Every central coboundary then is a coboundary.

– In case X is a group we define the LOCALLY CENTRAL COBOUNDARIES as

$$B_{lc}(\underline{X}, \mathcal{Z}) = \{(\gamma_{ij}) \in Z(\underline{X}, \mathcal{Z}) \mid \text{ for each } x = (x_i) \in X \text{ we have} \tag{4}$$

$$\gamma_{ij} = \gamma_i(x) \cdot \gamma_j^{-1}(x) \text{ for } \gamma_i(x) \in C_{X_i}(x_i)\}. \tag{5}$$

– Two cocycles are said to be CENTRALLY EQUIVALENT,

$$(\gamma_{ij}) \sim (\gamma'_{ij}) \Longleftrightarrow \gamma_i \cdot \gamma_{ij} \cdot \gamma_j^{-1} = \gamma'_{ij} \text{ for } \gamma_i \in \mathcal{Z}(X_i). \tag{6}$$

(Here we mean the maps induced on the quotients.)

– The equivalence classes of the central cocycles modulo central coboundaries are denoted by

$$\check{H}_c(\underline{X}, \mathcal{Z}(\underline{X})).$$

This is a pointed set with the class of the central coboundaries being the distinguished point.

This definition should be compared with the Čech cohomology of a topological space with respect to an open covering.

5. Locally inner automorphisms

We now assume that $X = \underline{X} = lim.proj.(X_i) \in \mathcal{K}$ where $\mathcal{K}$ is either the category of groups or of rings. We assume that $N_i = Ker(X \longrightarrow X_i)$ is a characteristic subobject in X.

DEFINITION 5.1. We denote by $Aut_{li}(X)$ the group of locally inner automorphisms, i. e. of those automorphisms $\rho \in Aut(X)$ which induce inner automorphisms $\rho_i = conj(\gamma_i(\rho))$ on each X_i. These give rise to coboundaries

$$\gamma(\rho) = (\gamma_{ij}(\rho)) = (\gamma_i(\rho) \cdot \gamma_j{}^{-1}(\rho)) \text{ with } \gamma_{ij} \in Z(X_{ij}).$$

LEMMA 5.2. *Let a central cocycle* $\gamma = (\gamma_{ij}) \in Z(\underline{X}, \mathcal{Z})$ *be given.*

1. *There exists* $\rho \in Aut_{li}(X)$ *with* $\gamma(\rho) = \gamma$ *if and only if* γ *is a coboundary. We will denote this automorphism by* $\rho(\gamma)$.
2. *If* γ *is a coboundary then* $\rho = \rho(\gamma)$ *is inner if and only if* γ *is a central coboundary.*
3. *For groups we have in addition: let* γ *be a coboundary. Then* $\rho = \rho(\gamma) \in Aut_c(X)$ *if and only if* γ *is a locally central coboundary.*
4. *Let* γ *and* δ *be two coboundaries. Then the corresponding automorphisms* $\rho(\gamma)$ *and* $\rho(\delta)$ *differ by an inner automorphism if and only if* γ *and* δ *are centrally equivalent.*

Proof of 1. Assume that γ is a coboundary, i. e. $\gamma_{ij} = \gamma_i \cdot \gamma_j^{-1}$ for $\gamma_i \in X_i$. We then define

$$\rho\colon X \longrightarrow X \text{ by } (x_i) \longmapsto (\gamma_i^{-1} \cdot x_i \cdot \gamma_i).$$

This is surely multiplicative and it remains to show that it maps X to X, i. e.

$$\gamma_i^{-1} \cdot x_i \cdot \gamma_i \equiv \gamma_j^{-1} \cdot x_j \cdot \gamma_j$$

if and only if

$$x_i \equiv (\gamma_i \cdot \gamma_j{}^{-1}) \cdot x_j \cdot (\gamma_i \cdot \gamma_j{}^{-1})^{-1};$$

but this is clear since $\gamma_i \cdot \gamma_j^{-1} = \gamma_{ij} \in Z(X_{ij})$. The other direction is obvious.

Proof of 2. The coboundary of an inner automorphism is the identity. Conversely, assume that ρ is given by conjugation with $\gamma_i \in X_i$, i. e.

$$\rho\colon (x_i) \longmapsto (\gamma_i^{-1} \cdot x_i \cdot \gamma_i),$$

and assume there are elements $\delta_i \in Z(X_i)$ with $\gamma_i \cdot \gamma_j^{-1} = \gamma_{ij} = \delta_i \cdot \delta_j^{-1}$. Then $x_0 := (\gamma_i^{-1} \cdot \delta_i) \in X$ and ρ is given by conjugation with x_0 and so ρ is inner.

Proof of 3. This is proved similarly as Part 2.

Proof of 4. This is also straightforward.

6. Projective limits of groups

We shall assume from now on that G is a finite solvable group. In order to set up an obstruction theory for $Ker(\Phi_R(G)) \neq 1$ we want to describe a finite group G as a projective limit with respect to certain families of normal subgroups $\{N_i\}_{1 \leq i \leq n}$. We always have an induced natural map:

$$G \longrightarrow lim.proj.(G/N_i) \simeq \tag{7}$$
$$\{(g_1, ..., g_n) \mid g_i \in G_i := G/N_i, g_i \equiv g_j \text{ in } G/(N_i \cdot N_j)\}. \tag{8}$$

We do not have a general criterion for when this map is an isomorphism. However, we have

LEMMA 6.1 ((KiRo,93)). *We have an isomorphism in Equation 7 provided* $\bigcap N_i = 1$ *and for every* $p \in \pi(G)$ *there is an index* i *with* $(p, |N_i|) = 1$.
This applies in particular for

$$N_i = O_{p_i'}(G) \textit{ where } \pi(G) = \{p_1, \ldots, p_n\}.$$

If G is the projective limit of the quotients $G_i = G/N_i$ then the projective limit

$$\Gamma_G := \Gamma_G(\{G_i\}) := lim.proj.(\mathbb{Z}G_i) \tag{9}$$

of the group rings $\mathbb{Z}G_i$ does not coincide with $\mathbb{Z}G$ in general. As a matter of fact,

$$\Gamma_G(\{G_i\}) = \{(\gamma_1, \ldots, \gamma_n) \mid \gamma_i \in \mathbb{Z}G_i, \gamma_i \equiv \gamma_j \text{ in } \mathbb{Z}G/(N_i \cdot N_j)\},$$

and over the rationals this is in general a relatively small proper quotient of $\mathbb{Q}G$.
The induced map $\sigma\colon \mathbb{Z}G \longrightarrow \Gamma(G)$ has the kernel

$$Ker(\sigma) = \bigcap_{i=1}^{n} I(G, N_i)$$

where $I(G, N_i)$ is the kernel of the natural map $\mathbb{Z}G \longrightarrow \mathbb{Z}G_i$. We should point out that G embeds into $\Gamma(G)$ and in many cases $\Gamma(G)$ determines G.
We have to introduce some MORE NOTATION. We know that G is the projective limit of the groups $G_i = G/O_{p_i'}(G)$ and we put

$$G_{ij} = G/(O_{p_i'}(G) \cdot O_{p_j'}(G)) \text{ with } K_{ij} = Ker(G_i \longrightarrow G_{ij})\,.$$

We denote the corresponding projective limit of the group rings $\mathbb{Z}G_i$ by $\Gamma_G(\mathcal{O})$ or simply by Γ if it leads to no confusion.
Assume that we are given a 'central cocycle of automorphisms'[1]

$$\rho = (\rho_{ij}), \text{ i. e. } \rho_{ii} = 1 \text{ and } \rho_{ij} = \rho_{ji}^{-1}\,,$$

where $\rho_{ij} \in Aut_c(G_{ij})$, the group of conjugacy class preserving automorphisms of G. Then

$$G(\rho) = \{(g_i) \in \prod_{1 \le i \le n} G_i \mid \rho_{ij}(g_i \cdot G_{ij}) = g_j \cdot G_{ij}\}$$

is a group.

THEOREM 6.2 ((KiRo,93)).

- *Assume that $\Gamma_G(\mathcal{O}) = \Gamma_H(\mathcal{O})$ as augmented algebras and that G is solvable. Then there exists a central cocycle of automorphisms ρ such that $H \simeq G(\rho)$.*
- *Moreover, $G(\rho) \simeq G$ if and only if there exist automorphisms $\rho_i \in Aut(G_i)$ – not necessarily in $Aut_c(G_i)$ – with*

$$\rho_{ij} = \rho_i \cdot \rho_j^{-1}\,.$$

Hence the obstruction is a question of lifting automorphisms.

[1] This should not be confused with 'central cocycle′ from Definition 4.1.

and so $(\gamma_{ij}^o)^2 = \delta_i^o \cdot (\delta_j^o)^{-1}$. Since δ_i^o is odd, there is a power n_i of δ_i^o such that $\delta_i^{n_i} \cdot ((\delta_j^o)^{n_i})^{-1} = \gamma_{ij}^o$. By replacing γ_i by $\gamma_i \cdot ((\delta_i^o)^{-n_i})$ the cocycle is now γ_{ij}^e.

The kernel of our homomorphism $\Phi\colon Out(G) \longrightarrow Out(\mathbb{Z}G)$ has an interesting interpretation in terms of $\widehat{G} = \prod_i G_i$, inside of which the group $G = lim.proj.(G_i)$ is naturally embedded.

LEMMA 8.2. *$Ker(\Phi) \neq 1$ if and only if there exist a group element $\widehat{g} \in N_{\widehat{G}}(G)$, $\widehat{g} \notin C_{\widehat{G}}(G) \cdot G)$ and a central unit $u \in \mathbb{Z}\widehat{G}$ such that $u \cdot \widehat{g} \in \mathbb{Z}G$.*

Proof. Let $\rho \in Ker(\Phi)$, $\rho \neq id$, be given on G_i as the conjugation by g_i. We set $\widehat{g} = (g_1, g_2, \ldots, g_n)$. Then obviously $\widehat{g} \in N_{\widehat{G}}(G) \backslash C_{\widehat{G}}(G) \cdot G$. Since $\rho \in Ker(\Phi)$ ρ it is also the conjugation by a unit $v \in \mathbb{Z}G \subset \mathbb{Z}\widehat{G}$. Hence $v \cdot \widehat{g}^{-1} = u$ is a unit in the center of $\mathbb{Z}\widehat{G}$. This proves the first direction.

The other direction is more general and will follow from the next lemma.

LEMMA 8.3. *Let G be a subgroup of the finite group $\widetilde{G}$ and assume that there exist a group element $\tilde{g} \in N_{\widetilde{G}}(G) \setminus C_{\widetilde{G}}(G) \cdot G$ and a central unit $u \in \mathbb{Z}\widetilde{G}$ such that $u \cdot \tilde{g} \in \mathbb{Z}G$. Then the restriction of the conjugation by $\tilde{g} \in \widetilde{G}$ to G is a non-trivial element in the kernel of $\Phi_{\mathbb{Z}}(G)$.*

Proof. Let $\rho \in Aut(G)$ be the conjugation with $\tilde{g} \in \widetilde{G}$. In $\mathbb{Z}G$ we have that conjugation by $v = u \cdot \tilde{g} \in \mathbb{Z}G$ has the same effect as conjugation by $\tilde{g}$, since conjugation by u is the identity. Since $\tilde{g} \in N_{\widetilde{G}}(G) \setminus C_{\widetilde{G}}(G) \cdot G$, the conjugation by $\tilde{g}$ is not inner on G.

Let us draw some elementary consequences from the hypotheses of the previous lemma which will be useful later.

NOTE 8.4. Let us be given $H \leq G$, an element $g \in N_G(H) \setminus C_G(H) \cdot H$ and a central unit

$$u = \sum_K u_K \cdot \widehat{K} \in \mathbb{Z}G \text{ such that } v := u \cdot g \in \mathbb{Z}H .$$

Here K stands for a conjugacy class and $\widehat{K} \in \mathbb{Z}G$ is its class sum. Taking augmentations we conclude that for every rational prime

p there must exist an element $x(p)$ in the support of u whose conjugacy class $K_{x(p)}$ has length relatively prime to p. This means that the centralizer $C_G(x(p))$ contains a Sylow p-subgroup of G.

Now we return to the situation of projective limits (cf. Lemma 8.3).

LEMMA 8.5. *Let $\widehat{g}$ and u be as in Lemma 8.3. For any prime p we can write the element $x(p)$ from Note 8.4 as $x(p) = (x_1, \ldots, x_n)$. Then $C_{\widehat{G}}(x(p)) = \prod_i C_{G_i}(x_i)$ and hence for each i we have $C_{G_i}(x_i) \supset P_i$ for a Sylow p-subgroup of G_i. In particular, when $p = p_i$ then $x_i \in Z(O_{p_i}(G_i))$.*

Proof. We have for $p = p_i$ that

$$x_i \in C_{G_i}(P_i) \subset C_{G_i}(O_{p_i}(G_i)) = Z(O_{p_i}).$$

9. Estimations on $Ker(\Phi_{\mathbb{Z}}(G))$

Let G be a finite solvable group. Recall that G is a projective limit of the system (G_i, ϕ_{ij}) where $G_i = G/O_{p_i'}(G)$ and Γ denotes the projective limit of the corresponding system of group rings $(\mathbb{Z}G_i, \phi_{ij})$. In this section we will estimate the kernel of the homomorphism

$$\Phi\colon Out(G) \longrightarrow Out(\mathbb{Z}G)$$

in terms of Γ.

We start with the observation that the image of the natural map $\tilde{\Phi}\colon Aut(G) \longrightarrow Aut(\mathbb{Z}G)$ is contained in the subgroup $Aut_{gr}(\mathbb{Z}G)$ consisting of those maps which preserve G. Let

$$Inn_{gr}(\mathbb{Z}G) = Aut_{gr}(\mathbb{Z}G) \cap Inn(\mathbb{Z}G),$$

the subgroup of conjugations by units of $\mathbb{Z}G$ which normalize G. Also, let $Out_{gr}(\mathbb{Z}G) = Aut_{gr}(\mathbb{Z}G)/Inn_{gr}(\mathbb{Z}G)$. Then the map Φ factors as

$$Out(G) \longrightarrow Out_{gr}(\mathbb{Z}G) \longrightarrow Out(\mathbb{Z}G).$$

Since the group G is naturally embedded in the ring Γ we can similarly define the groups $Aut_{gr}(\Gamma)$, $Inn_{gr}(\Gamma)$ and their quotient

$Out_{gr}(\Gamma)$. Notice that these groups project naturally to the corresponding groups of group-preserving automorphisms of the rings $\mathbb{Z}G_i$.
It is easy to see that a group-preserving automorphism of $\mathbb{Z}G$ induces a similar automorphism of Γ and, in fact, we have a homomorphism

$$\Phi' \colon Out_{gr}(\mathbb{Z}G) \longrightarrow Out_{gr}(\Gamma).$$

Let

$$\Phi_{gr} \colon Out(G) \longrightarrow Out_{gr}(\Gamma)$$

be the composition $\Phi' \circ \Phi$. In this way we obtain our first estimation:

$$Ker(\Phi) \leq Ker(\Phi_{gr}).$$

To estimate the group $Ker(\Phi)$ from below we will substitute the group $Out_{gr}(\Gamma)$ with a larger version.
To this end, let us have a closer look at the unit group $U(\Gamma)$ of the ring Γ. So far we have considered its two subgroups: G and $U_{gr}(\Gamma)$, the normalizer of G. We introduce an intermediate subgroup $U_{co}(\Gamma)$ of COSET UNITS. It consists of units $u = (u_1, \ldots, u_n) \in U_{gr}(\Gamma)$ such that the support of $u_i \in \mathbb{Z}G_i$ is contained in a single coset of $K_{ij} = Ker(\phi_{ij})$ in G_i for all $j = 1, \ldots, n$.
As cosets of a normal subgroup can be multiplied, $U_{co}(\Gamma)$ is a well defined group. It is easy to see that $G \leq U_{co}(\Gamma)$. Moreover, the subgroups K_{ij} are characteristic in G_i and hence $U_{co}(\Gamma)$ is preserved by any automorphism from $Aut_{gr}(\Gamma)$. It then follows that the subgroup $Inn_{co}(\Gamma)$, consisting of conjugations by the coset units, is normal in $Aut_{gr}(\Gamma)$ and we may form the group $Out_{co}(\Gamma) = Aut_{gr}(\Gamma)/Inn_{co}(\Gamma)$.
Now, the natural homomorphism $Aut(G) \longrightarrow Aut_{gr}(\Gamma)$ maps inner automorphisms of G to conjugations by coset units and hence we obtain a homomorphism $\Phi_{co} \colon Out(G) \longrightarrow Out_{co}(\Gamma)$. Our second estimation is given by the following

PROPOSITION 9.1.

$$Ker(\Phi_{co}) \leq Ker(\Phi).$$

Proof. Let $\rho \in Aut(G)$ be such that $\Phi_{co}([\rho]) = 1$. We must show that ρ is a restriction of an inner automorphism of $\mathbb{Z}G$.
We know that $\rho = conj(u)|_G$ for some coset unit $u = (u_1, \ldots, u_n)$

of Γ. Without loss of generality we can assume that each unit u_i has augmentation one. Let $\rho_i = conj(u_i)|_{G_i} \in Aut(G_i)$. From Corollary 7.3 we know that these are inner automorphisms and hence each u_i can be written as $g_i \cdot v_i$ for some group elements $g_i \in cG_i$ and central units $v_i \in \mathbb{Z}G_i$.
Let $\widehat{G} = \prod_{i=1}^n G_i$. Consider the element

$$v = v_1 \otimes \cdots \otimes v_n \in \otimes_{i=1}^n \mathbb{Z}G_i \simeq_{nat} \mathbb{Z}\widehat{G}.$$

Clearly, it is a central unit in $\mathbb{Z}\widehat{G}$ which maps to v_i under the natural projections $\mathbb{Z}\widehat{G} \longrightarrow \mathbb{Z}G_i$, due to the augmentation assumption. Consider also the element $\widehat{g} = (g_1, \ldots, g_n) \in \widehat{G}$.
We observe now that the element $\widehat{g} \cdot v$ of the ring $\mathbb{Z}\widehat{G}$ lies in fact in the subring $\mathbb{Z}G$.
To this end, let $t = (t_1, \ldots, t_n)$ be any element from the support of v. By construction, each t_i has to appear in the support of v_i. We must show that $\phi_{ij}(g_i \cdot t_i) = \phi_{ji}(g_j \cdot t_j)$ for all pairs i, j.
Recall that u is a coset unit, i.e. the support of u_i is contained in a single coset of K_{ij} in G_i. Clearly, the same is true about the shifted unit v_i. Therefore all elements from the support of v_i are mapped by ϕ_{ij} to the same fixed element of G_{ij}. But u is an element of Γ which means that $\phi_{ij}(u_i) = \phi_{ji}(u_j)$. Hence we obtain

$$\begin{aligned} &\phi_{ij}(g_i \cdot t_i) = \phi_{ij}(g_i) \cdot \phi_{ij}(t_i) = \phi_{ij}(g_i) \cdot \phi_{ij}(v_i) = \phi_{ij}(u_i) = \\ &\phi_{ji}(u_j) = \phi_{ji}(g_j) \cdot \phi_{ji}(v_j) = \phi_{ji}(g_j) \cdot \phi_{ji}(t_j) = \phi_{ji}(g_j \cdot t_j). \end{aligned}$$

Therefore $w = \widehat{g} \cdot v$ is a unit of $\mathbb{Z}G$. As v centralizes $\mathbb{Z}G$, the unit w normalizes G. On each factor G_i we have

$$conj(w)|_{G_i} = conj(g_i) = \rho_i.$$

Hence $conj(w)|_G$ is the restriction of $\prod \rho_i \in Aut(\widehat{G})$, i.e. it coincides with ρ. Consequently, ρ is the restriction of an inner automorphism of $\mathbb{Z}G$.

10. About the group $Ker(\Phi_{co}(G))$

Notice that the group $Ker(\Phi_{co}(G))$ is described purely in terms of Γ which is easier to handle than $\mathbb{Z}G$. Of course, if we can find a group with $Ker(\Phi_{co}(G)) \neq 1$ then its integral group ring will have non-trivial units normalizing G. In this section we show that in order to achieve this goal the group G must have a rather special structure of its 2-subgroups.

We are going to use again the projective system $(G_i = G/O_{p'}(G))$ for the solvable even group G. Recall that $K_{ij} = Ker(G_i \xrightarrow{\phi_{ij}} G_{ij})$. Let us also assume that the prime divisors of $|G|$ are naturally ordered, and so $p_1 = 2$.

PROPOSITION 10.1. *Suppose that the group G satisfies*

$$O_{2',2} \leq O_{2'} \cdot P_{p'} \text{ for all odd primes } p.$$

Then the homomorphism $\Phi_{co}(G)$ is injective.

Proof. Take any $\rho \in Aut(G)$ such that $\Phi_{co}([\rho]) = 1$. From the assumption it follows that $\rho = conj(u)|_G$ for a suitable coset unit $u = (u_1, \ldots, u_n) \in \Gamma$. We show that under the assumptions of the proposition, ρ must be inner.

To simplify the notation, we will denote the images of elements from G_i and G_j in G_{ij} by an overbar.

For each i we have $conj(u_i)|_{G_i} = conj(g_i)|_{G_i}$, for some $g_i \in G_i$. The elements (g_i) define a cocycle $\rho_{ij} = \overline{g}_i\overline{g}_j^{-1}$. Using Proposition 8.1 we can compose the homomorphism ρ with an inner automorphism to assure that all ρ_{ij} are involutions: $\rho_{ij}^2 = 1$.

Recall that for a coset unit u we have $\overline{u}_i = \overline{u}_j \in G_{ij}$, a group element. Hence $v_i = g_i \cdot u_i^{-1} \in Z(\mathbb{Z}G_i)$ and $\overline{v}_i = z_i \in Z(G_{ij})$. We have

$$\rho_{ij} = \overline{g}_i \cdot \overline{g}_j^{-1} = \overline{v}_i \cdot \overline{u}_i \cdot \overline{u}_j^{-1} \cdot \overline{v}_j^{-1} = z_i \cdot z_j^{-1}.$$

From Lemma 8.5 it follows that the support of v_i contains a p_i-element. As the whole support of v_i sits in a single coset of K_{ij}, all its elements map to $z_i \in G_{ij}$. In particular, z_i is a central p_i-element.

For $i, j > 2$ the primes p_i, p_j are odd and hence $z_i \cdot z_j^{-1}$ is of odd order. This is possible only if $\rho_{ij} = 1$. Thus we are left with the case $i = 1$, $j > 1$.

Here the assumption implies that $O_2(G_1) \leq \bigcap_{j\geq 2} K_{2j}$ which in turn guarantees that $z_1 = 1$. It easily follows that $\rho_{1j} = 1$, as well.

Hence we have $\overline{g}_i = \overline{g}_j$ for all i, j and so (g_i) defines a group element $g \in G$. Obviously, $\rho = conj(g)$.

11. More groups with $Ker(\Phi) = 1$

For any subgroup $H \le G$ we can define its 'Weil group' as the quotient $N_G(H)/C_G(H) =: W_G(H)$. Notice that this group acts naturally on H.
The following statement shows a construction of making groups with $Ker(\Phi) = 1$ larger.

PROPOSITION 11.1. *Let* $1 \longrightarrow H \longrightarrow G \longrightarrow \overline{G} \longrightarrow 1$ *be an exact sequence of finite groups. If* $Ker(\Phi_{\mathbb{Z}}(\overline{G})) = 1$ *and the groups* H *and* $W_G(H)$ *have odd orders then* $Ker(\Phi_{\mathbb{Z}}(G)) = 1$.
Proof. In our case $W_G(H) = G/C_G(H)$. Hence, for any subgroup $K \le H$, which is normal in G, the order of $W_G(K)$ is odd. Therefore we may proceed by induction with respect to the order of H. As H is solvable it contains a characteristic abelian p-subgroup A for some odd prime p. Clearly, the sequence

$$1 \longrightarrow H/A \longrightarrow G/A \longrightarrow \overline{G} \longrightarrow 1$$

satisfies the conditions of our proposition. By induction, we may assume that $Ker(\Phi_{\mathbb{Z}}(G/A)) = 1$. Therefore it is enough to prove our statement in the case when H is an abelian p-group.
Suppose that $[\rho] \in Ker(\Phi_{\mathbb{Z}}(G)) \le Out(G)$. Then ρ is a central automorphism and hence it preserves H and induces an automorphism $\overline{\rho}$ on $\overline{G}$.
Let $P \le G$ be a Sylow p-subgroup containing H. After composing ρ with a suitable inner automorphism we may assume that $\rho|_P = id_P$. Now, $\overline{\rho}$ is an inner automorphism, say $conj(x)$ which is the identity on $\overline{P}$ – the image of P in $\overline{G}$. Hence $x \in \overline{G}$ centralizes $\overline{P}$. Pick any element $y \in G$ which projects to x under the natural homomorphism. For any $z \in P$ we then have $y \cdot z \cdot y^{-1} = z \cdot a$ for some $a \in H$. But $H \le P$ and so y normalizes P.
Let us substitute ρ with $\rho \circ conj(y^{-1})$. Then $\overline{\rho} = id$ and $\rho|_P = conj(y^{-1})|_P$. As y clearly normalizes H, the automorphism $\rho|_H$ lies in the image of $W_G(H)$ and hence it is of odd order m.
Consider a new automorphism $\sigma = \rho^m$. Then $\overline{\sigma} = id$ as well as $\sigma|_H = id_H$. But then for any $g \in G$ we have $\sigma(g) = g \cdot a$ for some element $a \in H$. Let $k = o(a)$ be the order of a; it is again an odd number. Then $\sigma^k(g) = g \cdot a^k = g$, i.e. $\sigma^k = id$.
Now we only have to use the fact that $Ker(\Phi_{\mathbb{Z}}(G))$ is an elementary

abelian 2-group. In this group it follows from $[\sigma]^k = 1$ that $[\sigma] = 1$. But then $[\rho]^m = 1$ and hence $[\rho] = 1$, i.e. ρ is inner.

REMARK 11.2.

- The assumptions of the above proposition are satisfied in particular when the H-conjugacy classes of elements in H coincide with the G-conjugacy classes. This happens, for example, when H is central in G.
- If G is a nilpotent group then its Sylow subgroups are normal and clearly $W_G(P) = 1$ for each of them. Because $Ker(\Phi) = 1$ for p-groups, we obtain by an induction an easy proof that the same holds for all nilpotent groups.
- If G is a minimal example of a group with $Ker(\Phi) \neq 1$ then the center of G is a 2-group. Note that the trivial group is by definition a 2-group.

12. About $Ker(\Phi_{\mathbb{Z}}(G))$ for metabelian groups

In this section we consider the case of a finite metabelian group G. M. Hertweck has constructed an example of a (very complicated) metabelian group G with an automorphism $\rho \in Ker(\Phi_{\mathbb{Z}}(G))$, which is not inner. Here we shall discuss some situations when we have $Ker(\Phi_{\mathbb{Z}}(G)) = 1$.
Every automorphism, which belongs to $Ker(\Phi_{\mathbb{Z}}(G))$ automatically satisfies several conditions, most of which follow from Coleman's result. We collect them in a definition.

DEFINITION 12.1. Let ρ be an automorphism of a finite group G. We call ρ a COLEMAN AUTOMORPHISM, or a C-AUTOMORPHISM for short, when $\rho \circ \rho$ is inner, ρ preserves the conjugacy classes in G and for every Sylow p-subgroup $P < G$ we have $\rho|_P = conj(g)|_P$ for a suitable $g \in G$.

REMARK 12.2. Our interest in those automorphisms stems from the fact that all automorphisms $\rho \in Ker(\Phi_{\mathbb{Z}}(G))$ are C-automorphisms.

PROPOSITION 12.3. *Let G be a finite metabelian group. If a Sylow 2-subgroup of G is abelian, then every C-automorphism of G is inner.*

Proof. Let G be a metabelian group with an abelian Sylow 2-subgroup. Let B be an abelian normal subgroup of G group $A = G/B$ is also abelian. We first prove the following

CLAIM 12.4. *The restriction to B of every C-automorphism equals $conj(g)|_B$ for some $g \in G$.*

Proof of the claim. We proceed by induction on the number $r(B)$ of prime divisors of $|B|$. If $r(B) = 1$, then B is a p-group and the claim follows from the definition of a C-automorphism. Assume then that $r(B) > 1$. If $|B|$ is even let $p = 2$; otherwise pick for p any any prime dividing $|B|$. Write $B = B_p \times D$ where B_p is a Sylow p-subgroup of B. Then D is relatively prime to $2 \cdot p$.

Let $q\colon G \longrightarrow \overline{G} := G/B_p$ be the natural projection and let $\overline{\rho}$ be the induced automorphism of $\overline{G}$. By induction $\overline{\rho}|_{q(B)} = conj(\overline{g})|_{q(B)}$ for some $g \in G$. Hence for every $x \in D$ we have

$$(\rho \circ conj(g^{-1}))(x) \in B_p \cap D = \{1\},$$

i. e. $\rho|_D = conj(g)|_D$. Also, by definition, $\rho|_{B_p} = conj(g')|_{B_p}$ for some $g' \in G$.

Let us substitute for ρ a new C-automorphism $\rho \circ conj(g^{-1})$. Then $\rho|_D = id_D$ and $\rho|_{B_p} = con(h)|_{B_p}$ for some $h \in G$.

Consider the order of ρ. The square ρ^2 is inner; so if some odd power of ρ is inner, then ρ is inner and we are done. Hence, without loss of generality, we may assume that the order of ρ is a power of 2. Again after taking a suitable odd power of ρ we can assume that the element h is of order a power of 2.

We now consider the two cases $P = 2$ and $p \neq 2$.

If $p = 2$ let P be a Sylow 2-subgroup of G which contains the 2-element h. Clearly, $B_p \leq P$. As P is abelian by assumption, h centralizes B_p and hence $\rho|_B = id_B$; in particular, ρ is inner on B.

If p is odd we use the fact that ρ is central. Let $x \in B_p$ and $y \in D$ be any elements. Then there exists $z \in G$ depending on x and y such that $h \cdot x \cdot h^{-1} \cdot y = \rho(x \cdot y) = z \cdot (x \cdot y) \cdot z^{-1}$. It follows that

$$x^{-1} \cdot (h^{-1} \cdot z)^{-1} x \cdot (h^{-1} \cdot z) = y \cdot z^{-1} \cdot y^{-1} \cdot z \in B_p \cap D = \{1\}.$$

Hence $h^{-1} \cdot z \in C_G(x)$ and $z \in C_G(y)$. In other words, for any $x \in B_p$ and $y \in D$ we have $h \cdot C_G(x) \cap C_G(y) \neq \emptyset$.

Now, the action of G on B by inner automorphisms factors to an action of $A = G/B$. Let a be the image of h in A. Then for $x \in B_p$ and $y \in D$ we have $a \cdot C_A(x) \cap C_Y(y) \neq \emptyset$.

Let us factor $A = A_2 \times E$, where A_2 is the Sylow 2-subgroup of A. Then $C_A(b) = C_{A_2}(b) \times C_E(b)$ for all $b \in B$. As $a \in A_2$, we have $a \cdot C_{A_2}(x) \cap C_{A_2}(y) \neq \emptyset$ for all x and y.

From elementary representation theory it follows that when a 2-group A_2 acts on a $2'$-group M then there exists an element $m \in M$ such that $C_{A_2}(m) = C_{A_2}(M)$. Applying this remark to $M = B_p$ and $M = D$ respectively, we find elements $x_0 \in B_p$ and $y_0 \in D$ such that $C_{A_2}(x_0) = C_{A_2}(B_p)$ and $C_{A_2}(y_0) = C_{A_2}(D)$.

We obtain $a \cdot C_{A_2}(B_p) \cap C_{A_2}(D) \neq \emptyset$. Let $c \in A_2$ belong to this intersection. Then $a^{-1} \cdot c$ centralizes B_p and c centralizes D. Let $G \in G$ be any preimage of c. Then $\rho|B = conj(g)|_B$. In fact, for $x \in B_p$ we have

$$g{\cdot}x{\cdot}g^{-1} = c{\cdot}x{\cdot}x^{-1} = a{\cdot}(a^{-1}{\cdot}c){\cdot}x{\cdot}(a^{-1}{\cdot}C)^{-1}{\cdot}a^{-1} = a{\cdot}x{\cdot}a^{-1} = \rho(x),$$

and for $y \in D$ we have

$$g \cdot y \cdot g^{-1} = c \cdot y \cdot y^{-1} = Y = \rho(y).$$

In any case $\rho|_B$ is inner.

We are now ready to *prove our Proposition* . Let ρ be a C-automorphism of a metabelian group G. We can assume that $\rho|_B = id_B$ for an abelian normal subgroup $B < G$ with abelian quotient group $A := G/B$ and that the order of ρ is a power of 2. The automorphism ρ permutes the set of the Sylow 2-subgroups of G. As the order of ρ is a power of 2 and the number of permuted subgroups is odd, this permutation action has a fixed point. That is, ρ stabilizes a Sylow 2-subgroup P. Recall that ρ acts on P by conjugation with some $g \in G$. After taking a suitable power of ρ, we may assume that g is a 2-element. As g normalizes P, it belongs to P. As P is abelian, we get $\rho|_P = id_P$.

Obviously, ρ induces the identity on A. Hence $\delta(x) := \rho(x) \cdot x^{-1}$ is a 1-cocycle, $\delta\colon A \longrightarrow B$. Notice that for $k \geq 1$ we have

$\rho^k(x) \cdot x^{-1} = \delta^k(x)$. In fact, for $k = 1$ this is obvious. Assume that $k > 1$ and that the relation holds for $k-1$. Then

$$\rho^k(x) \cdot x^{-1} = \rho(\rho^{k-1}(x) \cdot x^{-1}) \cdot (\rho(x) \cdot x^{-1}) = \rho(\delta^{k-1}(x)) \cdot \delta(x) = \delta^{k-1}(x),$$

as $\delta^{k-1}(x) \in B$ and $\rho|_B = id_B$. In particular, δ takes values in the Sylow 2-subgroup B_2 of B. Thus $[\delta] \in H^1(A, B_2)$.
As $\delta|_P \equiv 1$, the class $[\delta]$ restricts to 0 in $H^1(A_2, B_2)$. However, the restriction map is injective. Thus δ is a coboundary and ρ is inner.

The above statement can not be extended to all metabelian groups as the next example shows.

EXAMPLE 12.5. There exists a C-automorphism of a finite metabelian group G which is not inner.

Let A be the additive group of upper triangular (2×2) matrices over $\mathbb{F}_2$. Thus $A \simeq C_2^3$. We define the action of A on two groups, $B_3 := C_2^4$ and $B_3 := C_3$.
Let the homomorphism $\phi_2 \colon A \longrightarrow GL_4(\mathbb{F}_2) = Aut(B_2)$ be given by

$$\phi_2(M) = \begin{pmatrix} I & M \\ 0 & I \end{pmatrix},$$

for $M \in A$, where I is the (2×2) identity matrix. Also let

$$\phi_3 \colon A \longrightarrow \{\pm 1\} = Aut(B_3) \, : \, \begin{pmatrix} a & b \\ 0 & c \end{pmatrix} \longrightarrow (-1)^{a+c}.$$

Let $B = B_2 \times B_3$ and $\phi = \phi_2 \times \phi_3 \colon A \longrightarrow Aut(B)$.
We define $G := B \rtimes_\phi A \simeq (C_2^4 \times C_3) \rtimes_\phi C_2^3$. Thus $|G| = 2^7 \cdot 3 = 384$. Let $g = \begin{pmatrix} 1 & 0 \\ 0 & 0 \end{pmatrix} \in A < G$. Consider that automorphism $\rho \colon G \longrightarrow G$ given by $\rho|_{D_2} = conj(g)|_{D_2}$, $\rho|_{B_3} = id_{B_3}$ and $\rho|_{A_2} = id_{A_2}$.
It is easy to see that ρ is a well defined automorphism of G and $\rho^2 = id_G$. Let us check that ρ is central.
Consider the subgroup $C := Ker(\phi_3) < A$. Then

$$C = \{\begin{pmatrix} a & b \\ 0 & a \end{pmatrix}\} \simeq C_2^2.$$

Notice that, although $g \notin C$, for every vector $w \in \mathbb{F}_2^2$ there exists an element $g_w \in C$ such that $g \cdot w = g_w \cdot w$. In fact

$$\begin{aligned} g \cdot (0,0) &= \begin{pmatrix} 0 & 0 \\ 0 & 0 \end{pmatrix} \cdot (0,0) \text{ and } g \cdot (1,0) = \begin{pmatrix} 1 & 0 \\ 0 & 1 \end{pmatrix} \cdot (1,0) \\ g \cdot (0,1) &= \begin{pmatrix} 0 & 0 \\ 0 & 0 \end{pmatrix} \cdot (0,1) \text{ and } g \cdot (1,1) = \begin{pmatrix} 0 & 1 \\ 0 & 0 \end{pmatrix} \cdot (1,1). \end{aligned} \tag{10}$$

This observation can be translated to a property of the action of A an B as follows. We can represent the elements of B_2 as pairs $b = (v, w)$ of vectors from $\mathbb{F}_2^2$. Then for $a \in A$ we have

$$a \cdot b \cdot a^{-1} = a \cdot (v + a \cdot w, w).$$

In particular, for every $b = (v, w) \in B_2$ there exists an element $g_b := g_w \in A$ such that $g \cdot b \cdot g^{-1} = g_b \cdot b \cdot g_b^{-1}$ and g_b centralizes B_3. Take any elements $x \in B_2$, $y \in B_3$ and $a \in A$. Then

$$\rho(x \cdot y \cdot a) = g \cdot x \cdot g^{-1} \cdot y \cdot a = g_x \cdot x \cdot g_x^{-1} \cdot y \cdot a = g_x \cdot (x \cdot y \cdot a) \cdot g_x^{-1},$$

i. e. $\rho(x \cdot y \cdot a)$ is conjugate to $x \cdot y \cdot a$.
The map ρ restricted to the unique Sylow 3-subgroup $B_3 < G$ is the identity. The group G has three Sylow 2-subgroups: $P_y := B_2 \rtimes y \cdot A \cdot y^{-1}$ for $y \in B_3$. We have $\rho|_{P_y} = conj(y^{-1} \cdot g)$. In fact for $x \in B_2$ and $a \in A$ holds

$$\begin{aligned} &\rho(x \cdot (y \cdot a \cdot y^{-1}) = (g \cdot x \cdot g^{-1}) \cdot (y \cdot a \cdot y^{-1}) = \\ &(g \cdot x \cdot g^{-1}) \cdot (y^{-2} \cdot a \cdot y^2) = (g \cdot x \cdot g^{-1}) \cdot (y^{-2} \cdot (g \cdot a \cdot g^{-1}) \cdot y^2) = \\ &((g \cdot x \cdot g^{-1})) \cdot ((y^{-2} \cdot g) \cdot a \cdot (y^{-2} \cdot g)^{-1}) = \\ &(g \cdot x \cdot g^{-1}) \cdot (y^{-1} \cdot g \cdot y) \cdot a(y^{-1} \cdot g)^{-1} = \\ &y^{-1} \cdot g \cdot x \cdot (y^{-1} \cdot g)^{-1} \cdot (y^{-1} \cdot g) \cdot (y \cdot aq \cdot y^{-1}) \cdot (y^{-1} \cdot g)^{-1} = \\ &(y^{-1} \cdot g) \cdot (x \cdot (y \cdot a \cdot y^{-1})) \cdot (y^{-1} \cdot g)^{-1}. \end{aligned} \tag{11}$$

Hence ρ is inner when restricted to any Sylow subgroup of G.
We have just checked that ρ is a C-automorphism. It remains to show that ρ is not inner. To this end, suppose that $\rho = conj(z)$ for some $z \in G$. It follows that $z \in C_G(B_3)$ and $z \cdot g^{-1} \in C_G(B_2)$. After a projection onto A we get an element $\overline{z} \in A$ such that

$\overline{z} \in C_A(B_2) = C$ and $\overline{z} \cdot g^{-1} \in C_A(B_2) = \{1\}$. It follows that $g \in C$ – a contradiction.

13. Acknowledgments

This paper was written during the first author's visit at Mathematisches Institut B, Stuttgart University in 1997 [2]. He is deeply grateful to the Institute for providing excellent working conditions and creating a very friendly atmosphere.

References

Co,64. Coleman, D.: On the modular group ring of a p–group, Proc. AMS, 15 (1964), 511–514.

Gr,82. Gross, F.: Automorphisms which centralize a Sylow p-subgroup, J. Algebra 77, 202-233 (1982).

He,97. Hertweck, M.: A counterexample to the isomorphism problem for integral group rings of finite groups, to appear in Comptes Rendu, Paris.

JaMa,87. Jackowski, S., Marciniak, Z.S.: Group automorphisms inducing the identity map on cohomology, J. of Pure and Appl. Algebra 44 (1987), 241–250.

KiRo,93. Kimmerle, W., Roggenkamp, K.W.: A Sylow-like theorem for integral group rings of finite solvable groups, Archiv d. Math. 60 (1993), 1–6.

Kr,87. Krempa, J.: see the remark in (JaMa,87).

Ma,95. Mazur, M.: On the isomorphism problem for integral group rings of infinite groups, Expo. Math. 13, No. 5 (1995), 433–445.

RoSc,85. Roggenkamp, K. W., Scott L. L.: A counterexample to the Zassenhaus conjecture, MS 1985, A strong answer to the isomorphism problem for finite p-solvable groups with a normal p-subgroup containing its centralizer MS 87. See (RoTa,92)

RoTa,92. Roggenkamp, K. W., Taylor, M. V.: Groups rings and class groups, DMV-Seminar 18, Birkhäuser, Basel 1992.

RoZi,92. Roggenkamp, K. W., Zimmermann, A.: On the isomorphism problem for integral group rings of finite groups, Arch. Math. 59 (1992), 534–544.

RoZi,95.1. Roggenkamp, K. W., Zimmermann, A.: Outer automorphisms may become inner in their integral group rings, J. Pure Appl. Algebra 103 (1995), 91–99.

[2] The paper is only published now, since a referee of the journal to which we submitted the paper first had made minor objections over a period of three years. We felt not being treated objectively and gave up with that journal.

RoZi,95.2. Roggenkamp, K. W., Zimmermann, A.: A counterexample for the isomorphism problem of polycyclic groups, J. Pure Appl. Algebra 103 (1995), 101–103.

Sc,87. Scott, L.L.: Recent progress on the isomorphism problem, Proceedings of Symposia in Pure Mathematics 47 (1987), 269-273.

REPRESENTATION THEORY OF SEMISIMPLE HOPF ALGEBRAS

S. MONTGOMERY (smontgom@math.usc.edu)*
University of Southern California,
Los Angeles,
CA 90089-1113

1. Introduction

Let H be a finite-dimensional semisimple Hopf algebra over a field k. We usually assume that k is algebraically closed of characteristic zero. In these notes we discuss the representation theory of such Hopf algebras, analogous to classical results about the representation theory of finite groups. We give fairly complete proofs of many of the results, since a number of the original proofs have been simplified, particularly by H.-J. Schneider and M. Lorenz.

We begin with a more general discussion of finite-dimensional Hopf algebras and their properties, and prove the semisimplicity criterion in terms of the antipode proved by Larson and Radford. We then specialize to the semisimple case, proving the orthogonality of characters and what is called the "Class Equation" for Hopf algebras. We then give a number of applications of the class equation, related to the classification of Hopf algebras and the dimensions of irreducible modules. We also discuss the Grothendieck ring $K_0(H)$ and its relation to twistings of Hopf algebras. Finally we consider the Schur indicator of an irreducible representation.

* The author was partially supported by NSF grant DMS 98-02086. She also thanks the organizers of this NATO workshop in Constanta, Romania for an interesting meeting. In addition she thanks the UNAM Mathematics Institute in Morelia, Mexico, for its hospitality in April 2000 where a preliminary version of these lectures were given.

K.W. Roggenkamp and M. Ştefănescu (eds.), Algebra - Representation Theory, 189–218.

Throughout H is a finite-dimensional Hopf algebra over k, with multiplication $m : H \otimes H \to H$, unit $u : k \to H$ (given by $u(1_k) = 1_H$), comultiplication $\Delta : H \to H \otimes H$, counit $\varepsilon : H \to k$, and antipode $S : H \to H$. We abbreviate $\Delta(h)$ using the so-called "sigma notation"; that is, $\Delta h = \sum h_1 \otimes h_2$. For example, let $H = kG$, the group algebra of the finite group G. H becomes an algebra using the group operation, and a Hopf algebra by defining $\Delta x = x \otimes x$, $\varepsilon(x) = 1$, and $S(x) = x^{-1}$, for any $x \in G$. These maps are extended linearly to kG.

Since H is finite-dimensional, its k-linear dual H^* is also a Hopf algebra, with structure maps given by appropriate transposes of those of H. That is, $m_{H^*} = (\Delta_H)^*$, $u_{H^*} = (\varepsilon_H)^*$, $\Delta_{H^*} = (m_H)^*$, $\varepsilon_{H^*} = (u_H)^*$, and $S_{H^*} = (S_H)^*$. Thus multiplication in H^* is given by *convolution*, that is

$$(f * g)(h) = m_{H^*}(f \otimes g)(h) = (f \otimes g)(\Delta h) = \sum f(h_1)g(h_2),$$

for all $f, g \in H^*$ and $h \in H$. Similarly comultiplication in H^* is given by $\Delta f(h \otimes l) = (m_H)^* f(h \otimes l) = f(hl)$, for all $f \in H^*$ and $h, l \in H$. In particular $(kG)^*$ is also a Hopf algebra: convolution becomes usual pointwise multiplication of functions, and the comultiplication can be given more explicity in terms of the dual basis $\{p_x | x \in G\}$ of $(kG)^*$: that is, $\Delta p_x = \sum_{y \in G} p_y \otimes p_{y^{-1}x}$.

There are several other Hopf algebras associated to H. H^{op} denotes the Hopf algebra with opposite multiplication to H but the same comultiplication, and H^{cop} is the Hopf algebra with the same multiplication as H, but opposite comultiplication (equivalently, H^* has opposite multiplication). We note that the antipode of both H^{op} and H^{cop} is S_H^{-1}, since the defining property of S_H is that $S * id = id * S = \varepsilon$, where $*$ denotes the convolution product in $\mathrm{End}_k(H)$.

In general we will say a Hopf algebra is *non-trivial* if it is not isomorphic to kG or $(kG)^*$ for any G. In any Hopf algebra, $G(H) = \{0 \neq h \in H | \Delta h = h \otimes h\}$ is the set of *group-like* elements in H; $kG(H)$ is a Hopf subalgebra of H isomorphic to the group algebra of $G(H)$. Knowing $G(H)$, or even whether it is non-trivial, is extremely useful.

A fundamental property of the representation theory of Hopf algebras is that, similarly to the case of groups, the tensor product of H-modules is again an H-module. That is, let $\mathcal{C} = {}_H\mathcal{M}$ denote the category of left H-modules. If $V, W \in \mathcal{C}$, then $V \otimes W \in \mathcal{C}$, via $h \cdot (v \otimes w) := \Delta(h) \cdot (v \otimes w) = \sum h_1 \cdot v \otimes h_2 \cdot w$, for all $h \in H, v \in V, w \in W$. This tensor product is associative and has an "identity" $I = V_0$, where $V_0 = kv_0$ is the *trivial* module, that is $h \cdot v_0 = \varepsilon(h)v_0$. Thus in fact $\mathcal{C}$ is a *monoidal* category. Moreover for each $V \in \mathcal{C}$, the dual module $V^* \in \mathcal{C}$ also, via $(h \cdot f)(v) := f(Sh \cdot v)$ for all $h \in H, f \in H^*$ and $v \in V$; this will imply that the category $\mathcal{C}$ is *rigid*; see (Ksl).

More generally, for $V, W \in \mathcal{C}$, also $Hom_k(V, W) \in \mathcal{C}$ via $(h \cdot f)(v) := \sum h_1 \cdot (f(Sh_2 \cdot v)) \in W$; V^* is the case when $W = V_0$.

Conversely if A is a k-algebra and ${}_A\mathcal{M}$ is a rigid monoidal category, then A can be given the structure of a Hopf algebra. Thus these elementary properties of the tensor product of H-modules can be considered as the defining properties of a Hopf algebra.

Note that it would be very useful to know that for any finite-dimensional $V \in \mathcal{C}$, $V^{**} \cong V$. This will happen if $S^2 = id$, or more generally if S^2 is an inner automorphism of H. The order of S^2 is considered Section 2.

We record here a few elementary facts about representations and characters. If V is a left H-module, we write $\phi_V : H \to End(V)$ for the representation of H on V. When V is finite-dimensional, we let χ_V denote the character of the representation; that is, $\chi_V(h)$ is the trace of $\phi_V(h)$, for any $h \in H$. Note that $\chi_V \in H^*$. The following lemma is an easy exercise.

LEMMA 1.1. *Let V, W be left H-modules. Then as left H- modules,*

(1) $(V \otimes W)^* \cong W^* \otimes V^*$,

(2) $V \otimes W^* \cong Hom(W, V)$, and

(3) $Hom_H(V, W) \cong Hom(V, W)^H$, where for any H-module V, $V^H := \{v \in V \mid h \cdot v = \varepsilon(h)v \ \forall h \in H\}$, the H- invariants of V.

If also V and W are finite-dimensional, then

(4) $\chi_{V \otimes W} = \chi_V * \chi_W$ and $\chi_{V^*} = \chi_V \circ S_H = S_{H^*}\chi_V$.

Some general references on Hopf algebras are (M) and (Sch1).

2. Frobenius algebras and semisimplicity

In this section we consider the relationship between semisimplicity and the square of the antipode S, as well as some useful trace formulas.

To say that a Hopf algebra H is *semisimple* simply means that it is semisimple as an algebra. The next results, due to Larson and Radford, were in fact conjectured by Kaplansky in 1975 (Kap).

THEOREM 2.1. *(LR1) Assume that k has characteristic 0. Then $S^2 = id$ if and only if both H and H^* are semisimple.*

THEOREM 2.2. *(LR2) Assume that k has characteristic 0. Then H is semisimple if and only if H^* is semisimple.*

Recently Etingof and Gelaki (EG2) have shown that the "if" direction of Theorem 2.1 also holds in characteristic $p > 0$; Theorem 2.2 is trivially false in that case (consider $H = kG$ such that p divides $|G|$).

Here we give fairly complete proofs of the two results of Larson and Radford. The original proofs of Theorems 2.1 and 2.2 were quite complicated. The first major simplification of the proof of Theorem 2.1 was given by (Sch1); he used properties of Frobenius algebras in an essential way. We follow the general outline of Schneider's proof, although with a few further small simplifications, including an idea in (Kad); again, whenever possible, Hopf computations are replaced by working in more general Frobenius algebras. We also give a new shorter proof of Theorem 2.2.

Recall that a k-algebra A is a *Frobenius algebra* if there exists an associative, non-degenerate bilinear form $(\ , \ ) : A \otimes A \to k$. There are a number of equivalent formulations of an algebra being Frobenius. To state these, note that A^* is always a left A-module via the transpose of right multiplication in A. That is, for $a \in A$ and $f \in A^*$, $a \rightharpoonup f \in A^*$ via $(a \rightharpoonup f)(b) := f(ba)$, for all $b \in A$. Similarly A^* is a right A-module via $(f \leftharpoonup a)(b) := f(ab)$. We sometimes write $a \rightharpoonup f = af$ and $f \leftharpoonup a = fa$, when there is no ambiguity.

LEMMA 2.3. *(CR, §9A,B) Let A be a finite-dimensional algebra. Then the following are equivalent:*

(1) A is a Frobenius algebra;

(2) there exists $f \in A^*$ such that the map $\Phi : {}_AA \to {}_AA^*$, via $a \mapsto (a \rightharpoonup f)$, is a left A-module isomorphism;

(3) there exists $f \in A^*$ such that the map $\Psi : A_A \to A^*{}_A$, via $a \mapsto (f \leftharpoonup a)$, is a right A-module isomorphism;

(4) there exists $f \in A^*$ and $r_i, l_i \in A$, $i = 1,...n$, where $n =$dim(A), such that for any $a \in A$, $a = \sum_i r_i f(l_i a) = \sum_i f(ar_i)l_i$.

The set (f, r_i, l_i) is called a *Frobenius system* for A. Given such a system, a bilinear form on A is given by $(a, b) := f(ab)$, and similarly f can be reconstructed from the bilinear form. Moreover the elements $\{r_i, l_i\}$ form a dual basis for $(\ ,\ )$ in the usual sense; that is, $(l_i, r_j) = \delta_{i,j}$. Of course these dual bases are not unique. However the element $e := \sum_i r_i \otimes l_i$ is uniquely determined by the bilinear form (more generally, this is true for any pair of dual bases for a non-degenerate bilinear form on a vector space).

We first review some basic properties of finite-dimensional Hopf algebras.

A *left integral* in H is an element $\Lambda \in H$ such that $h\Lambda = \varepsilon(h)\Lambda$ for all $h \in H$; similarly $\Gamma \in H$ is a *right integral* if $\Gamma h = \varepsilon(h)\Gamma$, for all $h \in H$. We let $\int_H^l$ denote the space of left integrals, and $\int_H^r$ the space of right integrals. We will use the notational convention that in H^*, a left integral is denoted by λ and a right integral by γ.

If $\int_H^l = \int_H^r$, then H is called *unimodular*; in that case we simply write $\int_H$ for the space of integrals. For example, both kG and $(kG)^*$ are unimodular, for $\Lambda = \Gamma = \sum_{g \in G} g$ in kG and $\lambda = \gamma = p_1$ in $(kG)^*$.

The following result of Larson and Sweedler guarantees that Hopf algebras are always Frobenius.

THEOREM 2.4. *(LS) (M, 2.1.3, 2.2.1) Let H be any (finite-dimensional) Hopf algebra. Then:*

(1) $\int_H^l$ and $\int_H^r$ are each one-dimensional;

(2) the antipode S of H is bijective;

(3) H^* is a cyclic left and right H-module, via $\rightharpoonup$ and $\leftharpoonup$;

(4) H is a Frobenius algebra;

(5) ("Maschke's theorem") H is semisimple $\iff \varepsilon(\Lambda) \neq 0$ for $0 \neq \Lambda \in \int_H^l \iff \varepsilon(\Gamma) \neq 0$ for $0 \neq \Gamma \in \int_H^r$. Moreover when H is semisimple, H is unimodular.

In the proof of the theorem, the cyclic generators of H^* as a left and right H-module are precisely the left (or right) integrals in H^*. Moreover, an associative, non-degenerate, bilinear form on H is given by $0 \neq \lambda \in \int_{H^*}^l$ via $(h,l) := \lambda(hl)$ for all $h, l \in H$; another such bilinear form is given by $0 \neq \gamma \in \int_{H^*}^r$ via $(h,l)' := \gamma(hl)$.

Even when H is not unimodular, it will still be true that $\Lambda h \in k\Lambda$, since $\int_H^l$ is one-dimensional by Theorem 2.4. Thus there exists $\alpha \in G(H^*)$ such that $\Lambda h = \alpha(h)\Lambda$ for all $h \in H$; α is called the *left modular function* for H. Similarly $\beta = \alpha^{-1}$ is the *right modular function* for H. For the Hopf algebra H^*, its left and right modular functions can be identified with elements $a, b \in H \cong H^{**}$.

The next lemma is well-known. The second part is the analog of a known result for Haar integrals.

LEMMA 2.5. *Choose* $\lambda \in \int_{H^*}^l$.

(1) Let $t \in H$ *be such that* $\lambda \leftharpoonup t = \varepsilon$. *Then* $t \in \int_H^r$ *and* $\lambda(t) = 1$.

(2) Let a *be the left modular element for* H^* *in* H. *Then* $a\lambda = a \rightharpoonup \lambda \in \int_{H^*}^r$; *similarly* $\lambda a \in \int_{H^*}^r$.

Moreover for any $\Gamma \in \int_H^r$, *we have* $(a \rightharpoonup \lambda)(\Gamma) = \lambda(\Gamma)$.

Proof. (1) For any $h \in H$,

$$\lambda \leftharpoonup th = (\lambda \leftharpoonup t) \leftharpoonup h = \varepsilon \leftharpoonup h = \varepsilon(h)\varepsilon = \lambda \leftharpoonup (\varepsilon(h)t)$$

and thus $t \in \int_H^r$. Now using that $1 \in H$ is the counit of H^*,

$$\lambda(t) = \lambda(t1) = (\lambda \leftharpoonup t)(1) = \varepsilon(1) = 1.$$

(2) Since $\lambda \in \int_{H^*}^l$, for any $f \in H^*$, $f\lambda = f(1)\lambda$ and $\lambda f = f(a)\lambda$. Thus for any $h \in H$, using that a is group-like,

$$((a \rightharpoonup \lambda)f)(h) = (a \rightharpoonup (\lambda(a^{-1} \rightharpoonup f))(h) =$$
$$(\lambda(a^{-1} \rightharpoonup f))(ha) = (a^{-1} \rightharpoonup f)(a)\lambda(ha) =$$
$$f(aa^{-1})\lambda(ha) = f(1)\lambda(ha) = f(1)(a \rightharpoonup \lambda)(h).$$

Thus $(a\lambda)f = f(1)(a\lambda)$ and so $a\lambda$ is a right integral in H^*. The argument for λa is similar.

Now if $\Gamma \in \int_H^r$, then $a\lambda(\Gamma) = \lambda(\Gamma a) = \lambda(\Gamma\varepsilon(a)) = \lambda(\Gamma)$, using that $\varepsilon(a) = 1$ since $a \in G(H)$.

Returning to arbitrary Frobenius algebras, note that any Frobenius algebra A has a natural automorphism η determined by the bilinear form, via $(a,b) = (b,\eta(a))$, for all $a,b \in A$; equivalently $f(ab) = f(b\eta(a))$. η is called the *Nakayama automorphism* of A relative to $(\, , \,)$.

A useful technique is to compare different pairs of dual bases, and different forms of the Nakayama automorphism. The next lemma goes back to work of the 1950's and 1960's; for a modern reference, see (Kad, Theorem 1.6).

LEMMA 2.6. *If A is a Frobenius algebra with Frobenius system (f, r_i, l_i), then any other Frobenius system for A is of the form $(d \rightharpoonup f, r_i d^{-1}, l_i)$ for some unit $d \in A$. The Nakayama automorphism for the new Frobenius system for A is given by $\eta'(a) = d\eta(a)d^{-1}$.*

This lemma was used by Kadison to further shorten Schneider's proof of Theorem 2.10 below. We also use it, in a slightly different way.

The next result was proved under an additional hypothesis in (OS). The general case is in (Sch1).

PROPOSITION 2.7. *Let H be any (finite-dimensional) Hopf algebra. Choose $\lambda \in \int_{H^*}^{l}$ and $\Gamma \in \int_{H}^{r}$ such that $\lambda(\Gamma) = 1$. Then $(\lambda, S\Gamma_1, \Gamma_2)$ is a Frobenius system for H. Moreover the Nakayama automorphism η for λ is given by $\eta(h) = \beta \rightharpoonup S^2(h)$, where $\beta \in H^*$ is the right modular function for H.*

Note that such a choice of λ and Γ can be made using the fact, from Theorem 2.4, that $H^* = \lambda \leftharpoonup H$: choose $\Gamma \in H$ such that $\lambda \leftharpoonup \Gamma = \varepsilon$. Then $\Gamma \in \int_H^r$ and $\lambda(\Gamma) = 1$ by Lemma 2.5.

Using the facts above about Frobenius algebras, we see:

COROLLARY 2.8. *(Sch1) Choose $\gamma \in \int_{H^*}^{r}$ and $\Gamma \in \int_{H}^{r}$ such that $\gamma(\Gamma) = 1$.*

(1) A Frobenius system for H is $(\gamma, S^{-1}\Gamma_2, \Gamma_1)$. Moreover the Nakayama automorphism η' for γ is given by $\eta'(h) = S^{-2}(h) \leftharpoonup \beta$.

(2) Another Frobenius system for H is $(\gamma, (S\Gamma_1)a^{-1}, \Gamma_2)$, where $a \in H$ is the left modular element for H^*. The Nakayama automorphism is also given by $\eta'(h) = a(\beta \rightharpoonup S^2(h))a^{-1}$.

Proof. Part (1) is simply Proposition 2.7 restated for H^{cop}.

To see (2), choose λ as in the Proposition such that $\lambda(\Gamma) = 1$. Now by Lemma 2.5, $a\lambda \in \int_{H^*}^r$, and so $a\lambda = c\gamma$, for some $c \in k$. But $a\lambda(\Gamma) = \lambda(\Gamma) = 1 = \gamma(\Gamma)$, and thus $a\lambda = \gamma$. The corollary now follows by Lemma 2.6 with $d = a$.

Corollary 2.8 enables one to compute the Nakayama automorphism explicitly. We consider a non-trivial example.

EXAMPLE 2.9. Let H be the Taft Hopf algebra of dimension n^2. That is,

$$H = k\langle g, x \mid g^n = 1,\ x^n = 0,\ xg = \omega gx,\ g \in G(H),\ \Delta x = x \otimes 1 + g \otimes x\rangle,$$

where ω is a primitive n^{th} root of 1. Then H is not unimodular, since $\Lambda = (1+g+\cdots+g^{n-1})x^{n-1}$ and $\Gamma = x^{n-1}(1+g+\cdots g^{n-1})$. Since H is self-dual, the left and right integrals in H^* are given similarly. The modular functions on H are given by $\alpha(x) = \beta(x) = 0$ and $\alpha(g) = \omega^{-1}$, $\beta(g) = \omega$. Thus by Proposition 2.7, the Nakayama automorphism for λ is given by $\eta(g) = \omega g$ and $\eta(x) = \omega x$; by Corollary 2.8, the Nakayama automorphism for γ is $\eta'(g) = \omega g$ and $\eta'(x) = x$.

We remark that this example is important for several reasons. First, if $n = p$, a prime, then the only known non-semisimple Hopf algebras over $\mathbb{C}$ of dimension p^2 are of this form. Second, in quantum groups, one can construct Lusztig's "restricted" quantum enveloping algebra $u_q(sl_2)'$, when q is a $2n^{th}$ root of 1, by amalgamating two copies of H in a suitable way.

Comparing the two forms of the Nakayama automorphism for γ in Lemma 2.7 enabled Schneider to give a shorter proof of the following crucial result of Radford.

THEOREM 2.10. *(R) Let H be any finite-dimensional Hopf algebra, with right modular elements $\beta \in H^*$ and $b \in H$. Then for all $h \in H$,*

$$S^4(h) = b(\beta^{-1} \rightharpoonup h \leftharpoonup \beta)b^{-1}.$$

Proof. Since b and β are group-like, $S^2(b) = b$ and S^2 commutes with $\beta \rightharpoonup$. By Corollary 2.8, using $b = a^{-1}$,

$$\eta'(h) = S^{-2}(h) \leftharpoonup \beta = b^{-1}(\beta \rightharpoonup S^2(h))b.$$

Applying S^2 and conjugating by b, we see that $b(h \leftharpoonup \beta)b^{-1} = \beta \rightharpoonup S^4(h)$.

Finally apply $\beta^{-1} \rightharpoonup$ to see that $S^4(h) = b(\beta^{-1} \rightharpoonup h \leftharpoonup \beta)b^{-1}$.

Now when H and H^* are semisimple, $b = 1$ and $\beta = \varepsilon$ by Theorem 2.4, part (3). Thus $S^4 = id$. To see that in fact $S^2 = id$ in this situation reduces to showing that -1 cannot be an eigenvalue for S^2 in $End(A)$. To prove this we will need some elementary facts about traces.

Recall that for any finite-dimensional vector space V, $V^* \otimes V \cong End_k(V)$ via $(\phi \otimes v)(w) := \phi(w)v$. Under this isomorphism, if $Tr_V : End(V) \to k$ is the usual trace map, then $Tr_V(\phi \otimes v) = \phi(v)$.

LEMMA 2.11. *Let A be a Frobenius algebra with Frobenius system (f, r_i, l_i). Let $e \in A$ with $e^2 = ce$, for $c \in k$, and let $F \in End_k(eA)$. Then*

$$cTr_{eA}(F) = \sum_i f(F(el_i)r_i).$$

Proof. For any $x \in A$, $ex = \sum_i f(exr_i)l_i$. Thus $e^2x = \sum_i f(exr_i)el_i$, and so

$$cF(ex) = F(cex) = F(e^2x) = \sum_i f(exr_i)F(el_i).$$

Now using the isomorphism $V^* \otimes V \cong End(V)$ above with $V = eA$, we see that cF corresponds to $\sum_i f(-r_i) \otimes F(el_i)$. Thus $cTr_{eA}(F) = \sum_i f(F(el_i)r_i))$.

Now for any H, the left regular representation of H on itself is given by $H \to End(H)$, via $h \mapsto L_h$, where L_h denotes left multiplication by h. Let χ_H denote the character of this representation, that is $\chi_H(h) = Tr_H(L_h)$. We need some elementary properties of χ_H.

LEMMA 2.12. *Let $\chi_H = Tr_H$ be as above. Then*

(1) $\chi_H(\Theta) = \varepsilon(\Theta)$, for all $\Theta \in \int_H^L \cup \int_H^r$;

(2) $S^2_{H^*}(\chi_H) = \chi_H)$;

(3) $\chi_H^2 = dim(H)\chi_H$.

Proof. (1) First choose $\Lambda \in \int_H^l$ and consider L_Λ, left multiplication by Λ on $V = H$. Now $L_\Lambda(h) = \alpha(h)\Lambda$ for all $h \in H$, and thus under the correspondence $V \otimes V^* \cong End(V)$ above, we see that L_Λ corresponds to $\alpha \otimes \Lambda$. Thus $Tr_H(L_\Lambda) = \alpha(\Lambda)$. But clearly $\alpha(\Lambda) = \varepsilon(\Lambda)$.

Similarly if $\Gamma \in \int_H^r$, then L_Γ corresponds to $\varepsilon \otimes \Gamma$ and so $Tr_H(L_\Gamma) = \varepsilon(\Gamma)$, proving (1).

(2) follows from the fact that S_H^2 is an algebra automorphism of H. For, $S_{H^*}^2(\chi_H)(h) = \chi_H(S_H^2(h)) = \chi_H(h)$ for any $h \in H$, since χ_H is the trace of the left regular representation.

To see (3), consider H as a left H-module under left multiplication, and let H_0 denote the vector space H but with the trivial left H-action. Then $H \otimes H_0 \cong H \otimes H$ as left H-modules, via $\phi : h \otimes v \mapsto \sum h_1 \otimes h_2 v$ (note ϕ is bijective since ϕ^{-1} is given by $h \otimes v \mapsto h_1 \otimes Sh_2 v$).

PROPOSITION 2.13. *(L) Let H be a finite-dimensional Hopf algebra, and as before choose $\lambda \in \int_{H^*}^l$ and $\Gamma \in \int_H^r$ with $\lambda(\Gamma) = 1$. Then:*

(1) $Tr_H(S^2) = \varepsilon(\Gamma)\lambda(1)$;

(2) if $S^2 = id$, then $dim(H) = \varepsilon(\Gamma)\lambda(1)$ and $\chi_H = \varepsilon(\Gamma)\lambda$. Consequently if also $dim(H) \neq 0$ in k, then $\lambda = \frac{\lambda(1)}{dim(H)}\chi_H$.

Proof. Apply Lemma 2.11 to H with $e = 1$ along with the fact that $(\lambda, S\Gamma_1, \Gamma_2)$ is a Frobenius system (Proposition 2.7) to see that for $F \in End(H)$, $Tr_H(F) = \sum \lambda(F(\Gamma_2 S\Gamma_1)$.

Now to prove (1), use $F = S^2$ in this expression:

$$Tr(S^2) = \sum \lambda(S^2(\Gamma_2)S(\Gamma_1)) = \sum \lambda(S(\Gamma_1 S(\Gamma_2))$$
$$= \lambda(S(\varepsilon(\Gamma)1) = \varepsilon(\Gamma)\lambda(1).$$

Here we have used the property of S that $id * S = \varepsilon$, applied to Γ.

For (2), assume $S^2 =$ id. Then $dim(H) = \varepsilon(\Gamma)\lambda(1)$ follows directly from (1). For the next equality, first note that if $S^2 = id$, then $S = S^{-1}$ and thus for all $h \in H$, $\sum h_2 S(h_1) = \varepsilon(h)$. Thus, again using $e = 1$ but now with $F = L_h$, we see

$$\chi_H(h) = Tr_H(L_h) = \sum \lambda(h\Gamma_2 S(\Gamma_1)) = \varepsilon(\Gamma)\lambda(h).$$

Thus $\chi_H = \varepsilon(\Gamma)\lambda$. The expression for λ now follows by multiplying by $\lambda(1)$ and dividing by $dim(H)$.

We require one more fact.

PROPOSITION 2.14. *(LR1) Choose* $\gamma \in \int_{H^*}^r$ *and* $\Lambda \in \int_H^l$ *such that* $\gamma(\Lambda) = 1$. *Then*

$$Tr_{H^*}(S_{H^*}^2) = \varepsilon(\Lambda)\gamma(1) = dim(H)Tr_{H^*}(S^2|_{\chi_H H^*}).$$

Proof. The first equality is simply Corollary 2.13(1) applied to H^*. For the second, we apply Lemma 2.11 to H^* with $e = \chi_H$, $c = \dim(H)$, and $F = S^2|_{\chi_H H^*}$, using the fact that $(\Lambda, S\gamma_1, \gamma_2)$ is a Frobenius system for H^*. By Lemma 2.12(3), $e^2 = ce$. Thus, $dim(H)Tr_{H^*}(S^2|_{\chi_H H^*}) = \sum \Lambda(S^2(\chi_H\gamma_2)S\gamma_1) = \sum \Lambda(\chi_H(S^2\gamma_2)S\gamma_1)$ $= \sum \Lambda(\chi_H S(\gamma_1 S\gamma_2)) = \Lambda(\chi_H\gamma(1)) = \gamma(1)\chi_H(\Lambda)$.

Now using Lemma 2.12(1), $\chi_H(\Lambda) = \varepsilon(\Lambda)$. This finishes the proof.

We are now able to prove the first Larson-Radford theorem.

Proof. (*of Theorem 2.1*) First, if $S^2 = id$, then by Proposition 2.13(1), both $\varepsilon(\Gamma) \neq 0$ and $\lambda(1) \neq 0$. Thus H and H^* are semisimple by Theorem 2.4(5).

We may thus assume that both H and H^* are semisimple. Then they are also unimodular, and so by Theorem 2.10, $S^4 = id$, as noted after that theorem. Thus $S_{H^*}^2$ has only 1 and -1 as possible eigenvalues on H^*. Let $\{\mu_j,\ j = 1, \ldots, n = dim(H)\}$ be the eigenvalues of $S_{H^*}^2$ on H^* and let $\{\eta_i,\ i = 1, \ldots, m\}$ be the eigenvalues of $S^2|_{\chi_H H^*}$. By Proposition 2.14, $\sum_{j=1}^n \mu_j = n(\sum_{i=1}^m \eta_i)$.

Now $\sum \mu_j = Tr_{H^*}(S^2) \neq 0$ by Proposition 2.13(1), and thus it is an integer divisible by n. Thus $|\sum_{j=1}^n \mu_j| = n$. But at least one $\mu_j = 1$, since $S^2(\varepsilon) = \varepsilon$. Thus all $\mu_j = 1$ and $S^2 = \text{id}$.

We can now also prove the second Larson-Radford theorem. We use the following elementary fact about matrices over $\mathbb{C}$.

LEMMA 2.15. *Let* k *be algebraically closed of characteristic 0 and let* $A = M_n(k)$, *the algebra of* $n \times n$ *matrices over* k. *Let* $T \in Aut_k(A)$ *be such that* $T^m = id$, *for some* $m > 0$. *Then* $Tr_A(T)$ *is a non-negative real number.*

Proof. (*of Theorem 2.2*) By extending the base field, we may assume k is algebraically closed. Now assume H is semisimple.

Let $T = S^2$; T is an automorphism of H since S is an anti-automorphism. By Theorem 2.10, T has finite order, since $\alpha = \varepsilon$ and b has finite order (group-like elements in any finite-dimensional Hopf algebra have finite order, since distinct group-like elements are linearly independent).

Now $H = \oplus M_{n_i}(k) = \oplus A_i$, so T permutes the $\{A_i\}$. We claim that $Tr_H(T) \neq 0$. Now if $T(A_i) = A_j$, for $j \neq i$, let $r > 0$ be minimal with $T^r(A_i) = A_i$. Letting $B = \oplus_{s=0}^{r-1} T^s(A_i)$, we see that $Tr_H(T|_B) = 0$. If T stabilizes A_i, then $T \in Aut_k(A_i)$ and so by Lemma 2.15, $Tr_H(T|_{A_i}) \geq 0$. Thus $Tr_H(T) \geq 0$.

However, H has a one-dimensional summand, say A_0, generated by $\int_H$. Since T is an automorphism of A_0, $T|_{A_0} = id$ and so $Tr_H(T|_{A_0}) = 1$. Thus $Tr_H(T) > 0$.

We are now done by Proposition 2.13(1): $Tr_H(T) = \varepsilon(\Lambda)\lambda(1)$ for $\Lambda \in \int_H^l$ and $\gamma \in \int_{H^*}^r$. Thus $\gamma(1) \neq 0$ and so H^* is semisimple by Maschke's theorem (Theorem 2.4(5)).

3. Character Theory and the Class Equation

We assume from now on that H is semisimple and that k has characteristic 0. Thus $S^2 = \mathrm{id}$ by Theorems 2.1 and 2.2, and so we may assume that $V^{**} = V$ for any finite-dimensional H-module V. Let $V_0, V_1, \ldots, V_m$ be a complete set of irreducible left H-modules, where $V_0 = kv_0$ denotes the trivial module as in §1. Note that $V_0 \cong k\Lambda$ for any $0 \neq \Lambda \in \int_H$.

Let $n_i = dim_k(V_i)$ and let $\phi_i : H \to End(V_i) \cong M_{n_i}(k)$ be the ith irreducible representation of H, with character χ_i. n_i is called the *degree* of χ_i (and of ϕ_i). Note that we may write $\chi_H = \sum_{i=0}^m n_i \chi_i$.

It is easy to see that V_i^* is also an irreducible left H-module; we sometimes write χ_{i^*} for $\chi_{v_i^*} = S\chi_i$.

Since H is semisimple, any H-module is a sum of the V_i, with multiplicities. In particular

$$V_i \otimes V_j \cong \sum_l m_{ij}(l) V_l \tag{3.0}$$

where the multiplicities $m_{ij}(l)$ are non-negative integers.

DEFINITION 3.1. The *character algebra* $\mathcal{R}(H)$ of H is the k-span in H^* of all the characters on H.

$\mathcal{R}(H)$ is spanned by $\{\chi_0, \ldots, \chi_m\}$, and in fact they form a basis.

We note that for $f \in H^*$, $f \in \mathcal{R}(H)$ if and only if $f(hl) = f(lh)$ for all $h, l \in H$. This is clear if $f \in \mathcal{R}(H)$; the converse follows since H is a sum of matrix rings and the only linear functional on matrices with this property is the trace, up to a scalar multiple.

Since Δ_{H^*} is the transpose of multiplication m_H, this condition may be rewritten to say that $f \in \mathcal{R}(H)$ if and only if $\Delta_{H^*} f = \sum f_1 \otimes f_2 = \sum f_2 \otimes f_1$; that is, f is a "cocommutative element" of H^*.

The analog of orthogonality of characters is true (an old result of Larson). First, we define a bilinear form on H^*, generalizing the one for functions on groups, as in (Se).

DEFINITION 3.2. Choose $\Lambda \in \int_H$ with $\varepsilon(\Lambda) = 1$. For $\phi, \psi \in H^*$, let

$$(\phi|\psi) := (\phi * S\psi)(\Lambda) = \sum_{\Lambda} \phi(\Lambda_1) S\psi(\Lambda_2) = \tilde{\Lambda}(\phi * S\psi),$$

where $\tilde{\Lambda}$ is the element corresponding to Λ in $H^{**} \cong H$ and $*$ denotes the convolution product in H^*.

THEOREM 3.3. *(L) (Orthogonality of Characters) Let H be semisimple. Then $(\chi_i|\chi_j) = \delta_{ij}$.*

Proof. Since $V_0 \cong k\Lambda$, $\chi_0(\Lambda) = \varepsilon(\Lambda)$ and $\chi_i(\Lambda) = 0$ for $i > 0$. The relation (3.0), replacing V_j by V_j^*, gives the corresponding relation on characters $\chi_i * \chi_{j^*} = \sum_l m_{ij^*}(l)\chi_l$. Applying this expression to Λ gives

$$\sum \chi_i(\Lambda_1) S\chi_j(\Lambda_2) = \sum_l m_{ij^*}(l)\chi_l(\Lambda) = m_{ij^*}(0)\varepsilon(\Lambda).$$

Now $m_{ij^*}(0)$ is the multiplicity of the trivial representation in $V_i \otimes V_j^*$. By Lemma 1.1(2)(3), this is the same as the multiplicity of the trivial representation in $Hom_H(V_j, V_i)$. Thus $m_{ij^*}(0) = dim(Hom_H(V_j, V_i))$. By Schur's Lemma, $Hom_H(V_j, V_i) = 0$ if $i \neq j$ and $Hom_H(V_i, V_i) \cong k$. Thus $m_{ij^*}(0) = \delta_{ij}$, proving the theorem.

As an application of the orthogonality theorem, one may prove a number of relations on the multiplicities in 3.0. We give one example, from (NR), in (3) of the next lemma. The present proof follows the classical one for groups, and seems shorter.

COROLLARY 3.4. *(1)* $(\ |\ )$ *is a symmetric* S*-invariant form on* $\mathcal{R}(H)$.

(2) For any i, j, l, $m_{ij}(l) = (\chi_l | \chi_i \chi_j)$.

(3) For any i, j, l, *the following multiplicity relation holds:* $m_{jl}(i) = m_{li^*}(j^*) = m_{il^*}(j)$.

Proof. (1) It suffices to prove for the χ_i, as they form a basis of $\mathcal{R}(H)$. It is clear from Theorem 3.3 that the form is symmetric on the χ_i. The S-invariance also follows from Theorem 3.3, since $S\chi_V = \chi_{V^*}$ and $V_i^* \cong V_j^* \iff V_i \cong V_j$.

(2) This follows immediately from the Theorem and (3.0).

(3) We use (1) and (2). Note that it follows from the definition that $(\phi|\psi) = (\phi S(\psi)|\varepsilon)$. Now

$$\begin{aligned} m_{jl}(i) &= (\chi_i|\chi_j\chi_l) = (\chi_i S(\chi_j\chi_l)|\varepsilon) = (\chi_i S\chi_l S\chi_j|\varepsilon) \\ &= (S(\chi_l S\chi_i)S\chi_j|\varepsilon) = (\chi_l S\chi_i|S\chi_j) = m_{li^*}(j^*). \end{aligned}$$

The second relation is similar.

The proof of the next lemma is close to an old argument for group algebras.

LEMMA 3.5. *(Kac, Z)* $\mathcal{R}(H)$ *is a semisimple algebra with involution* $*$.

Proof. The antipode S is an involution on $\mathcal{R}(H)$, since $S^2 = \mathrm{id}$ and $S\chi_i = \chi_{i^*}$. It suffices to prove semisimplicity for $\mathcal{R}_{\mathbb{Q}}(H)$, the $\mathbb{Q}$-linear span of the χ_i, since $\mathcal{R}(H)$ is obtained by extending the base field.

We first claim that for any $0 \neq \lambda \in \mathcal{R}_{\mathbb{Q}}(H)$, $\lambda\lambda^* \neq 0$. For, write $\lambda = \sum_i a_i \chi_i$, where $a_i \in \mathbb{Q}$. Then

$$\begin{aligned} \lambda\lambda^* &= \sum_{ij} a_i a_j \chi_i \chi_{j^*} = \sum a_i a_j (\delta_{ij}\varepsilon + \sum_{l \geq 1} m_{ij^*}(l)\chi_l) \\ &= (\sum_i a_i^2)\varepsilon + \sum_{l \geq 1} b_l \chi_l \end{aligned}$$

for some $b_l \in \mathbb{Q}$. Thus $\lambda\lambda^* \neq 0$.

Now if $\mathcal{R}_{\mathbb{Q}}(H)$ is not semisimple, its Jacobson radical $J \neq 0$ is nilpotent. Choose $0 \neq \lambda \in J$. Then by the above, $0 \neq \gamma = \lambda\lambda^* \in J$. But $\gamma^* = \gamma$ and so again by the above computation, $\gamma^2 \neq 0$. Repeating this, we see that $\gamma^t \neq 0$ for all t, a contradiction. Thus $J = 0$ and we are done.

The following theorem, known as the "Class Equation" for Hopf algebras, was shown by G. I. Kac for Hopf C^*-algebras and in general by Y. Zhu.

THEOREM 3.6. *(Kac, Z) Let H be a finite-dimensional semisimple Hopf algebra over an algebraically closed field of characteristic 0. Let $\{e_0, e_1, \ldots, e_m\}$ be a complete set of primitive orthogonal idempotents in $\mathcal{R}(H)$, where e_0 is an integral in H^*. Then for each i, $dim(e_iH^*)$ divides $dim(H)$, and*

$$dim(H) = 1 + \sum_{i=1}^{m} dim(e_iH^*)$$

We give Lorenz' recent representation-theoretic proof of the theorem (Lo). The first lemma is well-known; see for example (CR, p. 204), although the present proof seems shorter.

LEMMA 3.7. *Let k be a field of characteristic 0 and let A be a split semisimple k-algebra. That is, $A = \oplus_{j=1}^{d} A_i$, $A_i \cong M_{n_i}(k)$. Assume $\langle\ |\ \rangle$ is a non-degenerate symmetric associative bilinear form on A and let $\{a_j, b_j\}$, $j = 1, \ldots, n$ be dual bases for A with respect to $\langle\ |\ \rangle$. For each i, set*

$$f_i = \sum_{j=1}^{n} tr_i(a_j)b_j,$$

where tr_i is the usual matrix trace in A_i. If $\tilde{e}_i$ is the primitive central idempotent corresponding to A_i, then $f_i = \alpha_i\tilde{e}_i$, for some $0 \neq \alpha_i \in k$.

Proof. Since $\langle\ |\ \rangle$ is symmetric and associative, $f(a) := \langle 1|a\rangle$ satisfies $f(ab) = f(ba)$, for all $a, b \in A$. It follows that for some scalars $\gamma_i \in k$, $< a|b >= \sum_{i=1}^{d} \gamma_i tr_i(ab)$, where the $\gamma_i \neq 0$ since the form is non-degenerate. One set of dual bases is given by $\{\gamma_i^{-1}e^i_{lm}, e^i_{ml}\}$, where for each i, $\{e^i_{lm}\}$ is a set of matrix units for the ith summand $M_{n_i}(k)$. But has been noted earlier, for any two pair of dual bases $\{a_j, b_j\}$ and $\{c_r, d_r\}$ for A with respect to the form $\langle\ |\ \rangle$, it is always true that $\sum_j a_j \otimes b_j = \sum_r c_r \otimes d_r$.

It now follows that

$$f_i = \sum_{j,l,m} \gamma_j^{-1} tr_i(e^j_{lm}) e^j_{ml} = \sum_m \gamma_i^{-1} tr_i(e^i_{mm}) e^i_{mm} = \gamma_i^{-1} \tilde{e}_i.$$

Note that $\alpha_i = \gamma_i^{-1}$.

To apply this lemma, we need a new form on $\mathcal{R}(H)$, since the form in 3.2 is not associative.

DEFINITION 3.8. *Choose* $\Lambda \in \int_H$ *with* $\varepsilon(\Lambda) = 1$. *For* $\phi, \psi \in \mathcal{R}(H)$, *let*

$$\langle \phi | \psi \rangle := (\phi * \psi)(\Lambda) = \tilde{\Lambda}(\phi * \psi),$$

where as in 3.2, $\tilde{\Lambda} \in H^{**}$ *corresponds to* Λ.

The two forms are closely related: $\langle \phi | \psi \rangle = (\phi | S\psi)$. But now, $\langle\ |\ \rangle$ is clearly associative on $\mathcal{R}(H)$; moreover it is symmetric since $(\ |\ )$ is symmetric and S-invariant (alternatively, $\tilde{\Lambda}(\phi\psi) = \tilde{\Lambda}(\psi\phi)$ since $\tilde{\Lambda} \in \mathcal{R}(H)$ by Proposition 2.13(2)). In addition, it follows from Theorem 3.3 that $\{S\chi_j, \chi_j\}$ are dual bases for $\langle\ |\ \rangle$.

We can now finish the proof of the Class Equation.

Proof. (*of Theorem 3.6*) (Lo) Choose $\lambda \in \int_{H^*}$ with $\lambda(1) = 1$. By Proposition 2.13(2), $\lambda = (1/dim(H))\chi_H \in \mathcal{R}(H)$, so in fact we may choose $e_0 = \lambda$.

Consider $A = \mathcal{R}(H)$ and fix $e = e_i$ a primitive idempotent in A. Set $d := \frac{dim(H)}{dim(eH^*)} \in \mathbb{Q}$. It suffices to show $d \in \mathcal{A}$, the ring of algebraic integers in k, for then $d \in \mathbb{Z}$.

Let $\tilde{e} = \tilde{e}_i$ be the central primitive idempotent of A to which e belongs. Then $\tilde{e}\mathcal{R}(H) \cong M_{m_i}(k) \cong (e_i\mathcal{R}(H))^{(m_i)}$ and so $dim(\tilde{e}H^*) = m_i dim(eH^*)$. Thus $d = \frac{m_i(dim(H^*))}{dim(\tilde{e}H^*)}$.

Define $f_i := \sum_j tr_i(S\chi_j)\chi_j$. Then by the previous lemma, $f_i = \alpha_i \tilde{e}_i$.

We first claim that $d = \alpha_i$. Now by Proposition 2.13(2) applied to H^*, with Λ as in 3.8, $Tr_{H^*} = (dim(H))\Lambda$. Thus $dim(\tilde{e}H^*) = Tr_{H^*}(L_{\tilde{e}}) = (dim(H))\tilde{e}(\Lambda) = (dim(H))\alpha_i^{-1} f_i(\Lambda)$. But

$$f_i(\Lambda) = \sum_j tr_i(S\chi_j)\chi_j(\Lambda) = tr_i(\varepsilon)\varepsilon(\Lambda) = m_i.$$

Here we have used again (as in the proof of Theorem 3.3), that $\chi_0 = \varepsilon$ and that $\chi_j(\Lambda) = \delta_{j0}\varepsilon(\Lambda)$. Thus $dim(\tilde{e}H^*) = dim(H)\alpha_i^{-1}m_i$. It follows by the second formula for d that $d = \alpha_i$. Thus $f_i = d\tilde{e}_i$.

Finally we show that f_i is integral over $\mathbb{Z}$. This will finish the proof of the theorem, for then, since $\tilde{e}$ is an idempotent, $f_i^r = d^r\tilde{e}$ for all $r \geq 0$ and so d is integral over $\mathbb{Z}$.

Consider the ring $\mathcal{R}_{\mathbb{Z}}(H) := \sum_i \mathbb{Z}\chi_i \subset \mathcal{R}(H)$. Since the $\{\chi_i\}$ are $\mathbb{Z}$-independent by orthogonality, $\mathcal{R}_{\mathbb{Z}}(H)$ is a finite free $\mathbb{Z}$-module, and so any $x \in \mathcal{R}_{\mathbb{Z}}(H)$ is integral over $\mathbb{Z}$. If $\pi_i : \mathcal{R}(H) \to M_{m_i}(k)$ is the ith irreducible representation of $\mathcal{R}(H)$, then also $\pi_i(x)$ is integral over $\mathbb{Z}$, as are all of its eigenvalues, and so $tr_i(x)$ is integral over $\mathbb{Z}$.

In particular, letting $x = \chi_j$, we see that $tr_i(\chi_j) \in \mathcal{A}$, the ring of algebraic integers, for all j. Thus $f_i \in \sum_j \mathcal{A}\chi_j \subset \mathcal{A} \otimes_{\mathbb{Z}} \mathcal{R}_{\mathbb{Z}}(H)$, which is also integral over $\mathbb{Z}$.

When $H = kG$, the theorem reduces to the usual class equation for finite groups. For, $f \in \mathcal{R}(H) \iff f(ghg^{-1}) = f(h)$, $\forall g \in G \iff f$ is constant on the conjugacy classes of G. Thus $e_i = \sum_{g \in C_i} p_g$, where C_i is the ith conjugacy class. Then $dim(e_iH^*) = |C_i|$ and this divides $|G|$.

When $H = (kG)^*$, Theorem 3.6 specializes to Frobenius' classical theorem that the degree of an irreducible G-module divides $|G|$. For in this case, $H \cong k^{(|G|)}$ and so there are $|G|$ linearly independent characters. Thus $\mathcal{R}(H) = H^* = kG$, and the e_i are just the primitive idempotents of kG and the e_iH^* are the irreducible kG-modules.

The analog of Frobenius' theorem for Hopf algebras was conjectured by Kaplansky in 1975 and remains open. That is,

CONJECTURE 3.9. *(Kap) Let H be a finite-dimensional semisimple Hopf algebra over an algebraically closed field of characteristic 0. Then for any irreducible module V of H, $dim(V) \mid dim(H)$.*

We discuss this conjecture further in the next section.

4. Applications of the Class Equation

We assume throughout that k is algebraically closed of characteristic 0. Our first application of the Class Equation is the one for

Clearly $1_D = \varepsilon_H \bowtie 1_H$. For simplicity we write $h = (\varepsilon \bowtie h)$ and $f = (f \bowtie 1)$. Note that both $H \cong \varepsilon \bowtie H$ and $H^{*cop} \cong H^{*cop} \bowtie 1$ as Hopf algebras.

For example, if $H = kG$, and $g, g', h, h' \in G$, then multiplication is

$$\begin{aligned}(p_g \bowtie h)(p_{g'} \bowtie h') &= \sum p_g(h \rightharpoonup p_{g'} \leftharpoonup h^{-1}) \bowtie hh' \\ &= \sum p_g p_{hg'h^{-1}} \bowtie hh'.\end{aligned}$$

Thus $D(H) \cong (kG)^* \# kG$ as algebras, a skew group ring over $(kG)^*$.

Drinfel'd shows that $D(H)$ is quasi-triangular, with "R-matrix" given as follows: let $\{h_i, f_i\}$ be dual bases of H and H^* under the usual evaluation map. Then

$$R = \sum_i (\varepsilon_H \bowtie h_i) \otimes (f_i \bowtie 1_H) \in D(H) \otimes D(H).$$

We write $R = \sum_i h_i \otimes f_i$ for short. By R^{21} we mean the flip map τ applied to R; that is, $R^{21} = \tau(R) = \sum f_i \otimes h_i$. The reader is referred to (Ksl) or (M) for definitions and properties of quasi-triangular Hopf algebras. We note that an important property of any quasi-triangular Hopf algebra (H, R) is that H is "almost cocommutative"; that is, $\tau \circ \Delta(h) = R\Delta(h)R^{-1}$ for all $h \in H$. Consequently $V \otimes W \cong W \otimes V$ for any $V, W \in {}_H\mathcal{M}$ (see (M, 10.1.2)) and thus $\mathcal{R}(H)$ is a commutative ring.

We require several known results:

THEOREM 4.4. *(R2)(M, 10.3.12) If $0 \neq \Gamma \in \int_H^r$ and $0 \neq \lambda \in \int_{H^*}^l$, then $D(H)$ is unimodular and $\lambda \bowtie \Gamma \in \int_{D(H)}$.*

LEMMA 4.5. *(R2, Dr) Let $D = D(H)$ and consider $F : D^* \to D$ given by $p \mapsto (1 \otimes p)(R^{21}R) = \sum_{ij}(f_j \bowtie h_i)p(f_i \bowtie h_j)$. Then F is a vector space isomorphism which takes $\mathcal{R}(D)$ to $Z(D)$, the center of $D(H)$.*

Moreover $F(pq) = F(p)F(q)$ for all $p \in D^*, q \in \mathcal{R}(D)$.

In fact (R2, Dr) only note that F is multiplicative on $\mathcal{R}(D)$; Schneider shows that one of the arguments can be chosen arbitrarily.

Proof. (*of Theorem 4.3*) (Schneider (Sch2)) Again write D for $D(H)$ for the sake of simplicity. Let $\bar{e}$ be a primitive (central) idempotent of $Z(D)$ and let e be the primitive idempotent of $\mathcal{R}(D)$ such that $F(e) = \bar{e}$. Then by the lemma, $F(D^*e) = F(D^*)\bar{e} = D\bar{e}$. Thus $dim(D^*e) = dim(D\bar{e})$.

Now by the Class Equation 3.6, $dim(D^*e)$ divides $dim(D) = (dim(H))^2$. But $D\bar{e}$ is a full matrix summand of D, corresponding to the irreducible module V of D. Thus $dim(D\bar{e}) = (dim(V))^2$, and so $dim(V) \mid dim(H)$.

COROLLARY 4.6. *(EG) Assume H is any semisimple quasitriangular Hopf algebra, and let V be an irreducible H-module. Then $dim(V)$ divides $dim(H)$.*

Proof. This follows from Theorem 4.3 and the known (and fairly easy) fact that a Hopf algebra H is quasitriangular if and only if the map $D(H) \to H$, given by $f \bowtie h \mapsto h$, is an algebra morphism.

More generally, Schneider proves a generalization of this result to what are called "factorizable" Hopf algebras (Sch2).

The Corollary shows that Kaplansky's conjecture 3.9 holds if H is quasi-triangular. Earlier, two other results about the conjecture were known:

(1) (NR) show that if $dim(V) = 2$, then $2 \mid dim(H)$.

(2) (MW) show that if H is semisolvable, that is, if H has a normal series such that each Hopf quotient in the series is commutative or cocommutative, then Conjecture 3.9 holds.

It would be very useful in classifying Hopf algebras if the conjecture were known. In fact Etingof and Gelaki used 4.6 to prove that any (semisimple) Hopf algebra of dimension pq, where p and q are primes, is a group algebra or its dual (EG3). It had been proved earlier in (GW) that such a result held, provided Conjecture 3.9 was true for these Hopf algebras.

Recently a shorter proof of the pq result has been given by Natale (Na), although still using Theorem 4.3.

5. The Grothendieck ring and $K_0(H)$

The usual Grothendieck group $G(H)$ becomes a ring using tensor products of modules (an observation going back to Serre). When H is semisimple, all modules are projective and so $G(H) = K_0(H)$ and has as a basis the isomorphism classes of the irreducible modules. It follows that $K_0(H) \cong \mathcal{R}_{\mathbb{Z}}(H)$ as rings.

We first discuss here work of Nikshych on the relationship of the the ring structure of K_0 to the isomorphism class of H. To do this we need the notion of (dual) *cocycle twists.* Such twists were introduced by Drinfel'd (Dr2) and by Reshetikhin (Re) in their work on quasi-Hopf algebras.

Consider a Hopf algebra H and an invertible element $\Omega \in H \otimes H$. H^{Ω} is called a *cocycle deformation*, or *twist* of H with respect to Ω if:

(i) $H^{\Omega} = H$ as an algebra

(ii) H^{Ω} has comultiplication $\Delta^{\Omega}(h) := \Omega(\Delta h)\Omega^{-1}$

One can write down necessary and sufficient conditions on Ω for Δ^{Ω} to be coassociative; such an element Ω is called a *pseudo-cocycle.* An important special case of this is when Ω is an ordinary Hopf 2-cocycle on H^*; this means that

$$(\Omega \otimes 1)[(\Delta \otimes id)(\Omega)] = (1 \otimes \Omega)[(id \otimes \Delta)(\Omega)] \tag{5.0}$$

See (5.2) below for the usual (dual) form of this relation.

A nice connection between twisting and $K_0(H)$ has been made by Nikshych.

THEOREM 5.1. *(N) Let H_1 and H_2 be two finite-dimensional semisimple Hopf algebras over $\mathbb{C}$. Then H_1 is a twist of H_2 by a (dual) pseudo-cocycle if and only if $K_0(H_1) \cong K_0(H_2)$ as ordered rings with involution with a marked element.*

Here the marked element can be taken to be χ_H.

However, having isomorphic K_0-rings is not sufficient to guarantee that two Hopf algebras have isomorphic categories of representations. For, consider the non-commutative (semisimple) Hopf algebras of dimension 8. According to (Ma1), there are exactly three such Hopf algebras, namely the two group algebras kD_4 and

kQ of the dihedral group and the quaternion group, and H_8, the Kac-Paljutikin non-commutative, non-cocommutative Hopf algebra of dimension 8, constructed in (KacP) (we consider this example in more detail in the next section). Clearly they are all isomorphic as algebras, since $k^{(4)} \oplus M_2(k)$ is the only possibliity which has a one-dimensional summand. It is not difficult to see that their K_0-rings are isomorphic, and thus by Theorem 5.1 the three Hopf algebras are twists of each other by a pseudo-cocycle. However it is shown in (TY) that these three Hopf algebras have different categories of representations, considered as monoidal categories.

Nevertheless, $K_0(H)$ is a useful invariant of H. Kashina has classified the Hopf algebras of dimension 16 by first classifying all the possible K_0-rings for such algebras (Ksh). The situation is much more complicated than that for dimension 8; in fact there are (coincidentally) exactly 16 non-commutative, non-cocommutative (semisimple) Hopf algebras of dimension 16, with exactly seven isomorphism classes of K_0-rings.

One might hope that if $K_0(H)$ is commutative, then H is a twist of a group algebra. However this is false; the next example was shown to us by G. Mason and R. Ng.

EXAMPLE 5.2. We consider the Drinfel'd double $D = D(\mathbb{C}S_3)$. Since D is quasi-triangular, $\mathcal{R}(D)$ (and $K_0(D)$) is commutative, as noted in Section 4. However D is not a twist of a group algebra. To see this, we recall that for any finite group G, the irreducible representations of $D(\mathbb{C}G)$ arise as follows: choose one element g from each conjugacy class of G, consider the centralizer $C(g)$ of g in G, and let V be an irreducible representation of $C(g)$. Then the induced module $\mathbb{C}G \otimes_{C(g)} V$ becomes an irreducible D-module (Ms), and all irreducible D-modules are of this form. Using this one can check that $D(\mathbb{C}S_3)$ has exactly two 1-dimensional irreducible representations. However, there is no group G of order 36 with this property (such a group would have $[G : G'] = 2$, and consequently the Sylow 3-subgroup P of G' is actually normal in G. But then G/P is abelian of order 4, a contradiction). Thus D cannot be isomorphic to $\mathbb{C}G$ as an algebra, so is certainly not a twist of it .

If we restrict the situation to twists by cocycles, then much more can be said about equivalence. If $A = B^J$, where J is a cocycle, then the comodule categories are monoidally equivalent. The converse holds when A is finite-dimensional. Thus the three non-commutative dimension 8 Hopf algebras are not (dual) cocycle twists of one another. A result for infinite-dimensional H is proved in (EG4).

The dual situation, at least for ordinary cocycles, has also been studied; see for example (Do)(Tk). We recall that $\sigma : H \otimes H \to k$ is a *Hopf 2-cocycle* if it is convolution invertible, $\sigma(h,1) = \sigma(1,h) = \varepsilon(h)$, and for all $h, l, m \in H$,

$$\sum \sigma(l_1, m_1)\sigma(h, l_2 m_2) = \sigma(h_1, l_1)\sigma(h_2 l_2, m). \tag{5.2}$$

We may then form the twisted Hopf algebra H_σ. In this case the coalgebra structure is unchanged but the algebra structure is twisted by σ. That is, H_σ has new multiplication

$$h \cdot l := \sum \sigma(h_1, l_1) h_2 l_2 \sigma^{-1}(h_3, l_3),$$

where here we have used the iterated summation notation $(id \otimes \Delta)\Delta(h) = \Sigma h_1 \otimes h_2 \otimes h_3$.

See also (Sca), (Ma4).

6. The Schur Indicator

Again we assume H is semisimple over an algebraically closed field of characteristic 0. Just as in the representation theory of finite groups, we may define a *Schur indicator* $\nu(\chi)$ for any irreducible character χ of H. As before choose $\Lambda \in \textstyle\int_H$ with $\varepsilon(\Lambda) = 1$. Then we define

$$\nu(\chi) := \sum_{(\Lambda)} \chi(\Lambda_1 \Lambda_2).$$

When $H = kG$, let $\Lambda = \frac{1}{|G|}\sum_{g \in G} g$. Then $\nu(\chi) = \frac{1}{|G|}\sum \chi(g^2)$, which is the usual definition for groups (Se).

THEOREM 6.1. *(LM) Let H be a semisimple Hopf algebra over an algebraically closed field k. If k has characteristic $p \neq 0$, assume in addition that $p \neq 2$ and that H^* is semisimple. Then for Λ and $\nu(\chi)$ as above, and any irreducible character $\chi \in Irr(H)$, the following properties hold:*

(1) $\nu(\chi) = 0$, *1 or -1*, $\forall \chi \in Irr(H)$,

(2) $\nu(\chi) \neq 0$ *if and only if* $V_\chi \cong V_\chi^*$. *Moreover* $\nu(\chi) = 1$ *(respectively* -1*) if and only if* V_χ *admits a symmetric (resp. skew-symmetric) non-degenerate bilinear* H*-invariant form.*

(3)Considering $S \in End(H)$, $TrS = \sum_{\chi \in Irr(H)} \nu(\chi)\chi(1_H)$.

In fact (LM) prove a more general result for arbitrary split semisimple algebras with involution. After stating that result, we will use it to prove Theorem 6.1 in the case of charactersitic 0.

THEOREM 6.2. *Let* A *be a finite-dimensional split semisimple algebra over* k*, and write* $A = \oplus_{i=1}^d M_{n_i}(k)$*. Assume that* k *has characteristic* $\neq 2$ *and that each* $n_i \neq 0$ *in* k*. Assume that* A *has a* k*-involution* S*. Let* $V_1, \ldots, V_d$ *be the distinct irreducible modules for* A *and let* $\chi_1, \ldots \chi_d$ *be the corresponding irreducible characters. Also let* $\{a_r, b_r\}$ *be a pair of dual bases with respect to some symmetric bilinear associative non-degenerate form* $\langle \mid \rangle$ *on* A*. Then the scalars*

$$\mu(\chi_i) := \frac{n_i}{\chi_i(\sum_j a_j b_j)} \chi_i(\sum_r S(a_r)b_r)$$

satisfy the follwing properties:

1. $\mu(\chi_i) = 0$*,1 or -1, for all* $\chi_i \in Irr(A)$.
2. $\mu(\chi_i) \neq 0$ *if and only if* $V_i \cong V_i^*$*. Also* $\mu(\chi_i) = 1$ *(respectively -1) if and only if* V_i *admits a symmetric (resp. skew-symmetric) non-degenerate form such that* $S|_{A_i}$ *is the adjoint of the form, where* A_i *is the ith summand of* A*.*
3. $Tr_A S = \sum_{\chi \in Irr(A)} \mu(\chi)\chi(1_A)$.

The proof of this theorem involves a careful analysis of various trace functions, in the spirit of some of our previous arguments. In particular it is shown in the proof of Theorem 2.7 that if $\langle a|b \rangle = \sum_i \gamma_i tr_i(ab)$, then $\chi_i(\sum_j a_j b_j) = \gamma_i^{-1} n_i^2$;

Proof. (of Theorem 3.1.) If k has characteristic 0, then by Theorem 2.1 and 2.2, $S^2 = id$ and both H and H^* are semisimple. Thus S is an involution on H and we may apply Theorem 2.7 to H.

We next show that $\mu(\chi) = \nu(\chi)$. Choose $\lambda \in \int_{H^*}$ such that $\lambda(1) = \dim H$ and $\Lambda \in \int_H$ with $\varepsilon(\Lambda) = 1$. Then by Proposition 2.13, $\lambda(\Lambda) = 1$ and

$$\lambda = \sum_{\chi_i \in Irr(H)} n_i \chi_i = \chi_H.$$

Define a bilinear form $\langle\ |\ \rangle$ on H via

$$\langle a|b\rangle = \lambda(ab),$$

for all $a, b \in H$. It is clear that $\langle\ |\ \rangle$ is a non-degenerate associative symmetric bilinear form on H, and by Proposition 2.7, $\{S(\Lambda_1), \Lambda_2\}$ is a pair of dual bases with respect to $\langle\ |\ \rangle$.

Now in the bilinear form $\langle\ |\ \rangle$ above, $\gamma_i = n_i$. Thus using the remark before the proof, $\chi_i(\sum_j a_j b_j) = n_i$. Using the dual bases above,

$$\mu(\chi_i) = \chi_i(\sum_j S(a_j)b_j) = \chi_i(\sum_{(\Lambda)} S^2(\Lambda_1)\Lambda_2) = \chi_i(\sum_{(\Lambda)} \Lambda_1\Lambda_2) = \nu(\chi_i),$$

from the definition of $\nu(\chi)$ above.

Thus in order to finish the proof of Theorem 1 we only have to show that if $i = i^*$ then the bilinear form on V_i is H-invariant. However this is trivial:

$$\sum_{(h)}(h_1 \cdot v, h_2 \cdot w) = \sum_{(h)}(v, S(h_1)h_2 \cdot w) = \varepsilon(h)(v, w)$$

since S is the adjoint map with respect to the form.

This invariant can be computed for the representations of various Hopf algebras; in fact Kashina reports that some of the work in (Ksh) could be simplified by using Schur indicators rather than $K_0(H)$ (see (Ksh2)).

As an example, we study in more detail the unique non-commutative, non-cocommutative Hopf algebra H_8 of dimension 8 (KacP), mentioned in the last section.

EXAMPLE 6.3. *The Hopf algebra H_8*

We give generators and relations for $H = H_8$ as given in (Ma1):

$$H = k\langle x, y, z \mid x^2 = y^2 = 1, xy = yx, zx = yz,$$
$$zy = xz, z^2 = \frac{1}{2}(1 + x + y - xy)\rangle$$

The coalgebra structure is given by $x, y \in G(H)$ and

$$\Delta(z) = \frac{1}{2}[(1 + y) \otimes 1 + (1 - y) \otimes x](z \otimes z).$$

Thus $K := k\langle x, y\rangle \cong k(\mathbb{Z}_2 \times \mathbb{Z}_2)$ and K is a normal Hopf subalgebra of H. Then $H/HK^+ = k\langle \overline{z}\rangle$, and we see that $\Delta(\overline{z}) = \overline{z} \otimes \overline{z}$ and $\overline{z}^2 = 1$, so that $k\langle \overline{z}\rangle \cong k\mathbb{Z}_2$. Thus we can write H as the extension

$$K \cong k(\mathbb{Z}_2 \times \mathbb{Z}_2) \hookrightarrow H \xrightarrow{\pi} k\mathbb{Z}_2 = k(\overline{z})$$

where $\pi(x) = \pi(y) = 1$ and $\pi(z) = \overline{z}$. For computations, it is easier if we identify K with K^*; that is, we may write a basis for K as $\{e_{ij} \mid i, j = 0, 1\}$, where $e_{ij}(x^k y^l) = \delta_{ik}\delta_{jl}$. For example, with this notation,

$$\Delta(z) = \Big(\sum_{i,j,p,r=0}^{1} e_{ij} \otimes e_{pr}\Big)(z \otimes z).$$

We now compare the Schur indicators of the representations of H_8 with that of the other non-commutative Hopf algebras of dimension 8, that is the two group algebras kD_4 and kQ. Let χ denote the character of the unique 2-dimensional irreducible representation of each algebra. It is known that for D_4, $\nu(\chi) = +1$ and that for Q, $\nu(\chi) = -1$.

For H_8, an integral is given by $\Lambda = \frac{1}{8}\sum_{ijl=0}^{1} x^i y^j z^l$. Using this, or rather the $\{e_{ij}\}$ above, a direct computation shows that $\nu(\chi) = +1$ (we thank Kashina for the computation). Thus the representation theory of H_8 is closer in some sense to that of the dihedral group D_4 than to that of Q, although their categories of representations are different.

Note that the result of (TY) mentioned in the previous section says that H_8 has no non-trivial cocycle (or dual cocycle) deformations.

The Schur indicator is computed for a number of other examples of Hopf algebras, in (KMM). The examples considered there are all extensions of the dual of a group algebra by another group algebra, as is H_8 above.

References

CR. C. W. Curtis and I. Reiner, *Methods of Representation Theory*, Vol I, Wiley-Interscience, New York, 1981.

Do. Y. Doi, Braided bialgebras and quadratic bialgebras, *Comm. Alg* 12 (1993), 1731-1749.

Dr. V. G. Drinfel'd, On almost cocommutative Hopf algebras, *Leningrad Math J.* 1 (1990), 321-342.

Dr2. V. G. Drinfel'd, Quasi-Hopf algebras, *Leningrad Math J.* 1 (1990), 1415-1457.

EG. P. Etingof and S. Gelaki, Some properties of finite-dimensional semisimple Hopf algebras, *Math. Res. Lett.* 5 (1998), 191-197.

EG2. P. Etingof and S. Gelaki, On finite-dimensional semisimple and cosemisimple Hopf algebras in positive characteristic, *Inter. Math. Res. Not.* 16 (1998), 851-864.

EG3. P. Etingof and S. Gelaki, Semisimple Hopf algebras of dimension pq are trivial, *J. Algebra* 210 (1998), 664-669.

EG4. P. Etingof and S. Gelaki, On cotriangular Hopf algebras, preprint 1999.

GW. S. Gelaki and S. Westreich, On semisimple Hopf algebras of dimension pq, *Proc. Amer. Math. Soc.* 128 (2000), 2829-2831.

Kac. G.I. Kac, Certain arithetic properties of ring groups, *Functional Anal. Appl* 6 (1972), 158-160.

KacP. G. I. Kac and V. G. Paljutkin, Finite ring-groups, *Trudy Moscov. Mat. Ob.* 15 (1966) 224-261 (Russian).

Kad. L. Kadison, *New Examples of Frobenius Extensions*, University Lecture Notes 14, Amer. Math. Soc., Providence, 1999.

Kap. I. Kaplansky, *Bialgebras*, University of Chicago Lecture Notes in Math, 1975.

Ksh. Y. Kashina, Classification of semisimple Hopf algebras of dimension 16, *J. of Algebra* 232 (2000), 617-663.

Ksh2. Y. Kashina, On semisimple Hopf algebras of dimension 2^n, *Algebras and Representation Theory*, to appear.

KMM. Y. Kashina, G. Mason, and S. Montgomery, Computing the Schur indicator for some finite-dimensional Hopf algebras, in preparation.

Ksl. C. Kassell, *Quantum Groups*, GTM 155, Springer-Verlag, New York-Berlin, 1995.

KM. T. Kobayashi and A. Masuoka, A result extended from groups to Hopf algebras, *Tsukuba J. Math* 21 (1997), 55-58.

L. R. G. Larson, Characters of Hopf algebras, *J. of Algebra* 17 (1971), 352-368.

LR1. R. Larson and D. Radford, Semisimple cosemisimple Hopf algebras, *Amer. J. Math* 109 (1987), 287-295.

LR2. R. Larson and D. Radford, Finite-dimensional cosemisimple Hopf algebras in characteristic 0 are semisimple, *J. Algebra* 117 (1988), 267-289.

LS. R. Larson and M. E. Sweedler, An associative orthogonal bilinear form for Hopf algebras, *Amer. J. Math* 91 (1969), 75-93.

LM. V. Linchenko and S. Montgomery, A Frobenius-Schur theorem for Hopf algebras, *Algebras and Representation Theory* 3 (2000), 347-355.

Lo. M. Lorenz, On the class equation for Hopf algebras, *Proc. Amer. Math. Soc.* 126 (1998), 2841-2844.

Ms. G. Mason, The quantum double of a finite group and its role in conformal field theory, Groups '93 Galway/St. Andrews, Vol 2, 405-417, *LMS Lecture Notes* 212, Cambridge University Press, Cambridge, 1995.

Ma1. A. Masuoka, Semisimple Hopf algebras of dimension 6, 8, *Israel J. Math.* 92 (1995), 361-373.

Ma2. A. Masuoka, The p^n theorem for semisimple Hopf algebras, *Proc. Amer. Math. Soc.* 124 (1996), 735-737.

Ma3. A. Masuoka, Self-dual Hopf algebras of dimension p^3 obtained by extension, *J. Algebra* 178 (1995), 791-806.

Ma4. A. Masuoka, Cocycle deformations and Galois objects for some cosemisimple Hopf algebras of finite dimension, *AMS Contemporary Math*, to appear.

M. S. Montgomery, *Hopf Algebras and their Actions on Rings*, CBMS Lectures, Vol. 82, AMS, Providence, RI, 1993.

MW. S. Montgomery and S. Witherspoon, Irreducible Representations of Crossed Products, *J. Pure and Applied Algebra* 129 (1998), 315-326.

Na. S. Natale, On semisimple Hopf algebras of dimension pq^2, *J. Algebra* 221 (1999), 242-278.

NR. W. Nichols and M. B. Richmond, The Grothendieck group of a Hopf algebra, *J. Pure and Appl. Algebra*, 106 (1996), 297-306.

N. D. Nikshych, K_0-rings and twisting of finite dimensional semisimple Hopf algebras, *Comm. Algebra* 26 (1998) 321-342; Erratum: *Comm. Algebra* 26 (1998) 1347.

OS. U. Oberst and H.-J. Schneider, Über Untergruppen endlicher algebraischer Gruppen, *Manuscripta Math.* 8 (1973), 217-241.

R. D. Radford, The order of the antipode of a finite-dimensional Hopf algebra is finite, *Amer. J. Math* 98 (1976), 333-355.

R2. D. Radford, Minimal quasi-triangular Hopf algebras, *J. Algebra* 157 (1993), 285-315.

Re. N. Reshetikhin, Multiparameter quantum groups and twisted quasi-Hopf algebras, *Letters Math.Physics* 20 (1990), 331=335.

Sca. P. Schauenburg, Hopf bigalois extensions, *Comm. Algebra* 24 (1996), 3797-3825.

Sch1. H.-J. Schenider, *Lectures on Hopf algebras*, Trabajos Mat. 31/95, IMAF, Universidad Nac. Cordoba, Argentina, 1995. Available at www.mate.uncor.edu/andrus.

Sch2. H.-J. Schneider, Some properties of factorizable Hopf algebras, *Proc. Amer. Math. Soc.*, to appear.

Se. J. P. Serre, *Linear Representations of Finite Groups*, Springer-Verlag, Berlin and New York, 1977.

TY. D. Tambara and S. Yamagami, Tensor categories with fusion rules of self-duality for abelian groups, *J. Algebra* 209 (1998), 692-707.

Tk. M. Takeuchi, Cocycle deformations of coordinate rings of quantum matrices, *J. Algebra* 189 (1997), 23-33.

V. L. Vainerman, 2-cocycles and twisting of Kac algebras, *Comm. Math Physics* 191 (1998), 697-721.

Z. Y. Zhu, Hopf algebras of prime dimension, *Internat. Math. Res. Not.* (1994), 53-59.

SEMIGROUP COHOMOLOGY AND APPLICATIONS

B. V. NOVIKOV
(boris.v.novikov@univer.kharkov.ua)
Kharkov National University,
Dept. of Math. and Mech.
Svobody sq., 4,
Kharkov 68077 Ukraine

This article is a survey of the author's research. It consists of three sections concerned three kinds of cohomologies of semigroups. Section 1 considers 'classic' cohomology as it was introduced by Eilenberg and MacLane. Here the attention is concentrated mainly on semigroups having cohomological dimension 1. In Section 2 a generalization of the Eilenberg–MacLane cohomology is introduced, the so-called 0-cohomology, which appears in applied topics (projective representations of semigroups, Brauer monoids). At last Section 3 is devoted to further generalizing: partial cohomology defined and discussed in it are used then for calculation of the classic cohomology for some semigroups.

I am indebted to Prof. K. Roggenkamp and Prof. M. Ştefanescu for the support of my participation in the Workshop.

1. EM-cohomology

In this section we deal with the Eilenberg–MacLane cohomology of semigroups [3] (shortly EM-cohomology). Its definition is the same as for groups:

$$H^n(S, A) = \mathrm{Ext}^n_{\mathbf{Z}S}(\mathbf{Z}, A).$$

Here S is a semigroup, A a left S-module (i. e. a left module over the integral semigroup ring $\mathbf{Z}S$), $\mathbf{Z}$ is considered as a trivial S-module (i. e. $xa = a$ for all $x \in S$, $a \in \mathbf{Z}$); $H^n(S, A)$ is called *a nth cohomology group of S* (with the coefficient module A).

K.W. Roggenkamp and M. Ştefănescu (eds.), Algebra - Representation Theory, 219–234.

Another definition (equivalent to preceding one) of $H^n(S,A)$ is following. Denote by $C^n(S,A)$ the group of all maps

$$f : \underbrace{S \times \ldots \times S}_{n \text{ times}} \to A$$

(a group of n-cochains); a coboundary homomorphism $\partial^n : C^n(S,A) \to C^{n+1}(S,A)$ is given by the formula

$$\begin{aligned} \partial^n f(x_1,\ldots,x_{n+1}) &= x_1 f(x_2,\ldots,x_{n+1}) \\ &+ \sum_{i=1}^{n} (-1)^i f(x_1,\ldots,x_i x_{i+1},\ldots,x_{n+1}) \\ &\qquad +(-1)^{n+1} f(x_1,\ldots,x_n) \end{aligned} \tag{1}$$

Then $\partial^n \partial^{n-1} = 0$, i. e.

$$\mathrm{Im}\partial^{n-1} = B^n(S,A) \text{ (a group of coboundaries)}$$

$$\subseteq \mathrm{Ker}\partial^n = Z^n(S,A) \text{ (a group of cocycles)}$$

and cohomology groups are defined as

$$H^n(S,A) = Z^n(S,A)/B^n(S,A).$$

It is worth noting two simple properties of the semigroup cohomology [1] which are useful below.

(1) Adjoin to a given semigroup S an extra element 1 and extend the operation (multiplication) of S to $S^1 = S \cup \{1\}$ by

$$\forall s \in S \quad s \cdot 1 = 1 \cdot s = s, \qquad 1 \cdot 1 = 1.$$

Then S^1 becomes a monoid (i. e. a semigroup with an identity element) and is called *a semigroup with an adjoint identity.* Every S-module turns naturally into a (unitary) S^1-module and we have

$$H^n(S^1,A) \cong H^n(S,A), \qquad n \geq 0.$$

(2) If S possesses a zero element 0 then $H^n(S,A) = 0$ for each S-module A and for all $n \geq 1$. In particular, for any semigroup S we can define *the semigroup S^0 with an adjoint zero* (analogously to S^1); then we have $H^n(S^0,A) = 0$ $(n \geq 1)$.

The semigroup EM-cohomology has not so wide applications as the cohomology of groups. Nevertheless it is interesting for homologists at least as a model for testing homological methods. The problem of describing semigroups having cohomological dimension 1 is such an example. This problem has its own story.

Cohomological dimension (c. d.) of a semigroup S is the maximal integer n such that $H^n(S, A) \neq 0$ for some S-module A. There are many reasons to study algebraic objects of c.d. 1 (see, e. g., [4]).

It is an easy exercise to prove that both a free group and a free semigroup (or a free monoid) have c.d. 1 [3]. For groups the converse is true — this is the well-known Stallings–Swan theorem [2]. So a group has cohomological dimension one if and only if it is free.

Now, what about semigroups?

First, as we have mentioned above, every semigroup with 0 has c.d. ≤ 1. This is the main reason why we have to confine ourselves to considering cancellative semigroups only.

Second, a free group is not free as a semigroup. So even for cancellative semigroups the Stallings–Swan theorem doesn't hold.

B. Mitchell [13] has shown that a so-called partially free monoid (the free product of a free group and a free monoid) has c.d. 1. He has supposed that if c. d.$S = 1$ for S cancellative then S is partially free.

In [16] I have built the first counter-example for Mitchell conjecture:

$$S = \langle a, b, c, d \mid ab = cd \rangle \tag{2}$$

and in [17] I have formulated a 'weakened Mitchell conjecture'. It turned out true:

THEOREM 1. [22] *Every cancellative semigroup of c.d. 1 can be embedded into a free group.*

In the proof of this theorem the passage from the homological language to the semigroup one is realized by the following lemma (which may be helpful not only for semigroups).

Let A be a left module over an arbitrary ring R,

$$\dots \xrightarrow{d_2} P_2 \xrightarrow{d_1} P_1 \xrightarrow{d_0} P_0 \longrightarrow 0$$

its projective resolution. Evidently, d_n may be considered as a $(n+1)$-dimensional cocycle with values in the R-module $\operatorname{Im} d_n$.

LEMMA 2. *The cocycle $d_n \in Z^{n+1}(A, \operatorname{Im} d_n)$ is a coboundary iff the projective dimension of A is not greater than n.*

Applying Lemma 2 to a bar-resolution of the S-module $\mathbf{Z}$ we obtain the next property of a cancellative semigroup S with c.d. 1. Consider a graph with the elements of S as vertices and with the pairs $(a, b) \in S \times S$ such that $aS \cap bS \neq \emptyset$ as edges. Then every circuit of this graph is triangulable.

This property allows to prove that S can be embedded into a group (the latter turns out free by the Stallings–Swan theorem). Note that the converse assertion is certainly not true: there is a subsemigroup of a free semigroup, which has c.d. 1, while c.d. of its anti-isomorphic is equal to 2 (see the examples below). A nice answer is only obtained in the commutative case [19]: the c.d. of a commutative cancellative semigroup is equal to 1 if and only if this semigroup can be embedded into $\mathbf{Z}$.

So a new problem arises: to describe subsemigroups of a free group having c.d. 1. This question seems rather difficult even if we restrict ourselves to subsemigroups of a free semigroup. The following results are taken out of [24].

Let S be a subsemigroup of a free semigroup F. Further development of the proof of Theorem 1 gives us the next assertion:

THEOREM 3. *Let c.d. $S = 1$ and $aS \cap S \neq \emptyset \neq Sa \cap S$ for some $a \in F \setminus S$. There exists such $x \in aS \cap S$ that $aS \cap S \subset xF^1$.*

This theorem allows to build a lot of subsemigroups of F having c.d. > 1.

EXAMPLE 1. Let a, p, q, r are different elements of F such that:

1) $\min(|a|, |p|, |q|, |r|) = |a|$ ($|a|$ denotes the length of the word a),
2) p and q begin with different letters.

Then the subsemigroup $S = \langle p, q, r, ap, aq, ra\rangle$ has c.d. > 1.

From Theorem 3 a solution of the proposed problem for left ideals is obtained:

PROPOSITION 4. *A left ideal of a free semigroup has c.d. 1 iff it is free.*

COROLLARY 5. *Every proper two-sided ideal of a free semigroup has c.d. > 1.*

For principal right ideals the situation is similar:

PROPOSITION 6. *A principal right ideal of a free semigroup has c.d. 1 iff it is free.*

However for arbitrary right ideals the analog of Proposition 4 is not true:

EXAMPLE 2. Let $F = \langle a, b \rangle$ be a free semigroup. Then $R = \{b, aba\}F^1$ is not free but c. d.$R = 1$.

By the way, from these results a counter-example to another conjecture follows. Yu. Drozd supposed that for any $S \in F$ either S or the antiisomorphic to S has c.d. 1. Consider the principal left ideal $L = F^1 aba$ in a free semigroup $F = \langle a, b \rangle$. It is not free since its generators aba, $(ab)^2$, $(ab)^2 a$, $(ab)^3$ obey the relation

$$(ab)^2 \cdot (ab)^2 a = (ab)^3 \cdot aba.$$

By Proposition 4 c.d. $L > 1$. Of course its antiisomorphic $R = abaF^1$ is not free too and c.d. $R > 1$ by Proposition 6. Hence the pair (L, R) gives a counter-example to the conjecture.

2. 0-cohomology

In order to see how 0-cohomology appears let us try to define a projective representation of a semigroup.

Let K be a field, $K^\times$ its multiplicative group, n a positive integer, $M(n, K)$ the semigroup of all $n \times n$ matrices over K. Define an equivalence: for $A, B \in M(n, K)$

$$A \sim B \iff \exists \lambda \in K^\times \; A = \lambda B.$$

Then $\sim$ is a congruence on the semigroup $M(n, K)$ and we can consider a factor semigroup $PM(n, K) = M(n, K)/\sim$, 'the projective linear semigroup'.

Like for groups we call *a projective representation* of a given semigroup S a homomorphism $\Gamma : S \to PM(n, K)$.

Fix an element in each $\sim$-class. Then Γ induces a map $\Gamma' : S \to M(n, K)$. Now we can redefine a projective representation of S: it is a map $\Gamma' : S \to M(n, K)$ such that

1) $\Gamma'(x)\Gamma'(y) = 0 \Longleftrightarrow \Gamma'(xy) = 0$,

2) $\Gamma'(x)\Gamma'(y) = \Gamma'(xy)\rho(x, y)$, where $\rho : S \times S \to K^\times$ is a partial function defined on the subset $\{(x, y) \mid \Gamma'(xy) \neq 0\}$.

Certainly ρ yields the equation

$$\rho(x, y)\rho(xy, z) = \rho(x, yz)\rho(y, z)$$

for $\Gamma'(xyz) \neq 0$ and can be used as the corresponding 2-cocycle (like a factor system in Group Theory) excepting its partiality. Therefore we must anew define suitable cohomology as follows.

Let S be an arbitrary semigroup with a zero. An Abelian group A is called *a 0-module* over S, if an action $(S \setminus \{0\}) \times A \to A$ is defined which satisfies for all $s, t \in S \setminus \{0\}$, $a, b \in A$ the following conditions:

$$s(a + b) = sa + sb,$$

$$st \neq 0 \Longrightarrow s(ta) = (st)a.$$

A n-dimensional 0-cochain is a partial n-place map from S to A which is defined for all n-tuples $(s_1, \ldots, s_n)$, such that $s_1 \cdot \ldots \cdot s_n \neq 0$. A coboundary homomorphism is given like for the usual cohomology by the formula (1). The equality $\partial^n \partial^{n-1} = 0$ is valid too. We denote

$$\mathrm{Im}\partial^{n-1} = B_0^n(S, A) \text{ (a group of 0-coboundaries)}$$

$$\subseteq \mathrm{Ker}\partial^n = Z_0^n(S, A) \text{ (a group of 0-cocycles)}$$

and *0-cohomology groups* are defined as

$$H_0^n(S, A) = Z_0^n(S, A)/B_0^n(S, A).$$

Note that for a semigroup $T^0 = T \cup \{0\}$ with an adjointed zero

$$H_0^n(T^0, A) \cong H^n(T, A),$$

so the 0-cohomology may be considered as a generalization of the Eilenberg–MacLane cohomology.

Properties of 0-cohomology are not considered here since they follow from the properties of partial cohomologies (see Section 3).

Before returning to the projective representations we need a semigroup-theoretic construction, the so-called semilattice of groups [6].

Let Λ be a semilattice (i. e. a partially ordered set in which every two elements λ, μ have the greatest lower bound $\lambda\mu$) and let $\{G_\lambda \mid \lambda \in \Lambda\}$ be a family of disjoint groups. For each pair $\lambda, \mu \in \Lambda$ such that $\lambda \geq \mu$, let $\varphi^\lambda_\mu : G_\lambda \to G_\mu$ be a homomorphism. Suppose that

1) φ^λ_λ is identical for every $\lambda \in \Lambda$,

2) $\varphi^\lambda_\mu \varphi^\mu_\nu = \varphi^\lambda_\nu$ for all $\lambda \geq \mu \geq \nu$.

Define a multiplication on the set $T = \bigcup_{\lambda\in\Lambda} G_\lambda$ by the rule: if $x \in G_\lambda$, $y \in G_\mu$

$$xy = (\varphi^\lambda_{\lambda\mu} x)(\varphi^\mu_{\lambda\mu} y).$$

Then T becomes a semigroup which is called *a semilattice of groups.*

Now return to the projective representations. Recall [8] that if S is a group then one defines an equivalence on the set of the factor systems of S (which corresponds to the equivalence of projective representations); the factor set by this equivalence is a group $\mathrm{Sch}(S, K)$ which is called *a Schur multiplicator* and describes (in some sense) all projective representations of S over K. It is well-known that $\mathrm{Sch}(S, K) \cong H^2(S, K^\times)$, where $K^\times$ is considered as a trivial S-module.

What will be for semigroups? In this situation $\mathrm{Sch}(S, K)$ is not a group (more exactly, it becomes an inverse semigroup). Let Λ be a semilattice of all two-sided ideals of S (including $\emptyset$ and S) with respect to the inclusion and the union as a greatest lower bound. Then restriction of 0-cochains induces homomorphisms

$$\varphi^I_J : H^n_0(S/I, K^\times) \longrightarrow H^n_0(S/J, K^\times)$$

for ideals $I \subseteq J$ and we have a semilattice of groups

$$\bigcup_{I\in\Lambda} H^n_0(S/I, K^\times)$$

(here for $I = \emptyset$ we set $H^n_0(S/\emptyset, K^\times) = H^n(S, K^\times)$). The next assertion was proved in [15]:

THEOREM 7. *For every semigroup S and every field K*

$$\mathrm{Sch}(S, K) \cong \bigcup_{I\in\Lambda} H^2_0(S/I, K^\times)$$

Note that even if $0 \notin S$ we have to use 0-cohomology for describing of $\mathrm{Sch}(S, K)$.

Another application of the 0-cohomology appears in connection with the Brauer monoid. This notion was introduced by Haile, Larson and Sweedler [9], [10] while they studied the so-called strongly primary algebras (a generalization of central simple ones). I shall not give their original definition which is rather complicated. But it turned out that the Brauer monoid can be defined in terms of the 0-cohomology [20]. To do it one must introduce a new notion, a modification of a group.

By *a modification* $G(*)$ of a group G we mean a semigroup on the set $G^0 = G \cup \{0\}$ with an operation $*$ such that $x * y$ is equal either to xy or to 0, while

$$0 * x = x * 0 = 0 * 0 = 0$$

and the identity of G is the same for the semigroup $G(*)$.

In other words, to obtain a modification, one must erase the contents of some inputs in the multiplication table of G and insert there zeros so that the new operation would be associative.

Note some general properties of modifications. First, a modification of G satisfies the weak cancellation condition: from $x*z = y*z \neq 0$ it follows $x = y$ and analogously for left cancellation. Second, let U be a subgroup of all invertible elements in $G(*)$. Then its complement $I = G(*) \setminus U$ is a two-sided ideal. One can show that if G is finite, I is nilpotent.

Let $S = G(*)$ and $T = G(\star)$ be modifications of G. It is clear that $S \cap T = G(\circ)$ is a modification too, where

$$x \circ y \neq 0 \Longleftrightarrow x * y \neq 0 \neq x \star y.$$

We write $S \prec T$ if $x \star y = 0$ implies $x * y = 0$ for all $x, y \in G$. Obviously, all modifications of G constitute a semilattice $M(G)$: the greatest lower bound in it is $S \cap T$.

Each G-module A can be turned into a 0-module over a modification S in a natural way. Moreover, if $S \prec T$ then each 0-module over T is transformed into a 0-module over S. Therefore for $S \prec T$ a homomorphism is defined

$$\varphi_S^T : H_0^n(T, A) \longrightarrow H_0^n(S, A)$$

and we obtain a semilattice of groups $\bigcup_{S\in M(G)} H_0^n(S,A)$.

In particular, let L be a finite-dimensional normal extension of a field K with the Galois group G. Then $L^\times$ is a G-module. We define *a (relative) Brauer monoid* as

$$\mathrm{Br}(G,L) = \bigcup_{S\in M(G)} H_0^2(S,L^\times)$$

(the adjective 'relative' will be omitted since in this article relative Brauer monoids are only considered).

In the case when operation $*$ is defined in such a way that $x*y = xy$ for $x, y \neq 0$, we have

$$H_0^2(S,L^\times) \cong H^2(G,L^\times),$$

so the Brauer group is a subgroup of the Brauer monoid.

One can hope that the Brauer monoid will be useful. For example, it is well-known that the Brauer group is trivial for any finite field whereas the Brauer monoid is not trivial for each non-trivial field extension.

The Brauer monoid classifies strongly primary algebras over a field like the Brauer group classifies division algebras.

The use of 0-cohomology allows us to split the study of the Brauer monoid into two problems:

1) describing all modifications of a given finite group,
2) computing 0-cohomology of a modification.

Both of them seem rather difficult, especially the first. Its solution is unknown even for cyclic groups. In [21] some class of modifications of simple cyclic groups is described. It implies that the number of modifications of the group $\mathbf{Z}_p$ is $O(p^2)$. All multiplication tables of the modifications $S_1, \ldots, S_{15}$ of $\mathbf{Z}_5$ (up to automorphisms of the group) are shown in Table I; for $\mathbf{Z}_7$ their number equals 145.

As to the second problem, the initial step in solving it may consist in eliminating the influence of invertible elements of modifications on the structure of the Brauer monoid. Some results in this direction were obtained in [12], [20].

As above let G be the Galois group of a finite-dimensional extension L/K, $S = G(*)$ its modification, U the subgroup of

TABLE I. The modifications of $\mathbf{Z}_5$

$\mathbf{S}_1$	a^1	a^2	a^3	a^4
a^1	0	0	0	0
a^2	0	a^4	0	a^1
a^3	0	0	a^1	0
a^4	0	a^1	0	0

$\mathbf{S}_2$	a^1	a^2	a^3	a^4
a^1	0	0	0	0
a^2	0	a^4	0	0
a^3	0	0	a^1	0
a^4	0	0	0	0

$\mathbf{S}_3$	a^1	a^2	a^3	a^4
a^1	0	0	0	0
a^2	0	0	0	a^1
a^3	0	0	a^1	a^2
a^4	0	a^1	a^2	a^3

$\mathbf{S}_4$	a^1	a^2	a^3	a^4
a^1	0	0	0	0
a^2	0	0	0	a^1
a^3	0	0	a^1	0
a^4	0	a^1	0	0

$\mathbf{S}_5$	a^1	a^2	a^3	a^4
a^1	0	0	0	0
a^2	0	0	0	a^1
a^3	0	0	a^1	0
a^4	0	0	0	0

$\mathbf{S}_6$	a^1	a^2	a^3	a^4
a^1	0	0	0	0
a^2	0	0	0	0
a^3	0	0	a^1	a^2
a^4	0	0	a^2	0

$\mathbf{S}_7$	a^1	a^2	a^3	a^4
a^1	0	0	0	0
a^2	0	0	0	0
a^3	0	0	a^1	0
a^4	0	a^1	0	0

$\mathbf{S}_8$	a^1	a^2	a^3	a^4
a^1	0	0	0	0
a^2	0	0	0	0
a^3	0	0	a^1	0
a^4	0	0	a^2	0

$\mathbf{S}_9$	a^1	a^2	a^3	a^4
a^1	0	0	0	0
a^2	0	0	0	0
a^3	0	0	0	a^2
a^4	0	0	a^2	0

$\mathbf{S}_{10}$	a^1	a^2	a^3	a^4
a^1	0	0	0	0
a^2	0	0	0	0
a^3	0	0	0	a^2
a^4	0	0	a^2	a^3

$\mathbf{S}_{11}$	a^1	a^2	a^3	a^4
a^1	0	0	0	0
a^2	0	0	0	0
a^3	0	0	0	0
a^4	0	a^1	0	0

$\mathbf{S}_{12}$	a^1	a^2	a^3	a^4
a^1	0	0	0	0
a^2	0	0	0	0
a^3	0	0	0	0
a^4	0	a^1	0	a^3

$\mathbf{S}_{13}$	a^1	a^2	a^3	a^4
a^1	0	0	0	0
a^2	0	0	0	0
a^3	0	0	0	0
a^4	0	0	a^2	0

$\mathbf{S}_{14}$	a^1	a^2	a^3	a^4
a^1	0	0	0	0
a^2	0	0	0	0
a^3	0	0	0	0
a^4	0	0	0	0

$\mathbf{S}_{15}$	a^1	a^2	a^3	a^4
a^1	0	0	0	0
a^2	0	0	0	0
a^3	0	0	0	0
a^4	0	0	0	a^3

invertible elements of S, P the subfield of all U-fixed elements: $P = \{a \in L \mid Ua = a\}$. The inclusion $U \hookrightarrow S$ induces a homomorphism

$$\psi : H_0^2(S, L^\times) \longrightarrow H^2(U, L^\times)$$

We shall study this homomorphism in the situation when U is a normal subgroup of S (i. e. $x * U = U * x$ for all $x \in S$). Then the factor semigroup S/U is well-defined.

Further, if $U \triangleleft S$ then $P^\times$ is a 0-module over S/U. The inclusion $P^\times \hookrightarrow L^\times$ and the epimorphism $S \to S/U$ induce a homomorphism

$$\chi : H_0^2(S/U, P^\times) \longrightarrow H_0^2(S, L^\times)$$

THEOREM 8. *Let $U \triangleleft S$. Then the sequence*

$$0 \longrightarrow H_0^2(S/U, P^\times) \xrightarrow{\chi} H_0^2(S, L^\times) \xrightarrow{\psi} H^2(U, L^\times)$$

is exact.

COROLLARY 9. *If the field L is finite then*

$$H_0^2(S, L^\times) \cong H_0^2(S/U, P^\times)$$

Therefore, for finite fields the problem is reduced to the computation of 0-cohomology of a nilpotent 0-cancellative semigroup $(S/U) \setminus \{1\}$. I believe that such an algorithm can be built.

At last note that W. Clark [5] used 0-cohomology (however with trivial 0-modules only) for investigation of some matrix algebras.

3. Partial cohomologies

0-Cohomology has one more application: for calculating of EM-cohomology. However from this point of view it is worth once more to generalize our construction.

One can ask: what would be if we considered partial maps as cochains, starting from an arbitrary subset $W \subseteq S$, not necessary from $S \setminus \{0\}$? It was shown in [17] that this question is reduced to the following particular case.

Let a semigroup S be generated by a subset W with defining relations of the form $xy = z$ for some $x, y, z \in W$. Such a W

will be called *a root* of S. We denote by W_n a set of all n-tuples $(x_1, \ldots, x_n)$ such that $x_i x_{i+1} \cdot \ldots \cdot x_j \in W$ for all $1 \leq i \leq j \leq n$. Every map from W_n to a S-module A is called *a partial n-dimensional cochain* of W or *a W-cochain* with values in A. n-Dimensional W-cochains form an Abelian group $C^n(S, W, A)$ for $n > 0$. We set $C^0(S, W, A) = A$, and if $W_n = \emptyset$ then $C^n(S, W, A) = 0$. The coboundary homomorphism is given by the same formula (1); the corresponding *partial cohomology groups* (or *W-cohomology groups*) are denoted by $H^n(S, W, A)$.

It is clear that we obtain EM-cohomology if $W = S$. Reducing 0-cohomology to a partial one looks more complicated: if S is a semigroup with 0, $W = S \setminus \{0\}$, then we generate a new semigroup $T = \langle W \rangle$ with the operation $*$ and defining relations of the form $u * v = w$, where $u, v, w \in W$ and $uv = w$ in S. Then W is a root in T and

$$H_0^n(S, A) \cong H^n(T, W, A).$$

Having a presentation of a semigroup S one can easily build some of its roots.

EXAMPLE 3. Let $S = \langle a, b, c, d \mid ab = cd \rangle$ (see (2)). Then

$$W = \{a, b, c, d, x = ab\}, \quad W_2 = \{(a, b), (c, d)\}, \quad W_3 = \emptyset$$

and

$$S = \langle a, b, c, d, x \mid ab = x, cd = x \rangle$$

How are $H^n(S, W, A)$ and $H^n(S, A)$ connected?

The embedding $W \hookrightarrow S$ induces homomorphisms

$$\theta_W^n : H^n(S, A) \longrightarrow H^n(S, W, A),$$

PROPOSITION 10. [18] *If W is a root of a semigroup S then θ_W^n is an isomorphism for $n \leq 1$ and a monomorphism for $n = 2$.*

Generally speaking, θ_W^2 can be non-surjective (by the way it means that partial cohomology ought not be a derived functor in the category of S-modules).

Proposition 10 enables us to use partial cohomology for calculating 1-dimensional EM-cohomology of semigroups (and getting some information about 2-dimensional one) in the case when one succeeds to find a 'good' root in a given semigroup. For instance, consider the semigroup S from Example 3. Define for each $f \in Z^2(S, W, A)$ the 1-dimensional W-cochain h by

$$h(s) = \begin{cases} f(a, b), & \text{if } s = a, \\ f(c, d), & \text{if } s = c, \\ 0, & \text{otherwise.} \end{cases} \tag{3}$$

Then $f = \partial h$, so $H^2(S, W, A) = 0$. Since θ^2_W is injective, $H^2(S, A) = 0$ too.

To study θ^n for $n > 1$ we need some new definitions.

Let S be a semigroup, W be its root. A decomposition $x = x_1 \ldots x_k$ ($x_i \in W$) of an element $x \in S \setminus W$ is called *reduced* if $x_i x_{i+1} \ldots x_j \notin W$ for each i, j, $1 \leq i < j \leq k$. We mean that a reduced decomposition of an element $x \in W$ is its decomposition into the product of one multiplier. A root W is said to be *canonic* if each element $x \in S$ has the unique reduced decomposition.

For example, the set of all element of S is a canonic root.

A root W is called a *J-root* if $xy = x$, $yz = z$ implies $xz \in W$ for all $x, y, z \in W$.

THEOREM 11. [18, 23] *If W is a canonic J-root of S, then θ^n_W are isomorphisms for all $n \geq 0$.*

As above, we can use Theorem 11 for calculating EM-cohomology in higher dimensions. For example, if $S = T * U$ is the free product of semigroups T and U, then $W = T \cup U$ is its canonic J-root and $W_n = T_n \cup U_n$. Thus, we get

$$H^n(S, A) \cong H^n(S, W, A) \cong H^n(T, A) \oplus H^n(U, A)$$

for every S-module A. Below we consider less trivial examples.

EXAMPLE 4. Let $S = \langle a, b_1, b_2, \ldots \mid aP = Q \rangle$ be such a semigroup that the words P and Q do not contain the letter a. Denote $W = F \cup \{a\}$, where $F = \langle b_1, b_2, \ldots \rangle$ is a subsemigroup of S. Then W is a canonic J-root. The fact that F is free facilitates the calculation of W-cohomology of S; so we obtain $H^2(S, A) = 0$ for every X-module A (that is c.d. $S = 1$).

EXAMPLE 5. The semigroup

$$S^{\mathrm{op}} = \langle a, b_1, b_2, \dots \mid Pa = Q\rangle,$$

is antiisomorphic to S (see Example 4). Like for S, the subset $W = F \cup \{a\}$ is a canonic J-root. However c.d. $S^{\mathrm{op}} = 2$. Besides, $H^2(S^{\mathrm{op}}, A) \cong A/B$, where

$$B = PA + \sum_i \left(\frac{\partial P}{\partial b_i} - \frac{\partial Q}{\partial b_i}\right) A;$$

here $\frac{\partial}{\partial b}$ is an analog of the Fox' derivative [7] adapted to semigroups in [18].

Consider one more pair of antiisomorphic semigroups.

EXAMPLE 6. Let U be an arbitrary semigroup,

$$T = \langle U, p \mid Up = p\rangle \quad (p \notin U)$$

This notation means that T is generated by its subsemigroup U and by an element $p \notin U$ and is defined by relations of the form

$$u \cdot v = uv, \quad u \cdot p = p \quad (u, v \in U)$$

The subset $W = U \cup \{p\}$ turns out a canonic J-root. We get c.d. $T = 1$ and $H^1(T, A) \cong A/(p-1)A$ for every T-module A.

EXAMPLE 7. Now consider the semigroup

$$T^{\mathrm{op}} = \langle U, p \mid pU = p\rangle,$$

antiisomorphic to T. Its EM-cohomology is much more complicated:

PROPOSITION 12. *Let A be a T^{op}-module, A_1 be its additive group considered as a trivial T^{op}-module. The homomorphisms $\psi^n : H^n(T^{\mathrm{op}}, A) \to H^n(U, A)$ induced by the embedding $U \hookrightarrow T^{\mathrm{op}}$ are inserted into the long exact sequence*

$$0 \to H^0(T^{\mathrm{op}}, A) \xrightarrow{\psi^0} H^0(U, A) \to H^0(U, A_1) \to H^1(T^{\mathrm{op}}, A) \xrightarrow{\psi^1} \dots$$
$$\dots \to H^n(T^{\mathrm{op}}, A) \xrightarrow{\psi^n} H^n(U, A) \to H^n(U, A_1) \to \dots \quad (4)$$

By the last two examples we can build a semigroup T such that c.d. $T = 1$ and c.d. $T^{\mathrm{op}} = \infty$. To do it take the additive group of the ring $\mathbf{Z}_9$ as A and its multiplicative group as U. The action of U on A coincides with the multiplication in $\mathbf{Z}_9$. Then $H^n(T^{\mathrm{op}}, A) \cong \mathbf{Z}_3$ for $n > 1$.

It is worth to add that the notion of a canonic root can be applied to algorithmic problems. For instance, with its help a new family of semigroups with solvable word problem was obtained [11].

References

[1] N. Bernstein. *On the cohomology of semigroups.* Dissert. Abstrs., **25**(1965), N1, 6644-6645.

[2] K. S. Brown. *Cohomology of Groups.* Springer-Verlag, 1982.

[3] H. Cartan and S. Eilenberg. *Homological Algebra.* Princeton, 1956.

[4] C. C. Cheng and B. Mitchell. *DCC posets of cohomological dimension one.* J. Pure and Appl. Algebra, **13**(1978), N2, 125-137.

[5] W. E. Clark. *Cohomology of semigroups via topology with an application to semigroup algebras.* Commun. Algebra, **4**(1976), 979-997.

[6] A. H. Clifford and G. B. Preston. *The Algebraic Theory of Semigroups.* vol.1, 2, AMS Math. Surveys, 1964, 1967.

[7] R. H. Crowell and R. H. Fox. *Introduction to Knot Theory.* Ginn & Co, 1967.

[8] C. W. Curtis and I. Reiner. *Representation Theory of Finite Groups and Associative Algebras.* Willey & Sons, N.-Y., L., 1967.

[9] D. E. Haile. *The Brauer monoid of a field.* J. Algebra, **81**(1983), N2, 521-539.

[10] D. E. Haile, R. G. Larson and M. E. Sweedler. *A new invariant for* **C** *over* **R***: almost invertible cohomology theory and the classification of idempotent cohomology classes and algebras by partially ordered sets with Galois group action.* Amer. J. Math., **105**(1983), N3, 689-814.

[11] O. S. Kashcheeva and B. V. Novikov. *Canonic subsets in semigroups.* Filomat (Yugoslavia), **12**(1998), N1, 21-27.

[12] V. V. Kirichenko and B. V. Novikov. *On the Brauer monoid for finite fields.* In: Proc. of 5th Internat. Conf. on Finite Fields and Appl. (to appear).

[13] B. Mitchell. *On the dimension of objects and categories.* I. *Monoids.* J. Algebra, **9**(1968), N3, 314–340.

[14] B. V. Novikov. *On 0-cohomology of semigroups.* In: "Theor. and Appl. Quest. of Diff. Equat. and Algebra". Kiev, Naukova dumka, 1978, 185-188 (in Russian).

[15] B. V. Novikov. *On projective representations of semigroups.* Doklady AN USSR, 1979, N6, 474-478 (in Russian).

[16] B. V. Novikov. *An counter-example to a Mitchell conjecture.* Trudy Tbiliss. matem. inst. AN GSSR, **70**(1982), 52-55 (in Russian).

[17] B. V. Novikov. *On partial cohomologies of semigroups.* Semigroup Forum, **28**(1984), N1-3, 355-364.

[18] B. V. Novikov. *Partial cohomologies and their applications.* Izv. vuzov. Matem., 1988, N11, 25-32 (in Russian). Translation in Soviet Math. (Iz. VUZ) v.32 (1988), N11, 38-48.

[19] B. V. Novikov. *Commutative cancellative semigroups of dimension 1.* Matem. Zametki, **48**(1990), N1, 148-149 (in Russian).

[20] B. V. Novikov. *On the Brauer monoid.* Matem. Zametki, **57**(1995), N4, 633-636 (in Russian). Translation in Math. Notes 57 (1995), no. 3-4, 440-442.

[21] B. V. Novikov. *On modification of the Galois group.* Filomat (Yugoslavia), **9**(1995), N3, 867-872.

[22] B. V. Novikov. *Semigroups of cohomological dimension 1.* J. Algebra, **204**(1998), 386-393.

[23] B. V. Novikov. *Partial cohomologies and canonic roots in semigroups.* Matem. studii (Ukraine), **12**(1999), N1, 7-14 (in Russian).

[24] B. V. Novikov. *On cohomological dimension of ideals of free semigroups.* In: Colloq. on Semigroups. July 17-21, 2000. Szeged. Abstracts, 19.

IN SEARCH FOR NOETHERIAN ALGEBRAS

JAN OKNIŃSKI
Institute of Mathematics,
Warsaw University
Banacha 2,
02-097 Warsaw, Poland

1. Introduction

The noetherian property of certain finitely presented algebras has been considered in a number of recent papers. These algebras arise in the contexts of set theoretic solutions of Yang-Baxter equation, of Grobner bases techniques, and of certain subsemigroups of abelian-by-finite groups and their algebras, (3),(4),(5),(6),(7),(14),(20).

Here we consider unital algebras R over a field K defined via semigroup relations, that is relations of the form $u = v$ where u and v are words in the generators of R. Such an algebra coincides with the semigroup algebra $K[S]$ of the semigroup (with 1) defined by the same presentation.

A natural question, which might seem hopeless in full generality, is whether one can characterize the noetherian property of R in terms of the underlying semigroup S. This usually provides us with new tools in the study of the structure of R.

There have been several papers attempting to characterize the noetherian property in some special classes of algebras $R = K[S]$. In particular, if R is commutative, this is equivalent to R being finitely generated (8). A very readable solution was obtained for any subsemigroup S of a polycyclic-by-finite group (17). Namely, the quotient group $G = SS^{-1}$ of S should contain normal subgroups $U \subseteq H$ such that $U \subseteq S$, $S \cap H$ is finitely generated, H has finite index in G and H/U is abelian. Such algebras turn out to be finitely

K.W. Roggenkamp and M. Ştefănescu (eds.), Algebra - Representation Theory, 235–247.

presented. Classes of noetherian algebras with quadratic relations of the special type $x_i x_j = x_k x_l$ were considered in (6),(7),(14),(20). Principal ideal rings of the form $K[S]$ have been characterized in (13). Some other partial results can be also found in (15),(23).

By a theorem of Anan'in, every finitely generated right noetherian algebra R satisfying a polynomial identity embeds into the matrix ring $M_n(F)$ over a field F, (1). Therefore, if $R = K[S]$, then S embeds into the multiplicative semigroup of $M_n(F)$ and its structure can be approached through the theory of linear semigroups, (24). In particular, S has a finite ideal chain of a very special nature. One of our main aims is to show that this is often the case if $K[S]$ is right noetherian. In order to discuss this, we need some preparation.

2. Auxiliary notions

Let G be a group, X, Y nonempty sets and $P = (p_{yx})$ a $Y \times X$ matrix with entries in $G^0 = G \cup \{0\}$. Consider all triples (g, x, y), where $g \in G^0, x \in X, y \in Y$, with all triples $(0, x, y)$ identified with 0. On this set we define an operation by the rule: $(g, x, y)(g', x', y') = (gp_{yx'}g', x, y')$. By $\mathcal{M}(G, X, Y, P)$ we denote the semigroup obtained in this way. So, elements of $\mathcal{M}(G, X, Y, P)$ can be viewed as $X \times Y$ matrices over G with at most one nonzero entry and with multiplication $A * B = APB$. P is called the sandwich matrix of $\mathcal{M}(G, X, Y, P)$. Rees's theorem asserts that every semigroup with zero and no nonzero proper ideals (every semigroup with no proper ideals) is of this form (of the form $\mathcal{M}(G, X, Y, P) \setminus \{0\}$) for a sandwich matrix over G^0 (over G, respectively) with no zero rows or columns. By a row (a column) of $M = \mathcal{M}(G, X, Y, P)$ we mean all nonzero (g, x, y) for a given fixed $x \in X$ ($y \in Y$, respectively). We denote these sets by R_x and L_y. The sets $M_{xy} = R_x \cap L_y$ are called $\mathcal{H}$-classes. Each of them has either zero multiplication or it is a group isomorphic to G, called a maximal subgroup of $\mathcal{M}(G, X, Y, P)$.

If $S \subseteq \mathcal{M}(G, X, Y, P)$ is a subsemigroup that intersects all $\mathcal{H}$-classes then it is called a uniform semigroup. It then follows that there exists a unique subgroup H of G and a sandwich matrix P' over H^0 such that $S \subseteq \mathcal{M}(H, X, Y, P')$ and if F is a maximal subgroup of $\mathcal{M}(H, X, Y, P')$ then $S \cap F$ generates F as a group. If each $S \cap F$ has a right and left group of quotients and S is uniform as a subsemigroup in two such semigroups $\mathcal{M}(H, X, Y, P')$ then they are isomorphic. So $\mathcal{M}(H, X, Y, P')$ can be viewed as a quotient semigroup of S and it is denoted by $\widehat{S}$. By cancellative parts of S we mean the intersections $S \cap F$ with maximal subgroups F of $\mathcal{M}(H, X, Y, P')$. If P has exactly one nonzero entry in each row and column and X is finite then $\mathcal{M}(G, X, Y, P)$ has also a presentation of the form $\mathcal{M}(G, X, X, I)$ where I is the identity matrix. So $K_0[S]$ has a structure of a generalized matrix ring. This case will be crucial for the results of our paper.

For an ideal I of S we denote by S/I the corresponding Rees factor and we identify the contracted semigroup algebra $K_0[S/I]$ with $K[S]/K[I]$.

We refer to (24) for the basic facts concerning these notions, while (12) is a standard reference for semigroup theory.

3. Noetherian algebras

It is known that every linear semigroup S has a finite ideal chain with factors that are nilpotent or uniform, (24). It turns out that a similar chain exists for a large family of semigroups S with $K[S]$ noetherian. If A is a subset of S then $\langle A \rangle$ denotes the subsemigroup of S generated by A. By $\mathrm{GK}(S)$ we mean the Gelfand-Kirillov dimension of S, or equivalently of $K[S]$, see (21).

We need the following extension of Lemma 1.1 in (18).

LEMMA 3.1. *Let S be a subsemigroup of a finitely generated abelian-by-finite group. Then S has a quotient group G and $G = SS^{-1} = SZ(S)^{-1}$, where $Z(S)$ is the center of S.*

Proof. The quotient group G exists by (23), Lemma 7.1. G has the largest torsion normal subgroup H since this is the same as the largest locally finite normal subgroup of G. Then $G_1 = G/H$ has no nontrivial finite normal subgroups. Define $S_1 \subseteq G_1$ as the image of S in G_1. Now $K[G_1]$ is a prime PI ring, whence by an argument as in (18) we get that $G_1 = S_1 Z(S_1)^{-1}$. Let Z be the inverse image of $Z(S_1)$ in S. Take any $s \in Z$. Then $sxH = xHs$ for every $x \in S$. So s acts by conjugation on the coset xH, and it follows that s^n acts as the identity map, where $n = |H|!$. This means that $s^n \in Z(S)$, since $x \in S$ is arbitrary. So Z is periodic modulo $Z(S)$. Therefore $SZ(S)^{-1} = SZ^{-1}$. But the image of SZ^{-1} in G_1 is equal to $S_1Z(S_1)^{-1} = G_1$, whence $G = SZ(S)^{-1}H$. Now, for every $a \in SZ(S)^{-1}$ there exists $b \in SZ(S)^{-1}$ such that $ab \in H$. Then $(ab)^n = 1$, so $SZ(S)^{-1}$ is a group. It follows that $G = SZ(S)^{-1}$. □

THEOREM 3.2. *Assume that S is a monoid of finite Gelfand-Kirillov dimension and such that for every $a, b \in S$ we have*

$$a\langle a,b\rangle \cap b\langle a,b\rangle \neq \emptyset \neq \langle a,b\rangle a \cap \langle a,b\rangle b.$$

If the semigroup algebra $K[S]$ is right and left noetherian then S has an ideal chain $S_1 \subseteq S_2 \subseteq \cdots \subseteq S_n = S$ such that if Z denotes any of $S_1, S_j/S_{j-1}$ then Z is either nilpotent or uniform. In the latter case, Z has the same finite number of rows and columns and the sandwich matrix can be chosen diagonal.

Proof. We will show that every homomorphic image of S either has a nonzero nilpotent ideal or a uniform ideal. Since S has the ascending chain condition on (one-sided) ideals, it will follow that a finite chain of the desired type exists.

So, suppose the assertion does not hold. Replacing S by a homomorphic image we may assume that S has no nonzero nilpotent or uniform ideals. We assume also that S has a zero element. Let $\overline{S}$ be the image of S in the algebra $K_0[S]/B(K_0[S])$ where $B(K_0[S])$ denotes the prime radical of $K_0[S]$. The latter algebra is right noetherian and semiprime, so it has a classical ring of right quotients Q, which is semisimple artinian. In particular, $\overline{S}$ embeds into Q, whence it is a skew linear semigroup (a multiplicative subsemigroup of a simple artinian algebra Q'). Moreover $\overline{S}$ has no nonzero nilpotent ideals. From the structure theorem of such semigroups (24) it follows that $\overline{S}$ has a uniform ideal $\overline{I}$. (In fact, the ideal of elements of the least nonzero rank, as matrices in Q', is a finite union of uniform ideals.) Because of the ascending chain condition on one-sided ideals in S, $\overline{I}$ has finitely many rows and columns. Let $I \subseteq S$ be the inverse image of $\overline{I}$ under the natural homomorphism $\phi : S \longrightarrow \overline{S}$. Then $\phi^{-1}(0)$ is an ideal of S which consists of periodic elements. Suppose $\phi^{-1}(0) \neq 0$. Since S has no nonzero nil ideals and has finitely many completely 0-simple principal factors by (23),

Corollary 12.2, it follows that S has a completely 0-simple ideal M. Then M is a uniform ideal of S. Hence we may assume that $\phi^{-1}(0) = 0$.

Clearly, I has a "matrix pattern" induced from that in $\overline{I}$. Namely, if $\overline{I} = \bigcup_{i,j} \overline{I}_{ij}$ (where i, j run through the appropriate finite sets) is the decomposition into $\mathcal{H}$-classes (with zero adjoined), then $I = \bigcup_{i,j} I_{ij}$ where $I_{ij} = \phi^{-1}(\overline{I_{ij}})$ is a 0-disjoint union. Clearly $I_{ij}I_{kl} \subseteq I_{il}$, as the same is true of the image $\overline{I}$ of I. If I_{ip}, I_{iq} have nonzero multiplication for some i and some p, q then the assumption on the 2-generated subsemigroups of S easily implies that $p = q$. Similarly one shows that every column of I has at most one component with non-zero multiplication. Therefore, the number of rows of I is equal to the number of columns and permuting the rows and columns of I we may assume that only the diagonal components have nonzero multiplication.

Let $\overline{T}$ be a cancellative part of $\overline{I}$. Denote by T the inverse image of $\overline{T}$ in S. Notice that $T \cup \{0\} = I_{ij}$ for some i, j and T is a subsemigroup of I. Let $R = \{a \in S : Ta \subseteq T\}$. It is known that there exists a subsemigroup $\widehat{I} = \mathcal{M}(H, X, Y, P)$ of Q' whose all $\mathcal{H}$-classes are intersected by $\overline{I}$ and $\widehat{S} = \widehat{I} \cup (\overline{S} \setminus \overline{I})$ (a disjoint union) forms a subsemigroup of Q', see (24), Lemma 3.11. Assume that $ax \in R$ and $a \in R$ for some $x \in S$. Then $bx \in T$ for some $b \in T$. Let $\overline{c}$ be the inverse of $\overline{b} \in \overline{T}$ in the maximal subgroup G of $\widehat{I}$ containing $\overline{T}$. Then $\overline{T}\overline{x} = \overline{T}\overline{c}\overline{b}\overline{x} \subseteq G \cap \overline{S} = \overline{T}$. It follows that $x \in R$. Therefore $K_0[R]$ is a direct summand of the left $K_0[R]$-module $K_0[S]$. Since $K_0[S]$ is right noetherian, also $K_0[R]$ is right noetherian. It is easy to see that $T \cup \{0\}$ is an ideal of R.

Let ρ be the relation defined on T by $a\rho b$ if $xa = xb$ for some $x \in T$. Since $Ty \cap Tz \neq \emptyset$ for every $y, z \in T$ by the assumption on S, one verifies easily that ρ is a congruence on T, so it is the least left cancellative congruence on T. The linear span $I(\rho)$ of all $a - b$, where $a, b \in T$ and $a\rho b$, is an ideal in $K_0[R]$.

Now $I(\rho) = \sum_{i=1}^{r}(a_i - b_i)K_0[R]$ for some $a_i, b_i \in T$. Choose $c_i \in T$ such that $c_i(a_i - b_i) = 0$. Then $c_1T \cap \cdots \cap c_rT \neq \emptyset$ because every two right ideals of T intersect nontrivially. So the left annihilator C of $I(\rho)$ in T is nonempty, hence an ideal of T. We have $CI(\rho) = 0$. Similarly one finds an ideal C' of T such that $I(\rho')C' = 0$ where ρ' is the least right cancellative congruence on

T. We put $E = C \cap C'$.

Let $a, b \in E$ be such that $az = bz$ for some $z \in E$ but $a \neq b$. So $(a-b)E = 0$. Clearly $a - b \in B(K[S])$ because $\overline{T}$ is cancellative. Also $(a-b)\langle a, b\rangle = 0$ and we know that there exist $x, y \in \langle a, b\rangle$ such that $xa = yb$. It follows that $a^n = b^n$ for some n and $a^n(a-b) = 0$. So $a\rho b$. By a symmetric argument we show that $\rho = \rho'$.

Hence $I(\rho) = I(\rho')$ and we also get $EI(\rho) = 0 = I(\rho)E$. We claim that E contains a cancellative ideal of T.

The cancellative semigroup $T' = T/\rho$ has a group of quotients G' which is of finite Gelfand-Kirillov dimension because $\mathrm{GK}(S) < \infty$ (9). It is easy to see that subgroups of G' lift to right congruences on R. So by the ascending chain condition in $K_0[R]$ it follows that G' is finitely generated. Hence it is nilpotent-by-finite (10). The natural map $T \longrightarrow \overline{T}$ factors through $f : T' \longrightarrow \overline{T}$ and the latter extends to a group homomorphism $G' \longrightarrow G = \overline{T}\overline{T}^{-1}$. If $\mathrm{char}(K) = 0$ then this is an isomorphism, otherwise its kernel F is a finite subgroup of G', see (23), Theorem 7.19. As $e\overline{R} \subseteq G$ where $e = e^2 \in G$, we may define $R_1 = f^{-1}(e\overline{R})$. Notice that $er = ere$ for $r \in \overline{R}$. Since $K_0[R]$ is right noetherian, so is its homomorphic image $K_0[e\overline{R}]$. As F is finite, from (15), Lemma 1.2, it follows that $K_0[R_1]$ is right noetherian. Therefore G' contains normal subgroups U and A such that A has finite index in G', A/U is abelian and $U \subseteq R_1$, (15).

Let E' be the image of E in T'. If T' is a group, then every $z \in E'$ satisfies $zT' = T'z$. On the other hand, if T' is not a group, then let $z' \in E'$ be chosen so that $Uz's' = Us'z'$ for every $s' \in T'$ (see Lemma 3.1). Then $T'z's' = T'Uz's' = T'Us'z' = T's'z'$. So $T'z'T' = (T')^2z'$ and by symmetry $(T')^2z' = z'(T')^2$. So we always have an element $z' \in E'$ of the latter type. Now $(T')^2z' = z'(T')^2$ means that $ET^2z = EzT^2$ and $T^2zE = zT^2E$, where $z \in E$ is an inverse image of z'.

Consider any $a, b \in EzT^2$ with $a\rho b$. Then we can write $a = e_1z, b = e_2z$ for some $e_1, e_2 \in E$. Hence $e_1\rho e_2$ because ρ is a cancellative congruence. Therefore $(e_1 - e_2)z = 0$ (since $z \in E$) which proves that EzT^2 is right cancellative. A symmetric argument allows us to show that $D = EzT^2 \cap T^2zE$ is a cancellative ideal of T.

Notice also that $ID^2I \cap T \subseteq D$. Indeed, if $ad_1d_2b \in T$ for some $a, b \in I$ and $d_i \in D$ then it easily follows from the matrix structure on I that $ad_1 \in TD$ and $d_2b \in DT$.

We can repeat the above argument for every inverse image T_k in I of a cancellative part of $\overline{I}$. In this way, we find a cancellative ideal D_k in T_k. So, consider the ideal $J = \bigcap_k ID_k^2I$ where the intersection runs over all such D_k. Then J has a matrix structure $J = \bigcup_{i,j=1}^n J_{ij}$ with all components with non-zero multiplication being 0-cancellative.

Next we deal in a similar way with the non-diagonal components of J. Fix some J_{ij}. Let $B = \sum(a-b)K_0[S]$ where summation runs over all $a, b \in J_{ij}$ such that $ta = tb \neq 0$ for some $t \in J$ (notice that $ca = cb \neq 0$ for some $c \in J_{ii}$ in this case). This is a finitely generated right ideal of $K_0[S]$. So $L = \{x \in J_{ii} | xB = 0\}$ is nonempty, whence it is a left ideal of J_{ii}. Using the assumption on 2-generated subsemigroups of S one verifies that L is an ideal of J_{ii}. As before, using Lemma 3.1 we can find an element $0 \neq z_i \in L$ and an ideal E_i of J_{ii} such that $z_iI_{ii}^2E_i = I_{ii}^2z_iE_i$. We denote the latter by F_i. Then $(JF_i^2J)_{ij} \subseteq J_{ii}F_iF_iJ_{ij} \subseteq F_iJ_{ij} = zI_{ii}E_iJ_{ij} \subseteq z_iJ_{ij}$. So, if $a - b \in B$, then $a = z_ix, b = z_iy$ for some $x, y \in J_{ij}$. Clearly $x - y \in B$ and hence $a = b$ because $z_i \in L$. Hence, again replacing J by a smaller ideal of S we may assume that $ta = tb \neq 0$ implies $a = b$, for $a, b, t \in J$. A symmetric argument allows us to assume that $at = bt$ implies $a = b$ for $a, b, t \in J$.

Let $P = \{a_1 + \cdots + a_n | 0 \neq a_i \in J_{ii}\}$. It is now clear that P is a multiplicatively closed set of regular elements of $K_0[J]$. Fix some nonzero elements $x \in J_{ji}, y \in J_{jj}, z \in J_{ij}$. If $0 \neq a \in J_{ij}$ and $0 \neq c \in J_{ii}$ then $ayx, czx \in J_{ii}$ and hence $ayxJ_{ii} \cap czxJ_{ii} \neq 0$ and it is nonempty. Hence so is $ayxJ_{ii}z \cap czxJ_{ii}z$ and $aJ_{jj} \cap cJ_{ij} \neq 0$ is also nonempty. This easily implies that P is a right Ore subset of $K_0[J]$. Localizing with respect to P we get a subsemigroup J_P of $(K_0[J])_P$. It inherits the matrix pattern and the $(1,1)$ component is the quotient group of $J_{11} \setminus \{0\}$ with zero adjoined. As every nonzero ideal of J_P intersects this group nontrivially, it follows that J_P is 0-simple, and consequently completely 0-simple. Therefore J is a uniform semigroup. This contradiction completes the proof of the theorem. □

Notice that every semigroup S satisfying an identity of the form $xwy = yvx$, where w, v are some words in x, y, fulfills the hypothesis on the 2-generated subsemigroups of S in the theorem above. For example Malcev nilpotent semigroups are of this type, see (24).

Surprisingly, Theorem 3.2 admits a quite general partial converse.

THEOREM 3.3. *Assume that S is a finitely generated monoid with an ideal chain $S_1 \subseteq S_2 \subseteq \cdots \subseteq S_n = S$ such that S_1 and every factor S_j/S_{j-1} is either nilpotent or uniform. If S has the ascending chain condition on right ideals, the sandwich matrix of every uniform factor is square and diagonal and $GK(S)$ is finite, then $K[S]$ is right noetherian.*

Proof. We may assume S has a zero element θ, adjoining it to S if necessary. Write $S_0 = \{\theta\}$. By induction on j we will show that $K[S/S_{n-j}]$ is right noetherian. With $j = n$ this will yield the assertion. Since S is a monoid, S/S_{n-1} cannot be nilpotent. So, it is a monoid which is uniform, whence it must be a cancellative monoid with zero adjoined. As it has the ascending chain condition on right ideals and has finite Gelfand-Kirillov dimension, it follows that $K_0[S/S_{n-1}]$ is right noetherian, see (15).

Suppose now that we have already shown that $K_0[S/S_i]$ is right noetherian, for some i. Consider $T = S/S_{i-1}$. First consider the case where $I = S_i/S_{i-1}$ is nilpotent. Assume we have already shown that $K_0[T/I^j]$ is noetherian for some j. Since I_j is a finitely generated right ideal of T and the $K_0[T]$- and $K_0[T/I^j]$-right module structures on $K_0[I^j/I^{j+1}]$ coincide, it follows that $K_0[I^j/I^{j+1}]$ is a noetherian right $K_0[T]$-module. Therefore $K_0[T/I^{j+1}]$ is a right noetherian algebra. It follows that $K_0[T]$ is right noetherian.

Next, assume that $I = S_i/S_{i-1}$ is uniform. By the assumptions on S we know that $I \subseteq \widehat{I} \simeq \mathcal{M}(G, X, X, P)$ for a group G, a finite set X and the identity $X \times X$ matrix P. Moreover every maximal subgroup of $\widehat{I}$ is generated by its intersection with I. Then the disjoint union $\widehat{T} = (T \setminus I) \cup \widehat{I}$ has a natural semigroup structure, extending that of T, which is defined as follows. For $a \in \widehat{I}$ and $s \in \widehat{T}$ choose an idempotent $e \in \widehat{I}$ such that $a = ae$ and elements $x, y \in I \cap e\widehat{I}e$ such that $xy = e$. This is possible because $I \cap (e\widehat{I}e \setminus \{0\})$ is easily seen to be a right Ore subsemigroup in the group $e\widehat{I}e \setminus \{0\}$. Then we put $as = (ax)(ys)$. The product sa is defined by a symmetric argument. Since G is isomorphic to a group generated by a subsemigroup of S and $\mathrm{GK}(S) < \infty$, we also have $\mathrm{GK}(G) < \infty$ by (9). One shows that G is finitely generated because S is finitely generated, see (24), Proposition 3.16. Then G is nilpotent-by-finite (10).

Now $K_0[I] \subseteq K_0[\widehat{I}]$ and the latter is naturally isomorphic to $M_n(K[G])$. Let $\phi : K_0[T] \longrightarrow M_n(K[G])$ be given by $x \mapsto xE$ where E is the identity of $K_0[\widehat{I}]$. Since E is a central idempotent in $K_0[\widehat{T}]$, as the identity of an ideal, this is a homomorphism. We view $K_0[\widehat{I}]$ as the Munn algebra of $\widehat{I}$, (23). It is easy to see that every matrix xE for $x \in T$ has at most one nonzero entry in each row and in each column (because $a(xE), (xE)a \in I$ for every $a \in I$). From (16) it follows that $K_0[\phi(T)]$ is right noetherian. (The ascending chain condition in T, the fact that G is finitely generated and nilpotent-by-finite and the result of Chin and Quinn on rings graded by polycyclic-by-finite groups, (25), Theorem 1.3.7, are used here.) Therefore the image $\phi(K_0[T])$ is right noetherian. Since I is a finitely generated right ideal of $\phi(T)$, it follows that $K_0[I]$ is a noetherian right $K_0[\phi(T)]$-module. But $K_0[T/I]$ is right noetherian by the induction hypothesis, so that $K_0[T]$ also is right noetherian. This completes the proof of the theorem. □

For example, if S is a finitely generated Malcev nilpotent monoid then $K[S]$ is right noetherian if S has the ascending chain condition on right ideals and has an ideal chain of type described in Theorem 3.3. (Since Malcev nilpotent groups are nilpotent, it follows that the proof remains valid; also the uniform factors have to be of the desired form, see (24).)

The same proof would also work for (left and right) noetherian property of $K[S]$ if we knew that all involved groups G are polycyclic-by-finite (17). While there are no known noetherian group algebras of other types, there is no natural general condition that would allow us to soften the requirement $\mathrm{GK}(S) < \infty$.

It is clear that, if $K_0[S/I]$ is right noetherian and I is a finite ideal of S, then $K[S]$ also is right noetherian. So, the conditions in the theorem above are far from necessary. On the other hand, let $S = \{1\} \cup I$ where $I = \mathcal{M}(G, X, Y, P)$ with $X = \{1\}, Y = \{1, 2\}$, $p_{11} = p_{21} = 1$ and G the infinite cyclic group. Then $K[S]$ is not right noetherian as the left annihilator of $K_0[I]$ is infinite dimensional. One can represent S as a subsemigroup of $M_2(\mathbf{Q})$ with I consisting of all matrices of the forms $\begin{pmatrix} 2^n & 0 \\ 0 & 0 \end{pmatrix}$ and $\begin{pmatrix} 2^n & 2^n \\ 0 & 0 \end{pmatrix}$, where n runs over the set of integers. In particular, this shows that one of the assertions of (26) is not correct. In the context of the noetherian property of $K[S]$ the ascending chain condition on right congruences has also been studied, (11),(19). This might be important for a possible extension of Theorem 3.3.

4. Some examples

There are many examples of algebras satisfying conditions of Theorem 3.3. One of the interesting classes consists of certain Rees factors of cancellative semigroups. In all these cases the corresponding sandwich matrices are diagonal, whence after a normalization, they can be assumed identity. It is easy to see that the ascending chain condition on right ideals of a cancellative semigroup S implies that a group of right quotients of S exists. The following result comes from (17).

PROPOSITION 4.1. *Let S be a cancellative monoid satisfying the ascending chain condition on right ideals such that the group of right quotients $H = SS^{-1}$ is polycyclic-by-finite. Let P be a prime ideal in $K[S]$. If $P \cap S \neq \emptyset$, then*

1. *there exists an ideal I of S and a cancellative right Ore subsemigroup T of I such that the image $\overline{I}$ of I in $K[S]/P$ is a uniform subsemigroup of some $J = \mathcal{M}(G, X, Y, P)$ in the quotient ring $Q_{cl}(K[S]/P)$ such that $1 = e_1 + \cdots + e_m$, for the nonzero idempotents e_i of J.*
2. *the image $R = K\{\overline{I}\}$ of $K[I]$ in $K[S]/P$ is a generalized matrix ring $R = \oplus_{i,j=1}^{m} R_{ij}$ with $R_{ij} = e_i Q_{cl}(K[S]/P)e_j \cap R$ and $R_{11} = K\{\overline{T}\}$.*
3. *the image $\overline{S}$ of S in $K[S]/P$ is in monomial form with respect to $e_1, \ldots, e_m$. That is, if $s \in \overline{S}$, then for every i (every j) we have $e_i s e_j \neq 0$ for at most one j (at most one i, respectively).*

If moreover H is torsion-free then every height one prime ideal P of $K[S]$ intersecting S is of the form $P = K[Q]$ for a minimal prime ideal Q of S.

Situation of the above type occurs in particular in noetherian PI domains considered in (7) and (14). This includes algebras $K\langle x_1, \dots, x_n\rangle$ defined by $\binom{n}{2}$ relations of type $x_j x_i = x_{i'} x_{j'}$ where $i, i' < j, j'$ with each monomial occurring exactly once and such that $x_j x_i - x_{i'} x_{j'}; i, j = 1, \dots, n$, form a Grobner basis of the corresponding ideal of the free algebra on $x_1, \dots, x_n$. The underlying semigroups and groups have been independently considered from a slightly different perspective in (3). Proposition 4.1 is crucial for the study of the prime spectrum of $K[S]$, and therefore for other properties of the algebra. Notice that height one prime ideals of $K[S]$ not intersecting S come from prime ideals in the group algebra $K[H]$ of the quotient group H of S. Both $K[S]$ and $K[H]$ are maximal orders. So the result of Brown provides us with a deep insight into the prime spectrum of $K[H]$ (2). In particular, every height one prime is generated by a normal element. We conclude with a simple example of a slightly different nature which can also be used to illustrate all of these features.

Example Let M be the monoid defined by the presentation $M = \langle x, y \mid x^2 = y^2\rangle$. Then $K[M]$ is a right and left noetherian PI domain and it is a maximal order (7). Moreover Mx^2 is the only minimal prime ideal of M, and the abelian group generated by xy, yx is of finite index in the group of quotients of M.

Any ideal of M contains a power of x^2. Since x^2 is central, any prime ideal of M contains x^2. Using matrix notation, we have

$$M \setminus Mx^2 = \{1\} \cup \begin{bmatrix} xy\langle xy\rangle^1 & \langle xy\rangle^1 x \\ y\langle xy\rangle^1 & yx\langle yx\rangle^1 \end{bmatrix}$$

where $\langle xy\rangle^1$ denotes the monoid generated by xy. Because

$$K[M \setminus Mx^2] = K + \begin{bmatrix} K[xy\langle xy\rangle^1] & K[\langle xy\rangle^1 x] \\ K[y\langle xy\rangle^1] & K[yx\langle yx\rangle^1] \end{bmatrix},$$

the generalized matrix ring techniques easily imply that $K_0[M/Mx^2] = K[M]/K[Mx^2]$ is a prime algebra. Therefore $K[Mx^2]$ is the only height one prime intersecting M nontrivially.

Acknowledgment. This work was supported by KBN research grant 2 P03A 030 18, Poland.

References

1. Anan'in, A.Z. (1989) An intriguing story about representable algebras, *Ring Theory 1989, Israel Math. Conf. Proc.*, pp. 31–38, Weizmann, Jerusalem.
2. Brown, K.A. (1985) Height one primes of polycyclic group rings, *J. London Math. Soc.* **(2) Vol. no. 32**, pp.426–438.
3. Etingof, P., Schedler, T. and Soloviev, A. (1999) Set-theoretical solutions of the quantum Yang-Baxter equation, *Duke Math. J.* **Vol. no. 100**, pp. 169–209.
4. Gateva–Ivanova, T. (1991) On the noetherianity of some associative finitely presented algebras, *J. Algebra* **Vol. no. 138**, pp. 13–35.
5. Gateva–Ivanova, T. (1994) Noetherian properties of skew polynomial rings with binomial relations, *Trans. Amer. Math. Soc.* **Vol. no. 343**, pp. 203–219.
6. Gateva–Ivanova, T. (1996) Skew polynomial rings with binomial relations, *J. Algebra* **Vol. no. 185**, pp. 710–753.
7. Gateva–Ivanova, T. and Van den Bergh, M. (1998) Semigroups of I-type, *J. Algebra* **Vol. no. 206**, pp. 97–112.
8. Gilmer, R. (1984) *Commutative Semigroup Rings.* Univ. Chicago Press, Chicago.
9. Grigorchuk, R.I. (1988) Cancellative semigroups of power growth, *Mat. Zametki* **Vol. no. 43**, pp. 305–319.
10. Gromov, M. (1981) Groups of polynomial growth and expanding maps, *Publ. Math. IHES* **Vol. no. 53(1)**, pp. 53–73.
11. Hotzel, E. (1975/76) Semigroups with maximal conditions, *Semigroup Forum* **Vol. no. 11**, pp. 337–362.
12. Howie, J.M. (1995) *Fundamentals of Semigroup Theory*, Oxford University Press, Oxford.
13. Jespers, E. and Okniński, J. (1996) Semigroup algebras that are principal ideal rings, *J. Algebra* **Vol. no. 183**, pp. 837–863.
14. Jespers, E. and Okniński, J. (1998) Binomial semigroups, *J. Algebra* **Vol. no. 202**, pp. 250-275.
15. Jespers, E. and Okniński, J. (1999) Noetherian semigroup algebras, *J. Algebra* **Vol. no. 218**, pp. 543–562.
16. Jespers, E. and Okniński, J. (1999) On a class of noetherian algebras, *Proc. Roy. Soc. Edinburgh* **Vol. no. 129A**, pp. 1185–1196.
17. Jespers, E. and Okniński, J. Submonoids of polycyclic-by-finite groups and their algebras, *Algebras and Representation Theory*, to appear.
18. Jespers, E. and Wang, Q. Height one prime ideals in semigroup algebras satisfying a polynomial identity, preprint.
19. Kozhukhov, I.B. (1980) On semigroups with minimal or maximal conditions on left congruences, *Semigroup Forum* **Vol. no. 21**, pp. 337–350.
20. Kramer, X.H. (2000) The Noetherian property in some quadratic algebras, *Trans. Amer. Math. Soc.* **Vol. no. 351**, pp. 4295–4323.

21. Krause, G.R. and Lenagan, T.H. (2000) *Growth of Algebras and Gelfand-Kirillov Dimension, Revised edition.* Graduate Studies in Mathematics 22. American Mathematical Society, Providence, RI.
22. McConnell, J.C. and Robson, J.C. (1987) *Noncommutative Noetherian Rings.* Wiley, New York.
23. Okniński, J. (1981) *Semigroup Algebras.* Marcel Dekker, New York.
24. Okniński, J. (1998) *Semigroups of Matrices.* World Scientific, Singapore.
25. Passman, D.S. (1989) *Infinite Crossed Products.* Academic Press, London.
26. Shoji, K. On semigroup rings, (1986) *Proc. 9th Symposium on Semigroups and Related Topics*, pp. 28–34. Naruto Univ, Naruto.

COHEN-MACAULAY REPRESENTATION

DORIN POPESCU (dorin@stoilow.imar.ro)
Institute of Mathematics,
University of Bucharest,
P.O.Box 1-764,
Bucharest 70700,
Romania

Abstract. We study the maximal Cohen-Macaulay modules and their ranks or minimal numbers of generators over hypersurfaces of type $Y_1^t+Y_2^3$, $Y_1^3+\cdots+Y_n^3$.

Introduction

Let K be a field, $Y = \{Y_1, \ldots, Y_n\}$ a set of indeterminates over K, and let $R = K[[Y]]/J$, J being an ideal of $K[[Y]]$. It would be nice to describe the finitely generated R-modules. Unfortunately this is hard even when R is artinian. In general we would be happy to describe only a set of finitely generated R-modules, the so called the *maximal Cohen-Macaulay* R-modules (in brief *MCM* R-modules). A nonzero finitely generated R-module M is MCM if depth M = dim R, in other words a Cohen-Macaulay module of maximal Krull dimension. Why the set of MCM R-modules? One reason is that these modules preserve many things from the artinian case. For example if R is an isolated singularity (i.e. R_p is regular for p a prime ideal $\neq (Y)$) then there exists almost split sequences and the first Brauer-Thrall conjecture holds in the frame of MCM R-modules, as happens in the artinian case. Another reason is that if R is a Cohen-Macaulay ring of Krull dimension d then the d-syzygy $\Omega_R^d(N)$ of any finitely generated R-module N is MCM, so each finitely generated R-module N can be aproximated modulo a part of its free resolution by a MCM R-module. In particular, the

K.W. Roggenkamp and M. Ştefănescu (eds.), Algebra - Representation Theory, 249–256.

Grothendieck group of R can be described using only the classes of MCM R-modules.

If R is a hypersurface ring, that is $J = (f)$ for a certain $0 \neq f \in (Y)$, then after Eisenbud (3) any MCM R-module M has a minimal free resolution of periodicity 2 which is completely given by a matrix factorization (φ, ψ), φ,ψ being square matrices over $K[[Y]]$ such that $\varphi\psi = \psi\varphi = fI_n$ for a certain positive integer n. So we would like to describe the matrix factorizations of all MCM R-modules.

Suppose that $f = Y_n^s + h$, $s \geq 2$, for $0 \neq h \in K[[Y_1, \ldots, Y_{n-1}]]$ and let $A = K[[Y_1, \ldots, Y_{n-1}]]/(h)$. Is it possible to describe MCM(R) using MCM(A)? A partial answer is given in the following:

Theorem 0.1 *(Knörrer (8)) Suppose $s = 2$ and char $K \neq 2$. Then*

i) For every MCM R-module N without free summands $\Omega_A^1(N/Y_nN) \cong N \oplus \Omega_R^1(N)$.

ii) Each indecomposable MCM R-module is a direct summand in a module of type $\Omega_R^1(M)$ for an indecomposable MCM A-module M.

The statement ii) follows easily from i) taking $M = N/Y_nN$ and it is clear that ii) can be used to describe inductively MCM(f) for special f. Unfortunately the condition $s = 2$ is too restrictive. What happens when $s > 2$? An important piece of help is given in the following:

Theorem 0.2 *(Yoshino (14)) If $s > 2$ and M is an indecomposable MCM A-module, then $\Omega_R^1(M)$ is an indecomposable MCM R-module.*

So the above theorem gives the possibility to describe some indecomposable MCM R-modules using MCM(A). However this is just a small part of the story. For example if $n = 2$, $h = Y_1^6$ and $s = 3$, then MCM(A) has just 6 indecomposable modules but MCM(R) is of infinite Cohen-Macaulay type ($f = Y_1^6 + Y_2^3$ is not a simple singularity!), that is there exists an infinite set of nonisomorphic indecomposable MCM R-modules.

A natural extension of 0.1 is given below.

Theorem 0.3 *((13)) Suppose s is not a multiple of char K. If N is a MCM R-module then $\Omega_R^1(N/Y_n^{s-1}N) \cong N \oplus \Omega_R^1(N)$.*

The next section presents some extensions of the above theorem together with some applications. The last section studies the MCM modules and their minimal numbers of generators over hypersurface singularities of type $Y_1^3 + \cdots + Y_n^3$.

1. Extensions and applications of Theorem 0.3

A strong extension of Theorem 0.3 is given below.

Theorem 1.1 *((6), (10)) Let r be a positive integer, $1 \leq r < n$, and suppose that $f = h + \sum_{i>n-r}^{n} Y_i^{s_i}$, $s_i \geq 2$, where $h \in K[[Y_1, \ldots, Y_{n-r}]]$ and the s_i are not multiples of char K. Then for every MCM R-module N we have*

$$\Omega_R^r(N/(Y_{n-r+1}^{s_{n-r+1}-1}, \ldots, Y_n^{s_n-1})N) \cong \oplus_{j=0}^{r} (\Omega_R^j(N))^{\binom{r}{j}}.$$

Actually behind this theorem there exists a general result which is given next.

Theorem 1.2 *((11)) Let (R, m) be a Noetherian local ring and let P, N be finitely generated R-modules such that $Tor_i^R(P, N) = 0$ for all $i \geq 1$. Suppose that $t = pd_R P < \infty$, let us say*

$$0 \to R^{\beta_t} \to R^{\beta_{t-1}} \to \cdots \to R^{\beta_0} \to P \to 0$$

is a minimal free resolution of P over R, its maps being $(\delta_i)_{0 \leq i \leq t}$. If $Ext_R^1(\delta_i \otimes_R N, -) = 0$ for all $1 \leq i \leq t$ then $\Omega_R^t(P \otimes_R N) \cong \oplus_{i=0}^{t} (\Omega_R^i(N))^{\beta_{t-i}}$.

Let $\tilde{R} = R/(Y_n^{s-1})$. An $\tilde{R}$-module F is a *lifting* (see (1)) of a MCM A-module M from A to $\tilde{R}$ if $F/(Y_n)F \cong M$ and $\mathrm{Tor}_i^{\tilde{R}}(A, F) = 0$ for $i \geq 1$. In other words F is a *deformation* of M to $\tilde{R}$. The interesting case of the above theorem is $P = R/a$ for an ideal a of R. Then the condition $\mathrm{Tor}_i^R(P, N) = 0$ for $i \geq 1$ says that N is a lifting of N/aN from R/a to R. In Theorem 1.1, $a = (Y_{n-r+1}^{s_{n-r+1}-1}, \ldots, Y_n^{s_n-1})$ is generated by a regular sequence and since N is a lifting of a MCM R/a-module from R/a to R we see that N is a MCM R-module. Certainly here $r = t$ and $\beta_{r-i} = \binom{r}{i}$ because we deal with the Koszul complex.

For applications it is useful to state the following easy consequence of Theorem 0.3.

Proposition 1.3 *Each indecomposable MCM R-module is a direct summand in a module of type $\Omega_R^1(F)$ for an indecomposable lifting F of a MCM A-module M from A to $\tilde{R}$.*

Some applications of Theorem 0.3 were obtained in the case when K is algebraically closed with char $K \neq 3$, $n = 2$ and $f = Y_1^t + Y_2^3$ for a positive integer t. When $t \leq 5$, f is a simple singularity and so MCM(R) is of finite Cohen-Macaulay type and the indecomposable MCM R-modules are well known (see e.g. 0.2). So it would be nice to have a complete description of MCM(R) for $t = 6$ or $t = 7$. Using Theorem 0.3 we may reduce the description of MCM(R) to the description of all liftings of $K[[Y_1]]/(Y_1^t)$-modules to $\tilde{R} = K[[Y_1, Y_2]]/(Y_1^t, Y_2^2)$ but this is still very complicated. However along these lines we have the following:

Theorem 1.4 *Let* $1 \leq i < j < t$ *be two positive integers and* $\mathcal{M}_{ij}$ *the class of all MCM* R*-modules* N *such that* N/Y_2N *is a direct sum of copies of* $K[[Y_1]]/(Y_1^i)$, $K[[Y_1]]/(Y_1^j)$. *Then*

i) If $i + j \neq t$ *then* $\mathcal{M}_{ij}$ *is of finite Cohen-Macaulay representation type and all indecomposable modules of* $\mathcal{M}_{ij}$ *are completely described by some matrix factorizations of size* 4 *(see (4)).*

ii) If $i + j = t$, $t > 5$, $t \neq 3i$ *then* $\mathcal{M}_{ij}$ *is of infinite Cohen-Macaulay representation type and in (12) there is given a list of matrix factorizations (not necessarily corresponding to indecomposable modules) such that each indecomposable module from* $\mathcal{M}_{ij}$ *is a direct summand in one defined by a matrix factorization from the list.*

Another possible application it is inspired by the following:

Question 1.5 (Herzog) *Let* B *be a Noetherian complete local ring. Is it possible to describe MCM(*$B[[X]]$*) using the description of MCM(*B*),* X *being an indeterminate?*

Though the above question seems very natural it is rather complicated. In fact if B is not regular then $B[[X]]$ is not an isolated singularity and so MCM($B[[X]]$) has no almost split sequences and the first Brauer-Thrall Conjecture does not hold. The easiest case $B = K[[Z]]/(Z^2)$ was considered in (2), (15) as an example of complete local ring where the first Brauer-Thrall Conjecture does not hold. Actually, it shows that the complete list of nonisomorphic, nonfree indecomposable modules from MCM($B[[X]]$) are the ideals $\{(X^i, Z)B[[X]] | i \geq 1\}$, $ZB[[X]]$. Since XB is the only nonfree indecomposable module from MCM(B) =Mod B we see that the condition $h \neq 0$ is really necessary in Theorems 0.1, 0.3.

Now, let $B = K[[Z]]/(Z^t)$. The classification of all MCM

$B[[X]]$-modules (these are free over $K[[X]]$ and the multiplication by Z is given by a square matrix T over $K[[X]]$) is equivalent with the classification of all square matrices T over $K[[X]]$ such that $T^t = 0$. Suppose we were be able to describe $\mathrm{MCM}(B[[X]])$; then using Theorem 0.3 like in the proof of Theorem 1.4 we might be able to find some families of MCM modules over $K[[X,Y,Z]]/(Y^3+Z^t)$, if char $K \neq 3$. Even the case $2 < t < 6$ has a certain interest.

2. Maximal graded Cohen-Macaulay modules over hypersurfaces of type $Y_1^3 + \cdots + Y_n^3$

Let $R_n = \mathbf{C}[Y_1, \ldots, Y_n]/(f_n)$, where $f_n = \sum_{i=1}^n Y_i^3$. C. Kahn (7) described "geometrically" the graded MCM R_3-modules using Atiyah's classification of vector bundles over elliptic curves (clearly a graded MCM R_3-module is free outside (Y) since f_3 is an isolated singularity). Using the ranks and the degrees of the bundles he gave a partition on the set of isomorpism classes of indecomposable graded MCM R_3-modules such that each slice is in bijection with the points of the curve $V(f_3) \subset \mathbf{P}^2_{\mathbf{C}}$. Moreover, Kahn also described the Auslander-Reiten quiver of $\mathrm{MCM}(R_3)$. Unfortunately, his method does not give the matrix factorizations of the indecomposable MCM R_3-modules so his description is somewhat abstract.

Let $\lambda = [\lambda_1 : \lambda_2 : \lambda_3]$ be a point on the curve $V(f_3) \subseteq \mathbf{P}^2_{\mathbf{C}}$ with equation $f_3 = 0$. After a suitable change of coordinates if necessary, we may suppose that $\lambda_3 \neq 0$ and so may take $\lambda_3 = 1$. Then we may define a MCM R_3-module M_λ by the matrix factorization $(\varphi_\lambda, \psi_\lambda)$, where

$$\varphi_\lambda = \begin{pmatrix} Y_1 - \lambda_1 Y_3 & Y_2 - \lambda_2 Y_3 \\ -(Y_2^2 + \lambda_2 Y_2 Y_3 + \lambda_2^2 Y_3^2) & Y_1^2 + \lambda_1 Y_1 Y_3 + \lambda_1^2 Y_3^2 \end{pmatrix},$$

$$\phi_\lambda = \begin{pmatrix} Y_1^2 + \lambda_1 Y_1 Y_3 + \lambda_1^2 Y_3^2 & -(Y_2 - \lambda_2 Y_3) \\ Y_2^2 + \lambda_2 Y_2 Y_3 + \lambda_2^2 Y_3^2 & Y_1 - \lambda_1 Y_3 \end{pmatrix}.$$

Proposition 2.1 *((9)) The correspondence $\lambda \to M_\lambda$ defines a bijection between the points of the curve $V(f_3)$ and the set of* *isomorphism classes of indecomposable nonfree, graded maximal Cohen-Macaulay modules over R_3 having two generators.*

It is not clear that this bijection is Kahn's bijection but this is very likely. A nice family of three generated graded MCM R_3-modules is Coker (φ_{uv}), $u, v \in \mathbf{C}^*$, $u^{-1} + v^{-1} = uv$ (see (9)), where

$$\varphi_{uv} = \begin{pmatrix} Y_1 & Y_2 & Y_3 \\ Y_2 & -v^{-1}Y_3 & -uY_1 \\ vY_3 & u^{-1}Y_1 & -Y_2 \end{pmatrix}.$$

Special three generated graded MCM R_3-modules are Coker (ϕ_ω), where $\omega \in \mathbf{C}$ satisfies $\omega^3 = 1$ and

$$\phi = \begin{pmatrix} 0 & -(Y_1 + Y_2) & Y_3 \\ Y_3 & 0 & Y_1 + \omega Y_2 \\ Y_1 + \omega^2 Y_2 & Y_3 & 0 \end{pmatrix}.$$

If $n > 3$ we know almost nothing about $\mathrm{MCM}(R_n)$. In this case we would be content to have an idea about the possible ranks or minimal numbers of generators of graded MCM R_n-modules. Let

$$\mu(A) = \min\{\mu(M) \mid M \text{ an indecomposable nonfree MCM } A\text{-module}\},$$

where $\mu(M)$ denotes the minimal number of generators of M.

It is easy to see the following:

Lemma 2.2 *((9)) For $n \geq 1$, $\mu(R_n) \geq n/2$.*

Herzog and Kühl (5) proved that if R is a Gorenstein local ring of multiplicity e and M is a MCM R-module then

$e/(e-1)\ \mathrm{rank}(M) \leq \mu(M) \leq e\ \mathrm{rank}(M)$.

Bruns (5) showed that if M is a MCM module over a hypersurface ring then $\mathrm{rank}(M) \geq (\dim R - 1)/2$. It follows that:

Lemma 2.3 *For $n \geq 1$, $\mu(R_n) \geq 3(n-2)/4$.*

The proof uses the above inequalities, since $e = 3$, $\dim R_n = n - 1$. Note that Lemma 2.3 is stronger than Lemma 2.2 when $n > 7$, but it is weaker for $n = 3$. It is easy to see that $\mu(R_2) = 1$, $\mu(R_3) = \mu(R_4) = 2$, $\mu(R_5) = 4$. There exist 3-generated standard MCM R_4-modules : take

$$\phi = \begin{pmatrix} Y_1 + Y_4 & Y_2 & Y_3 \\ Y_2 & -v^{-1}Y_3 & -u(Y_1 + \rho^2 Y_4) \\ vY_3 & u^{-1}(Y_1 + \rho Y_4) & -Y_2 \end{pmatrix},$$

where $\rho, u, v \in K \setminus \{0\}$ satisfy $\rho^3 = 1, \rho \neq 1$, $u^3 = \rho(1+\rho)$ and $v = -u/(1+\rho)$.

Since the description of μ seems to be not easy, the following theorem gives some idea about the jumps of μ.

Theorem 2.4 *((9)) Let $n \geq 3$ be such that $\mu(R_n) > \mu(R_{n-1})$. Then there exist families of graded indecomposable MCM modules over R_n of rank $\mu(R_{n-1})$ minimally generated by $2\mu(R_{n-1})$-elements.*

Corollary 2.5
$\mu(R_{n-1}) \leq \mu(R_n) \leq 2\mu(R_{n-1})$.

References

1. M. Auslander, S. Ding and Ø. Solberg, *Lifting and weak liftings of modules.* J. Algebra **156**(1993), 273-317.
2. R.O. Buchweitz, G.M. Greuel and F.O. Schreyer, *Cohen-Macaulay modules over hypersurface singularities II.* Invent. Math. **88** (1987), 165-182.
3. D. Eisenbud, *Homological algebra with an application to group representations.* Trans. American Math. Soc. **260** (1980), 35-64.
4. V. Ene and D. Popescu, *Steps in the classification of Cohen-Macaulay modules over singularities of type $X^t + Y^3$*, Algebra and Representation Theory, **2** (2) (1999), 35-64.
5. J. Herzog and M. Kühl, *Maximal Cohen-Macaulay modules over Gorenstein rings and Bourbaki sequences.* In *Commutative algebra and combinatorics.* (Kyoto, 1985), 65-92, Adv. Stud. Pure Math. **11**, North-Holland, Amsterdam-New York, 1987.
6. J. Herzog and D. Popescu, *Thom-Sebastiani problems for maximal Cohen-Macaulay modules.* Math. Ann. **309** (1997), 677-700.
7. C.P. Kahn, *Reflexive modules on minimally elliptic singularities.* Math. Ann. **285** (1989), 141-160.
8. H. Knörrer, *Cohen-Macaulay modules over hypersurface singularities I.* Invent. Math. **88**(1987), 153-164.
9. R. Laza, L. O'Carroll and D. Popescu, *Maximal Cohen-Macaulay modules over $Y_1^3 + \cdots + Y_n^3$ with few generators.*, Prepr. Bucharest, 2000.
10. L. O'Carroll and D. Popescu, *Free resolutions for deformations of maximal Cohen-Macaulay modules.* Comm. in Algebra, to appear.
11. L. O'Carroll and D. Popescu, *Splitting syzygies.* J. Algebra **228**(2000), 682-709.
12. G. Pfister and D. Popescu, *A family of Cohen-Macaulay modules over singularities of type $X^t + Y^3$.* Comm. Algebra, **27** (6) (1999), 2555-2572.
13. D. Popescu, *Maximal Cohen-Macaulay modules and their deformations.* An. Stiint. Constanta **2**(1994), 112-119.

14. Y. Yoshino, *Tensor product of matrix factorizations.* Nagoya Math. J. **152** (1998), 39-56.
15. Y. Yoshino, *Cohen-Macaulay modules over Cohen-Macaulay rings.* London Math. Soc. Lecture Notes Ser. **146**, Cambridge, 1990.

HEREDITARY ABELIAN CATEGORIES AND ALMOST SPLIT SEQUENCES

IDUN REITEN (idunr@math.ntnu.no)
Department of Mathematical Sciences
Norwegian University of Sciences and Technology

These notes are based on a series of three lectures given at the Advanced Study Institute (NATO ASI) in Algebra - Representation Theory in Constantza in August 2000.

The notes are centered around the topic of hereditary abelian k-categories, where k is a field which we for simplicity assume to be algebraically closed. That the abelian category $\mathcal{C}$ is a k-category means that the homomorphism groups are vector spaces over k and the composition of maps is bilinear over k. The category $\mathcal{C}$ is said to be hereditary if the Yoneda functor $\mathrm{Ext}^2_{\mathcal{C}}(,)$ vanishes. We shall in addition assume that $\mathcal{C}$ is Ext-finite, that is $\mathrm{Hom}_{\mathcal{C}}(A, B)$ and $\mathrm{Ext}^1_{\mathcal{C}}(A, B)$ are finite dimensional vector spaces over k for all A and B in $\mathcal{C}$. It is an interesting problem to classify such categories, under additional natural assumptions. A general classification theorem under the extra assumptions of $\mathcal{C}$ being noetherian and having Serre duality is given in (RV1). The purpose of these notes is to provide some background material for this work.

The first task is to provide basic examples of Ext-finite hereditary abelian k-categories, taken from different areas of mathematics. Then we discuss some additional properties satisfied for these examples. In particular we emphasize almost split sequences, and Chapter II is devoted to discussing the relationship between almost split sequences and Serre duality. Another important additional condition is having a tilting object. Such categories are of interest in the representation theory of finite dimensional algebras, as discussed in Chapter III.

K.W. Roggenkamp and M. Ştefănescu (eds.), Algebra - Representation Theory, 257–271.

Derived categories play an important role in the investigation of hereditary abelian k-categories, because of their simple description in this case, and since seemingly unrelated examples are connected via derived categories. We should also point out that Ext-finite hereditary abelian k-categories are important in noncommutative algebraic geometry (SV).

I had been asked to emphasize some of the important work of Maurice Auslander through my lectures. First this is done through discussing almost split sequences, as a natural additional property of hereditary abelian k-categories, as well as by discussing their relationship to Serre duality. Also Auslander's existence theorems for almost split sequences of Cohen–Macaulay modules, with the relationship to isolated singularities (A3), have interesting connections to hereditary categories in the two-dimensional case (RV1, RV2). Finally, the connection between AR-quivers and extended Dynkin diagrams for two-dimensional isolated hypersurface singularities (A2) is the basis for explaining a connection with hereditary path algebras for extended Dynkin quivers. These topics are treated in the first two chapters.

We shall freely use background material from (R, ARS), and refer to these books for original references.

I. Basic examples and connections

In this chapter we discuss some examples of hereditary abelian k-categories and give a brief introduction to the theory almost split sequences and AR-quivers (see (ARS, R)). We also stress the importance of derived categories in the study of hereditary categories, and explain different occurrances of extended Dynkin diagrams this way.

1 Examples from finite dimensional algebras. Let Λ be the path algebra kQ over k, of a finite quiver Q without oriented cycles. Since k is an algebraically closed field, the kQ are, up to Morita equivalence, exactly the finite dimensional hereditary k-algebras. In particular, the category $\operatorname{mod}\Lambda$ of finitely generated Λ-modules and the category $\operatorname{Mod}\Lambda$ of all Λ-modules are examples of hereditary abelian k-categories. Clearly the first example is Ext-finite, but the second one is not.

The hereditary k-algebras have played a prominent role in the representation theory of finite dimensional algebras. A finite dimensional k-algebra Λ is said to be of *finite (representation) type* if there is only a finite number of indecomposable modules in $\operatorname{mod}\Lambda$ up to isomorphism. For a finite connected quiver Q without oriented cycles, it is known, by an old theorem of Gabriel, that kQ is of finite type if and only if the underlying graph of Q is a Dynkin diagram $A_n, D_n, E_i\ (i = 6, 7, 8)$. The indecomposable algebras kQ of infinite type are divided into two classes: the tame ones, given by quivers whose underlying graphs are extended Dynkin diagrams $\tilde{A}_n, \tilde{D}_n, \tilde{E}_i\ (i = 6, 7, 8)$, and the wild ones, whose quivers correspond neither to Dynkin nor to extended Dynkin diagrams.

2 Geometric examples. Interesting examples of Ext-finite hereditary abelian k-categories appear also in algebraic geometry. The most basic example is the category $\operatorname{coh}\mathbb{P}^1(k)$ of coherent sheaves over the projective line over k. Note that unlike what is the case for the previous examples, there are no nonzero projective or injective objects in $\operatorname{coh}\mathbb{P}^1(k)$.

It is well known that, by the Serre description, $\operatorname{coh}\mathbb{P}^1(k)$ is equivalent to the quotient category

$$\operatorname{qgr}(k[X,Y]) = \operatorname{gr}(k[X,Y])/\text{finite length},$$

where $\operatorname{gr} k[X,Y]$ denotes the category of finitely generated graded $k[X,Y]$-modules with degree zero homomorphisms. Similarly, let $S = k[X,Y,Z]/(X^2 + Y^3 + Z^5)$ be $\mathbb{Z}$-graded via assigning X degree 15, Y degree 10 and Z degree 6. Then $\operatorname{qgr} S$ is also an Ext-finite hereditary abelian k-category. There is a whole class of related examples, obtained from two-dimensional commutative $\mathbb{Z}$-graded noetherian Cohen–Macaulay isolated singularities (see (GL, RV1)).

3 Almost split sequences. Let Λ be a finite dimensional k-algebra. A homomorphism $g\colon B \to C$ in $\operatorname{mod}\Lambda$ with C indecomposable, is said to be *right almost split* if it is not a split epimorphism, and for any nonisomorphism $t\colon X \to C$, where X is indecomposable, there is a map $s\colon X \to B$ with $gs = t$. The map $g\colon B \to C$ is *right minimal* if whenever $t\colon B \to B$ is such that $gt = g$ then t is an isomorphism. We say that g is *minimal right almost split* if it is right minimal and right almost split. The notions of left almost split, left minimal and minimal left almost split maps are defined in a dual way. An exact sequence $0 \to A \xrightarrow{f} B \xrightarrow{g} C \to 0$ is *almost split* if g is minimal right almost split, or equivalently, if f is minimal left almost split. If C is indecomposable nonprojective (or A is indecomposable noninjective) there is always an almost split sequence $0 \to A \to B \to C \to 0$, which is unique up to isomorphism. When P is indecomposable projective, there is a minimal right almost split map $f\colon \underline{r}P \to P$, where $\underline{r}$ denotes the radical of Λ, and a minimal left almost split map $g\colon I \to I/\operatorname{soc} I$, where $\operatorname{soc} I$ denotes the socle of I. This is summarized in saying that $\operatorname{mod}\Lambda$ has almost split sequences (see (ARS, R)).

The information provided by existence of almost split sequences can be recorded in what is called the AR-quiver. This is a (possibly infinite) quiver whose vertices are in one-one correspondence with the nonisomorphic indecomposable modules in $\operatorname{mod}\Lambda$, and the arrows are given by information from the almost split sequences (and minimal right and left almost split maps), as we shall illustrate through examples.

For $\operatorname{coh}\mathbb{P}^1(k)$, and related examples, we also have existence of almost split sequences, and hence an associated AR-quiver. The situation for $\operatorname{coh}\mathbb{P}^1(k)$ is a little simpler than for $\operatorname{mod}\Lambda$, since there are no nonzero projective or injective objects, so that an almost split sequence $0 \to A \to B \to C \to 0$ exists for any indecomposable object C (or for any indecomposable object A).

4 Examples of AR-quivers. Let Q be the quiver

$$\cdot_1 \xrightarrow{\alpha} \cdot_2 \xrightarrow{\beta} \cdot_3$$

and kQ the corresponding path algebra over k, which is easily seen to be isomorphic to the lower triangular matrix algebra $\begin{pmatrix} k & 0 & 0 \\ k & k & 0 \\ k & k & k \end{pmatrix}$.

There are three indecomposable projective modules P_1, P_2 and P_3 corresponding to the columns of the matrix, with three associated simple modules $S_1 = P_1/\underline{r}P_1$, $S_2 = P_2/\underline{r}P_2$ and $S_3 = P_3/\underline{r}P_3$. We write $P_1 = \begin{pmatrix} S_1 \\ S_2 \\ S_3 \end{pmatrix}$, $P_2 = \begin{pmatrix} S_2 \\ S_3 \end{pmatrix}$ and we have $P_3 = S_3$. There are totally six indecomposable modules, obtained as factor modules of the P_i, and we have the almost split sequences

$$0 \to S_3 \to P_2 \to S_2 \to 0,$$
$$0 \to P_2 \to S_2 \oplus P_1 \to P_1/S_3 \to 0,$$
$$0 \to S_2 \to P_1/S_3 \to S_1 \to 0.$$

In addition we have the minimal right almost split maps $P_2 \to P_1$ and $S_3 \to P_2$, and minimal left almost split maps $P_1 \to P_1/S_3$ and $P_1/S_3 \to S_1$. Note that $P_1, P_1/S_3$ and S_1 are the indecomposable injective modules. The AR-quiver is then the following, where the dotted lines indicate the correspondence between the end terms of almost split sequences.

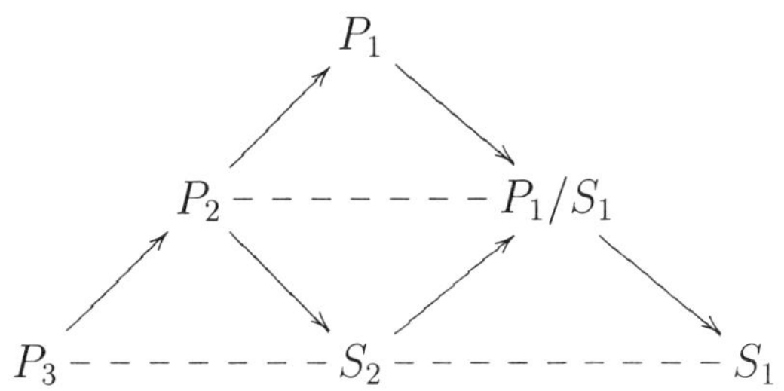

For coh $\mathbb{P}^1(k)$ the AR-quiver looks as follows. One connected component is of the form

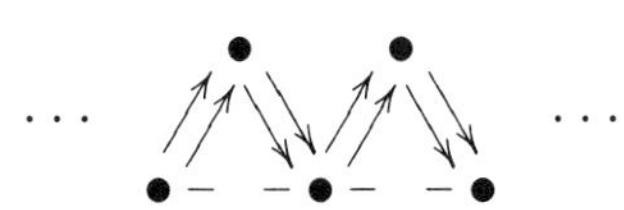

and in addition there is an infinite number of components of the form

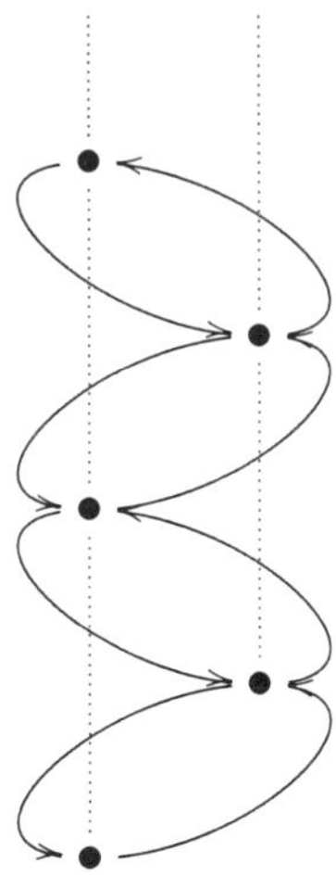

which are stable tubes of rank one.

5 Derived categories Let $\mathcal{C}$ be an Ext-finite hereditary abelian k-category, and denote by $D^b(\mathcal{C})$ the associated bounded derived category. One of the advantages of $\mathcal{C}$ being hereditary is that $D^b(\mathcal{C})$ has a particularly nice description which makes it easy to work with. The indecomposable objects of $D^b(\mathcal{C})$ are up to isomorphism just the stalk complexes $X[i]$ where X is an indecomposable object in $\mathcal{C}$ placed in degree $-i$. The maps are given by $\operatorname{Hom}(X[i], Y[j]) = \operatorname{Ext}_{\mathcal{C}}^{j-i}(X, Y)$ for X and Y in $\mathcal{C}$.

Considering derived categories is important in this context for establishing connections between seemingly unrelated categories. For the examples discussed before we have the following bounded derived categories:

If $\mathcal{C} = \operatorname{mod} kQ$ with Q the quiver $\cdot_1 \xrightarrow{\alpha} \cdot_2 \xrightarrow{\beta} \cdot_3$, we can illustrate $D^b(\mathcal{C})$ as follows:

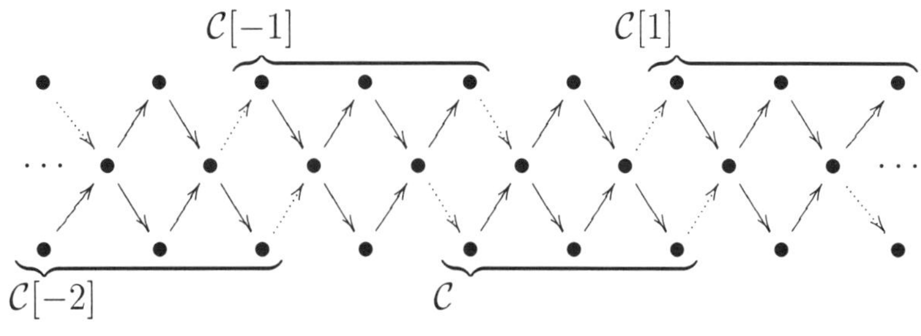

For $D^b(\operatorname{coh}\mathbb{P}^1(k))$ we have

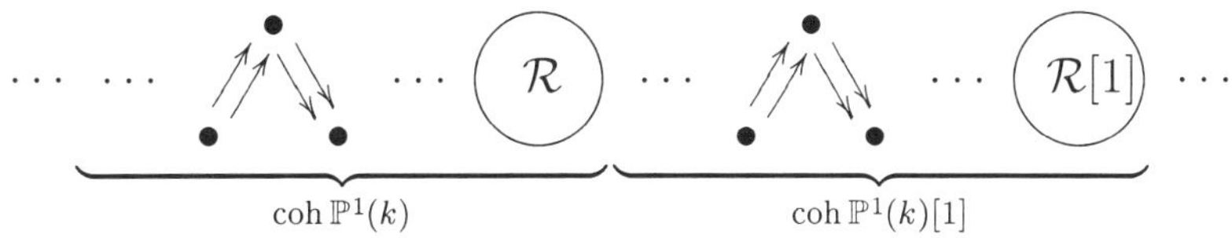

where $\mathcal{R}$ denotes the infinite family of stable tubes of rank one. It turns out that the path algebra $k(\cdot \rightrightarrows \cdot)$ has the same bounded derived category. Hence the geometric example $\operatorname{coh}\mathbb{P}^1(k)$ is related to this finite dimensional algebra through an equivalence of their bounded derived categories.

6 Hereditary categories and almost split sequences. Let R be a two-dimensional complete local (nonregular) commutative noetherian Cohen–Macaulay domain, and denote by $\mathrm{CM}(R)$ the category of finitely generated maximal Cohen–Macaulay R-modules. Auslander established an interesting connection between existence of almost split sequences for $\mathrm{CM}(R)$ and geometric properties. In fact, he proved, as a special case of a more general result, that $\mathrm{CM}(R)$ has almost split sequences if and only if R is an *isolated singularity*, that is the localizations R_p have finite global dimension for any nonmaximal prime ideal p in R (A3). Considering the quotient category $\operatorname{qgr} R$, then actually the same condition of isolated singularity characterizes when $\operatorname{qgr} R$ is hereditary (RV2).

There is also an analogous result in the graded case (AR3, AR4), which is more relevant for our purposes here. Let $S = k + S_1 + S_2 + \cdots + S_i + \cdots$ be a two-dimensional commutative $\mathbb{Z}$-graded (nonregular) Cohen–Macaulay domain, generated as a k-algebra by a finite number of homogenous elements, and denote by $\mathrm{CM}(\operatorname{gr} S)$ the category of finitely generated graded Cohen–Macaulay modules with degree zero homomorphisms. Then $\mathrm{CM}(\operatorname{gr} S)$ has almost split sequences if and only if S is an isolated singularity, which again is the case if and only if the abelian k-category $\operatorname{qgr} S$ is hereditary (RV1, RV2).

Note that $\mathrm{CM}(\operatorname{gr} S)$ is a subcategory both of $\operatorname{gr} S$ and of $\operatorname{qgr}(S)$. When S is an isolated singularity, then $\mathrm{CM}(\operatorname{gr} S)$ has almost split sequences, but the category gr S of finitely generated graded modules does not. However, when considering qgr(S) instead of gr S, we see that there is in fact an abelian category containing CM(gr S) which has almost split sequences (RV1), and even such that the almost split sequences in CM(gr S) stay almost split in qgr(S). It is interesting to note that it is a hereditary abelian category which accomplishes this.

7 Hereditary categories and extended Dynkin diagrams. Let R and S be as above, and assume in addition that they are simple isolated hypersurface singularities. Equivalently, when k has characteristic zero, these are the invariant rings $R = k[[X,Y]]^G$ and $S = k[X,Y]^G$ where G is a finite subgroup of the special linear group $SL(2,k)$. Then the AR-quivers for the Cohen–Macaulay modules are given by extended Dynkin diagrams, where the corresponding Dynkin diagram appears in the resolution graph of the singularity (see (A2) for the case of complete rings.) For example $S = k[X,Y]/(X^2 + Y^3 + Z^5)$ corresponds to the extended Dynkin diagram $\tilde{E}_8$. Associated with the graded ring S we have the hereditary abelian category qgr S. On the other hand, there are finite dimensional tame hereditary path algebras kQ corresponding to an extended Dynkin diagram. Actually, the hereditary categories qgr S and mod kQ have equivalent bounded derived categories. So through hereditary categories one detects a connection between the occurrance of extended Dynkin diagrams in the two situations.

8 Additional conditions. We have seen that several basic examples of hereditary abelian k-categories are Ext-finite and have almost split sequences, so these are natural conditions to impose. In the geometric context, Serre duality is a central condition, which is closely related to almost split sequences. The connection will be discussed in the next chapter. Other natural conditions are having a tilting object or having a finitely generated Grothendieck group, as well as being noetherian.

II. Almost split sequences and Serre duality.

In this chapter we discuss the connection between almost split sequences and Serre duality, from a historical point of view.

1 Origins. Almost split sequences were first introduced in the context of artin algebras, where existence was proved in the early seventies (AR1, AR2), via two different approaches. One approach used finitely presented functors, and the other one was based on a crucial formula relating Ext^1 and Hom. We here concentrate on the second approach. Specifically, if Λ is a finite dimensional k-algebra with A and B in $\mathrm{mod}\,\Lambda$, there is a functorial isomorphism $\mathrm{Ext}^1_\Lambda(A, \tau B) \simeq D\underline{\mathrm{Hom}}(B, A)$. Here

$$\underline{\mathrm{Hom}}(A, B) = \mathrm{Hom}_\Lambda(A, B)/P(A, B),$$

where $P(A, B)$ denotes the homomorphisms $f\colon A \to B$ which factor through some projective Λ-module. Further, D denotes the ordinary duality $D = \mathrm{Hom}_k(\ ,k)$ for finite dimensional k-algebras and τ denotes the correspondence $D\,\mathrm{Tr}$, where Tr is the transpose operation given by the exact sequences $P_1 \to P_0 \to C \to 0$ and $\mathrm{Hom}_\Lambda(P_0, \Lambda) \to \mathrm{Hom}_\Lambda(P_1, \Lambda) \to \mathrm{Tr}\,C \to 0$, where $P_1 \to P_0 \to C \to 0$ is a minimal projective presentation of C. For $A = B$ indecomposable, we then have the isomorphism $\mathrm{Ext}^1_\Lambda(A, \tau A) \simeq D\underline{\mathrm{Hom}}(A, A)$. Then $\underline{\mathrm{Hom}}(A, A)/m \simeq k$, where m is the maximal ideal of the local ring $\underline{\mathrm{Hom}}(A, A)$. This gives rise to a nonzero element in $D\underline{\mathrm{Hom}}(A, A) = \mathrm{Hom}_k(\underline{\mathrm{Hom}}(A, A), k)$. Via the isomorphism this gives an exact sequence $0 \to \tau A \to E \to A \to 0$, which turns out to be almost split (AR2).

The concept of Serre duality, which is a central concept in algebraic geometry, came up in the fifties in connection with sheaf cohomology. The following formulation is appropriate for our purposes: For a nonsingular projective space X over k there is a functorial isomorphism $\mathrm{Ext}^n(A, FB) \simeq D\,\mathrm{Hom}(B, A)$, where $F\colon \mathrm{coh}\,X \to \mathrm{coh}\,X$ is an equivalence of categories on the coherent sheaves over X, given by tensoring with the canonical sheaf. Note that for $n = 1$ the two formulas are similar.

2 Cohen–Macaulay modules and vector bundles. Existence theorems for almost split sequences were proved in more general contexts than finite dimensional algebras (A1), also in the seventies. Again the existence was deduced from formulas similar to the one cited above for finite dimensional k-algebras. An important special case was the category $\mathrm{CM}(R)$ of maximal Cohen–Macaulay modules for a complete local noetherian commutative Cohen–Macaulay isolated singularity R. Motivated by the connection between existence of almost split sequences and concepts and results from algebraic geometry, analogous existence theorems were proved in the context of graded maximal Cohen–Macaulay modules $\mathrm{CM}(\mathrm{gr}\, S)$ over $\mathbb{Z}$-graded rings $S = k + S_1 + S_2 + \cdots + S_i + \cdots$ (AR3, AR4). In the formula $\mathrm{Ext}^1(A, \tau B) \simeq D\underline{\mathrm{Hom}}(B, A)$ the modules A and B lie in $\mathrm{CM}(R)$, respectively $\mathrm{CM}(\mathrm{gr}\, S)$, and τ and D are defined in a different way (A1, AR3, AR4).

When X is a nonsingular projective curve, the category $\mathrm{coh}\, X$ of coherent sheaves over X is equivalent to $\mathrm{qgr}\, S$ for some $\mathbb{Z}$-graded two-dimensional noetherian commutative Cohen–Macaulay isolated singularity S, and the subcategory $\mathrm{vect}\, X$ of vector bundles is equivalent to $\mathrm{CM}(\mathrm{gr}\, S)$. In the mid eighties existence of almost split sequences for $\mathrm{coh}\, X$ was proved on one hand by using existence for $\mathrm{CM}(\mathrm{gr}\, S)$ (AR3), and on the other hand it was deduced from the Serre duality formula by Schofield.

3 Derived categories. After Happel's discovery of the connection between tilting theory and derived categories in the early eighties (H1), it became important for the representation theory of finite dimensional algebras to investigate derived categories more closely. Generalizing the notion of almost split sequences Happel introduced the concept of Auslander–Reiten triangles (almost split triangles) in the derived category $D^b(\Lambda)$ for finite dimensional k-algebra Λ. He proved existence of almost split triangles if and only if Λ has finite global dimension (H2). Also in this setting the proof was based on a formula relating Ext^1 and Hom. More specifically, we have $\mathrm{Ext}^1(A, \tau B) \simeq D\,\mathrm{Hom}(B, A)$ for A and B in $D^b(\Lambda)$, where $D = \mathrm{Hom}_k(\ , k)$ and $\tau \colon D^b(\Lambda) \to D^b(\Lambda)$ is an equivalence induced by the equivalence $N \colon \mathcal{P} \to \mathcal{I}$ from the projective to the injective Λ-modules given by $N(P) = D\,\mathrm{Hom}_\Lambda(P, \Lambda)$.

Motivated by the geometric setting, a general definition of Serre duality was given in the late eighties (BK). If $\mathcal{C}$ is an abelian category, then $D^b(\mathcal{C})$ is said to have Serre duality if there is an equivalence of categories $F\colon D^b(\mathcal{C}) \to D^b(\mathcal{C})$ such that we have a functorial isomorphism $\operatorname{Ext}^1(A, FB) \simeq D\operatorname{Hom}(B, A)$ for all A and B in $D^b(\mathcal{C})$. With this definition the usual geometric definition is covered, and it follows from the above discussion that the same holds for $D^b(\Lambda)$ when Λ is a finite dimensional algebra of finite global dimension. For the categories $D^b(\mathcal{C})$ (which are Ext-finite) it is proved in (RV1) that we have existence of almost of almost split triangles if and only if we have Serre duality. Note that the formulation in (BK) and the result in (RV1) are in the more general setting of triangulated categories. When $n > 1$ the category $\operatorname{coh} X$ of coherent sheaves for a nonsingular projective space of dimension n does not have almost split sequences, even though it has Serre duality. But the connection is recovered on the level of $D^b(\operatorname{coh} X)$.

4 Hereditary abelian categories. Let $\mathcal{C}$ be an Ext-finite hereditary abelian k-category. We say that $\mathcal{C}$ has Serre duality if $D^b(\mathcal{C})$ has Serre duality in the above definition (RV1). Because of the close connection between almost split sequences in $\mathcal{C}$ and almost split triangles in $D^b(\mathcal{C})$ in the hereditary case, we get a close connection between existence of almost split sequences and Serre duality for $\mathcal{C}$. In fact $\mathcal{C}$ has Serre duality if and only if $\mathcal{C}$ has almost split sequences and there is a one-one correspondence between indecomposable projective objects P in $\mathcal{C}$ and indecomposable injective objects I in $\mathcal{C}$ such that $P/\operatorname{rad} P \simeq \operatorname{soc} I$ (RV1). Here $\operatorname{rad} P$ is the unique maximal subobject of P and $\operatorname{soc} I$ denotes the socle of I.

5 The formula and existence of almost split sequences. In most of the situations where almost split sequences exist there is a formula of the type discussed above explaining the existence. There are however existence theorems for almost split sequences in subcategories of $\operatorname{mod}\Lambda$ for an artin algebra Λ, where we do not know if there is always a corresponding formula.

III. Hereditary categories with tilting object

The Ext-finite hereditary abelian k-categories with the additional property of having a tilting object are important in the representation theory of finite dimensional algebras. In this chapter we give a brief account of this topic, including some background material on tilting theory (see (R)).

1 Tilting theory. A module T over a finite dimensional hereditary k-algebra Λ is said to be a *tilting module* if $\operatorname{pd}_\Lambda T \leq 1$ (where $\operatorname{pd}_\Lambda T$ denotes the projective dimension of T), $\operatorname{Ext}^1_\Lambda(T,T)=0$ and there is an exact sequence

$$0 \to \Lambda \to T_0 \to T_1 \to 0$$

with T_0 and T_1 summands of a finite direct sum of copies of T. The last condition is equivalent to requiring that $\operatorname{Hom}_\Lambda(T,C)=0=\operatorname{Ext}^1_\Lambda(T,C)$ implies $C=0$. The algebras of the form $\Gamma=\operatorname{End}_\Lambda(T)^{\mathrm{op}}$, where Λ is a finite dimensional hereditary k-algebra, are by definition the *tilted* k-algebras, which have played a central role in the representation theory of finite dimensional algebras. An important feature of tilting theory is the connection with torsion theories. The tilting module T gives rise to a torsion pair $(\mathcal{T},\mathcal{F})$ in $\operatorname{mod}\Lambda$, where the objects of the torsion class $\mathcal{T}$ are the factors of finite direct sums of copies of T. The subcategory $\mathcal{F}$ is the associated torsionfree class, whose objects are the C in $\operatorname{mod}\Lambda$ such that $\operatorname{Hom}(\mathcal{T},C)=0$. There is also an associated torsion pair $(\mathcal{X},\mathcal{Y})$ in $\operatorname{mod}\Gamma$, together with induced equivalences $\operatorname{Hom}_\Lambda(T,\ )\colon \mathcal{T}\to\mathcal{Y}$ and $\operatorname{Ext}^1_\Lambda(T,\ )\colon \mathcal{F}\to\mathcal{X}$. In fact, there is even induced an equivalence of derived categories $D^b(\Lambda)\to D^b(\Gamma)$ (see (H1)).

2 Quasitilted algebras. Let $\mathcal{C}$ be an Ext-finite hereditary abelian k-category, and assume that $\mathcal{C}$ has some tilting object T, that is an object T satisfying $\operatorname{Ext}^1_{\mathcal{C}}(T,T)=0$, and whenever $\operatorname{Hom}_{\mathcal{C}}(T,C)=0=\operatorname{Ext}^1_{\mathcal{C}}(T,C)$, then $C=0$. The algebras of the form $\Gamma=\operatorname{End}_{\mathcal{C}}(T)^{\mathrm{op}}$ for a tilting object T in $\mathcal{C}$, which are clearly finite dimensional over k, are called *quasitilted* algebras (HRS). By definition the class of quasitilted algebras contains the class of tilted algebras. Also in this more general setting there is the same connection with torsion pairs, and the same equivalences of categories. The quasitilted algebras also contain another important class of algebras, the canonical algebras of Ringel, as well as other examples, so the investigation of quasitilted algebras provides a unified treatment. An attractive property is that the quasitilted algebras have a nice homological characterization as the algebras of global dimension at most two, where every indecomposable module has projective or injective dimension at most one.

3 Tilting with respect to torsion pairs. Let T be a tilting Λ-module, where Λ is a finite dimensional hereditary k-algebra, and let $\Gamma = \operatorname{End}_\Lambda(T)^{\mathrm{op}}$. It is then possible to reconstruct Λ, up to Morita equivalence, as $\Lambda \simeq \operatorname{End}_\Gamma(D(T))^{\mathrm{op}}$. Here the right Γ-module T is a tilting module, and then by definition the dual module $D(T)$ is what is called a cotilting module. When $\mathcal{C}$ is not equivalent to some such $\operatorname{mod}\Lambda$, it can not be reconstructed from $\Gamma = \operatorname{End}_{\mathcal{C}}(T)^{\mathrm{op}}$ in the same way. However, there is a more general procedure, where we can construct $\operatorname{mod}\Gamma$ from $\operatorname{mod}\Lambda$ by performing a tilting procedure inside $D^b(\Lambda)$ with respect to the torsion pair $(\mathcal{T}, \mathcal{F})$ associated with the tilting module T (HRS). More generally, we can construct $\operatorname{mod}\Gamma$, where $\Gamma = \operatorname{End}_{\mathcal{C}}(T)^{\mathrm{op}}$, from an Ext-finite hereditary abelian k-category $\mathcal{C}$ with tilting object T this way and, more importantly, we can construct $\mathcal{C}$ from $\operatorname{mod}\Gamma$ by a similar procedure.

The hereditary abelian category $\mathcal{C} = \operatorname{coh}\mathbb{P}^1(k)$ has a tilting object $T = \mathcal{O} \oplus \mathcal{O}[1]$, where $\mathcal{O}$ denotes the structure sheaf. Then $\Gamma = \operatorname{End}_{\mathcal{C}}(T)^{\mathrm{op}}$ is isomorphic to the path algebra $k(\cdot \rightrightarrows \cdot)$. So the derived equivalence between $\mathcal{C}$ and Γ can be explained through existence of a tilting object. As a special case of the above discussion, $\mathcal{C}$ can be constructed from $\operatorname{mod}\Gamma$ via a tilting procedure, which informally amounts to "moving" a torsionfree class from left to right within the derived category. A rough illustration is as follows:

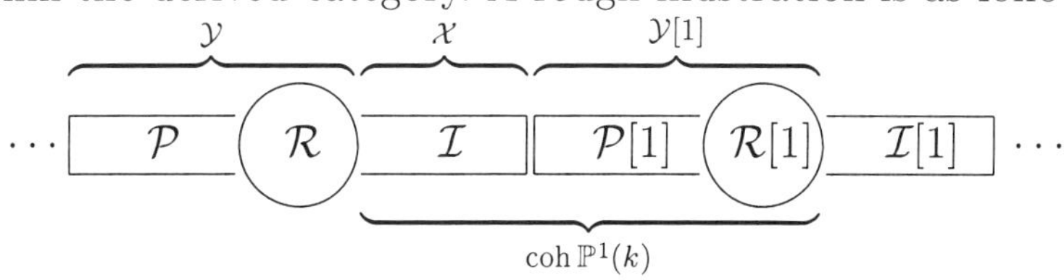

Here $\mathcal{P}$ denotes the preprojective modules for Γ, $\mathcal{I}$ the preinjective modules, and $\mathcal{R}$ the regular ones. $(\mathcal{X}, \mathcal{Y})$ is a torsion pair in $\operatorname{mod}\Gamma$ with respect to which we do the tilting. Note that only the indecomposable objects are included in the picture.

4 Classification. It is an important problem to classify the Ext-finite hereditary abelian k-categories $\mathcal{C}$. For those having a tilting object this is of interest for the study of quasitilted algebras. For the case of k being algebraically closed, a classification of the $\mathcal{C}$ with tilting object has recently been given by Happel (H3). It was shown earlier by Lenzing (L) that the only (indecomposable) noetherian examples are the mod Λ for a finite dimensional hereditary k-algebra Λ, and the categories coh $\mathbb{X}$ of coherent sheaves on a weighted projective line $\mathbb{X}$ in the sense of Geigle–Lenzing (GL). It was shown in (H3), as was supported by previous results, that the noetherian ones are the only ones up to derived equivalence.

References

A1. M. Auslander, *Functors and morphisms determined by objects*, Proc. Conf. on Representation Theory, Philadelphia, Lecture Notes in Pure and Applied Math. 37, Marcel Dekker, New York -Basel, **1978**, 1–244.

A2. M. Auslander, *Rational singularities and almost split sequences*, Trans. Amer. Math. Soc. 293, **1986**, 511–531.

A3. M. Auslander, *Isolated singularities and existence of almost split sequences*, in: Proceedings ICRA IV Ottawa, Lecture Notes in Math. 1178 Springer-Verlag, Berlin - New York, **1986**.

AR1. M. Auslander and I. Reiten, *Stable equivalence of dualizing R-varieties*, Adv. in Math., Vol. 12, No. 3, **1974**, 306–366.

AR2. M. Auslander and I. Reiten, *Representation theory of artin algebras III, Almost split sequences*, Comm. in Alg., Vol. 3, No. 3, **1975**, 239–294.

AR3. M. Auslander and I. Reiten, *Almost split sequences in dimension two*, Adv. in Math., Vol. 66, No. 1, **1987**, 88–118.

AR4. M. Auslander and I. Reiten, *Almost split sequences for $\mathbb{Z}$-graded rings, in: Singularities, representations of algebras and vector bundles*, (Lambrecht 1985) Lecture Notes in Math. 1273, Springer, Berlin–New York, **1987**, 232–243.

ARS. M. Auslander, I. Reiten, S. O. Smalø, *Representation theory of artin algebras*, Cambridge Univ. Press, **1995**.

BK. Bondal and I. Kapranov, *Representable functors, Serre functors and reconstructions*, Izv. Akad. Nauk SSSR Ser. Mat. 53, **1989**, no. 6, 1183–1205.

GL. W. Geigle and H. Lenzing, *A class of weighted projective curves arising in representation theory of finite dimensional algebras, in: Singularities, representations of algebras and vector bundles*, Springer Lecture Notes 1273, **1987**, 265–297.

H1. D. Happel, *Triangulated categories in the representation theory of finite dimensional algebras*, London Math. Soc. Lecture Notes Series 119, **1988**.

H2. D. Happel *Auslander-Reiten triangles in derived categories of finite dimensional algebras*, Proc.Amer.Math.Soc. 112, **1991**, no. 3, 641–648.

H3. D. Happel, *A characterization of hereditary categories with tilting object*, Invent. Math. (to appear).

HR1. D. Happel and I. Reiten, *An introduction to quasitilted algebras*, An. St. Univ. Ovidius Constantza, Vol. 4, **1996**, f. 2, pp. 137–149.

HRS. D. Happel, I. Reiten and S. O. Smalø, *Tilting in abelian categories and quasitilted algebras*, Memoirs of the AMS, Vol. 575, Amer. Math. Soc., **1996**.

L. H. Lenzing, *Hereditary noetherian categories with a tilting complex* Proc. Amer. Math. Soc. 125, **1997**, 1893–1901.

RV1. I. Reiten and M. Van den Bergh, *Noetherian hereditary abelian categories satisfying Serre duality*, preprint.

RV2. I. Reiten and M. Van den Bergh, *Grothendieck gorups and tilting objects*, Journal of Algebra and Representation theory, (60th birthday volume for K. W. Roggenkamp), to appear.

R. C. M. Ringel, *Tame algebras and integral quadratic forms*, Lecture Notes in Math. 1099, Springer–Verlag, Berlin–Heidelberg–New York, **1984**.

SV. T. Stafford and M. Van den Bergh, *Noncommutative curves and noncommutative surfaces*.

THE ABELIAN DEFECT GROUP CONJECTURE: SOME RECENT PROGRESS

JEREMY RICKARD
School of Mathematics
University of Bristol
Bristol BS8 1TW
United Kingdom

1. Background and the basic conjecture

1.1. BLOCK THEORY

We start by summarizing the basics of the block theory of finite groups.

Let G be a finite group. We shall consider representations of G over a field k of characteristic p and (to relate the representation theory over k with the ordinary character theory) over a complete discrete valuation ring $\mathcal{O}$ whose residue field is k and whose field of fractions K has characteristic zero. We shall assume that both k and K are 'large enough' (i.e., contain enough roots of unity) that we do not need to worry about rationality questions. Technically, we want these fields to be **splitting fields** for the groups concerned.

Classical representation theory tells us that the group algebra KG is a product of matrix rings $M_n(K)$, so its module theory is very simple: there are a finite number of simple modules, one for each irreducible character, and every other module is a direct sum of simple modules.

If p divides the order of G, the group algebra kG has a much more complicated structure, but often it will have a direct product decomposition

$$kG \cong kA_0 \times kA_1 \times \cdots \times kA_n.$$

K.W. Roggenkamp and M. Ştefănescu (eds.), Algebra - Representation Theory, 273–299.

If we choose this decomposition to be as fine as possible then the algebras kA_i are called **block (algebras)** of kG. A basic result is that the group algebra $\mathcal{O}G$ has a corresponding decomposition

$$\mathcal{O}G \cong \mathcal{O}A_0 \times \mathcal{O}A_1 \times \cdots \times \mathcal{O}A_n$$

into block algebras, where $kA_i = k \otimes_{\mathcal{O}} \mathcal{O}A_i$. The notation we use $(kA_i, \mathcal{O}A_i)$ is chosen so that the correspondence between blocks of kG and $\mathcal{O}G$ is evident: A_i in isolation has no meaning.

Each indecomposable kG-module M is annihilated by all but one block of kG: we say it 'belongs' to this block. If M is the trivial module k (G acts trivially), then this block is called the **principal block**. We shall choose our notation so that this is kA_0. Similarly, $\mathcal{O}A_0$ is called the principal block of $\mathcal{O}G$. $KA_i = K \otimes_{\mathcal{O}} \mathcal{O}A_i$ is a product of matrix rings corresponding to ordinary characters of G, so every irreducible character also 'belongs' to a block.

To each block is associated a conjugacy class of p-subgroups of G called the **defect groups** of the block. These are the minimal subgroups D such that every module for the block is a direct summand of one induced from D. A simple example to keep in mind is that the defect groups of the principal block are just the Sylow p-subgroups. If the defect group of a block is the trivial group, then the block is just a matrix ring $M_n(k)$.

Brauer's First Main Theorem ((3, Theorem 6.2.6), for example) gives a natural 1-1 correspondence between the blocks of kG with defect group D and the blocks of $kN_G(D)$ with defect group D. This is called the **Brauer correspondence**. In the simple example of the principal block, where D is a Sylow p-subgroup of G, the Brauer correspondent block of $kN_G(D)$ is again the principal block.

The character theory of G and the representation theory of kG are related by a matrix of integers, called the **decomposition matrix**. Given a character χ of G, it is always possible to find an $\mathcal{O}G$-module M whose character is χ, and one can then form the kG-module $\overline{M} = M \otimes_{\mathcal{O}G} kG$. Now M and $\overline{M}$ are by no means unique up to isomorphism, but the simple composition factors of $\overline{M}$, and their multiplicities, are independent of the choice of M. The decomposition matrix is a matrix, with rows indexed by the irreducible characters and columns by the isomorphism classes of simple kG-modules, with the entry in the row indexed by χ and the column indexed by a simple kG-module S recording how many composition factors of $\overline{M}$ are isomorphic to S.

If χ belongs to a particular block, so will all the composition factors of $\overline{M}$, so each block has a decomposition matrix, and if there are n blocks, then the decomposition matrix D of the group will be built out of matrix blocks as follows, where $D_1, \ldots, D_n$ are the decomposition matrices of the individual blocks.

$$D = \begin{pmatrix} D_1 & 0 & \cdots & 0 \\ 0 & D_2 & \cdots & 0 \\ \vdots & \vdots & \ddots & \vdots \\ 0 & 0 & \cdots & D_n \end{pmatrix}$$

The decomposition matrix has another interpretation. Since there is a natural bijection between isomorphism classes of simple kG-modules and indecomposable projective kG-modules, sending a simple module to its projective cover, the columns of the decomposition matrix are also indexed by the isomorphism classes of indecomposable projective kG-modules. If P is such a module, then P can be lifted (uniquely up to isomorphism) to an indecomposable projective $\mathcal{O}G$-module $\tilde{P}$ with $P \cong \tilde{P} \otimes_{\mathcal{O}G} kG$. Then the corresponding column of the decomposition matrix records the multiplicities of the irreducible characters as constituents of the character of $\tilde{P}$.

It follows that the matrix $C = D^{\text{tr}}D$, the **Cartan matrix** of kG, whose rows and columns are indexed by the isomorphism classes of simple kG-modules, records the multiplicities of composition factors of the indecomposable projective kG-modules.

The basics of block theory can be found in any introductory text on modular representation theory. Some relatively recent examples are the books by Alperin (1), Benson (3) and Curtis and Reiner (6),(5).

1.2. BLOCK THEORY: SOME EXAMPLES

Here we will describe two examples, one very simple, the other more complicated, that will serve to illustrate much of the material that we shall describe in this article.

1.2.1. *A_4 in characteristic 2*

In the simpler example, we shall take G to be the alternating group A_5, of order 60, and take k to be a field of characteristic two: in order that k and K are splitting fields, we shall take k to be the finite field $GF(4)$ of order four, although of course any field containing $GF(4)$ would do just as well.

Since we are working in characteristic 2, we must consider the Sylow 2-subgroup P of G, which is the Klein 4-group $C_2 \times C_2$. The normalizer $H = N_G(P)$ is the alternating group A_4, embedded in G in the obvious way.

There are two blocks of kG, the principal block kA (whose defect group is P, or, more precisely, the defect *groups* are are the conjugates of P in G) and a second block 'of defect zero'; i.e., with the trivial group as defect group.

There are five ordinary irreducible characters of G, which we shall label χ_1, χ_{3_a}, χ_{3_b}, χ_4 and χ_5. The subscript refers to the degree of the character, so that χ_5 has degree 5 and χ_1 is the trivial character of degree 1. The extra labels a and b on χ_{3_a} and χ_{3_b} are arbitrary labels just serving to distinguish two different characters of degree 3.

The only irreducible character that belongs to the block of defect zero is χ_4, the others belonging to the principal block.

Up to isomorphism, kG has four simple modules, the trivial module $k = V_1$ and three other modules V_{2_a}, V_{2_b} and V_4. As with the irreducible characters, the subscript indicates the dimension of the module. Only V_4 belongs to the non-principal block, the other three belonging to the principal block.

Listing the irreducible characters in the order $\chi_1, \chi_{3_a}, \chi_{3_b}, \chi_5, \chi_4$ and the simple kG-modules in the order k, V_{2_a}, V_{2_b}, V_4, the decomposition matrix is

$$\begin{pmatrix} 1 & 0 & 0 & 0 \\ 1 & 1 & 0 & 0 \\ 1 & 0 & 1 & 0 \\ 1 & 1 & 1 & 0 \\ 0 & 0 & 0 & 1 \end{pmatrix},$$

the decomposition matrix of the principal block being

$$\begin{pmatrix} 1 & 0 & 0 \\ 1 & 1 & 0 \\ 1 & 0 & 1 \\ 1 & 1 & 1 \end{pmatrix}$$

and the decomposition matrix of the non-principal block being just the 1×1 matrix (1). (More generally, any block of defect zero has just one irreducible character and one isomorphism class of simple modules, with the single entry of the decomposition matrix being 1.)

The Cartan matrix of the principal block of kG is

$$\begin{pmatrix} 4 & 2 & 2 \\ 2 & 2 & 1 \\ 2 & 1 & 2 \end{pmatrix}.$$

The subgroup $H = N_G(P) \cong A_4$ has four irreducible characters, ϕ_1, ϕ_{1_a}, ϕ_{1_b} and ϕ_3, all of which belong to the principal block kB, which in this case is the only block.

There are three isomorphism classes of simple $kN_G(P)$-module, all one-dimensional: the trivial module $k = U_1$, U_{1_a} and U_{1_b}.

The decomposition matrix of $kN_G(P)$ (or equivalently, the principal block kB) is

$$\begin{pmatrix} 1 & 0 & 0 \\ 0 & 1 & 0 \\ 0 & 0 & 1 \\ 1 & 1 & 1 \end{pmatrix},$$

and the Cartan matrix is

$$\begin{pmatrix} 2 & 1 & 1 \\ 1 & 2 & 1 \\ 1 & 1 & 2 \end{pmatrix}.$$

Notice that the principal block kA of kG and its Brauer correspondent, the group algebra $kN_G(P) = kB$, have the same number (four) of irreducible characters and the same number (three) of isomorphism classes of simple modules. However, the two blocks are not isomorphic, as the simple modules have different dimensions, and do not have equivalent module categories, as the Cartan matrices (which are invariants of the module categories up to simultaneous permutation of rows and columns) are different.

1.2.2. *J_1 in characteristic 2*

For the more complicated example, we shall take G to be the smallest Janko group J_1, a sporadic simple group of order 175560. Again, $k = GF(4)$ is a splitting field.

G has an elementary abelian Sylow 2-group $P = C_2 \times C_2 \times C_2$ of order 8, whose normalizer $H = N_G(P)$ is of order 168, with $N_G(P)/P$ being the semi-direct product of a cyclic group C_7 by C_3. $k = GF(4)$ is also a splitting field for this group.

There are seven blocks of kG: the principal block kA, five blocks of defect zero and one with cyclic defect group C_2.

There are 15 irreducible characters. Eight of these, χ_1, χ_{77_a}, χ_{77_b}, χ_{77_c}, χ_{133_a}, χ_{133_b}, χ_{133_c} and χ_{209}, belong to the principal block. One of the characters of degree 77, which we shall choose to be χ_{77_a} takes rational values, with χ_{77_b} and χ_{77_c} being algebraically conjugates of one another. Similarly, one of the characters of degree 133, which we choose to be χ_{133_a} is rational, with χ_{133_b} and χ_{133_c} an algebraically conjugate pair.

Up to isomorphism, kG has 11 simple modules, of which five belong to the principal block kA, these being the trivial module $k = V_1$, V_{20}, V_{56_a}, V_{56_b} and V_{76}.

The decomposition matrix of kA is

$$\begin{pmatrix} 1&0&0&0&0\\ 1&0&0&0&1\\ 1&1&1&0&0\\ 1&1&0&1&0\\ 1&1&1&1&0\\ 1&0&0&1&1\\ 1&0&1&0&1\\ 1&1&1&1&1 \end{pmatrix},$$

and the Cartan matrix is

$$\begin{pmatrix} 8 & 4 & 4 & 4 & 4 \\ 4 & 4 & 3 & 3 & 1 \\ 4 & 3 & 4 & 2 & 2 \\ 4 & 3 & 2 & 4 & 2 \\ 4 & 1 & 2 & 2 & 4 \end{pmatrix}.$$

As in the previous example, $kN_G(P)$ has only one block, the principal block kB.

$N_G(P)$ has eight irreducible characters: the trivial character ϕ_1, ϕ_{1_a}, ϕ_{1_b}, ϕ_{3_a}, ϕ_{3_b}, ϕ_{7_a}, ϕ_{7_b} and ϕ_{7_c}, where ϕ_{7_a} is rational, and the other six non-trivial characters are in three complex conjugate pairs.

There are five isomorphism classes of simple $kN_G(P)$-modules: the trivial module $k = U_1$, U_{1_a}, U_{1_b}, U_{3_a} and U_{3_b}.

The decomposition matrix is

$$\begin{pmatrix} 1 & 0 & 0 & 0 & 0 \\ 0 & 1 & 0 & 0 & 0 \\ 0 & 0 & 1 & 0 & 0 \\ 0 & 0 & 0 & 0 & 1 \\ 0 & 0 & 0 & 1 & 0 \\ 1 & 0 & 0 & 1 & 1 \\ 0 & 1 & 0 & 1 & 1 \\ 0 & 0 & 1 & 1 & 1 \end{pmatrix},$$

and the Cartan matrix is

$$\begin{pmatrix} 2 & 0 & 0 & 1 & 1 \\ 0 & 2 & 0 & 1 & 1 \\ 0 & 0 & 2 & 1 & 1 \\ 1 & 1 & 1 & 4 & 3 \\ 1 & 1 & 1 & 3 & 4 \end{pmatrix}.$$

Notice that, just as in the first example, the principal block kA of kG and its Brauer correspondent $kB = kH$ have the same number (eight) of irreducible characters and the same number (five) of isomorphism classes of simple modules. But for the same reason as before (the Cartan matrices are different), the module categories of the two blocks are not equivalent.

1.3. DERIVED CATEGORIES AND STABLE MODULE CATEGORIES

Let kA be a block algebra. Then we'll denote by $\mathrm{mod}(kA)$ the category of finitely generated (left) modules for kA. This is an **abelian category**: inter alia, module homomorphisms have kernels and cokernels that satisfy all the familiar properties. We'll be looking at some other categories based on $\mathrm{mod}(kA)$ that are not abelian.

Firstly, the (bounded) **derived category** $D^b(\mathrm{mod}(kA))$ of $\mathrm{mod}(kA)$ has as its objects the bounded chain complexes

$$X := \cdots \xrightarrow{d^X_{n+2}} X_{n+1} \xrightarrow{d^X_{n+1}} X_n \xrightarrow{d^X_n} X_{n-1} \xrightarrow{d^X_{n-1}} \cdots$$

of finitely generated kA-modules. By 'bounded' we mean that $X_n \cong 0$ for all but finitely many n. The morphisms are obtained from the chain maps by adjoining inverses to chain maps

$$X \to Y$$

that induce an isomorphism

$$H_n(X) \cong H_n(Y)$$

in homology in every degree n. For example, if X is **acyclic** (i.e., $H_n(X) \cong 0$ for every n), then the zero map from X to the zero chain complex is inverted, so $X \cong 0$ in the derived category. More generally, isomorphic objects of the derived category have isomorphic homology, but the converse is not true in general.

We shall not go into the technical details of the construction of the derived category here, but a helpful way of thinking of the derived category is the slogan, which can be made precise:

The derived category is the category of chain complexes modulo acyclic complexes.

Most maps in the derived category do not have kernels and cokernels in the categorical sense, so it is not an abelian category. However, it has a structure that is in many ways even better: it is a **triangulated category**. We shall not go into details here, but there are two important aspects of this structure. Firstly, a 'shift' which is an automorphism of the category: given any chain complex

X we can form a new chain complex $X[1]$ with $X[1]_n = X_{n-1}$ and $d_n^{X[1]} = -d_{n-1}^X$, and we can iterate this to define $X[t]$ for any integer t. Secondly, the concept of '(distinguished) triangles' or 'cofibre sequences': these are diagrams

$$X \to Y \to Z \to X[1]$$

that play an analogous role in a triangulated category to that played by short exact sequences in an abelian category.

We can regard $D^b(\text{mod}(kA))$ as an extension of the module category: we can embed $\text{mod}(kA)$ into $D^b(\text{mod}(kA))$ by sending a module M to the complex with $M_0 = M$ and $M_n = 0$ for $n \neq 0$. This is a 'full embedding': the maps between modules are exactly the same as those between the corresponding complexes.

Moreover, if we regard the module category as a subcategory of the derived category in this way, the Ext groups have a simple description. If M and N are modules and $t > 0$ is an integer, then there is a natural isomorphism

$$\text{Hom}_{D^b(\text{mod}(kA))}(M, N[t]) \cong \text{Ext}^t_{kA}(M, N),$$

and if $t < 0$ then

$$\text{Hom}_{D^b(\text{mod}(kA))}(M, N[t]) \cong 0.$$

Of course, if $t = 0$ then

$$\text{Hom}_{D^b(\text{mod}(kA))}(M, N[t]) \cong \text{Hom}_{kA}(M, N).$$

Of course this all makes sense if we replace kA with $\mathcal{O}A$, so we can also define the bounded derived category $D^b(\text{mod}(\mathcal{O}A))$ of finitely generated $\mathcal{O}A$-modules.

There are now many published introductions to derived categories and triangulated categories. For example, Weibel's book (17), or the recent lecture notes of König and Zimmermann (9).

Another, related, triangulated category is the **stable module category** $\underline{\text{mod}}(kA)$. Here the objects are finitely generated kA-modules, and the morphisms are equivalence classes of module homomorphisms, where two maps $M \to N$ are equivalent if their difference factors through a projective module (a property of block algebras that is important here is that a module is projective if and

The objects of $D^b(\mathrm{mod}(kA))$ that are isomorphic to bounded complexes of finitely generated projective modules can be characterized by natural properties inside the derived category: namely, they are the objects X for which, for every other object Y of $D^b(\mathrm{mod}(kA))$, $\mathrm{Hom}_{kA}(X, Y[t])$ is zero for all but finitely many integers t. Therefore the stable module category, which is in many ways easier to understand, is determined up to equivalence by the derived category.

1.4. THE BASIC CONJECTURE

Much recent work on the modular representation theory of general finite groups has been concerned with various conjectures concerning the relationship between blocks and their Brauer correspondents (and other related blocks of subgroups). These conjectures often take a simpler form if we insist that the defect group be abelian. For example, Alperin's Weight Conjecture (2), when applied to blocks with abelian defect group, is as follows.

CONJECTURE 1.1. *Let G be a finite group, let $\mathcal{O}A$ be a block algebra of $\mathcal{O}G$ with abelian defect group D, and let $\mathcal{O}B$ be the Brauer correspondent block of $\mathcal{O}N_G(D)$. Then*

(i) kA and kB have the same number of (isomorphism classes of) simple modules, and

(ii) KA and KB have the same number of irreducible characters.

In fact, the two statements are equivalent.

This conjecture is still open in general, but it has been verified in a very large number of special cases.

Of course, the conclusion would follow if $\mathcal{O}A$ and $\mathcal{O}B$ were isomorphic algebras, or, more generally, if the module categories $\mathrm{mod}(\mathcal{O}A)$ and $\mathrm{mod}(\mathcal{O}B)$ were equivalent (in which case $\mathcal{O}A$ and $\mathcal{O}B$ are said to be **Morita equivalent**). Occasionally this is the case, but by no means always. We must therefore look for a weaker structural connection between the blocks.

The derived category (and, to a lesser extent, the stable module category) provide such a connection: Broué (4) made the following conjecture (the Abelian Defect Group Conjecture).

CONJECTURE 1.2. *Let G be a finite group, let $\mathcal{O}A$ be a block algebra of $\mathcal{O}G$ with abelian defect group D, and let $\mathcal{O}B$ be the Brauer correspondent block of $\mathcal{O}N_G(D)$. Then the derived categories $D^b(\mathrm{mod}(\mathcal{O}A))$ and $D^b(\mathrm{mod}(\mathcal{O}B))$ are equivalent as triangulated categories.*

The main reason for stating this for $\mathcal{O}A$ rather than kA is to allow us to consider character-theoretic consequences. Broué showed that if $D^b(\mathrm{mod}(\mathcal{O}A))$ and $D^b(\mathrm{mod}(\mathcal{O}B))$ are equivalent as triangulated categories, then there is a **perfect isometry** I between the groups of virtual characters (i.e., the groups of class functions generated by the irreducible characters belonging to $\mathcal{O}A$ and $\mathcal{O}B$): this means that I is an isomorphism such that for every irreducible character χ, $\pm I(\chi)$ is an irreducible character, and that I restricts to an isomorphism between the groups of class functions generated by characters of projective modules for $\mathcal{O}A$ and $\mathcal{O}B$. This last condition can be translated into an arithmetic condition on character values. The existence of a perfect isometry has been verified in a large number of cases: for example, using the classification of finite simple groups, Fong and Harris (7) have proved this for principal blocks of all groups with abelian Sylow p-subgroups for $p = 2$ and (more recently) $p = 3$.

In particular, the number of irreducible characters belonging to a block $\mathcal{O}A$ is an invariant of the derived category $D^b(\mathrm{mod}(\mathcal{O}A))$, so Alperin's Weight Conjecture (for blocks with abelian defect group) would be a consequence of Broué's Conjecture.

Also, $D^b(\mathrm{mod}(kA))$, and hence $\underline{\mathrm{mod}}(kA)$, is determined by $D^b(\mathrm{mod}(\mathcal{O}A))$, and so another consequence of Broué's Conjecture is the following.

CONJECTURE 1.3. *Let G be a finite group, let kA be a block algebra of kG with abelian defect group D, and let kB be the Brauer correspondent block of $kN_G(D)$. Then the stable module categories $\underline{\mathrm{mod}}(kA)$ and $\underline{\mathrm{mod}}(kB)$ are equivalent as triangulated categories.*

This is of interest because an equivalence of stable module categories is often easier to prove than an equivalence of derived categories. Incidentally, it is hard to recover the number of simple modules from the stable module category (although it is conjectured that this is possible), but if the last conjecture were true, then Alperin's Weight Conjecture (for blocks with abelian defect group) would be a consequence.

1.5. PERFECT ISOMETRIES: SOME EXAMPLES

We shall now return to the examples of Section 1.2

1.5.1. A_5

In this example, a perfect isometry is given by the mapping

$$\begin{aligned} \chi_1 &\mapsto \phi_1 \\ \chi_{3_a} &\mapsto -\phi_{1_a} \\ \chi_{3_b} &\mapsto -\phi_{1_b} \\ \chi_5 &\mapsto -\phi_3 \end{aligned}$$

Note that verifying that this is a perfect isometry is a simple calculation with the decomposition matrices. We need to check that the character of any projective $\mathcal{O}G$-module is sent to an integer linear combination of characters of projective $\mathcal{O}H$-modules. For example, the projective cover of the trivial module has character

$$\chi_1 + \chi_{3_a} + \chi_{3_b} + \chi_5,$$

which is sent to

$$\phi_1 - \phi_{1_a} - \phi_{1_b} - \phi_3,$$

which is

$$[P(U_1)] - [P(U_{1_a})] - [P(U_{1_b})],$$

where $P(M)$ denotes the projective cover of a module M, and $[P(M)]$ denotes its character.

1.5.2. J_1

Here, a perfect isometry is given by

$$\begin{aligned}
\chi_1 &\mapsto \phi_1\\
\chi_{77_a} &\mapsto -\phi_{3_a}\\
\chi_{77_b} &\mapsto \phi_{1_a}\\
\chi_{77_c} &\mapsto \phi_{1_b}\\
\chi_{133_a} &\mapsto -\phi_{3_b}\\
\chi_{133_b} &\mapsto -\phi_{7_b}\\
\chi_{133_c} &\mapsto -\phi_{7_c}\\
\chi_{209} &\mapsto -\phi_{7_a}
\end{aligned}$$

2. Refinements of the conjecture

2.1. THEORETICAL CONDITIONS FOR EQUIVALENCES

Let us give some theoretical conditions that guarantee an equivalence of module categories, stable module categories, and derived categories. These are all very similar.

The first is classical. If $\mathcal{O}A$ and $\mathcal{O}B$ are Morita equivalent blocks, then there is an $\mathcal{O}A$-$\mathcal{O}B$-bimodule ${}_{\mathcal{O}A}X_{\mathcal{O}B}$, which is finitely generated and projective as a left $\mathcal{O}A$-module and as a right $\mathcal{O}B$-module, such that, if $X^\vee$ is the $\mathcal{O}$-linear dual $\mathrm{Hom}_{\mathcal{O}}(X, \mathcal{O})$ of X, which is an $\mathcal{O}B$-$\mathcal{O}A$-bimodule, then

$$X \otimes_{\mathcal{O}B} X^\vee \cong \mathcal{O}A$$

as $\mathcal{O}A$-bimodules, and

$$X^\vee \otimes_{\mathcal{O}A} X \cong \mathcal{O}B$$

as $\mathcal{O}B$-bimodules.

If a bimodule X with these properties exists, then the equivalence of module categories is easy to describe, as

$$? \otimes_{\mathcal{O}A} X : \mathrm{mod}(\mathcal{O}A) \to \mathrm{mod}(\mathcal{O}B)$$

is an equivalence of categories, with inverse given by

$$? \otimes_{\mathcal{O}B} X^\vee : \mathrm{mod}(\mathcal{O}B) \to \mathrm{mod}(\mathcal{O}A).$$

The bimodule X has to be projective over $\mathcal{O}A$ and $\mathcal{O}B$, as otherwise the functors described above would not be exact, and so could not possibly be equivalences of categories.

Of course, there is a similar condition for Morita equivalence of blocks over k.

Adjusting the condition on X slightly, and bearing in mind the slogan

The stable module category is the module category modulo projective modules.

we get a similar (this time sufficient) condition for **stable equivalence** of blocks (i.e., equivalence of stable module categories.

Suppose there is an kA-kB-bimodule ${}_{kA}X_{kB}$, which is finitely generated and projective as a left kA-module and as a right kB-module, such that, if $X^\vee$ is the k-linear dual $\mathrm{Hom}_k(X, k)$ of X, which is an kB-kA-bimodule, then

$$X \otimes_{kB} X^\vee \cong kA \oplus P$$

as kA-bimodules for some projective kA-bimodule P, and

$$X^\vee \otimes_{kA} X \cong kB \oplus Q$$

as kB-bimodules for some projective kB-bimodule Q. Then there is an equivalence of (triangulated) categories

$$? \otimes_{kA} X : \underline{\mathrm{mod}}(kA) \to \underline{\mathrm{mod}}(kB)$$

with inverse

$$? \otimes_{kB} X^\vee : \underline{\mathrm{mod}}(kB) \to \underline{\mathrm{mod}}(kA).$$

This time the reason that the bimodule X has to be projective over $\mathcal{O}A$ and $\mathcal{O}B$ is that otherwise the functors described above would not even be well-defined on the stable module categories, as they would not send projective modules to projective modules.

A stable equivalence of this form is called a **stable equivalence of Morita type**, because of its similarity to the case of Morita equivalence.

For **derived equivalence** (i.e., equivalence of derived categories) there is also a similar condition. This time bear in mind the slogan

The derived category is the category of chain complexes modulo acyclic complexes.

Suppose there is a bounded complex X of $\mathcal{O}A$-$\mathcal{O}B$-bimodules that are all finitely generated and projective as left $\mathcal{O}A$-modules and as right $\mathcal{O}B$-modules, such that, if $X^\vee$ is the $\mathcal{O}$-linear dual $\mathrm{Hom}_\mathcal{O}(X, \mathcal{O})$ of X, which is a bounded complex of $\mathcal{O}B$-$\mathcal{O}A$-bimodules, then

$$X \otimes_{\mathcal{O}B} X^\vee \cong \mathcal{O}A \oplus S$$

as complexes of $\mathcal{O}A$-bimodules, where S is an acyclic complex, and

$$X^\vee \otimes_{\mathcal{O}A} X \cong \mathcal{O}B \oplus T$$

as complexes of $\mathcal{O}B$-bimodules, where T is an acyclic complex. Then there is an equivalence of (triangulated) categories

$$? \otimes_{kA} X : D^b(\mathrm{mod}(kA)) \to D^b(\mathrm{mod}(kB))$$

with inverse

$$? \otimes_{kB} X^\vee : D^b(\mathrm{mod}(kB)) \to D^b(\mathrm{mod}(kA)).$$

If the terms of X were not projective as $\mathcal{O}A$-modules and as $\mathcal{O}B$-modules, then the tensor product functors would not preserve acyclicity of complexes, and so would not induce functors between derived categories.

Finally, the slogan

The stable module category is the derived category modulo bounded complexes of projective modules.

suggests a different sufficient condition for stable equivalence.

Suppose there is a bounded complex X of kA-kB-bimodules that are all finitely generated and projective as left kA-modules and as right kB-modules, such that, if $X^\vee$ is the k-linear dual $\mathrm{Hom}_k(X, k)$ of X, which is a bounded complex of kB-kA-bimodules, then

$$X \otimes_{kB} X^\vee \cong kA \oplus S \oplus P$$

as complexes of kA-bimodules, where S is an acyclic complex and P is a complex of projective bimodules, and

$$X^\vee \otimes_{kA} X \cong kB \oplus T \oplus Q$$

as complexes of kB-bimodules, where T is an acyclic complex and Q is a complex of projective bimodules. Then there is an equivalence of (triangulated) categories

$$? \otimes_{kA} X : \underline{\mathrm{mod}}(kA) \to \underline{\mathrm{mod}}(kB)$$

with inverse

$$? \otimes_{kB} X^{\vee} : \underline{\mathrm{mod}}(kB) \to \underline{\mathrm{mod}}(kA).$$

This generalizes the earlier condition, which we can recover by insisting that the complex X be a single bimodule. In fact, if there is a complex inducing a stable equivalence, then there is a single bimodule inducing an isomorphic equivalence. However, as we shall see, it might be that there is a complex with more than one term that is in some sense more natural than any single bimodule.

2.2. SPLENDID EQUIVALENCES

When Broué introduced perfect isometries and the Abelian Defect Group Conjecture, he noted that perfect isometries do not occur in isolation. Restricting our attention, for simplicity, to principal blocks, in examples where the principal blocks of G and $H = N_G(P)$ have a perfect isometry between them, there are also 'compatible' perfect isometries between the principal blocks of the 'local' subgroups $C_G(Q)$ and $C_H(Q)$ for all subgroups Q of P. Broué formalized this 'compatibility' in the definition of an **isotypy**. It turns out that it is much easier to check the existence of an isotypy than to check the existence of a perfect isometry alone. In virtually all the cases where a perfect isometry is known to occur, there is known to be an isotypy.

For principal blocks, the notion of a **splendid equivalence** was defined in (14), giving a definition of a compatible family of equivalences of derived categories corresponding to the notion of an isotypy. This has since been extended to other blocks by Harris, Linckelmann and, most generally, by Puig (12).

In this section, we describe the theory of splendid equivalences. For simplicity, we shall restrict our attention to principal blocks.

First recall that a complex X is **chain contractible** if there is a graded homomorphism $h : X \to X$ of degree one (i.e., a sequence of homomorphisms $h_n : X_n \to X_{n+1}$) such that

$$\mathrm{id}_X = hd + dh$$

(i.e., for each n, $\mathrm{id}_{X_n} = h_{n-1}d_n^X + d_{n+1}^X h_n$). This is a stronger condition than being acyclic, and is equivalent to requiring that X be a direct sum of complexes of the form

$$\cdots \to 0 \to M \to M \to 0 \to \ldots$$

for various modules M, in various degrees, where the differentials $M \to M$ are isomorphisms.

Let G be a finite group, and H a subgroup that contains a Sylow p-subgroup P of G. For example, this is the case in the situation of the Abelian Defect Group Conjecture for principal blocks, if we take $H = N_G(P)$. Let $\mathcal{O}A$ and $\mathcal{O}B$ be the principal blocks of $\mathcal{O}G$ and $\mathcal{O}H$ respectively. Then a bounded complex X of finitely generated $\mathcal{O}A$-$\mathcal{O}B$-bimodules is called a **splendid tilting complex** if

$$X \otimes_{\mathcal{O}B} X^\vee \cong \mathcal{O}A \oplus S$$

as complexes of $\mathcal{O}A$-bimodules, where S is a chain contractible complex, and

$$X^\vee \otimes_{\mathcal{O}A} X \cong \mathcal{O}B \oplus T$$

as complexes of $\mathcal{O}B$-bimodules, where T is a chain contractible complex, and that each bimodule X_n is (considered as a $\mathcal{O}[G \times H]$-module by letting

$$x.(g,h) = g^{-1}xh$$

for $x \in X_n$, $g \in G$, $h \in H$) a direct summand of a direct sum of induced $\mathcal{O}[G \times H]$-modules of the form

$$\mathrm{Ind}_{\Delta Q}^{G\times H}\mathcal{O}$$

for various subgroups $Q \leq P$. Here ΔQ is the diagonal copy

$$\{(q,q) \in G \times H : q \in Q\}$$

of Q in $G \times H$, and $\mathcal{O}$ denotes the trivial $\mathcal{O}[\Delta Q]$-module.

Note that the condition on X_n implies that it is projective as a left $\mathcal{O}A$-module and as a right $\mathcal{O}B$-module, so a splendid tilting complex always induces a derived equivalence as in the previous section; such a derived equivalence is called a **splendid equivalence**.

Of course, we can also make the analogous definition over k instead of $\mathcal{O}$. If X is a splendid tilting complex for $\mathcal{O}A$ and $\mathcal{O}B$, then $X \otimes_{\mathcal{O}} k$ will be a splendid tilting complex for kA and kB.

Any functor between module categories preserves chain contractibility of complexes, whereas only an exact functor preserves acyclicity. The main reason for strengthening the conditions on $X \otimes_{\mathcal{O}B} X^{\vee}$ and $X^{\vee} \otimes_{\mathcal{O}A} X$ to involve chain contractibility is that there is one particular non-exact functor that we shall want to apply to a splendid tilting complex.

In slightly more generality than we need, let Γ be a finite group, and let $R \leq \Gamma$ be a p-subgroup. Then the **Brauer construction with respect to** R is the functor

$$?(R) : \mathrm{mod}(k\Gamma) \to \mathrm{mod}(kN_{\Gamma}(R))$$

given by

$$M(R) = M^R / \sum_{R'<R} \mathrm{Tr}_{\mathrm{R'}}^{\mathrm{R}}(\mathrm{M}^{\mathrm{R'}}).$$

Here M^R denotes the fixed points of R on M, and $Tr_{R'}^{R}$ denotes the relative trace map, which sends $m \in M^{R'}$ to

$$\sum_r mr \in M^R,$$

where r runs over a set of representatives for the cosets $R' \backslash R$. The action of $N_{\Gamma}(R)$ on $M(R)$ is induced by the action on M^R, so in fact, $M(R)$ is a module for $N_{\Gamma}(R)/R$, since R acts trivially.

If M is a permutation module (i.e., $M = k\Omega$ has a basis Ω that is fixed setwise by Γ), then $M(R)$ is isomorphic to the $kN_{\Gamma}(R)$-module $k[\Omega^R]$ whose basis is the set of fixed points of R acting on Ω. We shall only apply the Brauer construction to direct summands of permutation modules, so this simpler description of $M(R)$ will suffice. However, this description does not make it clear that $M \mapsto M(R)$ is a functor, since it appears to depend on the choice of a basis Ω (and, indeed, the simpler description would not give a functor if R were not a p-group), whereas it is clear that the original definition defines a functor.

The most important property of splendid equivalences is given in the following theorem, the special case of (14, Theorem 4.1) that applies to the situation of the Abelian Defect Group Conjecture for principal blocks.

THEOREM 2.1. *Suppose G has an abelian Sylow p-subgroup P and $H = N_G(P)$. Let kA and kB be the principal blocks of kG and kH respectively, and let ${}_{kA}X_{kB}$ be a splendid tilting complex. Then for every subgroup $Q \leq P$, the complex $X(\Delta Q)$ is a splendid tilting complex for the principal blocks of the centralizers $C_G(P)$ and $C_H(P)$.*

Note that the conclusion makes sense, since

$$C_G(Q) \times C_H(Q) = C_{G\times H}(\Delta Q)$$

is a subgroup of $N_{G\times H}(\Delta Q)$, and so we can regard $X(\Delta Q)$ as a complex of $kC_G(Q)$-$kC_H(Q)$-bimodules.

This theorem gives the promised notion of a 'compatible family' of derived equivalences.

The theorem is not true if we replace k with $\mathcal{O}$, as the Brauer construction does not behave so nicely for modules over $\mathcal{O}$. However, this is not a problem, because of the following theorem (14, Theorem 5.2).

THEOREM 2.2. *Suppose X is a splendid tilting complex for kA and kB. Then there is a splendid tilting complex $\tilde{X}$ for $\mathcal{O}A$ and $\mathcal{O}B$, unique up to isomorphism, such that*

$$X \cong \tilde{X} \otimes_{\mathcal{O}} k.$$

Thus we can perform the construction of 'local' splendid equivalences over k using the Brauer construction, and then lift them to $\mathcal{O}$.

2.3. ROUQUIER'S GLUEING OF SPLENDID TILTING COMPLEXES

Recently, Rouquier (16, 15) has proved a partial converse to Theorem 2.1. We shall not state his results in the greatest possible generality, but in particular he showed that if, for each non-trivial subgroup $Q \leq P$, one had a splendid tilting complex X_Q for the principal blocks of $kC_G(Q)$ and $kC_H(Q)$ (in the situation of the Abelian Defect Group Conjecture, say), and if these complexes satisfied suitable compatibility conditions and all the differentials (which are maps between direct summands of permutation modules) were induced by maps between permutation modules preserving the permutation bases, then one could 'glue' these complexes to produce a complex X of kG-kH-bimodules such that, for each non-trivial $Q < P$,

$$X_Q \cong X(\Delta Q).$$

Now, it is not reasonable to expect that X should necessarily be a splendid tilting complex for the principal blocks kA and kB of kG and kH: since every Brauer functor $?(\Delta Q)$ sends all projective bimodules to zero, the most one would hope for is that X should be a splendid tilting complex 'up to projective terms'. In the following sense, that is exactly what happens.

THEOREM 2.3 (Rouquier). *The complex X described above induces a stable equivalence between kA and kB. More precisely,*

$$X \otimes_{kB} X^{\vee} \cong kA \oplus S \oplus P$$

as complexes of kA-bimodules, where S is a chain contractible complex and P is a complex of projective bimodules, and

$$X^{\vee} \otimes_{kA} X \cong kB \oplus T \oplus Q$$

as complexes of kB-bimodules, where T is a chain contractible complex and Q is a complex of projective bimodules.

Rouquier uses this, in (15), to prove that if G is any group whose Sylow p-subgroup P is isomorphic to $C_p \times C_p$, then the principal blocks of kG and $kN_G(P)$ are stably equivalent. It is still very much an open question whether Broué's Conjecture is true for such blocks: i.e., whether they are derived equivalent.

Rouquier's results give some hope of an inductive proof of the Abelian Defect Group Conjecture: the major missing step being a procedure for 'lifting' Rouquier's stable equivalences to splendid equivalences.

2.4. GLUEING IN THE EXAMPLES

2.4.1. A_5

Here Rouquier's procedure gives nothing new. For every non-trivial subgroup $Q \leq P$, $C_G(Q) = C_H(Q) = P$, so the identity functor is a local splendid equivalence for every Q, the local splendid tilting complex being the single bimodule ${}_{kP}kP_{kP}$. Rouquier's glueing procedure produces the kA-kB-bimodule ${}_{kA}kA_{kB}$, which amounts to saying that restriction induces a stable equivalence between kA and kB, a well-known fact.

In (14), a splendid tilting complex is exhibited for this example, having the form

$$X = \cdots \to 0 \to M \to kA \to 0 \to \ldots,$$

where M is a projective bimodule. As expected, this agrees with the result of Rouquier's glueing procedure up to projective terms.

2.4.2. J_1

This example is more interesting. For most non-trivial subgroups $Q \leq P$, we have $C_G(Q) = C_H(Q) = P$, so that kP is a 'local' splendid tilting complex. However, there is one conjugacy class of subgroups $Q \cong C_2$ for which

$$C_G(Q) \cong Q \times A_5$$

but

$$C_H(Q) \cong Q \times A_4.$$

Here we can use the previous example to construct a splendid tilting complex $X \otimes_k kQ$, where X is the splendid tilting complex for the principal blocks of A_5 and A_4.

The compatibility conditions needed for Rouquier's glueing procedure are trivially satisfied in this case, so we get a complex Y inducing a stable equivalence between kA and kB.

3. Lifting stable equivalences to derived equivalences

As we have pointed out, if we had a general procedure for lifting stable equivalences to splendid equivalences for Brauer correspondent blocks with abelian defect group, then we would be close to a proof of Broué's Conjecture. No such procedure exists at present, but substantial progress has been made recently, starting with an idea of Okuyama (11), based on a theorem of Linckelmann (10).

Linckelmann's theorem states that if a stable equivalence of Morita type between two blocks (over k) sends simple modules to simple modules, then the two blocks are actually Morita equivalent.

Okuyama's method is to use this as follows.

Suppose there is a stable equivalence of Morita type between kA and kB, and suppose we can find a derived equivalence between kB and some k-algebra C such that the composition

$$\underline{\mathrm{mod}}(kA) \equiv \underline{\mathrm{mod}}(kB) \equiv \underline{\mathrm{mod}}(C)$$

(of the original stable equivalence with the one induced by the derived equivalence) send simple kA-modules to simple C-modules. Then by Linckelmann's theorem, kA and C are Morita equivalent, and so, since kB and C are derived equivalent, so are kA and kB.

At first sight, this doesn't appear to be much progress, since we seem to have replaced the problem of proving kA and kB to be derived equivalent with a similar problem of proving kB and C derived equivalent. However, so long as we know the images of the simple kA-modules under the stable equivalence with kB, we can in principle apply Okuyama's method with no further knowledge of the block kA. This is a huge help, since in a typical case of Broué's conjecture, the structure of the block kA is complicated and poorly understood, whereas the much smaller block kB is much better understood and has a much simpler structure.

Okuyama (11) used this idea to prove several cases of Broué's Conjecture, especially for $p = 3$ and groups with Sylow p-subgroup $C_3 \times C_3$. More recently he has used this method to give a general proof for $PSL(2, p^n)$ in characteristic p.

Building on Okuyama's method, the following theorem (13) gives a strategy for determining the algebra C.

THEOREM 3.1. *Suppose that kB is a block algebra, and that $X_1, \ldots, X_n$ are objects of $D^b(\mathrm{mod}(kB))$ such that*

(a) $\mathrm{Hom}(X_i, X_j[t]) = 0$ *for* $1 \leq i, j \leq n$ *and* $t < 0$,

(b)

$$\mathrm{Hom}(X_i, X_j) = \begin{cases} 0 & \text{if } i \neq j \\ k & \text{if } i \neq j, \end{cases}$$

and

(c) $X_1, \ldots, X_n$ *generate* $D^b(\mathrm{mod}(kB))$ *as a triangulated category.*

Then there is a k-algebra C and a derived equivalence from C to kB that takes the simple C-modules to $\{X_1, \ldots, X_n\}$.

Note the resemblance of condition (b) to Schur's Lemma.

To combine this theorem with Okuyama's method, suppose we have a stable equivalence of Morita type from kA to kB. Then if we can find objects $X_1, \ldots, X_n$ of $D^b(\mathrm{mod}(kB))$ that satisfy the conditions of the theorem and such that the images of $X_1, \ldots, X_n$ under the natural functor

$$D^b(\mathrm{mod}(kB)) \to \underline{\mathrm{mod}}(kB)$$

are isomorphic to the images of the simple kA-modules under the original stable equivalence, then Linckelmann's theorem shows that kA and C are Morita equivalent, and so kA and kB are derived equivalent.

3.1. LIFTING A STABLE EQUIVALENCE TO A DERIVED EQUIVALENCE FOR J_1

Here we sketch how the procedure of the previous section can be carried out in our example of the Janko group J_1, with the help of the computer algebra package Magma. Broué's Conjecture was originally proved in this example (using different methods, but also with the help of computer calculations) by Okuyama.

First, we must calculate the images of the simple kA-modules under the stable equivalence given by the complex Y obtained in Section 2.4. This complex has two non-zero terms, and calculation with Magma shows that, modulo projective modules,

$$\begin{aligned} V_1 \otimes_{kA} Y &= \cdots \to 0 \to U_1 \to \ldots \\ V_{20} \otimes_{kA} Y &= \cdots \to 0 \to B_0 \to \ldots \\ V_{56_a} \otimes_{kA} Y &= \cdots \to C_1 \to C_0 \to \ldots \\ V_{56_b} \otimes_{kA} Y &= \cdots \to D_1 \to D_0 \to \ldots \\ V_{76} \otimes_{kA} Y &= \cdots \to E_1 \to E_0 \to \ldots, \end{aligned}$$

where $\dim_k(B_0) = 18$, $\dim_k(C_0) = \dim_k(D_0) = 8$, $\dim_k(C_1) = \dim_k(D_1) = 28$, $\dim_k(B_0) = 12$ and $\dim_k(B_1) = 56$, and all the non-zero differentials are surjective.

Denoting the projective cover of the simple kB-module U_{1_a} by P_{1_a} and so on, further calculation with Magma shows that it is possible to 'adjust' the complexes above by introducing extra projective modules to produce complexes with fairly small homology groups.

$$\begin{array}{llllllllllll} X_1 &= 0 \to & 0 & \to & 0 & \to & 0 & \to & U_1 & \to 0 \\ X_2 &= 0 \to & 0 & \to & P_{1_a} \oplus P_{1_b} & \to & P_{3_b} & \to & B_0 & \to 0 \\ X_3 &= 0 \to & P_{1_b} & \to & P_{3_b} & \to & C_1 & \to & C_0 & \to 0 \\ X_4 &= 0 \to & P_{1_a} & \to & P_{3_b} & \to & D_1 & \to & D_0 & \to 0 \\ X_5 &= 0 \to & P_{3_b} & \to & P_{1_a} \oplus P_{1_b} \oplus P_{3_a} \oplus P_{3_b} & \to & E_1 & \to & E_0 & \to 0 \end{array}$$

Here X_3, X_4 and X_5 all have zero homology in all degrees except degrees one and two (1-dimensional, isomorphic to the trivial module U_1) and three (4-dimensional, but different in each case).

Only X_2 has homology groups that are at all large. The degree one homology is 7-dimensional, and the degree two homology is 11-dimensional.

Calculation with Magma verifies that these objects satisfy the conditions of the theorem, proving that Broué's Conjecture is true for this group. Checking that homomorphisms between the objects vanish is simplified by a theorem of Holloway (8) that allows one to do the calculation in the chain homotopy category rather than the derived category.

More complicated calculations of the same sort have recently been carried out by Holloway (8): in particular, he has proved that the Abelian Defect Group Conjecture is true for the sporadic Hall-Janko group (of order 604800) in characteristic 5.

References

1. J. L. Alperin. *Local representation theory.* Cambridge University Press, Cambridge, 1986. Modular representations as an introduction to the local representation theory of finite groups.
2. J. L. Alperin. Weights for finite groups. In *The Arcata Conference on Representations of Finite Groups (Arcata, Calif., 1986)*, pages 369–379. Amer. Math. Soc., Providence, RI, 1987.
3. D. J. Benson. *Representations and cohomology. I.* Cambridge University Press, Cambridge, second edition, 1998. Basic representation theory of finite groups and associative algebras.
4. Michel Broué. Isométries parfaites, types de blocs, catégories dérivées. *Astérisque*, (181-182):61–92, 1990.
5. Charles W. Curtis and Irving Reiner. *Methods of representation theory. Vol. II.* John Wiley & Sons Inc., New York, 1987. With applications to finite groups and orders, A Wiley-Interscience Publication.
6. Charles W. Curtis and Irving Reiner. *Methods of representation theory. Vol. I.* John Wiley & Sons Inc., New York, 1990. With applications to finite groups and orders, Reprint of the 1981 original, A Wiley-Interscience Publication.
7. Paul Fong and Morton E. Harris. On perfect isometries and isotypies in finite groups. *Invent. Math.*, 114(1):139–191, 1993.
8. Miles Holloway. Ph.D. thesis, Bristol. In preparation.
9. Steffen König and Alexander Zimmermann. *Derived equivalences for group rings.* Springer-Verlag, Berlin, 1998. With contributions by Bernhard Keller, Markus Linckelmann, Jeremy Rickard and Raphaël Rouquier.
10. Markus Linckelmann. Stable equivalences of Morita type for self-injective algebras and p-groups. *Math. Z.*, 223(1):87–100, 1996.
11. Tetsuro Okuyama. Some examples of derived equivalent blocks of finite groups. Preprint, Hokkaido, 1998.
12. Lluís Puig. *On the local structure of Morita and Rickard equivalences between Brauer blocks.* Birkhäuser Verlag, Basel, 1999.
13. Jeremy Rickard. Equivalences of derived categories for symmetric algebras. Preprint, Bristol, 2000.
14. Jeremy Rickard. Splendid equivalences: derived categories and permutation modules. *Proc. London Math. Soc. (3)*, 72(2):331–358, 1996.
15. Raphaël Rouquier. Block theory via stable and rickard equivalences. Preprint, Paris, 2000.
16. Raphaël Rouquier. Gluing p-permutation modules. Preprint, Paris, 1998.
17. Charles A. Weibel. *An introduction to homological algebra.* Cambridge University Press, Cambridge, 1994.

2-DIMENSIONAL ORDERS AND INTEGRAL HECKE ORDERS

KLAUS W. ROGGENKAMP
(roggenkamp@mathematik.uni-stuttgart.de)
Fakultät Mathematik,
Universität Stuttgart
Pfaffenwaldring 57,
D–70550 Stuttgart, Germany

1. Introduction

These Notes are organized as follows:

SECTION I: Introduction to 2-dimensional orders, The Higman ideal and Tits' deformation theorem.

SECTION II: Hecke orders and their blocks with cyclic defect and tree orders.

SECTION III: Classification of Cohen-Macaulay modules over 2-dimensional tree-orders.

SECTION IV: Extensions between Cohen-Macaulay Specht-modules of Hecke orders of the symmetric group (Soriano's Lemma).

We shall mainly consider Cohen-Macaulay $\mathbb{Z}[q]$-orders Λ in a separable K-algebra A i. e. Λ is a unitial subring of A, finitely generated and projective over R and spanning A.

K.W. Roggenkamp and M. Ştefănescu (eds.), Algebra - Representation Theory, 301–349.

2. 2-dimensional orders

Unless otherwise stated we shall assume that R IS A REGULAR INTEGRAL DOMAIN OF KRULL DIMENSION 2 – in particular, R is noetherian.

2.1. THE ARITHMETICAL SURFACE

We shall fix some basic notation, which will be used throughout the paper:

NOTATION 2.1.

1. *We consider the* ARITHMETIC SURFACE $\mathbb{Z}[q]$ *and put*

$$[n] = \sum_{i=0}^{n-1} q^i \text{ for } n \in \mathbb{N}, \ [n](s) = \sum_{i=0}^{n-1} s^i \text{ for } s \text{ in an ring } S .$$

2. *The* MAXIMAL IDEALS *in* $\mathbb{Z}[q]$ *are of the form (cf. (Mum; 88))*

$$\mathfrak{m} = \langle f, p \rangle ,$$

where $f \in \mathbb{Z}[q]$ *is irreducible modulo the rational prime* p.

3. *The importance of the polynomial ring* $\mathbb{Z}[q]$ *lies in the* TWO DIFFERENT TYPES OF MINIMAL PRIME IDEALS [1], *which are principal,* $\mathbb{Z}[q]$ *being factorial.*
The ARITHMETIC PRIME IDEALS *are of the form* $< f(q) >$ *for a primitive irreducible polynomial* $f(q) \in \mathbb{Z}[q]$. *As quotient field of the residue ring* EVERY *algebraic number field occurs, so that in some sense the ring* $\mathbb{Z}[q]$ *incorporates all of algebraic number theory. From these ideals we get the integral theory.*
The GEOMETRIC PRIME IDEALS [2] *are of the form* $< p >$ *for a rational prime* p, *and the quotient ring is* $\mathbb{F}_p[q]$, *the polynomial ring, which in turn maps onto* EVERY *finite field. From these ideals we get the modular theory.*

4. *For a rational prime* p *let* $\Phi_{p^n}(q)$ *be the* MINIMUM POLYNOMIAL OF A PRIMITIVE p^n-TH ROOT OF UNITY θ_{p^n}.

---

[1] This means height one primes in our case.
[2] The name was suggested to me by Claus M. Ringel.

5. *We denote by*

$$R_{p^n} := \mathbb{Z}[\theta_{p^n}]$$

the ring of integers in the p^n-th cyclotomic number field.

6. *If p_1 is another (it could be the same) rational prime, then we denote by $\Phi_{p^n,p_1}(q)$ the minimum polynomial of θ_{p^n} over the prime field of characteristic p_1, $\mathbb{F}_{p_1}$.*
According to the above observations, the ideals

$$\mathfrak{m}_{p^n,p_1} := \langle\, \Phi_{p^n,p_1}(q), p_1 \,\rangle$$

are then maximal for $p \neq p_1$.

We list some facts:

CLAIM 2.2.

(i) As ideals we have

$$\langle\, q-1, [n] \,\rangle = \langle\, q-1, n \,\rangle$$
$$\langle\, q-1, p \,\rangle = \langle\, q-1, \Phi_{p^n}(q) \,\rangle\,.$$

The latter is a maximal ideal with residue field $\mathbb{F}_p$.

(ii) Assume that m and n are relatively prime natural numbers, then

$$\langle\, [m], [n] \,\rangle = \mathbb{Z}[q]\,.$$

(iii) Let p_1 and p_2 be different rational primes.
Then $\langle\, \Phi_{p_1^{n_1}}, \Phi_{p_2^{n_2}} \,\rangle = \mathbb{Z}[q]$.

(iv) Let $\theta = \theta_{p^n}$ be a primitive p^n-th root of unity, and let $\mathbb{Z}[q]_{p^n} = \mathbb{Z}[q][\theta_{p^n}]$.
The ideal $\langle\, \theta_{p^n} - q, q-1 \,\rangle = \langle\, \theta_{p^n} - 1, q-1 \,\rangle$ is maximal with residue field $\mathbb{F}_p$, lying above $\langle\, p, q-1 \,\rangle$.

(v) For a maximal ideal $\mathfrak{m}$ in an integral domain S we denote by $S_{\mathfrak{m}}$ the localization at $\mathfrak{m}$ and by $\widehat{S_{\mathfrak{m}}}$ completion at $\mathfrak{m}$. We shall use a similar notation for S-modules. For a rational prime number p we have an injection

$$\mathbb{Z}_p \subset \mathbb{Z}[q]_{\langle\, q-1, [p] \,\rangle}\,.$$

Proof **(i)**: We have $[n] = f(q) \cdot (q-1) + n$ for some polynomial $f(q)$ as follows from calculating modulo $q-1$. Similarly the second assertion is verified.
(ii) Indeed, we can write $a \cdot m = b \cdot n + 1$ for a, $b \in \mathbb{N}$ possibly after exchanging the role of m and n. Then

$$\sum_{i=0}^{a} q^{i \cdot m} \cdot [m] - q \cdot (\sum_{i=0}^{b} q^{i \cdot n} \cdot [n]) = 1 .$$

(iii) This is an immediate consequence of (i) and (ii).
(iv) We obviously have

$$\langle \theta_{p^n} - q, q-1 \rangle \supseteq \langle \theta_{p^n} - 1, q-1 \rangle,$$

however,

$$\mathbb{Z}[q]_{p^n} / \langle \theta_{p^n} - 1, q-1 \rangle = \mathbb{F}_p .$$

(v) Because of the above if suffices to show $\widehat{\mathbb{Z}_{<p>}} \subset \widehat{\mathbb{Z}[q]_{\langle q-1,p \rangle}}$. We have an injection $\langle p^n \rangle \subset \langle q-1, p \rangle^n$, which in turn induces an injection $\mathbb{Z}/\langle p^n \rangle \longrightarrow \mathbb{Z}[q]/\langle q-1, p \rangle^n$, since $\widehat{\mathbb{Z}_{<p>}} \subset \prod_{n \in \mathbb{N}} \mathbb{Z}/\langle p^n \rangle$ and similarly for $\widehat{\mathbb{Z}[q]_{\langle q-1,p \rangle}}$. The statement follows. *q.e.d.*

REMARK 2.3. In the theory of the Hecke orders of the symmetric group and also for the other Coxeter groups, the relevant quotients of $\mathbb{Z}[q]$ are

1. modulo $q-1$ this gives rise to the integral representation theory;
2. modulo Φ_{p^n}, i. e. specializing q to p^n-th roots of unity and there only those roots of unity whose order divides the group order. These are in some sense singularities, since the cyclotomic polynomials appear in the amalgamations;
3. the rational prime ideals, and here only those which divide the group order. These give rise to the modular representation theory;
4. at the other roots of unity and prime ideals resp. the reduction modulo the corresponding ideals is semi-simple, even separable (cf. Section 2.1 Corollary 2.13).

Because of this the ring $\mathbb{Z}[q]$ is not so important, but rather the ring – keeping in mind the statement of Claim 2.2

$$\begin{aligned}\mathbb{Z}^q &:= \cap_{<p>\in max(\mathbb{Z})}\ \mathbb{Z}[q]_{<[p],q-1>} \\ &= \mathbb{Q}[q,q^{-1}] \cap (\cap_{<p>\in max(\mathbb{Z})}\ \widehat{\mathbb{Z}[q]_{<[p],q-1>}})\,.\end{aligned}$$

NOTE 2.4. We point out that q is invertible in $\mathbb{Z}[q]_{<[p],q-1>}$ for every prime p, and hence q is invertible in $\mathbb{Z}^q$.

The advantage here is that we have a global local principle.

2.2. THE HIGMAN IDEAL

Let us fix some more NOTATION:

DEFINITION 2.5.

1. *Let R be a regular integral domain of Krull-dimension 2 (e. g. $\mathbb{Z}[q]$) with field of fractions K and let Λ be a Cohen-Macaulay R-order in the separable K-algebra A, i. e. Λ is a unitial subring of A, finitely generated and projective over R and spanning A. We write $\Lambda^e := \Lambda \otimes_R \Lambda^{op}$ for the* ENVELOPING ORDER. *It is – A being separable – a Cohen-Macaulay R-order in the separable K-algebra $A^e := A \otimes_K A^{op}$* [3].
2. *We denote by Λ-mod the category of finitely generated left Λ-modules, by ${}_{CM}\Lambda$ are those left Λ-modules which are R-projective of finite rank and by Λ-lat the Λ-lattices, i. e. the finitely generated R-torsion-free Λ-modules.*
3. *With the enveloping order Λ^e we have associated the* AUGMENTATION SEQUENCE

$$\mathbb{E}_\Lambda\ :\ 0 \longrightarrow I(\Lambda) \longrightarrow \Lambda^e \xrightarrow{\ \mu\ } \Lambda \longrightarrow 0\,,$$

where μ is induced from the multiplication $\lambda_1 \otimes \lambda_2 \longrightarrow \lambda_1 \cdot \lambda_2$ and the left Λ^e-ideal $I(\Lambda)$ is generated by

$$\{\lambda \otimes 1 - 1 \otimes \lambda \,|\, \lambda \in \Lambda\}.$$

This sequence is split both as sequence of left and of right Λ-modules, but not as a sequence of bimodules.

[3] Λ^{op} and A^{op} are the opposite rings.

4. *For a Λ-bimodule B, the first* HOCHSCHILD COHOMOLOGY GROUP $H^1(\Lambda, B)$ *is defined as quotient the cocycles modulo the coboundaries, where we define a* 1-COCYCLE *and a* 1-COBOUNDARY *resp. as*

$$\rho : \Lambda \longrightarrow B\,,\ \rho(\lambda_1, \lambda_2) = \lambda_1 \cdot \rho(\lambda_2) + \rho(\lambda_1) \cdot \lambda_2 \text{ and}$$
$$\rho : \lambda \longrightarrow \lambda \cdot b_0 - b_0 \cdot \lambda \text{ for a fixed } b_0 \in B \text{ resp. }.$$

We observe that there is a natural isomorphism

$$H^1(\Lambda, B) := Ext^1_{\Lambda^e}(\Lambda, B)\,.$$

5. *The* HIGMAN IDEAL $\mathbb{H}(\Lambda)$ *of Λ is by definition the annihilator in R of the* HOCHSCHILD 1-COHOMOLOGY $H^1(\Lambda, -)$, *i. e.*

$$\mathbb{H}(\Lambda) := \cap_B\, ann_R(H^1(\Lambda, B))\,,$$

where B runs over all finitely generated R-projective left Λ^e-modules.
6. *The order Λ is said to be* SEPARABLE *if the sequence $\mathbb{E}_\Lambda$ is split as Λ^e-sequence, i. e. Λ is projective as left Λ^e-module, or equivalently as Λ-bimodule.*

We collect some WELL KNOWN FACTS as

CLAIM 2.6.

1. *The category ${}_{CM}\Lambda$ of left Cohen-Macaulay Λ-modules is closed under taking syzygies.*
2. *Let M, $N \in {}_{CM}\Lambda$ and let $\wp$ of R be a prime ideal, then there is a natural isomorphism*

$$R_\wp \otimes_R Hom_\Lambda(M, N) \simeq_{nat} Hom_{\Lambda_\wp}(R_\wp \otimes_R M, R_\wp \otimes_R N)\,,$$

which extends to the derived functors. A similar statement holds for the completion.

3. *Given a finitely generated torsion-free Λ-module X, we can associate with it in a natural way a Cohen-Macaulay Λ-module*

$${}_{CM}X := Hom_R(Hom_R(X, R), R)$$

the COHEN-MACAULAY-FICATION *of* X. *The* EVALUATION FUNCTOR *gives a natural injection of* Λ*-modules* $\mu_X : X \longrightarrow {}_{CM}X$. *Both modules span the same* K*-module.*

4. *Let* $N \in {}_{CM}\Lambda$ *and let* $M \in \Lambda$*-lat. Then* $Hom_\Lambda(M, N)$ *is* R*-projective of finite rank.*

Proof Let us just indicate the last statement: Choose a Λ-projective resolution of M

$$0 \longrightarrow \Omega^1(M) \xrightarrow{\alpha} P_M \xrightarrow{\beta} M \longrightarrow 0,$$

and note that $\Omega^1_M \in {}_{CM}\Lambda$ as kernel of the projective cover of a lattice. Applying $Hom_\Lambda(-, N)$ we get the exact sequence

$$0 \to Hom_\Lambda(M, N) \xrightarrow{(\beta, N)^0} Hom_\Lambda(P_M, N) \xrightarrow{(\alpha, N)^0} Hom_\Lambda(\Omega^1_M, N).$$

Since Λ and N are Cohen-Macaulay we conclude that $Hom_\Lambda(P_M, N)$ is Cohen-Macaulay. On the other hand, $Hom_\Lambda(\Omega^1_M, N)$ is surely an R-lattice and hence so is $Im((\alpha, N)^0)$. As syzygy of a lattice $Hom_\Lambda(M, N)$ is finitely generated R-projective. The next claim shows the importance of the Higman ideal for the study of Cohen-Macaulay Λ-lattices.

CLAIM 2.7.

1. $\mathbb{H}(\Lambda) \neq 0$.
2. $Ext^1_\Lambda(M, N) \simeq_{nat} H^1(\Lambda, Hom_R(M, N))$ *for* M, $N \in {}_{CM}\Lambda$.
3. $\mathbb{H}(\Lambda) = \bigcap_{\mathfrak{m} \in max(R)} \mathbb{H}(\Lambda_\mathfrak{m})$.
4. $\widehat{R}_\mathfrak{m} \otimes_{R_\mathfrak{m}} \mathbb{H}(\Lambda_\mathfrak{m}) \simeq \mathbb{H}(\widehat{\Lambda_\mathfrak{m}})$.

Proof We have the AUGMENTATION SEQUENCE $\mathbb{E}_\Lambda$ from Equation 3. Since both Λ^e and Λ are R-projective of finite rank, so is I_Λ. We write $\underline{Hom_\Lambda(M, N)}$ for $Hom_\Lambda(M, N)$ modulo the HOMOMORPHISMS FACTORING VIA PROJECTIVE MODULES. For B a left Λ^e-module we have

$$\{\mathbb{E}_\Lambda \phi \,:\, \phi \in \underline{Hom_{\Lambda^e}(I_\Lambda, B)}\} \simeq Ext^1\Lambda^e(\Lambda, B).$$

Hence we have for $r \in R$ the relation $r \in Ext^1\Lambda^e(\Lambda, B) = 0$ for all B if and only if $\mathbb{E}_\Lambda \cdot r = 0$.
Let $\mathcal{H}(\Lambda) := ann_R(Ext_{\Lambda^e}(\Lambda, I_\Lambda)$. Then we have seen by the above argument that $\mathcal{H}(\Lambda) \subseteq \mathbb{H}(\Lambda)$. On the other hand, I_Λ is a bimodule, which occurs in the intersection for $\mathbb{H}(\Lambda)$ thus $\mathcal{H}(\Lambda) \supseteq \mathbb{H}(\Lambda)$. Hence we have actually proved:

$$\mathbb{H}(\Lambda) = ann_R(H^1(\Lambda, I_\Lambda)), \tag{1}$$

which also shows that $\mathbb{H}(\Lambda) \neq 0$, A^e being separable.
The second statement is proved as in (CaEi; 56) Ch. IX, 4.4 (cf. also (RoDy; 70), III, 4.11). It heavily depends on the fact that M is R-projective. *q.e.d.*

DEFINITION 2.8. *We write $dim(\Lambda)$ for the homological dimension of Λ as Λ^e-module.*

Let us quote from (CaEi; 56) Ch. IX, 7.1.

PROPOSITION 2.9. *Let S be a commutative R-algebra. Then*

$$dim(S \otimes_R \Lambda) \leq dim(\Lambda).$$

2.3. SEPARABLE ORDERS

NOTE 2.10. Since Λ is separable if and only if $\Lambda_\mathfrak{m}$ is separable for every $\mathfrak{m} \in max(R)$ if and only if $\widehat{\Lambda}_\mathfrak{m}$ is separable for every $\mathfrak{m} \in max(R)$ – observe that the Higman-ideal commutes with localization and completion – , we shall *assume from now on* that R is *complete* with maximal ideal $\mathfrak{m}$ and residue field $\mathfrak{k}$.

Let us collect some OBVIOUS PROPERTIES OF SEPARABLE ORDERS.

CLAIM 2.11. *Let Λ be a separable Cohen-Macaulay R-order,*

1. *then $\Lambda/(\mathfrak{m} \cdot \Lambda)$ is a separable $\mathfrak{k}$-algebra,*
2. *more generally, for every ideal $\mathfrak{a}$ of R the $R/\mathfrak{a}$-algebra $\Lambda/(\mathfrak{a} \cdot \Lambda)$ is separable,*
3. *and moreover, $rad(\Lambda) = \mathfrak{m} \cdot \Lambda$.*

Proof The first two statements follow directly from Proposition 2.9. The last statement is an immediate consequence of the second, since a separable algebra over a field is semi-simple. *q.e.d.*

We next turn to an important characterization of separable orders.

CLAIM 2.12. *The Cohen-Macaulay R-order Λ in a separable K-algebra A is separable if and only if $\Lambda/\mathfrak{m}\cdot\Lambda$ is separable if and only if $\Lambda/\mathfrak{a}\cdot\Lambda$ is separable for every ideal $\mathfrak{a}$ of R.*

Proof Let us assume that $\overline{\Lambda} := \Lambda/\mathfrak{a}\cdot\Lambda$ is separable. Reducing further modulo $\mathfrak{m}$, we may assume that $\mathfrak{a} = \mathfrak{m}$. We now consider the augmentation sequence (3) reduced modulo $\mathfrak{m}$, which gives the augmentation sequence $\mathbb{E}_{\overline{\Lambda}}$ of $\overline{\Lambda}$, since all modules in the augmentation sequence are R-projective. Now $\overline{\Lambda}$ is separable and so there is a homomorphism $\tilde{\phi}\colon \overline{\Lambda} \longmapsto (\overline{\Lambda})^e \simeq \overline{\Lambda^e}$ such that $\pi_{\overline{\Lambda}} \circ \tilde{\phi} = \overline{\epsilon}$ is an idempotent in $\overline{\Lambda^e}$. By the method of lifting idempotents, there is an idempotent ϵ in Λ^e reducing to $\overline{\epsilon}$. We identify ϵ with the idempotent endomorphism $\epsilon\colon \Lambda^e \longmapsto \Lambda^e$ and consider the map $\mu_\epsilon := \mu_\Lambda \downarrow_{\Lambda^e\cdot\epsilon}\colon \Lambda^e\cdot\epsilon \longmapsto \Lambda$. Reducing modulo $\mathfrak{m}$ we see that μ_ϵ is an isomorphism modulo $\mathfrak{m}$. But then μ_ϵ must be surjective by Nakayama's lemma. Since the R-rank of $\Lambda^e\cdot\epsilon$ is the same as the R-rank of Λ, we conclude that μ_ϵ must be an isomorphism, hence the augmentation sequence is split and Λ is separable. *q.e.d.*

We actually have proved much more.

COROLLARY 2.13. *The Cohen-Macaulay R-order Λ in a finite-dimensional (not necessarily separable) K-algebra A is separable if and only if $\Lambda/\mathfrak{m}\cdot\Lambda$ is separable (and consequently A is separable) if and only if $\Lambda/\mathfrak{a}\cdot\Lambda$ is separable for every ideal $\mathfrak{a}$ of R.*

DEFINITION 2.14. *Let B be a separable F-algebra for a field F and let $\widetilde{F}$ be its algebraic closure. The* NUMERICAL INVARIANTS *of B are the dimensions and the number of the non-isomorphic simple $(\widetilde{F}\otimes_F B)$-modules.*

We shall use the next observation over and over again. Let $\mathfrak{a}$ be an ideal in R and denote by $(\overline{})$ reduction modulo $\mathfrak{a}$.

CLAIM 2.15. *Let P and Q be indecomposable projective Λ-modules. Then $\overline{Hom_\Lambda(P,Q)} = Hom_{\overline{\Lambda}}(\overline{P},\overline{Q})$. In particular, $Hom_\Lambda(P,Q)=0$ if and only if $Hom_{\overline{\Lambda}}(\overline{P},\overline{Q})=0$.*

Proof Since R is complete, we have the Krull-Schmidt theorem available, and so $P = \Lambda \cdot \epsilon_P$ and $Q = \Lambda \cdot \epsilon_Q$ for idempotents ϵ_p and ϵ_Q resp. Now $Hom_\Lambda(P,Q) = \epsilon_P \cdot \Lambda \cdot \epsilon_Q$ is R-projective, and the result follows. *q.e.d.*

LEMMA 2.16. *Assume that Λ is a basic, connected [4], separable Cohen-Macaulay R-order in the separable K-algebra A. Then*

1. *A is a skew field,*
2. *A and $\overline{\Lambda} := \Lambda/\mathfrak{m}\cdot\Lambda$ have the same numeric invariants.*

Proof Since Λ is separable, it is a projective Λ^e-module, and the center of Λ is given by $Hom_{\Lambda^e}(\Lambda,\Lambda)$. Hence we may apply Claim 2.15 to conclude that central idempotents of $\overline{\Lambda}$ lift to central idempotents of Λ – note that this heavily uses that Λ is separable. Since $\overline{\Lambda}$ is semi-simple, we may assume that $\overline{\Lambda}$ is simple. However, for projective Λ-modules P and Q we have $P \simeq Q$ if and only if $\overline{P} = \overline{Q}$, and so Λ is local, since it was assumed to be basic.
If A is not a skew field there is a non-trivial idempotent $\epsilon \in A$. Then the Cohen-Macaulay-fication ${}_{CM}(\Lambda\cdot\epsilon)$ of $\Lambda\cdot\epsilon$ is a finitely generated Cohen-Macaulay Λ-module, which is then projective, Λ being separable. But this is a contradiction since Λ was assumed to be local, and since the Krull-Schmidt theorem holds. Thus A is a skew field as is $\overline{\Lambda}$. Let now L be a finite extension field of K with S the integral closure of R in L such that L is a splitting field for A and $S/rad(S)$ is a splitting field for $\overline{\Lambda}$. We now consider $\Lambda^S := S \otimes_R \Lambda$. Invoking again Claim 2.15 we conclude that A and $\overline{\Lambda}$ have the same numerical invariants. *q.e.d.*

COROLLARY 2.17 (Tits' Deformation Theorem (CR; 87)). *Let R be a regular integral domain of dimension 2, with K field of fractions K. Assume we are given a homomorphism $f\colon R \longmapsto F$ of R into a field F. Let Λ be an R-algebra with finite basis [5]. Let $\Lambda^K = K \otimes_R \Lambda$*

[4] I. e. Λ is indecomposable as ring.
[5] It is enough to require that Λ is R-projective of finite type.

and denote by Λ_f *the specialized algebra* $F \otimes_R \Lambda$ *(via* f*). If both* Λ^K *and* Λ_f *are separable, the they have the same numerical invariants.*

Proof Let $\wp$ be the kernel of f. It is a prime ideal in R, since $Im(f)$ is an integral domain. We may assume that F is the field of fractions of $R/\wp$. We denote by $R_\wp$ the localization of R at $\wp$ and by $\widehat{R_\wp}$ the completion of R at $\wp$. Then $F = R_\wp/\wp \cdot R_\wp = \widehat{R_\wp}/\wp \cdot \widehat{R_\wp}$. Moreover, $< \wp >$ is the maximal ideal in $R_\wp$. K is still the field of fractions of $R_\wp$. Similar statements hold for the completion.
Since the numerical invariants arise in a big field, we may assume that:

1. R is complete with maximal ideal $\mathfrak{m}$, the Cohen-Macaulay R-order Λ has a separable ring of quotients and $\overline{\Lambda} := \Lambda/\mathfrak{m} \cdot \Lambda$ is separable.

With Claim 2.12 we conclude that Λ is separable. Now Lemma 2.16 gives the desired result. *q.e.d.*

CLAIM 2.18. *Assume that* R *is a complete regular local domain of dimension* 2 *with field of fractions* K*. If* Λ *is a basic connected Cohen-Macaulay* R*-order in a separable* K*-algebra* A*, then* $\Lambda = S$ *is a (commutative) unramified extension of* R*.*

Proof The result is true for orders over rings of Dimension 1. Since d has dimension 2 we choose a prime ideal $\wp$ such that $R/\wp$ is of dimension one. Then the ring $\Lambda/\wp \cdot \Lambda = \overline{S}$ is an unramified (commutative) extension of $R/\wp$. Consequently, $\Lambda/\mathfrak{m} \cdot \Lambda$ is a then a commutative field (cf. Lemma 2.16). Again the claims 2.12 and 2.15 and the arguments at the beginning of the proof of Lemma 2.16 come into play and imply that Λ is commutative. *q.e.d.*

NOTE 2.19. We point out that most of the results in this section are valid for R a regular integral domain of dimension $d > 0$. Since we are mainly interested in the arithmetic surface, we have only formulated them for dimension 2.

3. 2-dimensional tree orders and their Cohen-Macaulay module

As above R is a regular complete local integral domain of dimension 2 with field of quotients K, maximal ideal $\mathfrak{m}$ and residue field $\mathfrak{k}$.

3.1. TWO-DIMENSIONAL BRAUER TREE ORDERS

In this section we recall the definition of the BRAUER TREE ORDERS which we have sometimes also called GREEN ORDERS. Let us be given a finite tree T with vertex set V and edge set E [6]. Such a tree can be embedded into the plane. This embedding gives rise to a numbering of the local edges at each vertex $v \in V$, i. e. the edges starting at v.

In order to define a tree order to the given tree T we need the following ingredients:

- To each vertex v we associate a local Cohen-Macaulay R-order Δ_v in a separable K-algebra with a regular principal ideal [7] $\langle \delta_v \rangle = \delta_v \cdot \Delta_v = \Delta_v \cdot \delta_v$.
- An R-algebra $\overline{\Delta}$ and for each vertex an isomorphism $\phi_v : \Delta_v/\langle \delta_v \rangle \longrightarrow \overline{\Delta}$.
- To each vertex a GENERALIZED HEREDITARY order, which is by definition of the form

$$\mathbb{H}_v = \mathbb{H}(\Delta_v, \delta_v, n_v) := \begin{pmatrix} \Delta_v & \Delta_v & \Delta_v & \cdots & \Delta_v & \Delta_v \\ \langle\delta_v\rangle & \Delta_v & \Delta_v & \cdots & \Delta_v & \Delta_v \\ \langle\delta_v\rangle & \langle\delta_v\rangle & \Delta_v & \cdots & \Delta_v & \Delta_v \\ & & & \ddots & & \\ \cdots & \cdots & \cdots & & \cdots & \cdots \\ \langle\delta_v\rangle & \langle\delta_v\rangle & \langle\delta_v\rangle & \cdots & \Delta_v & \Delta_v \\ \langle\delta_v\rangle & \langle\delta_v\rangle & \langle\delta_v\rangle & \cdots & \langle\delta_v\rangle & \Delta_v \end{pmatrix}_{n_v},$$

where n_v is the valency of the vertex v.

---

[6] For the more general notion of the ORDER TO A GRAPH WITH TRUNCATED EDGES, which occurs in the inductive description of cellular orders and of of blocks with cyclic defect of group rings and Hecke orders, we refer to (Ro; 00, 1). We also point out that the definition makes sense in a much wider generality.

[7] An ideal I in Λ is said to be REGULAR, provided $K \otimes_R \Lambda$.

– The local edge at the j-th position at v corresponds to the j-th column of $\mathbb{H}_v$, say $M_{v,j}$, $1 \leq j \leq n_v$.

The TREE ORDER $\Lambda = \Lambda(T, \{\Delta_v, \phi_v\}_{v\in V})$ is constructed as a subring of $\prod_{v\in V} \mathbb{H}_v$: For an edge from the *local edge* (v,i) to the *local edge* (w,j) the i-th diagonal element in $\mathbb{H}_v$ and the j-th diagonal element in $\mathbb{H}_w$ is replaced by the pull-back

$$\Delta_v \xrightarrow{(\delta_v,\delta_w)} \Delta_w ,$$

where we use the following notation for the pull-back

$$\begin{array}{ccc} \Delta_v \xrightarrow{(\delta_v,\delta_w)} \Delta_w & \xrightarrow{\alpha_v} & \Delta_v \\ \alpha_w \downarrow & & \downarrow \phi_v \\ \Delta_w & \xrightarrow{\phi_w} & \overline{\Delta} \end{array} .$$

DEFINITION 3.1. *We say that a Cohen-Macaulay R-order Λ is an order with a* TREE RESOLUTION, *provided there exists a tree $T = T(V,E)$ embedded into the plane, provided the following conditions are satisfied:*

1. *There is a bijection between the edges and the indecomposable projective Λ-modules, $e \longmapsto P_e$, $e \in E$. Moreover, P_e is also injective in the category of Cohen-Macaulay Λ-modules.*
2. *The local edges to each vertex correspond to irreducible* [8] *indecomposable Cohen-Macaulay Λ-modules, $(v,j) \longmapsto M_{v,j}$ for $1 \leq j \leq n_v b$, $v \in V$, which are rationally isomorphic. There are no homomorphisms between modules corresponding to different vertices.*
3. *If P_e corresponds to the edge $(v,i) \overset{e}{\text{———}} (w,j)$, then there is a projective resolution*

$$0 \longrightarrow M_{v,i-1} \longrightarrow P_e \longrightarrow M_{w,j} \longrightarrow 0 .$$

We point out that the above exact sequences give rise to "Green's" walk around the Brauer tree for blocks with cyclic defect.
The next results describes explicitly the orders with a tree resolution.

[8] This means that M does not have a submodule of smaller R-rank; equivalently $K \otimes_R M$ is simple.

THEOREM 3.2 ((Ro; 00, 1)). *Assume that Λ is a Cohen-Macaulay R-order with a tree resolution to the tree $\mathcal{T}$. Then Λ is a tree order to the tree $\mathcal{T}$.*

The result is proved by induction on the number of edges of $\mathcal{T}$ and involves also trees with truncated edges. We just sketch the argument in case the tree has the form $\bullet$ —— $\bullet$. In this case we have the projective resolutions

$$0 \longrightarrow M_1 \xrightarrow{\phi_1} P \xrightarrow{\psi_1} M_2 \longrightarrow 0 \quad \text{and}$$
$$0 \longrightarrow M_2 \xrightarrow{\phi_2} P \xrightarrow{\psi_2} M_1 \longrightarrow 0 \quad .$$

Since $K \otimes_R P \simeq K \otimes_R M_1 \oplus K \otimes_R M_2$ is the direct sum of two non-isomorphic modules (because of our assumption), we conclude that we have a pull-back diagram

$$\begin{array}{ccc} P & \xrightarrow{\psi_2} & M_1 \\ {\scriptstyle\psi_1}\downarrow & & \downarrow{\scriptstyle\kappa_1} \\ M_2 & \xrightarrow{\kappa_2} & \overline{M} \end{array} .$$

From the above sequences it follows that $Ker(\kappa_i) \simeq_{\iota_i} M_i$. The composition $\delta_i : M_i \xrightarrow{\iota_i} Ker(\kappa_i) \xrightarrow{\widetilde{\phi_i}} M_i$ is then an endomorphism in $\Delta_i := End_\Lambda(M_i)$ – we have used the notation $\widetilde{\phi_i}$ for the inclusion $Ker(\kappa_i) \longrightarrow M_i$. Then $\overline{M} \simeq M_i/M_i \cdot \delta_i$. Since $Hom_\Lambda(M_i, M_j) = 0$ for $i \neq j$, we conclude that

$$\Lambda = P = \{(\alpha, \beta) \,|\, \alpha \in \Delta_1 \,,\, \beta \in \Delta_2 \,:\, \alpha = \beta \text{ on } \overline{M}\}$$

is a pull-back. We then write for the above $\Lambda \simeq \Delta_1 \overset{(\delta_1,\delta_2)}{\text{——}} \Delta_2$, which is the tree order to the tree $\bullet$ —— $\bullet$ with respect to (Δ_i, δ_i) and the isomorphism $\Delta/\langle\, \delta_i \,\rangle \simeq \overline{\Delta}$.
The exact sequences

$$0 \longrightarrow \langle\, \delta_j \,\rangle \longrightarrow \Lambda \xrightarrow{\alpha_i} \Delta_i \longrightarrow 0$$

for $i \neq j$ and $i, j \in \{1, 2\}$ show that $\langle\, \delta_j \,\rangle$ is a Cohen-Macaulay module as kernel of a projective resolution of a torsion-free module. However, $\langle\, \delta_j \,\rangle \simeq \Delta_j$ and hence Δ_j is a Cohen-Macaulay R-module for $j = 1, 2$.

3.2. COHEN-MACAULAY MODULES FOR TREE ORDERS

Let R be a complete local integral domain of Krull dimension $n \geq 2$ and let Λ be an R-order in a separable K-algebra A. A description of the Cohen-Macaulay Λ-modules seems to be out of reach for $n = 2$, since in general the epimorphic images of Cohen-Macaulay modules are not Cohen-Macaulay.
A good substitute seem to be those Cohen-Macaulay modules M which have a filtration with irreducible Cohen-Macaulay modules as sections, i. e. there is a chain of Cohen-Macaulay submodules

$$0 \subset M_1 \subset M_2 \subset \cdots \subset M_n = M$$

such that the modules M_i/M_{i-1} are irreducible Cohen-Macaulay Λ-modules occurring in a prescribed order (cf. below).
In case $n = 1$ the situation for tree orders is very lucid, since in this case the extension groups between the irreducible lattices are isomorphic to the residue field, and so there is essentially only one extension, the projective cover sequence.
Let us assume now that $n = 2$ and that Λ is a tree order with $\Delta_v = R$ and $\delta_v = \pi$ a height one – and hence principal – prime ideal. In this case the extension groups are of the form $\overline{R} := R/\langle\pi\rangle$. The structure of the category of filtered Cohen-Macaulay Λ-modules now depends heavily on the arithmetical structure of $\overline{R}$.
Let us turn to the situation $R = \mathbb{Z}[q]_{<p,q-1>}$ is the completion of the arithmetic surface at the maximal ideal $\langle\, p, q-1\,\rangle$. The ring $\overline{R}$ is then a subring of the ring of p-adic integers R_p in the algebraic number field $K_p =: \mathbb{Q}[q]_{<p,q-1>}/\langle\pi\rangle$.
In case $\overline{R} = R_p$, i. e. $\overline{R}$ is a complete discrete rank one valuation ring, the only indecomposable filtered Cohen-Macaulay modules are then given as follows:
We start the *walk around the tree* from an end vertex. The walk then gives rise to a numbering of the rational idempotents and also to the projective resolution of the unique irreducible Cohen-Macaulay module M_0 corresponding to the chosen end vertex. The only irreducible Cohen-Macaulay Λ-modules are the syzygies $\Omega^i(M_0)$, $0 \leq i \leq s$, where $s+1$ is the number of vertices in our tree. The construction of the remaining Cohen-Macaulay Λ-modules is

done as follows: We calculate $Ext^1_\Lambda(\Omega^{i-1}(M_0), \Omega^i(M_0)) \simeq \overline{R}$ and so the non-isomorphic exact sequences are given as extensions

$$\mathbb{E}_\nu : 0 \longrightarrow \Omega^i(M_0) \longrightarrow X_\nu \longrightarrow \Omega^{i-1}(M_0) \longrightarrow 0,$$

which correspond to the element $\rho^\nu \in Ext^1_\Lambda(\Omega^{i-1}(M_0), \Omega^i(M_0))$, with $\langle \rho \rangle = rad(\overline{R})$. It turns out that $\{X_\nu \mid \nu = 0, 1, 2, \cdots\}$ are non-isomorphic indecomposable Cohen-Macaulay modules.

The **main result** is that these are all indecomposable filtered Cohen-Macaulay Λ-modules. Note that these depend on the chosen starting point in the tree.

In case $\overline{R} \neq R_p$ we can find in the category of filtered Cohen-Macaulay Λ-modules modules of arbitrarily large R-rank. This case will be considered below.

Luckily, in all the examples I know of blocks of cyclic defect of Hecke orders and of deformations of blocks with defect p we are in the first situation, i. e. $\overline{R} = R_p$.

We shall give here – as a demonstration – a description of the filtered Cohen-Macaulay modules for the localized Hecke order of the dihedral groups of order $p \cdot 2$ for an odd prime p. One can also take the description as a pull-back below as a definition.

So let $\mathcal{H}_p = \mathcal{H}_{D_p} \otimes_{\mathbb{Z}[q,q^{-1}]} \mathbb{Z}[q,q^{-1}]_{\langle q-1,p \rangle}$ be the completion of the Hecke order $\mathcal{H}_{D_p}$ of the dihedral group of order $2 \cdot p$ for an odd prime p at the maximal ideal $\langle q-1, p \rangle$. Then $\mathcal{H}_p$ is a tree order which is described as follows (cf. (Ro; 00, 2)):

Let $R = \mathbb{Z}[\theta_p]^{C_2}$ be the ring of the integers in the fixed field under the cyclic group of order two of the p-th cyclotomic field and define $R_2 = R \otimes_\mathbb{Z} \mathbb{Z}[q,q^{-1}]_{\langle q-1,p \rangle}$. In R_2 we have the prime ideal $\langle \pi \rangle = \langle (q-\theta_p) \cdot (q-\theta_p^{-1}) \rangle$. We observe that $\overline{R_2} = R_2/\langle \pi \rangle \simeq \mathbb{Z}_{\langle p \rangle}[\theta_p]$ is regular of dimension one, i. e. a rank one valuation ring. The order Λ_2 is defined as

$$\Lambda_2 =: \begin{pmatrix} R_2 & R_2 \\ \langle \pi \rangle & R_2 \end{pmatrix}$$

We put $\Lambda_1 = \mathbb{Z}[q,q^{-1}]_{\langle q-1,p \rangle} = \Lambda_3$ and denote by $\Phi_p(q)$ the p-th cyclotomic polynomial. The ring $\mathcal{H}_p$ is then obtained as the pull-back

$$\begin{array}{ccc} \mathcal{H}_p & \longrightarrow & \Lambda_1 \times \Lambda_3 \\ \downarrow & & \phi \downarrow \\ \Lambda_2 & \xrightarrow{\psi} & \mathbb{Z}_{\langle p \rangle}[\theta_p] \times \mathbb{Z}_{\langle p \rangle}[\theta_p] \end{array},$$

where ϕ is reduction modulo $\langle \Phi_p(q) \rangle \times \langle \Phi_p(q) \rangle$, ψ is reduction modulo the ideal

$$\begin{pmatrix} \langle \pi \rangle & R_2 \\ \langle \pi \rangle & \langle \pi \rangle \end{pmatrix}.$$

We label the rational central primitive idempotents of A, the algebra spanned by $\Lambda = \mathcal{H}_p$ as (e_1, e_2, e_3), where e_i is the unit element in Λ_i. The non-isomorphic irreducible Cohen-Macaulay $\mathcal{H}_p$-modules are the modules:

$$\{N_1 = \Lambda_1\,,\ M_1 = \begin{pmatrix} R_2 \\ \langle \pi \rangle \end{pmatrix}\,,\ M_2 = \begin{pmatrix} R_2 \\ R_2 \end{pmatrix}\,,\ N_3 = \Lambda_3\}\,.$$

Green's walk around the Brauer tree (cf. (Gr; 74)) gives the projective resolutions

$$N_3 \longrightarrow P \xrightarrow{\alpha_3} Q \xrightarrow{\alpha_2} Q \xrightarrow{\alpha_1} P \longrightarrow N_3,$$

which is put together from the projective cover sequences

$$\begin{array}{l} \mathbb{E}_1 : 0 \longrightarrow M_1 \longrightarrow P \xrightarrow{\phi_1} N_3 \longrightarrow 0 \\ \mathbb{E}_2 : 0 \longrightarrow N_1 \longrightarrow Q \xrightarrow{\phi_2} M_1 \longrightarrow 0 \\ \mathbb{E}_3 : 0 \longrightarrow M_2 \longrightarrow Q \xrightarrow{\phi_3} N_1 \longrightarrow 0 \\ \mathbb{E}_4 : 0 \longrightarrow N_3 \longrightarrow P \xrightarrow{\phi_4} M_2 \longrightarrow 0 \end{array}.$$

We now look at the modules X, which have a (e_1, e_2, e_3)-COHEN-MACAULAY FILTRATION, i. e. a chain of Cohen-Macaulay submodules

$$X_1 \subseteq X_2 \subseteq X_3$$

such that $X_1 \simeq N_1^{(s_1)}$, $X_2/X_1 \simeq M_1^{(t_1)} \oplus M_2^{(t_2)}$ and $X_3/X_2 \simeq N_3^{(s_3)}$.

THEOREM 3.3. *The indecomposable non-isomorphic Cohen-Macaulay $\mathcal{H}_p$-modules having a (e_1, e_2, e_3)-filtration are given as:*

– *The irreducible Cohen-Macaulay $\mathcal{H}_p$-modules*

$$\{N_1, M_1, M_2, N_3\},$$

– *the modules* $\phi_1^{-1}((p^\nu + \pi) \cdot N_3)$, $\nu = 0, 1, \dots,$
– *the modules* $\phi_2^{-1}(((1-\theta_p)^\nu + \pi) \cdot M_1)$, $\nu = 0, 1, \dots$.

If we look at the modules X, which have a (e_2, e_1, e_3)-Cohen-Macaulay filtration, i. e.

$$X_1 \subseteq X_2 \subseteq X_3$$

such that $X_1 \simeq M_1^{(t_1)} \oplus M_2^{(t_2)}$, $X_2/X_1 \simeq N_1^{(s_1)}$ and $X_3/X_2 \simeq N_3^{(s_3)}$ we obtain

THEOREM 3.4. *The indecomposable non-isomorphic Cohen-Macaulay $\mathcal{H}_p$-modules having a (e_2, e_1, e_3)-filtration are given as:*

– *The irreducible Cohen-Macaulay $\mathcal{H}_p$-modules*

$$\{N_1, M_1, M_2, N_3\},$$

– *the modules* $\phi_1^{-1}((p^\nu + \pi) \cdot N_3)$, $\nu = 0, 1, \dots,$
– *the modules* $\phi_3^{-1}((p^\nu + \pi) \cdot N_1)$, $\nu = 0, 1, \dots$.

It is shown by W. Rump in (Ru; 00), that for a tree order every Cohen-Macaulay module has a filtration by primitive idempotents, numbered in a special way.

3.3. EXAMPLES OF COHEN-MACAULAY MODULES FOR GRAPH ORDERS

As pointed out above the reduction modulo principal prime ideals has to be done with great care. *We shall elaborate on two examples.* Let $\mathbb{Z}[q, q^{-1}]_{\langle \mathfrak{m} \rangle}$ be the completion with respect to a maximal ideal. We shall now consider the following two essentially different cases:

DEFINITION 3.5.

- *Let p be an odd rational prime and let $\pi_1 := \Phi_{p^n}(q)$ be the p^n-th cyclotomic polynomial. In this case we choose $\mathfrak{m} = \langle q-1, p\rangle$. We denote by θ_{p^n} a primitive p^n-th root of unity and put $R_1 := \mathbb{Z}[q,q^{-1}]_{\mathfrak{m}}$ the completion at $\mathfrak{m}$. In this case $\overline{R_1} := R_1/\langle\pi_1\rangle = \mathbb{Z}[\theta_{p^n}]_{\langle\theta_{p^n}-1\rangle}$ is a discrete rank one valuation ring with maximal ideal $\mathfrak{m}_1 := \langle\theta_{p^n}-1\rangle$, a principal prime ideal over $\langle p\rangle$.*
- *As second example let $\pi_2 := f_0(q)$ be a monic irreducible polynomial over $\mathbb{Z}_{\langle p\rangle}$ and let $\mathfrak{m}$ be a maximal ideal of $R_2 := \mathbb{Z}_{\langle p\rangle}[q,q^{-1}]$ such that $f_0(q)$ stays irreducible in the completion $R_2 := \mathbb{Z}_{\langle p\rangle}[q,q^{-1}]_{\mathfrak{m}}$. Assume that $\overline{R}_2 := R_2/\langle f_0(q)\rangle$ is not the ring of p-adic integers in $\mathbb{Q}_{\langle p\rangle}[q,q^{-1}]/\langle f_0(q)\rangle$. In this case $\overline{R_2}$ is a local $\mathbb{Z}_{\langle p\rangle}$-order which is not maximal, and so its maximal ideal $\mathfrak{m}_2 = \langle\alpha_1,\alpha_2,\cdots,\alpha_m\rangle$ is not cyclic, say it is minimally generated by $m \geq 2$ elements.*

We shall next consider the R_i graph orders to the graph $\bullet$ —— $\bullet$:

DEFINITION 3.6. *Let $\phi_i\colon R_i \longrightarrow \overline{R_i}$ be the reductions modulo π_i for $i = 1,2$. We define the Cohen-Macaulay R_i-order Λ_i as pullbacks:*

$$\begin{array}{ccc} \Lambda_i & \longrightarrow & R_i \\ \big\downarrow & & \big\downarrow{\scriptstyle \phi_i} \\ R_i & \xrightarrow{\phi_i} & \overline{R_i} \end{array}.$$

Let K_i be the quotient field of R_i and let A_i be the K_i algebra generated by R_i.

The structure of the Cohen-Macaulay Λ_i-modules is completely different in Case $i = 1$ and in Case $i = 2$.
Let us collect some easy facts:

NOTE 3.7.

- We have the projective resolutions

$$\mathbb{E}_i^l: \qquad 0 \longrightarrow \pi_i \cdot N_i \longrightarrow P_i \longrightarrow M_i \longrightarrow 0$$

and

$$\mathbb{E}_i^r : \quad 0 \longrightarrow \pi_i \cdot M_i \longrightarrow P_i \longrightarrow N_i \longrightarrow 0,$$

moreover, $Hom_{\Lambda_i}(M_i, N_i) = Hom_{\Lambda_i}(N_i, M_i) = 0$.

- $Ext^1_{\Lambda_i}(M_i, N_i) = \overline{\Omega_i} = Ext^1_{\Lambda_i}(N_i, M_i)$,
- the corresponding extensions are described as above.
- We shall write $CM^l(\Lambda_i)$ ($CM^r(\Lambda_i)$) for the category of Cohen-Macaulay Λ_i module having a Cohen-Macaulay filtration with respect to (e_1, e_2) ((e_2, e_1) resp.), where e_1 is the primitive idempotent of A_i with $e_1 \cdot N_i \neq 0$ and e_2 is the primitive idempotent of A_i with $e_2 \cdot M_i \neq 0$.

CLAIM 3.8. *Given an extension of Cohen-Macaulay Λ_i-modules*

$$0 \longrightarrow N_i \oplus N_i \longrightarrow E_i \longrightarrow M_i \longrightarrow 0\,.$$

- *If $i = 1$, then E_1 decomposes.*
- *If $i = 2$, then there exists an extension where E_2 is indecomposable.*

Proof Because of the hypotheses from Note 3.7 (cf. (Ro; 70), Chap. X) the module E_i decomposes if and only if the corresponding matrix

$$\mathbb{E}_i \in Ext^1_{\Lambda_i}(M_i, N_i^2)$$

decomposes under transformations of the form $A_i \cdot \mathbb{E}_i \cdot B_i$ with $A_i \in Gl(1, \overline{\Omega_i})$ and $B_i \in Gl(2, \overline{\Omega_i})$. The matrix $\mathbb{E}_i$ is given as $\mathbb{E}_i = (\alpha_i, \beta_i)$ with $\alpha_i, \beta_i \in \overline{\Omega_i}$.

We now consider our two cases:

CASE 1: Let $i = 1$. Then – if necessary after renumbering – there exists an element $\gamma_1 \in \overline{\Omega_1}$ with $\alpha_1 \cdot \gamma_1 = \beta_1$, since $\overline{\Omega_1}$ is a valuation ring. Hence E_1 decomposes.

CASE 2: If $i = 2$, then there are elements α_2 and β_2 such that none of them is a multiple in $\overline{\Omega_2}$ of the other. Hence the matrix (α_2, β_2) cannot be decomposed and so the middle term of the corresponding extension $(\alpha_2, \beta_2) \in Ext^1_{\Lambda_1}(M_1, N_1^{(2)})$ is indecomposable. *q.e.d.*

PROPOSITION 3.9.

- *In Case 1 let* $\overline{\tau}$ *be a generator of the radical of* $\overline{\Omega_1}$ *and let* τ *be a preimage in* Ω_1.
  The non-isomorphic indecomposable Cohen-Macaulay Λ_1*-modules in* $CM^l(\Lambda_1)$ *are the modules* E^l_ν *occurring in the middle term of the exact sequence* $\tau^\nu \cdot \mathbb{E}^l_1$ *for* $\nu = 0, 1, \ldots$ *and the two irreducible modules* M *and* N.
  The non-isomorphic indecomposable Cohen-Macaulay Λ_1*-modules in* $CM^r(\Lambda_1)$ *are the modules* E^r_ν *occurring in the middle term of the exact sequence* $\mathbb{E}^r_1 \cdot \tau^\nu$, *for* $\nu = 0, 1, \ldots$ *and the two irreducible modules* M *and* N.
  The indecomposable Cohen-Macaulay Λ_1 *modules in* $CM^l(\Lambda_1) \cap CM^r(\Lambda_1)$ *are thus the modules* P_1, M_1, N_1 [9].
- *In Case 2 there are indecomposable modules of arbitrarily large rank and with a* (e_1, e_2) *filtration.*

Proof Case 1: Let $X \in CM^l(\Lambda_1)$ be indecomposable. Since it has a Cohen-Macaulay (e_1, e_2)-filtration, we have an exact sequence

$$\mathbb{E}: \qquad 0 \longrightarrow X' \longrightarrow X \longrightarrow X'' \longrightarrow 0$$

with $e_1 \cdot X' = X'$ and $e_2 \cdot X'' = X''$. If N (M) is the only indecomposable Cohen-Macaulay $\Lambda_1 \cdot e_1$- $(\Lambda_1 \cdot e_1)$-module, we conclude that

$$X' \simeq N_1^{(s_1)} \text{ and } X'' \simeq M_1^{(s_2)}.$$

Hence the above sequence $\mathbb{E}$ corresponds to a matrix $E \in Mat(s_2 \times s_1, \overline{\Omega_1})$ which is operated on by $Gl(s_2, \overline{\Omega_1})$ from the left and by $Gl(s_2, \overline{\Omega_1})$ from the right. Since $\overline{\Omega_1}$ is a principal ideal domain we can reduce E under this operation to diagonal form, i. e. the sequence $\mathbb{E}$ decomposes unless $s_1 = 1 = s_2$. Thus $\mathbb{E} = \rho \cdot \mathbb{E}^l_1$ for some $\rho \in \overline{\Omega_1}$. Since modifying with a unit does not change the isomorphism class of X, we may replace ρ by τ^ν, provided ρ has value ν. The case $CM^r(\Lambda_1)$ is dealt with similarly.

---

[9] I do not know whether there are indecomposable Cohen-Macaulay Λ_1-modules of arbitrarily large rank which then cannot have a Cohen-Macaulay filtration.

Case 2: We generalize the example from above in Claim 3.8: Let $\mathfrak{m}_2 : \{\alpha, \beta, \cdots\}$. Then the matrix

$$\mathbb{E}_n := \begin{pmatrix} \alpha & \beta & 0 & \cdots & 0 & 0 \\ 0 & \alpha & \beta & 0 \cdots & 0 & 0 \\ \cdots & \cdots & \cdots & \cdots & \cdots & \cdots \\ 0 & \cdots & \cdots & 0 & 0 & \alpha \end{pmatrix}_n$$

gives an indecomposable filtered Cohen-Macaulay module as an exact sequence in $Ext^1_{\Lambda_2}(M_2^{(n)}, N_2^{(n)})$. *q.e.d.*

4. Hecke orders of Coxeter groups and their blocks with cyclic defect

4.1. HECKE ORDERS

DEFINITION 4.1.

1. *A finite* COXETER GROUP W *is a finite group with a presentation*

$$W = \langle\, s_i\,,\, 1 \leq i \leq n\, :\, (s_i \cdot s_j)^{m_{ij}} = 1\,,\ \textit{for all}\ 1 \leq i, j \leq n \,\rangle\,,$$

where $\{m_{ij}\}$ *are positive integers with*

$$m_{ii} = 1\,,\ m_{ij} > 1\ \textit{for}\ i \neq j\,,\ m_{ij} = m_{ji}\,.$$

The set of involutions $S = \{s_i\}_{1 \leq i \leq n\}}$ *is called a* FINITE COXETER SYSTEM.

2. *The* HECKE ORDER $\mathcal{H}_W$ *of the Coxeter group* W *with respect to the Coxeter system* S *is the* $\mathbb{Z}[q]$*-algebra generated by* $\Sigma := \{\sigma_i\}_{1 \leq i \leq n}$ *subject to the following relations:*

   a) The QUADRATIC RELATIONS*:*

$$\sigma_i^2 = (q-1) \cdot \sigma_i + q \cdot 1\ \textit{and}$$

   b) the HOMOGENEOUS RELATIONS

$$(\sigma_i \cdot \sigma_j)^{k_{ij}} = ((\sigma_j \cdot \sigma_i)^{k_{ij}}\ \textit{provided}\ m_{ij} = 2 \cdot k_{ij}\ \textit{and}$$
$$(\sigma_i \cdot \sigma_j)^{k_{ij}} \cdot \sigma_i = ((\sigma_j \cdot \sigma_i)^{k_{ij}} \cdot \sigma_j\ \textit{provided}\ m_{ij} = 2 \cdot k_{ij} + 1\,.$$

We quote a fact from (CR; 87) Section 66 ff:

THEOREM 4.2. *For $q = 1$ the above quadratic and homogeneous relations describe the Coxeter group W with Coxeter system S, and so we obtain as $\mathcal{H}_W/\langle\, q-1\,\rangle$ the integral group ring $\mathbb{Z}W$ of the Coxeter group W.*

THEOREM 4.3.

1. *$\mathcal{H}_W$ is a Cohen-Macaulay $\mathbb{Z}[q]$-algebra.*
2. *$\mathbb{Q}(q) \otimes_{\mathbb{Z}[q]} \mathcal{H}_W$ is a separable $\mathbb{Q}(q)$-algebra.*
3. *$\mathbb{Q}(q) \otimes_{\mathbb{Z}[q]} \mathcal{H}_W$ and $\mathbb{Q}W$ have the same numerical invariants.*

Proof We leave it as an exercise to show that $\mathcal{H}_W$ is a $\mathbb{Z}[q]$-free algebra.
Since W is a finite group the Higman ideal of $\mathbb{Z}G$ is $\mathbb{H}(\mathbb{Z}G) = \langle\, |G|\,\rangle$. We now choose a rational prime p, relatively prime to $|G|$, put $R := \mathbb{Z}[q]_{\langle p,(q-1)\,\rangle}$ and let $\Lambda := R \otimes_{\mathbb{Z}[q]} \mathcal{H}_W$. Then R is local with maximal ideal $\langle\, p, q-1\,\rangle$ and $\Lambda/\langle\, p\,\rangle{\cdot}\Lambda$ is separable and we may apply Corollary 2.13 to conclude that Λ is separable and thus also the algebra $\mathbb{Q}(q)\otimes_{\mathbb{Z}[q]} \mathcal{H}_W$ is separable. The statement about the numerical invariants is a consequence of Tits' Deformation Theorem 2.17.
q.e.d.

REMARK 4.4. For every rational prime p which is prime to $|W|$ the localized Hecke order

$$\mathbb{Z}[q]_{\langle p,(q-1)\,\rangle} \otimes_{\mathbb{Z}[q]} \mathcal{H}_W$$

is separable. Hence also

$$\mathbb{Q}[q]_{\langle\,(q-1)\,\rangle} \otimes_{\mathbb{Z}[q]} \mathcal{H}_W$$

is separable as order over the one-dimensional ring $\mathbb{Q}[q]_{\langle\,(q-1)\,\rangle}$. This means that the congruences which occur in $\mathbb{Q}[q]_{\langle\,(q-1)\,\rangle} \otimes_{\mathbb{Z}[q]} \mathcal{H}_W$ must be prime to $\langle\, q-1\,\rangle$.

5. Blocks with cyclic defect of Hecke orders

Let $\mathcal{H} = \mathbb{Z}[q]_{\mathfrak{m}} \otimes_{\mathbb{Z}[q]} \mathcal{H}_0$ be the completion of the Hecke order $\mathcal{H}_0$ of a finite Coxeter group W at the maximal ideal $\mathfrak{m} := \langle p, q-1 \rangle$ for a rational prime p. We shall be using the following facts:

LEMMA 5.1. *The reduction map modulo $q-1$ from $\mathcal{H}$ to the p-adic group ring $\mathbb{Z}_pW$ has the following properties:*

- *It lifts indecomposable projective $\mathbb{Z}_pW$ modules. (This follows since $\mathbb{Z}[q]_{\mathfrak{m}}$ is complete.)*
- *It maps the center of $\mathcal{H}$ onto the center of $\mathbb{Z}_pW$ (cf. (GR; 94))*
- *If $\mathbb{Q}$ is a splitting field for W or if W is a dihedral group of order $2 \cdot p^n$ for an odd prime p, then there is a bijection between the irreducible rational representations V of $\mathcal{H}$ and $\widetilde{V}$ of $\mathbb{Z}_pW$ such that the dimensions over the endomorphism ring are the same. (This is an extension of Tits' deformation theorem 2.17.)*
- *For the above family of Coxeter groups, under the reduction and with this identification of the rational irreducible representations the decomposition numbers of $\mathcal{H}$ and $\mathbb{Z}_pW$ coincide. (This follows from Tits' deformation theorem and the method of lifting idempotents.)*

From now on we shall always assume that W is either a dihedral group of order $2 \cdot p^n$ for an odd prime p or that $\mathbb{Q}$ is a splitting field for W. (I am positive that the results also hold for the other Coxeter groups.)

THEOREM 5.2 ((Ro; 00, 1)). *Let B be a block of $\mathbb{Z}_pW$.*

- *There is a block $\mathcal{B}$ of $\mathcal{H}$ reducing to B modulo $\langle q-1 \rangle$ and having the same decomposition numbers as B.*
- *If B is a block with cyclic defect and Brauer tree T, then $\mathcal{B}$ is a tree order over $\mathbb{Z}[q]_{\mathfrak{m}}$ with tree T.*
- *Moreover, except for possibly one vertex – the exceptional vertex – the rings Δ_v are commutative integral domains and moreover separable extensions of $\mathbb{Z}[q]_{\mathfrak{m}}$.*

Proof : Since the center of $\mathcal{H}$ maps *onto* the center of $\mathbb{Z}_pW$ the central primitive idempotents of $\mathbb{Z}_pW$ lift to central primitive idempotents of $\mathcal{H}$. Whence the statement for the blocks follows with Lemma 5.1.

Let now $\mathcal{B}$ be a block of $\mathcal{H}$ such that the reduced block B of $\mathbb{Z}_pW$ is a block with cyclic defect and hence a tree order with associated tree T. Since projective B-lattices lift under preservation of isomorphism to projective Cohen-Macaulay $\mathcal{B}$-modules, we may assume that both $\mathcal{B}$ and B are basic.

We note that T has only genuine edges. Moreover, only one vertex v_0 can possibly be the exceptional vertex to which we have associated a local order $\widetilde{\Delta_{v_0}}$. At all other vertices v, the rings $\widetilde{\Delta_v}$ are unramified extensions of $\mathbb{Z}_p$, i. e. they are separable $\mathbb{Z}_p$-orders, which are at the same time integral domains.

According to Lemma 5.1 the rational representations of $\mathcal{B}$ and those of B are in bijection. Since every tree has at least two ends we may choose an end v_1 in T such that the corresponding ring $\widetilde{\Delta_{v_1}}$ is unramified over $\mathbb{Z}_p$. Moreover, the projection of B into the rational component to v_1 is just $\widetilde{\Delta_{v_1}}$. Hence the corresponding projection of $\mathcal{B}$ is a separable $\mathbb{Z}[q]_{\mathfrak{m}}$-order Δ_{v_1}. This shows in particular that the irreducible lattice $\widetilde{M}$ corresponding to v_1 in B lifts to an irreducible Cohen-Macaulay module M for Δ_{v_1}. Since the projective resolution for B of $\widetilde{M}$ lifts to a projective resolution for $\mathcal{B}$ of M, we know that $\mathcal{B}$ is a Cohen-Macaulay order with a tree resolution (cf. Definition 3.1). Note that there are no truncated edges, and hence we conclude with Theorem 3.2 that the order $\mathcal{B}$ is a tree order with tree T. *q.e.d.*

NOTE 5.3. In case B has defect 1 the indecomposable Cohen-Macaulay $\mathcal{B}$-modules which have a Cohen-Macaulay filtration have been described in the previous section (cf. (Ro; 99, 2)).

COROLLARY 5.4. *Assume that* $W = S(n)$ *is the symmetric group on* n *letters with local Hecke order* $\mathcal{H}(n)_p = \mathbb{Z}[q]_{\langle q-1,p\rangle} \otimes_{\mathbb{Z}[q]} \mathcal{H}(n)$, *where* $\mathcal{H}(n)$ *is the Hecke order of* $S(n)$ *over* $\mathbb{Z}[q]$. *Let* B *be a block of* $\mathbb{Z}_pS(n)$ *with cyclic defect, and let* $\mathcal{B}$ *be the corresponding block of* $\mathcal{H}$, *then the tree of* $\mathcal{B}$ *is a stem and for all the vertices the rings* Δ_v *coincide with* $\mathbb{Z}[q]_{\langle q-1,p\rangle}$. *For the principal p-block* $\mathcal{B}_0$ *of* $\mathcal{H}(p)_p$ *of the* *symmetric group $S(p)$ we have for all the corresponding elements δ_v the description $\delta_v = [p]$.*

Proof Since $\mathbb{Q}$ is a splitting field for $S(n)$, a block with cyclic defect of $\mathbb{Z}_p S(n)$ has all the rings $\widetilde{\Delta_v} = \mathbb{Z}_p$. In particular the block has defect 1 only since there is no room for an exceptional vertex, which would be there for a block of defect at least p^2. Hence in the corresponding block we have $\Delta_v = \mathbb{Z}[q]_{\langle q-1,p \rangle}$.
We know from (So; 98) that for a rational component $Mat(\nu, \mathbb{Q}(q))$ with central primitive idempotent e we have that $c_e \cdot Mat(\nu, \mathbb{Z}[q]_{\langle q-1,p \rangle}) \subset \mathcal{H}(p)_p$, where c_e is a factor of the Poincaré polynomial $[p]!$ (cf. (Ge; 95)). However, in $\mathbb{Z}[q]_{\langle q-1,p \rangle}$ we have $[p]! = [p]$. In fact, we have always $[s] = s$ modulo $\langle q-1 \rangle$, and if in addition, for an integer s we have $(s,p) = 1$, then in $\mathbb{Z}_p$ the element s is a unit, and so $[s]$ is a unit in $\mathbb{Z}[q]_{\langle q-1,p \rangle}$. Thus $\delta_v = [p]$ for every $v \in V$, since it cannot be invertible. *q.e.d.*

We shall here explain only the results in case of the principal block $\mathcal{B}_0$ of the Hecke order of the symmetric group $S(p)$ on p elements.

The results apply equally well to blocks with cyclic defect of the Hecke order of any symmetric group $S(n)$. In fact, since $\mathbb{Q}$ is a splitting field for $\mathbb{Q}S(n)$, every non-maximal block with cyclic defect must have defect 1, since there is no room for a genuine exceptional vertex. Moreover, the Brauer tree of every such block is a stem, since "complex conjugation" leaves the block elementwise fixed.

The principal block $\mathcal{B}$ of $\mathbb{Z}_p S(p)$ has as Brauer tree T a stem with p vertices,

$$T : 1 —— 2 —— \cdots —— (p-1) —— p\,.$$

Let $\mathcal{B}$ be the basic order derived from $\mathcal{B}_0$. The vertices of T correspond to the different rational components of $\mathcal{B}$. With the vertex i we associate the Cohen-Macaulay $\mathbb{Z}[q]_{\langle q-1,p \rangle}$-order Λ_i, which is described as follows:

$$\Lambda_1 = \mathbb{Z}[q]_{\langle q-1,p \rangle}\,,$$
$$\Lambda_i = \begin{pmatrix} \mathbb{Z}[q]_{\langle q-1,p \rangle} & \mathbb{Z}[q]_{\langle q-1,p \rangle} \\ [p] \cdot \mathbb{Z}[q]_{\langle q-1,p \rangle} & \mathbb{Z}[q]_{\langle q-1,p \rangle} \end{pmatrix} \text{ for } 1 < i < p \text{ and}$$
$$\Lambda_p = \mathbb{Z}[q]_{\langle q-1,p \rangle}\,.$$

Then $\mathcal{B}$ is the Brauer tree order of the stem T with associated "hereditary orders" Λ_i. Recall that for $\Gamma = \prod_{k=1}^{p} \Lambda_k$ "*we identify in* Γ *the* (i,i)*-entry in* Λ_k *with the* (j,j)*-entry in* Λ_l *modulo* $[p]$", i. e. we replace these entries by the pull-back

$$\begin{array}{ccc} \mathbb{Z}[q]_{\langle q-1,p\rangle} \stackrel{[p]}{\text{---}} \mathbb{Z}[q]_{\langle q-1,p\rangle} & \longrightarrow & (\mathbb{Z}[q]_{\langle q-1,p\rangle})^{k}_{(i,i)} \\ \downarrow & & \downarrow{\scriptstyle mod([p])} \\ (\mathbb{Z}[q]_{\langle q-1,p\rangle})^{l}_{(j,j)} & \xrightarrow{mod([p])} & \mathbb{Z}[\theta_p] \end{array},$$

where θ_p is a primitive p-th root of unity.
The main structural result is then as follows:

THEOREM 5.5 ((Ro; 00, 1)). *The basic version of the principal block* $\mathcal{B}$ *of the Hecke order of the symmetric group* $S(p)$ *is described as follows as a subring of* Γ*:*

- Λ_1 *is identified modulo* $[p]$ *with the* $(1,1)$*-entry in* Λ_2.
- *The* $(2,2)$*-entry of* Λ_i *is identified modulo* $[p]$ *with the* $(1,1)$*-entry in* Λ_{i+1} *for* $2 \le i \le p-2$.
- *The* $(2,2)$*-entry of* Λ_{p-1} *is identified modulo* $[p]$ *with* Λ_p.

In order to describe the filtered Cohen-Macaulay $\mathfrak{B}$-modules we let $\{M_1^1, M_i^k, M_1^p\}$ for $i = 1,2$ and $2 \le k \le p-1$ be the indecomposable projective Γ-modules and let P_i for $1 \le i \le p-1$ be the indecomposable projective $\mathcal{B}$-modules. We then have projective resolutions:

$$\mathbb{E}_i : \quad 0 \longrightarrow 0 \longrightarrow M_1^i \longrightarrow P_i \longrightarrow M_1^{i+1} \longrightarrow 0,$$

for $1 \le i \le p-1$.
Using these sequences, we shall give a description of the indecomposable Cohen-Macaulay modules M for $\mathcal{B}$, which have a Cohen-Macaulay filtration:

$$0 = M_0 \subset M_1 \subset M_2 \subset \cdots \subset M_{p-1} \subset M_p = M,$$

such that M_i/M_{i-1} for $1 \le i \le p$ is a Cohen-Macaulay module for Λ_i.
The classification of these Cohen-Macaulay modules is given as follows:

THEOREM 5.6 ((Ro; 99, 2)(Ro; 00, 1)). *The indecomposable $\mathcal{B}_0$-modules with a Cohen-Macaulay filtration are given as follows:*

- *The $p-1$ projective indecomposable modules P_i for $1 \leq i \leq p-1$,*
- *the $2 \cdot (p-1)$ projective indecomposable Γ-modules M_i^k, and*
- *the $p-1$ countable series, which arise as the middle terms of the exact sequences $(1-\theta_p)^n \cdot \mathbb{E}_i$.*

The last statement needs a word of explanation: One computes that $Ext^1_{\mathcal{B}}(M_1^{i+1}, M_1^i) \simeq End_{\mathcal{B}}(M_1^i)/[p] \cdot End_{\mathcal{B}}(M_1^i) = \mathbb{Z}[\theta_p]$, and so the non-equivalent extensions are given by $(1-\theta_p)^n \cdot \mathbb{E}_i$.

5.1. SORIANO'S TRICK

For the so-called *cellular orders* over $\mathbb{Z}[q]$ M. Soriano has made an interesting observation which allows for example to describe the extension groups of the *Specht* modules for the Hecke orders of the symmetric groups. We first state Soriano's observation:

PROPOSITION 5.7 (sori00). *Let R be a commutative ring and let Λ be an R-algebra, which is finitely generated as R-module. Let M and N be left Λ-modules such that*

1. *M and N are R-projective finitely generated,*
2. *$Hom_\Lambda(M,N) = 0$,*
3. *$r \cdot Ext^1(M,N) = 0$ for a regular element [10] $r \in R$,*

then $Ext^1(M,N) \subseteq (R/r \cdot R)^n$ for some natural number n.

Proof Since $r \cdot R \simeq R$ we have a short exact sequence

$$0 \longrightarrow N \xrightarrow{\cdot r} M \longrightarrow M/r \cdot M \longrightarrow 0,$$

which gives rise to the long exact sequence:

$$0 \longrightarrow Hom_\Lambda(M,N) \xrightarrow{\cdot r} Hom_\Lambda(M,N) \longrightarrow Hom_\Lambda(M, N/r \cdot N)$$
$$\longrightarrow Ext^1_\Lambda(M,N) \xrightarrow{\cdot r} Ext^1_\Lambda(M,N)$$

[10] This means $r \cdot R \simeq R$

The hypotheses now imply that

$$Hom_\Lambda(M, N/r \cdot N) \simeq Ext^1_\Lambda(M, N).$$

However,

$$Hom_\Lambda(M, N/r \cdot N) \subseteq Hom_R(M, N/r \cdot N).$$

Since M and N are R-projective finitely generated over R, we conclude $Hom_R(M, N/r \cdot N) \subseteq (R/r \cdot R)^n$ as claimed. *q.e.d.*

As an application we get an important result for Hecke orders and the extension groups of Specht lattices. In order to formulate the result we have to give several definitions. In (KoXi; 96) S. König and C. Xi gave a definition of cellular algebras which is equivalent to the original one of J. J. Graham and G. I. Lehrer (GrLe; 96), but more suitable for our purposes.

DEFINITION 5.8 (König-Xi (KoXi; 96)). *Let Λ be an R-algebra, where R is a commutative Noetherian integral domain. Assume that there is an R-linear anti-involution ι on Λ. A two-sided ideal J of Λ is called a* CELL IDEAL *provided*

- *$\iota(J) = J$ and*
- *there exists a left Λ-ideal $\Delta \subset J$ such that*
  - *Δ is finitely generated and free over R and*
  - *there is an isomorphism of bimodules*

$$\alpha\colon J \longrightarrow \Delta \otimes_R \iota(\Delta), \tag{2}$$

making the following diagram commute

$$\begin{array}{ccc} J & \xrightarrow{\alpha} & \Delta \otimes_R \iota(\Delta) \\ \iota\downarrow & & \downarrow x\otimes y \longmapsto \iota(y)\otimes \iota(x) \\ J & \xrightarrow{\alpha} & \Delta \otimes_R \iota(\Delta) \end{array}$$

The algebra Λ is called CELLULAR *(with respect to the anti-involution ι) provided there exists a chain of two-sided Λ-ideals*

$$0 = J_0 \subset J_1 \subset J_2 \subset \cdots \subset J_n = A,$$

each of them fixed by ι, such that for each index $0 < i$ the quotient J_i/J_{i-1} is a cell ideal (with respect to ι) in Λ/J_{i-1}.
The irreducible Cohen-Macaulay Λ-modules Δ_i generating the cell ideals J_i/J_{i-1} are called SPECHT MODULES.

The importance of cellular algebras stems from the following examples.

1. Hecke algebras of type A or B or more generally Ariki-Koike algebras (cf. (GrLe; 96)) [11].
2. Brauer algebras (cf. (GrLe; 96)),
3. Temperly-Lieb algebras (cf. (GrLe; 96)),
4. q-Schur algebras (cf. (DiJa; 89)),
5. Jones' annular algebras (cf. (GrLe; 96)).

NOTE 5.9. The cell structure of a cellular R-order Λ gives automatically a numbering of the central primitive idempotents ϵ_i of the K-algebra A. Similarly as in case of the quasi-hereditary algebras, for a cellular Hecke order $\mathcal{H}$ the category $\mathcal{H}(\mathcal{S}$ of those Cohen-Macaulay $\mathcal{H}$-modules M which have a filtration by Specht modules) should play an important role in their representation theory.

$$0 = M_0 \subset M_1 \subset M_2 \subset \cdots \subset M_n = M$$

such that $M_i/M_{i-1} \simeq \Delta_i^{\nu_i}$.

The next result should be considered in this connection.-

THEOREM 5.10 (Soriano 2000(So; 00)). *Let $\mathcal{H}_W$ be the Hecke order of the Coxeter group W. Assume that the Higman ideal contains a monic polynomial $f(q) \in \mathbb{Z}[q]$. Then $Ext^1_{\mathcal{H}_W}(M,N)$ is finitely generated $\mathbb{Z}$-torsion-free for M and N Cohen-Macaulay modules of $\mathcal{H}_W$, which lie in different rational components.*
This is in particular true for W the symmetric group on n letters and M and N Specht lattices.

I have asked Marcos Soriano to add a more detailed version of his results as an appendix to this article (cf. Appendix: Generic Hecke Algebras of Type A).

[11] This includes in particular the Hecke order of the symmetric group.

6. Appendix: Generic Hecke Algebras of Type A.

Marcos Soriano[12]

After giving a brief historical account on Hecke rings we state some essential facts about generic Iwahori-Hecke algebras. In combination with a method of calculating certain Ext^1-groups, these results are used to derive some basic facts about the structure of extension groups between generic Specht modules.

6.1. PRELIMINARIES.

For any finite Coxeter group W with system of generating involutions S one can define a corresponding associative unital ring, the *generic Iwahori-Hecke algebra*. Recall that irreducible Coxeter groups[13] are classified and can be described in terms of (weighted) Dynkin diagrams.

The *symmetric group* on n letters Σ_n can be presented as such an irreducible finite reflection group corresponding to the diagram of type A with $n-1$ nodes. One might think of the system of generating involutions as corresponding to the permutations $s_i = (i, i+1)$, where $i \in \{1, \ldots, n-1\}$. For simplicity we shall restrict ourselves to this case:

DEFINITION 6.1. *The generic Iwahori-Hecke algebra* $\mathcal{H}_n$ *associated to the symmetric group* Σ_n *is the* $\mathbb{Z}[q, q^{-1}]$*-algebra generated by elements* T_{s_i}*, where* $i \in \{1, \ldots, n-1\}$*, satisfying the* ***homogeneous*** *or* ***braid*** *relations:*

$$\begin{aligned} T_{s_i} \cdot T_{s_{i+1}} \cdot T_{s_i} &= T_{s_{i+1}} \cdot T_{s_i} \cdot T_{s_{i+1}} && \forall\, i\,, \\ T_{s_i} \cdot T_{s_j} &= T_{s_j} \cdot T_{s_i} && \text{whenever } |j-i| > 1\,; \end{aligned}$$

and the ***quadratic*** *relations:*

$$T_{s_i}^2 = (q-1) \cdot T_{s_i} + q \;\Leftrightarrow\; (T_{s_i} - q)(T_{s_i} + 1) = 0\,.$$

---

[12] Partially supported by the DFG.
Author's address:
Fakultät Mathematik, Universität Stuttgart
Pfaffenwaldring 57, D–70550 Stuttgart, Germany
(soriano@mathematik.uni-stuttgart.de)

[13] This means they do not admit a proper direct product decomposition into smaller *Coxeter* groups.

One can picture the homogeneous relations in the following way, which makes the connection to Artin's braid group evident:

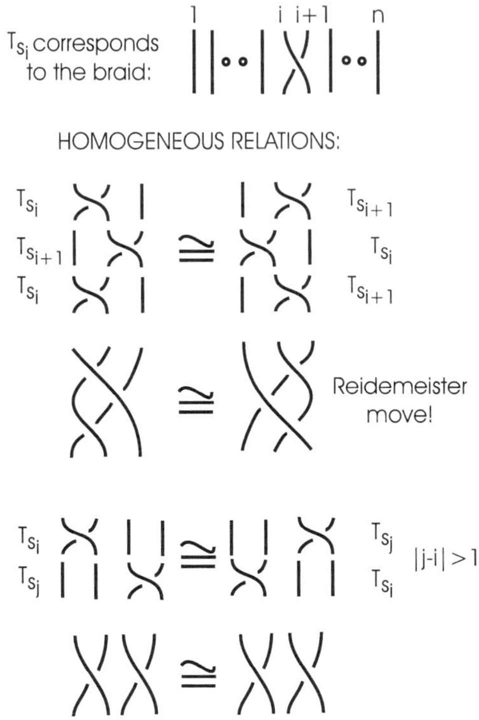

Following Jones (Jo87), one can think of representations of the Iwahori-Hecke algebra as being representations for the braid group, but where the canonical generators are only allowed to have two eigenvalues: -1 and q. This condition is imposed by the quadratic relations[14]. Note that we keep at least the freedom of choosing the second eigenvalue, q being a variable; this accounts for the word *generic* in the above definition. The ring of Laurent polynomials $\mathbb{Z}[q, q^{-1}]$, which we want to abbreviate by $\mathbb{Z}^q$, is the universal ring generated by one explicitly invertible element. So if we want the

[14] Therefore, one can think of the Iwahori-Hecke algebra $\mathcal{H}_n$ as the quotient of the semigroup ring over $\mathbb{Z}[q, q^{-1}]$ of the braid monoid of type A on $n-1$ generators defined by factoring out the quadratic relations (cf. (G-P00), Ch.4). One could even consider it as a quotient of the group ring over $\mathbb{Z}[q, q^{-1}]$ of Artin's braid group itself, since the canonical generators T_{s_i} are invertible in $\mathcal{H}_n$, due to the fact that q is a unit in $\mathbb{Z}[q, q^{-1}]$.

second eigenvalue to take a concrete value, we may do this by *specializing* q to be any invertible element in a commutative ring R. So, for example, if we let $q = 1 \in R$, we are just studying representations of the symmetric group Σ_n, still depending on the characteristic of R. Other interesting specializations arise when we let q be an n-th root of unity. The classical Hecke algebras first treated by Iwahori (Iw64) are those with $q = p^r$, where p is an integral prime. One should think here of q as corresponding to the index of the intersection $|B : B \cap B^{s_i}|$, where B is the (Borel) subgroup of upper triangular matrices within $\mathrm{GL}_n(\mathbb{F}_q)$ and s_i is the permutation matrix switching the canonical basis vectors e_i and e_{i+1}.

Remark 6.1. At this place we want to give a glance at the development of the concept of Hecke algebras, at least to justify the appearance of the names of Iwahori and Hecke in the definition above. We refer the interested reader to the monography (Kr90), where the theory of Hecke algebras is developed from a completely algebraic point of view, and where many historical comments are made.

The most general definition for Hecke algebras starts from a (semi)-group G and a subgroup S such that every double coset SgS contains only finitely many right and left cosets. Then a multiplication of double cosets is defined through counting arguments, generalizing the concept of group ring (which is just the case $S = 1$). Since its definition just involves counting, we obtain an algebra defined over the integers $\mathbb{Z}$, so we can pass to any other coefficient ring. Coming from groups, Hecke algebras are canonically equipped with an anti-involution (induced from group inversion) and with a symmetric bilinear form.

The origin of Hecke algebras lies in number theory and complex analysis, joint by the theory of modular forms. In his paper (He37), Hecke introduces in §2 the algebra corresponding to $S = \mathrm{SL}_2(\mathbb{Z})$ with respect to the monoid G consisting of all integer 2×2 matrices with positive determinant. Rather than introducing it as an abstract algebra as above, Hecke identifies it with the linear representation it induces on the complex space of entire elliptic modular forms. This space of functions is invariant under the action of the modular group S, this being the key fact allowing the definition of the corresponding ring of operators. The study of spaces invariant under the action of some subgroup is one of the main motivations for introducing Hecke algebras.

Shimura seems to be the first who gives a completely abstract definition of Hecke algebra (Sh59). The first appearance in a text book is in Shimura's book on automorphic forms (Sh71), p. 51–55. Other motivating examples for introducing the concept of Hecke algebra coming from the arithmetic of (even non-commutative) orders in number theory are given there.

In the mid sixties Hecke algebras enter representation theory. Chevalley groups give all but one (the alternating groups) infinite series of finite non-abelian simple groups. Trying to develop their theory from a small number of facts, the axiomatic of finite groups G with a BN-pair was developed. The finite groups $\mathrm{GL}_n(\mathbb{F}_q)$ with $q = p^r$ some prime power and B the Borel subgroup of upper triangular matrices are a prominent example. Groups with a BN-pair have a canonical decomposition into double cosets with respect to the subgroup B, the so-called Bruhat decomposition. This decomposition is surprisingly parametrized by a (Coxeter) group, the so-called Weyl group W. Hecke algebras provide an appropriate frame for linearization of the Bruhat decomposition, giving rise to families of algebras closely related to the group algebra of W. Iwahori (Iw64) and Matsumoto (Ma64) give for the first time a presentation of the Hecke algebras associated to Chevalley groups defined over a finite field. They even prove that these algebras, when defined over the complex numbers, are indeed isomorphic to the complex group algebra of W. This fact will be later on generalized by Lusztig and (for arbitrary parabolic subgroups) by Curtis–Iwahori–Kilmoyer. This algebra presentation is recovered in an exercise from (Bo68). Curtis seems to be the first to underline the importance of the concept of generic algebras for groups G having a BN-pair, since one obtains whole families of Hecke algebras and even the group algebra of W as specializations from a *single* generic algebra.

The importance of Hecke algebras for representation theory (especially, as seen above, for finite groups G of Lie type) is based on the following fact, which we describe rather roughly: the problem of classifying simple modules for G can often be divided into two parts: first, finding all (so-called cuspidal) representations necessary for a certain induction process, and second, decomposing induced modules from these cuspidal representations. Hecke algebras are the endomorphism algebras of these induced modules, and therefore simple modules for Hecke algebras parametrize the simple modules appearing in the decomposition of the induced modules. For introductory accounts the reader is refered to (C-R81), §11, and (C-R87), Ch. 8. For a recent survey on the latest developments we refer to the paper (DGHM99).

The other prominent applications of Hecke algebras are mainly in number theory, where they were first introduced. A famous example from the past years is contained in (W-T95), where the gap discovered in Wiles' proof of Fermat's Last Theorem is closed.

A rather unexpected application (at that time) in a completely different field from number theory or representation theory was the construction of polynomial invariants of knots and links by Jones in the late eighties via Hecke algebras of type A (cf. the picture of the braid relations). We refer to the paper (Jo87) for the connections of type A Hecke algebras to topology.

Next we list some essential facts about *generic* Iwahori-Hecke algebras needed later. Meanwhile there exists an extensive and accessible literature on the subject where the reader may find most of the missing proofs (e.g. (C-R87), Ch. 8, (G-P00), (Ma99)).

PROPOSITION 6.1 (Properties of $\mathcal{H}_n$).
① *$\mathcal{H}_n$ is $\mathbb{Z}^q$-free with a basis indexed by the group elements $w \in \Sigma_n$.*
② *We denote by $\mathbb{Q}(q)$ the function field in one variable q over the rationals; this is the field of fractions of $\mathbb{Z}^q$. Set $\mathbb{H}_n := \mathbb{Q}(q) \otimes_{\mathbb{Z}^q} \mathcal{H}_n$. Then $\mathbb{H}_n$ is a semisimple algebra and $\mathbb{Q}(q)$ is already a splitting field for it.*
③ *$\mathbb{H}_n$ is a symmetric $\mathbb{Q}(q)$-algebra. Following the theory of symmetric algebras, one can associate to any irreducible module of $\mathbb{H}_n$ with character ϕ an* ***invariant*** *$c_\phi \in \mathbb{Q}(q)$.*
④ *We define for any natural number n the polynomial $[n]$ with integral coefficients by: $[n] = \frac{q^n-1}{q-1} = q^{n-1} + \ldots + q + 1$. Note $[1] = 1$; generally $[n]$ equals the product of irreducible cyclotomic*

polynomials $\prod_{1\neq d|n}\Phi_d$. *We define the Poincaré polynomial* P_n *by*

$$P_n := \sum_{w\in\Sigma_n} q^{l(w)} \overset{!}{=} [n]! := [n]\cdot[n-1]\cdots[1] = \prod_{i=1}^{n}[i]\,.$$

⑤ *Define* $D_\phi := \frac{P_n}{c_\phi}$. *These invariants are called* ***generic degrees*** *and are known to belong to* $\mathbb{Z}[q]$, *i.e. they are polynomials with integer coefficients.*
⑥ *Although* $\mathbb{Z}^q$ *is* not *a principal ideal domain, one can realize all the irreducible* $\mathbb{H}_n$*-representations already as* $\mathbb{Z}^q$*-free modules of finite rank. A special set of* $\mathbb{Z}^q$*-free modules affording these representations is rather well known: the so-called* ***Specht*** *modules.*

Remark 6.2. **Ad ①:** Pick an element w, write it as a product of elementary transpositions $w = s_1\cdot s_2\cdots s_k$, with k as small as possible. Call such an expression for w *reduced* and denote its length by $l(w)$. The (standard) basis element T_w is given in terms of the above presentation as the product $T_{s_1}\cdot T_{s_2}\cdots T_{s_k}$. Observe that the unit element of the algebra $\mathcal{H}_n$ corresponds to the unit element of the group Σ_n.
This fact is due to Matsumoto, in (Ma64) he shows that the braid relations are sufficient for transforming into each other any two reduced expressions of an element w as a product of s_i's.
Ad ②: $\mathbb{H}_n$ is even isomorphic to the group ring of Σ_n over the field $\mathbb{Q}(q)$. Lusztig has given an *explicit* isomorphism. *But* note that the generic Iwahori-Hecke algebra is *not* isomorphic to the group ring $\mathbb{Z}^q\Sigma_n$ (cf. e.g. (Sh86), §2.2).
Ad ③: Using the above notation, one has a canonical pair of dual bases: $\{T_w\}_{w\in\Sigma_n}$ and $\{\frac{1}{q^{l(w)}}T_{w^{-1}}\}_{w\in\Sigma_n}$. Note that the elements of the second basis also belong to $\mathcal{H}_n$, since q is invertible in $\mathbb{Z}^q$.
The invariants c_ϕ are usually called *Schur elements.* They could be defined (rather than computed) as follows: pick a pair of dual bases $\{h_i\}$ and $\{h_i^\star\}$ of $\mathbb{H}_n$, and pick an irreducible matrix representation X of $\mathbb{H}_n$ with character ϕ. Then, summing up over the following product of matrix entries, one obtains:

$$\sum_i X(h_i)_{1,1}\cdot X(h_i^\star)_{1,1} = c_\phi\,.$$

One should note that c_ϕ does *not* depend on the pair of dual bases and the representation used for its computation. As for any sym-

metric semisimple algebra, one can construct a full set of primitive idempotents for $\mathbb{H}_n$ using the Schur elements: the idempotents can all be chosen to be of the form $e = \frac{1}{c_\phi} \cdot E$ for an element $E \in \mathcal{H}_n$, provided all irreducible representations of $\mathbb{H}_n$ can be realized over $\mathbb{Z}^q$, which is the case (see ⑥ below).
In type A the Schur elements correspond to the q-analogue of the product of hook lengths, as follows: irreducible $\mathbb{H}_n$-representations are indexed by partitions λ of n. Filling the hook lengths l_i into the corresponding Young diagram, the invariant c_λ equals (up to some q-power, which is a unit in $\mathbb{Z}^q$) the product of the polynomials $[l_i]$. For example, take $\lambda = (5,3,2)$, a partition of 10. Its Young diagram with the hook lengths filled in is (for combinatorial terms cf. e.g. (Ja78)):

7	6	4	2	1
4	3	1		
2	1			

Therefore, the principal ideal in $\mathbb{Z}^q$ generated by $c_{(5,3,2)}$ equals

$$\langle\, c_{(5,3,2)} \,\rangle = \langle\, [7] \cdot [6] \cdot [4] \cdot [2] \cdot [4] \cdot [3] \cdot [2] \,\rangle = \langle\, \Phi_7 \Phi_6 \Phi_4^2 \Phi_3^2 \Phi_2^5 \,\rangle\,.$$

For an introduction to the theory of symmetric algebras with special emphasis on Iwahori-Hecke algebras we refer to (Ge95), §1 or (G-P00), Ch.7.
Ad ④: One might think of P_n as a kind of Hilbert series of the algebra $\mathcal{H}_n$ when grading it by putting T_w at degree $l(w)$. This polynomial is analogous to the group order $n!$ of Σ_n. The appearance of cyclotomic polynomials might suggest that specializations of $\mathcal{H}$ at roots of unity may be interesting: in fact, exactly these specializations give rise to *non-semisimple* algebras (cf. e.g. (Ge95)).
Ad ⑤: The reason for calling D_ϕ a *degree* is as follows: $D_\phi(p^r)$ is indeed the degree of a (so-called unipotent) representation of the group $\mathrm{GL}_n(\mathbb{F}_{p^r})$. Thus, one may interpret the defining equation $D_\phi = \frac{P_n}{c_\phi}$ as a generalization of the hook formula for the dimension of the irreducible modules of the symmetric group Σ_n over fields of characteristic zero. Steinberg was the first to calculate the polynomials D_ϕ for type A in the mid-fifties (St51); Lusztig completed the task by computing them for all Weyl groups ((Lu77); see also

(Ca85), Ch.13, for closed formulas and tables).
Ad ⑥: There is one Specht module S_λ for $\mathcal{H}_n$ associated to any partition λ of n; these give a full set of pairwise non-isomorphic, irreducible $\mathbb{H}_n$-representations. There are several ways of deriving irreducible lattices for generic Iwahori-Hecke algebras: Dipper and James follow closely in (D-J86) the classical theory of Specht modules for the symmetric group Σ_n. Kazhdan and Lusztig gave already in (K-L79) a completely different construction introducing the new ideas of W-graphs and cell representations. These ideas have been axiomatized in the work of Graham and Lehrer (G-L96), where they introduce the concepts of cellular basis and cellular algebras. A ring theoretical interpretation of the concept of cellularity, following closely the inductive definition of quasi-hereditary algebras, was given by König and Xi in (K-X98). Murphy gave in (Mu92) a cellular basis of $\mathcal{H}_n$, completely different from the one constructed in (K-L79), which is combinatorially easier to describe and which extends easily to the q-Schur algebra, therefore proving in a unified way that $\mathcal{H}_n$ is cellular and that the q-Schur algebra is quasi-hereditary (cf. (Ma99) for an account on this theme).
It is important to note that the Specht modules S_λ are absolutely irreducible in the following sense: $\mathrm{End}_{\mathcal{H}_n}(S_\lambda) = \mathbb{Z}^q \cdot \mathrm{id}_{S_\lambda} \cong \mathbb{Z}^q$.
An easy proof of this fact can be derived from the cellularity of $\mathcal{H}_n$.

6.2. A METHOD FOR COMPUTING SOME Ext^1-GROUPS.

We want to present a method for computing some Ext^1-groups in fairly general terms, although we will only apply it to the Iwahori-Hecke algebra $\mathcal{H}_n$ and its Specht modules. The reader might keep this example in mind while further reading.

First let R be an integral domain with quotient field K. Since we do some kind of "integral" theory, it is important that R does *not* coincide with K, otherwise the method to be described lacks significance. We further consider an R–algebra Λ. We mainly have in mind R–orders[15] inside finite dimensional semisimple algebras, but

[15] By an R–order Λ we mean a *unital* subring of the K–algebra $A := K \otimes_R \Lambda$ containing a basis of A and being finitely generated as an R–module. It is sometimes strongly recommended to let Λ be projective as an R–module, but we shall not demand this.

this is not at all necessary. Finally, we let M and N be Λ–modules, and we note that we can consider them also as R–modules. In the course of the section we shall impose additional conditions on these modules.

Assume N is *R–torsion-free*, and let $f \neq 0$ be an element in R. Multiplication with f induces a short exact sequence of Λ–modules:

$$\mathbb{E} : 0 \longrightarrow N \xrightarrow{f\cdot} N \longrightarrow N/fN \longrightarrow 0 .$$

In order to exclude trivial cases one should take care that f does not act as an automorphism on N, i.e. $f{\cdot}N$ should be properly contained in N[16]. Observe N being R–torsion-free implies the injectivity of the map "$f\cdot$". Applying the covariant functor $\mathrm{Hom}_\Lambda(M, -)$ to the sequence $\mathbb{E}$ gives rise to the following long exact sequence:

$$\begin{aligned} 0 \longrightarrow \mathrm{Hom}_\Lambda(M, N) &\xrightarrow{f\cdot} \mathrm{Hom}_\Lambda(M, N) \longrightarrow \mathrm{Hom}_\Lambda(M, N/fN) \\ \longrightarrow \mathrm{Ext}^1_\Lambda(M, N) &\xrightarrow{f\cdot} \mathrm{Ext}^1_\Lambda(M, N) \longrightarrow \dots \end{aligned}$$

Observe every third map really coincides with multiplication by f on the underlying R–modules.

Suppose we would know *a priori* that f belongs to the R–annihilator of the module $\mathrm{Ext}^1_\Lambda(M, N)$. Then the fourth map in the sequence is the *zero* map: one may interpret this roughly by saying that "every extension of M by N comes from a homomorphism in $\mathrm{Hom}_\Lambda(M, N/fN)$".

Suppose further there are *no homomorphisms* at all between the modules M and N: then the first terms of the above long exact sequence collapse to

$$0 \longrightarrow \mathrm{Hom}_\Lambda(M, N/fN) \longrightarrow \mathrm{Ext}^1_\Lambda(M, N) \longrightarrow 0 ,$$

so these modules are even isomorphic. Since every homomorphism from M into N/fN factors over $f \cdot M$, we have even proved:

LEMMA 6.1. *Let Λ be an R-algebra, M and N Λ-modules and f an element in R satisfying:*

[16] In more general terms, one could already *start* with a short exact sequence in appropriate form.

① *N is R–torsion-free,*
② $\operatorname{Hom}_\Lambda(M,N)=0$,
③ $0\neq f\in \operatorname{Ann}_R(\operatorname{Ext}^1_\Lambda(M,N))$.
Then we have an isomorphism

$$\operatorname{Hom}_\Lambda(M/fM, N/fN) \xrightarrow{\sim} \operatorname{Ext}^1_\Lambda(M,N)\,.$$

We want to make some remarks on this lemma before we continue:
➤ Thinking for a moment of M and N being R–free, one can interpret the set $\operatorname{Hom}_\Lambda(M/fM, N/fN)$ as "matrices commuting only modulo f", or rather "morphisms up to f". This is what Künzer (for integers f, cf. (Ku99)) calls for obvious reasons *modular morphisms*. In this explicit *proto*form for matrices one can already find the lemma in the book (C-R62)[17]. One may also easily derive an *explicit* isomorphism between the two objects in question, but we will not be interested in doing so.
➤ One should also at least point out the connection to *matrix factorizations* in the theory of so-called Cohen-Macaulay modules for *hypersurfaces*, i.e. rings R of the form "regular local ring S modulo the principal ideal generated by f". Matrix factorizations are pairs of (square) matrices (ϕ,ψ) with entries in S satisfying

$$\phi\cdot\psi = f\cdot\mathbf{1} = \psi\cdot\phi\,,$$

therefore being obvious f-modular morphisms. They give rise to 2-periodic free resolutions of Cohen–Macaulay modules over the hypersurface and can actually be used to classify these modules (see e.g. (Yo90), Ch.7).

Example 6.1. ① Of course, when computing extension groups via Lemma 6.1, it does not matter which annihilating element is chosen. We demonstrate this by the following example.
We set $R:=\mathbb{Z}[X]$ and $\mathfrak{m}:=\langle p, X\rangle$, where p is a prime integer. Observe that $\mathfrak{m}$ is a maximal ideal in R, the quotient field $R/\mathfrak{m}$ being isomorphic to the prime field $\mathbb{F}_p$. Finally, $K:=\mathbb{Q}(X)$ denotes the field of quotients of R.

[17] This fact as well as the main idea of computing extension groups via *modular morphisms* was pointed out to me by M. Künzer, whom I want to thank here.

Now we define an R-order inside $K \times K$ by

$$\Lambda := \left(R \overset{\mathfrak{m}}{=\!=\!=} R\right) := \{\, (x, y) \in R \times R \,|\, x - y \in \mathfrak{m} \,\} \, .$$

One should note that Λ is *not* R-free. With π_1 and π_2 we denote the projections onto the first resp. the second component of Λ. Observe that we can then write Λ as the pullback along two copies of the quotient map from R to $\mathbb{F}_p$:

$$\begin{array}{ccccc}
 & & (\mathfrak{m}, 0) & \xrightarrow{\ \simeq\ } & \mathfrak{m}_1 \\
 & & \downarrow & & \downarrow \\
(0, \mathfrak{m}) & \hookrightarrow & \Lambda & \xrightarrow{\ \pi_1\ } & R_1 \, . \\
\downarrow \simeq & & \downarrow \pi_2 & & \downarrow \\
\mathfrak{m}_2 & \hookrightarrow & R_2 & \twoheadrightarrow & \mathbb{F}_p
\end{array}$$

The subscripts indicate which component of Λ operates on the corresponding modules. Thus we have a short exact sequence

$$\mathbb{E} \, : \, 0 \longrightarrow \mathfrak{m}_2 \xrightarrow{\ \iota\ } \Lambda \xrightarrow{\ \pi_1\ } R_1 \longrightarrow 0 \, .$$

As usual, by applying the functor $\mathrm{Hom}_\Lambda(-, \mathfrak{m}_2)$ to the sequence $\mathbb{E}$ and observing that $\mathbb{E}$ is a projective presentation of the Λ-module R_1, one easily obtains:

$$\mathrm{Ext}^1_\Lambda(R_1, \mathfrak{m}_2) \simeq \frac{\mathrm{End}_\Lambda(\mathfrak{m}_2)}{\mathrm{Hom}_\Lambda(\Lambda, \mathfrak{m}_2)} \simeq \frac{\mathrm{End}_\Lambda(\mathfrak{m}_2)}{\mathfrak{m}_2} \simeq \frac{R}{\mathfrak{m}} \simeq \mathbb{F}_p \, .$$

Thus, $\mathrm{Ext}^1_\Lambda(R_1, \mathfrak{m}_2) \simeq \mathbb{F}_p$, having obviously p-torsion. Since R_1 and $\mathfrak{m}_2$ give rise to two non-isomorphic simple $K \otimes_R \Lambda$-modules by extending the scalars to K, there are no non-zero homomorphisms between these modules, and Lemma 6.1 could be used for the computation of $\mathrm{Ext}^1_\Lambda(R_1, \mathfrak{m}_2)$. There are two obvious choices for the annihilating element: $f = p$ or $f = X$. The first choice makes p-torsion evident, in contrast with the second choice, which leads to the isomorphism:

$$\mathrm{Ext}^1_\Lambda(R_1, \mathfrak{m}_2) \simeq \mathrm{Hom}_\Lambda(R_1 / X \cdot R_1, \mathfrak{m}_2 / X \cdot \mathfrak{m}_2) \, .$$

sentation Theory with Applications to Finite Groups and Orders, Volume II. John Wiley & Sons, (1987).

DGHM99. RICHARD DIPPER, MEINOLF GECK, GERHARD HISS AND GUNTER MALLE, *Representations of Hecke Algebras and Finite Groups of Lie Type.* In "Algorithmic Algebra and Number Theory", B.H. Matzat, G.-M. Greuel, G. Hiss (Eds.), Springer Verlag, 332–378, (1999).

D-J86. RICHARD DIPPER AND GORDON JAMES, *Representations of Hecke algebras of general linear groups.* Proc. London Math. Soc. (3), **52**, 20–52, (1986).

Ge95. MEINOLF GECK, *Beiträge zur Darstellungstheorie von Iwahori–Hecke–Algebren.* Verlag der Augustinus Buchhandlung, Aachener Beiträge zur Mathematik, **11**, (1995).

G-P00. MEINOLF GECK AND GÖTZ PFEIFFER, *Characters of Finite Coxeter Groups and Iwahori–Hecke–Algebras.* Oxford Science Publications, London Mathematical Society Monographs, New Series, **21**, (2000).

G-L96. J. J. GRAHAM AND G. I. LEHRER, *Cellular algebras.* Invent. Math., **123**, 1–34, (1996).

He37. E. HECKE *Über Modulfunktionen und die Dirichletschen Reihen mit Eulerscher Produktentwicklung I.* Mathematische Annalen, **114**, 1–28, (1937).

Iw64. IWAHORI, N., *On the structure of the Hecke ring of a Chevalley group over a finite field.* J. Fac. Sci. Univ. Tokyo, **10**, 215–236, (1964).

Ja78. G. D. JAMES, *The Representation Theory of the Symmetric Groups.* Springer Verlag, Lecture Notes in Mathematics, **682**, (1978).

Jo87. V. F. R. JONES, *Hecke algebra representations of braid groups and link polynomials.* Annals of Mathematics, **126**, 335–388, (1987).

K-L79. DAVID KAZHDAN AND GEORGE LUSZTIG, *Representations of Coxeter Groups and Hecke Algebras.* Invent. Math., **53**, 165–184, (1979).

K-X98. STEFFEN KÖNIG AND CHANGCHANG XI, *On the structure of cellular algebras.* In "Algebras and modules II", Idun Reiten et al. (Eds.), CMS Conf. Proc., published by the AMS, **24**, 365–386, (1998).

Kr90. ALOIS KRIEG, *Hecke algebras.* Memoirs of the AMS, **435**, (1990).

Ku99. MATTHIAS KÜNZER, *Ties for the $\mathbb{Z}S_n$*, PhD. Thesis, Universität Bielefeld, http://www.mathematik.uni-bielefeld.de/~kuenzer/ (1999).

Lu77. GEORGE LUSZTIG, *Irreducible Representations of Finite Classical Groups.* Invent. Math., **43**, 125–175, (1977).

Ma99. ANDREW MATHAS, *Iwahori–Hecke Algebras and Schur Algebras of the Symmetric Group.* University Lecture Series of the AMS, **15**, (1999).

Ma64. MATSUMOTO, H., *Générateurs et rélations des groupes de Weyl generalisés.* C. R. Acad. Sci. Paris, **258**, 3419–3422, (1964).

Mu92. G. E. MURPHY, *On the Representation Theory of the Symmetric Groups and Associated Hecke Algebras.* J. Algebra, **152**, 492–513, (1992).

Sh86. SHI JIAN-YI, *The Kazhdan-Lusztig Cells in Certain Affine Weyl Groups.* Springer Verlag, Lecture Notes in Mathematics, **1179**, (1986).

Sh59. SHIMURA, G., *Sur les integrales attachées aux formes automorphes.* J. Math. Soc. Japan, **11**, 291–311, (1959).

Sh71. SHIMURA, G., *Introduction to the Arithmetic Theory of Automorphic Functions.* Iwanami Publishers and Princeton University Press, (1971).

St51. R. STEINBERG, *A geometric approach to the representations of the full linear group over a Galois field.* Trans. Amer. Math. Soc., **71**, 274–282, (1951).

W-T95. RICHARD TAYLOR AND ANDREW WILES, *Ring-theoretic properties of certain Hecke algebras.* Annals of Mathematics, **141**, 553–572, (1995).

Yo90. YUJI YOSHINO, *Cohen-Macaulay Modules over Cohen-Macaulay Rings.* Cambridge University Press, London Math. Soc. Lecture Note Series, **146**, (1990).

INVERTIBLE IDEALS AND NON-COMMUTATIVE GENERALIZATIONS OF REGULAR RINGS

WOLFGANG RUMP
(wolfgang.rump@ku-eichstaett.de)
Mathematisch-Geographische Fakultät,
Katholische Universität Eichstätt
Ostenstrasse 26-28,
D-85071 Eichstätt, Germany

Key words: invertible ideal, regular ring, valuation ring, non-commutative

2000 Mathematics Subject Classification: 16P40, 16S38

1. Introduction

Starting from the well-established theory of commutative noetherian rings [32, 22, 10, 41, 19], there have been found various lines of generalization into the vast and unknown field of non-commutative rings. McConnell and Robson [30] specify this program as the search for a unification of the two equally well-developed theories of commutative noetherian rings and non-commutative artinian rings, thinking, for instance, of non-commutative localization as a series of attempts [21, 42, 24] which illustrate the difficulties of that procedure.

From a geometric point of view, regular rings are the foremost object of study. Describing smooth varieties in the commutative case, they should retain some nice behaviour when the familiar province of commutativity is left behind. There are various equivalent descriptions of commutative regular rings, each suggesting a different route into the non-commutative area. The obvious choice, of course, is a failure: The class of rings with finite global dimension is far too large. However, it can be regarded as a first approximation.

K.W. Roggenkamp and M. Ştefănescu (eds.), Algebra - Representation Theory, 351–378.

Recall that an ideal P of a ring R is said to be *invertible* if the bimodule ${}_RP_R$ belongs to the Picard group $\mathrm{Pic}(R)$ of R. The occurrence of invertible ideals in connection with non-commutative regularity concepts will be discussed in the present article. We show that invertible ideals can be used to get a notion of a non-commutative Cohen-Macaulay ring that generalizes Cohen-Macaulay algebras. In a natural way, this definition of Cohen-Macaulayness gives rise to a concept of regularity which retains many features known from the commutative case. For a comparison with other recently developed regularity concepts, invertible ideals will be the connecting link.

For graded algebras over a field, Artin and Schelter [2] introduced a concept of regularity which led to 13 generic classes of rings in global dimension three. Artin, Tate, and Van den Bergh [3] gave a complete description of such three-dimensional regular algebras in case they are generated by elements of degree one, and recently, Stephenson [46, 47] extended the classification to the general case. According to their Hilbert function, Artin-Schelter regular algebras of global dimension 3 can be regarded as non-commutative analogues of the polynomial ring in three variables (each endowed with a degree $\geqslant 1$). Thus if the Hilbert function coincides with that of a polynomial ring with the ordinary grading, there is good reason to speak of a non-commutative $\mathbb{P}^2$.

A particularly interesting class of Artin-Schelter regular algebras can be described in terms of an elliptic plane curve E together with an automorphism σ. (Generically, σ will be either an involution or a translation with respect to the group law on E.) These algebras A contain a homogenous regular element g which is *normal* (i. e. $Ag = gA$) such that A/gA is the twisted homogenous coordinate ring $\mathcal{B}(E, \sigma, \mathcal{O}(1))$ of E (see [5]).

Using the results of Artin, Tate, and Van den Bergh [3, 4], Levasseur [27] proved that three-dimensional Artin-Schelter regular algebras satisfy a couple of conditions that could make up a candidate for a well-behaved regularity concept: He showed that they are Auslander-regular and Macaulay. Moreover, they are noetherian domains [3, 27, 46]. Conversely, let $A = k \oplus A_1 \oplus A_2 \oplus \cdots$ be a noetherian finitely generated graded k-algebra of polynomial growth, i. e. there is an integer $d \in \mathbb{N}$ such that $n^{-d} \dim(k \oplus \cdots \oplus A_n)$ lies between two positive constants for $n \gg 0$. If A is Auslander-regular, then A is Artin-Schelter regular. When a graded algebra A of polynomial growth has a homogenous regular normal element g, Levasseur [27] proves under some mild extra condition that A is Auslander-regular and Macaulay if and only if A/gA is.

Regular normal elements occurred in early investigations of universal enveloping algebras of nilpotent or solvable Lie algebras [29, 49]. They give rise to an obvious generalization of regular sequences and thus provide a concept of regularity introduced by Walker [49]. Not all of the Artin-Schelter regular algebras are regular in this sense, even if they are generated by elements of degree one. For such an algebra A, however, there is always a (possibly zero) normal element $g \in A$ such that A/gA is a twisted homogenous coordinate ring (see §2).

Every regular normal element g of a ring R generates an invertible ideal $Rg = gR$. We call it *principal* invertible. Unless R is commutative, invertible ideals need not be principal. By forming "R-sequences" $\{P_1, \dots, P_n\}$ such that $P_1 + \cdots + P_{i+1}$ is invertible modulo $P_1 + \cdots + P_i$ for $i < n$, we get a concept of *weak Cohen-Macaulayness* for semilocal rings R. The appropriate concept of dimension will be analogous to the Gelfand-Kirillov dimension, with the difference that we have to consider decreasing instead of increasing filtrations. If the $\{P_1, \dots, P_n\}$ are themselves invertible, we speak of a (strong) *Cohen-Macaulay ring* R. Accordingly, there are two concepts of regularity. Namely, we call R *weakly regular* if R is a weak Cohen-Macaulay ring with $\operatorname{gld} R = \dim R$, and (strongly) *regular* if R is a Cohen-Macaulay ring with $P_1 + \cdots + P_n = \operatorname{Rad} R$ for some R-sequence $\{P_1, \dots, P_n\}$. Of course, the weak version will turn out to be a consequence of the strong one.

For example, consider the universal enveloping algebra $\mathrm{U}(\mathfrak{g})$ for any finite dimensional nilpotent Lie algebra $\mathfrak{g}$. The localization of $\mathrm{U}(\mathfrak{g})$ at the ideal generated by $\mathfrak{g}$ is then weakly regular. Furthermore, every weakly regular ring R is Auslander-regular and Macaulay [39].

For strongly regular rings, stronger results are available. There is a complete description [38] of semiperfect regular rings in terms of certain local rings V which are obtained by a higher-dimensional analogue of a discrete valuation. We call these rings V *almost regular* since (strongly) regular local rings are of this type. Unless

$\dim V = 1$, almost regular rings need not be regular. A relationship to invariant rings of tori $\mathrm{GL}_1(k)^m$ will be explained in §5.

2. Homological concepts of regularity

As a basis for generalization, let us briefly review the commutative situation (cf., e. g., [41, 28]). Let R be a commutative noetherian local ring with maximal ideal $\mathfrak{m}$ and residue class field $\mathfrak{k}$. There are several concepts of dimension for R:

1. the Krull dimension $\dim R$;
2. the tangential dimension $\dim_{\mathfrak{k}} \mathfrak{m}/\mathfrak{m}^2$;
3. the depth $\operatorname{cdh} R$, i. e. the length of a maximal R-sequence in $\mathfrak{m}$;
4. the global dimension $\operatorname{gld} R$.

Recall that $a_1, \ldots, a_d \in \mathfrak{m}$ is said to be an *R-sequence* if a_i is regular (i. e. a non-zerodivisor) modulo $Ra_1 + \cdots + Ra_{i-1}$ for $i \in \{1, \ldots, d\}$. The depth $\operatorname{cdh} R$ (also called the *homological codimension* ([41]) is equal to the grade of $\mathfrak{k}$, that is, the smallest integer d with $\operatorname{Ext}_R^d(\mathfrak{k}, R) \neq 0$. Auslander and Buchsbaum [7] gave another characterization of $\operatorname{cdh} R$. They showed that $\operatorname{cdh} R$ coincides with the finitistic global dimension in the category R-**mod** of finitely generated R-modules, i. e. the supremum over the finite projective dimensions of finitely generated R-modules. Moreover, they proved the inequalities

$$\operatorname{cdh} R \leqslant \dim R \leqslant \dim_{\mathfrak{k}} \mathfrak{m}/\mathfrak{m}^2 \leqslant \operatorname{gld} R. \tag{1}$$

Notice that it makes no difference whether the supremum $\operatorname{gld} R$ of all projective dimensions is extended over R-**mod** (as in [7]) or over the category R-**Mod** of all R-modules. The ring R is said to be *Cohen-Macaulay* if $\operatorname{cdh} R = \dim R$, and *regular* if $\dim R = \dim_{\mathfrak{k}} \mathfrak{m}/\mathfrak{m}^2$, that is, if $\operatorname{cdh} R = \operatorname{gld} R$. By Serre's theorem, the latter already happens if $\operatorname{gld} R < \infty$.

Now let us turn to the non-commutative case. As above, R-**Mod** (R-**mod**) denotes the category of (finitely generated) left R-modules. For simplicity, we fix an algebraically closed field k of characteristic 0. If a k-algebra A is finitely generated, there exists a k-subspace V with $1 \in V$ such that $A = \bigcup_{n=0}^{\infty} V^n$. The *Gelfand-Kirillov dimension*

$$\operatorname{GKdim} A := \inf\{\nu \in \mathbb{R} \mid \dim V^n \leqslant n^\nu \text{ for almost all } n \in \mathbb{N}\} \tag{2}$$

of A does not depend on the particular choice of V. For example,

$$\operatorname{GKdim} k[x_1, \ldots, x_n] = n. \tag{3}$$

If M is a finitely generated A-module, then $\operatorname{GKdim} M$ is defined by a formula similar to (2). Now assume that A is *connected graded*, that is, $A = \bigoplus_{n=0}^{\infty} A_n$ with $A_0 = k$ and $A_m A_n \subset A_{m+n}$. Then V^n in formula (2) can be replaced by $\sum_{i=0}^{n} A_i$. Artin and Schelter [2] define A to be *regular* if the following are satisfied:

(AS1) $\operatorname{gld} A =: d < \infty$
(AS2) $\operatorname{GKdim} A < \infty$
(AS3) $\operatorname{Ext}_A^i(k, A) = 0$ for $i \neq d$, and $\operatorname{Ext}_A^d(k, A) \cong k$.

To distinguish between other regularity concepts, we speak of an *AS-regular* k-algebra A. Condition (AS3) is called the *AS-Gorenstein property*.

Artin and Schelter [2] added the condition that A is generated by A_1. For $d = 2$ one obtains the affine quantum planes $U_q = k\langle x, y\rangle/(f_q)$, $q \in k^\times \cup \{\infty\}$, with $f_q = xy - qyx$ for $q \in k^\times$, and $f_\infty = xy - (x + y)x$.

In case $d = 3$ let us briefly outline the approach in [2]. Assume that $x_1, \ldots, x_r$ are linearly independent generators of degree 1 satisfying homogenous relations $f_i(x_1, \ldots, x_r) = 0$, where $i \in \{1, \ldots, m\}$. With $f = (f_1, \ldots, f_m)^\top$ and $x = (x_1, \ldots, x_r)^\top$, the relations can be written in matrix form

$$f = Mx. \tag{4}$$

If homomorphisms are operating from the right, this gives a minimal free resolution

$$0 \to A^l \xrightarrow{\;y\;} A^m \xrightarrow{\;M\;} A^r \xrightarrow{\;x\;} A \twoheadrightarrow k \tag{5}$$

of the trivial left A-module k. By virtue of (AS3), the application of $\operatorname{Hom}_A(-, A)$ to (5) yields a minimal (graded) free resolution of the trivial right A-module:

$$0 \to A \longrightarrow A^r \longrightarrow A^m \longrightarrow A^l \twoheadrightarrow k_A. \tag{6}$$

Hence $l = 1$, and $m = r$. Therefore, the matrix y in (5) is a row vector with entries in A_1, differing from $x^\top$ just by a base change of A_1. So M can be modified such that (5) turns into a graded free resolution

$$0 \to A \xrightarrow{x^\top} A^r \xrightarrow{M} A^r \xrightarrow{x} A \twoheadrightarrow {}_A k \tag{7}$$

with

$$x^\top M = (Qf)^\top \tag{8}$$

for some $Q \in \mathrm{GL}_r(k)$. Since (7) is a graded resolution, the $f_1, \ldots, f_r$ have a common degree s, while M is a homogenous matrix of degree $s-1$. Condition (AS2) implies that there are only two possibilities for (r, s), namely, $(3, 2)$ and $(2, 3)$. The Hilbert function of A is that of a weighted polynomial ring in three variables x, y, z, endowed with degrees 1,1,1 (as usual) in the first case, and 1,1,2 in the second case. For example, the universal enveloping algebra of the Heisenberg Lie algebra (generated by x, y, z with $[x, y] = z$, and $[x, z] = [y, z] = 0$) is AS-regular of the second type. Artin and Schelter [2] obtained 7 generic types of AS-regular algebras with $r = 3$, and 6 types with $r = 2$.

A more geometric approach to AS-regular algebras is given in [3]. For example, let us consider the case $r = 3$. Then A is obtained from the tensor algebra $T = \bigoplus_{n=0}^{\infty} A_1^{\otimes n}$ by factoring out three homogenous relations $f_1, f_2, f_3 \in A_1^{\otimes 2}$. Regarding the f_i as bilinear forms on A_1^*, they have a zero set $\Gamma \subset \mathbb{P}^2 \times \mathbb{P}^2$. By virtue of (8), the projections of Γ to the first and second factor of $\mathbb{P}^2 \times \mathbb{P}^2$ are given by the same equation

$$\det M = 0, \tag{9}$$

regarding the entries of M (which lie in A_1) as linear forms on A_1^*. Unless $\det M$ is identically zero, equation (9) determines a cubic divisor E in $\mathbb{P}^2$. (In the first case, we put $E = \mathbb{P}^2$.) Then the relation $\Gamma \subset \mathbb{P}^2 \times \mathbb{P}^2$ gives rise to an automorphism

$$\sigma : E \to E. \tag{10}$$

Conversely, the subscheme $\Gamma \subset \mathbb{P}^2 \times \mathbb{P}^2$ can be recovered by the embedding $\pi : E \hookrightarrow \mathbb{P}^2$ and the automorphism (10). Furthermore, the pair (π, σ) can be used to define a twisted homogenous coordinate ring of E,

$$B := \bigoplus_{n=0}^{\infty} \mathrm{H}^0(E, L \otimes L^\sigma \otimes \cdots \otimes L^{\sigma^{n-1}}), \tag{11}$$

where L denotes the invertible sheaf $\pi^*\mathcal{O}(1)$, and $L^\sigma := \sigma^* L$. If $E = \mathbb{P}^2$, then B is isomorphic to A. Otherwise, there exists a (left and right) regular element $g \in A_3$ which is *normal*, (i. e. $Ag = gA$) such that ([3], Theorem 2)

$$A/gA \cong B. \tag{12}$$

In case E is a non-singular cubic, the regular normal element $g \in A_3$ is unique up to a constant factor ([4], Proposition 5.13). There is reason to consider g as a non-commutative version of the determinant $\det M$. Accordingly, $g = 0$ can be regarded as a quantum analog of (9), defining a quantization of E with "coordinate ring" (12).

For example, let us denote the homogenous coordinates in $\mathbb{P}^2$ by $(x : y : z)$, and let A be given by the matrix

$$M = \begin{pmatrix} cx & bz & ay \\ az & cy & bx \\ by & ax & cz \end{pmatrix}. \tag{13}$$

Then E is the zero-set of

$$\det M = (a^3 + b^3 + c^3)xyz - abc(x^3 + y^3 + z^3). \tag{14}$$

Assume that $abc \neq 0$ and $(3abc)^3 \neq (a^3 + b^3 + c^3)^3$. Then E is an elliptic curve, and in particular, the algebra A is AS-regular. The matrix Q in (8) is the identity matrix. Here the regular normal element $g \in A_3$ is central:

$$g = (a^3 - b^3)cy^3 + (b^3 - c^3)axyz + (c^3 - a^3)bzyx. \tag{15}$$

By means of the relations $Mx = 0$ it can be written in the form:

$$\begin{aligned} g =&(b - a)\left(a^3xzy + b^3yxz + c^3zyx - abc(x^3 + y^3 + z^3)\right) + \\ &a\left(a^3(xzy - yzx) + b^3(yxz - zxy) + c^3(zyx - xyz)\right) + \\ &ab(a^2 - ab + b^2)(xyz + yzx + zxy - zyx - yxz - xzy). \end{aligned}$$

If the variables x, y, z are made commutative, this element turns into $(b - a)\det M$. There exist AS-regular algebras with $b - a = 0$. In that case, the algebra A merely depends on $d := -\frac{c}{a}$, and the relations $Mx = 0$ become

$$dx^2 = yz + zy; \quad dy^2 = xz + zx; \quad dz^2 = xy + yx.$$

Let R be a commutative noetherian local domain. An R-algebra Λ is said to be a *Cohen-Macaulay order* over R if ${}_R\Lambda$ is a maximal Cohen-Macaulay module.

THEOREM 3. *A Cohen-Macaulay order Λ over a regular local ring R is Auslander-regular and Macaulay if and only if* $\operatorname{gld}\Lambda = \dim R$.

Proof. By [6], Chap. III, Proposition 3.3, and [33], Proposition 3.5, $\operatorname{gld}\Lambda = \dim R$ implies that Λ is smooth. Since Λ is FBN, Teo's result [48], Theorem 3.9, implies that Λ is Auslander-regular and Macaulay. Conversely, let Λ be Auslander-regular and Macaulay, hence smooth. Then the Macaulay property for ${}_\Lambda(\Lambda/\operatorname{Rad}\Lambda)$ implies

$$\dim R = \dim\Lambda = j(\Lambda/\operatorname{Rad}\Lambda) \leqslant \operatorname{pd}(\Lambda/\operatorname{Rad}\Lambda) \leqslant \operatorname{gld}\Lambda =: n.$$

On the other hand, $\operatorname{Ext}^n_\Lambda(\Lambda/\operatorname{Rad}\Lambda, \Lambda) \neq 0$ implies

$$\dim\Lambda \geqslant j(\operatorname{Ext}^n_\Lambda(\Lambda/\operatorname{Rad}\Lambda, \Lambda)) \geqslant n$$

since Λ is Auslander-regular. Thus dim $R = \operatorname{gld}\Lambda$. □

For a commutative noetherian ring R, Auslander-regularity is equivalent to finite global dimension [35], whereas the Macaulay property (with respect to ordinary dimension) says that R is an equidimensional Cohen-Macaulay ring, i. e. $R_\mathfrak{m}$ is Cohen-Macaulay with $\dim R_\mathfrak{m} = \dim R$ for every maximal ideal $\mathfrak{m}$. This is Nagata's definition of a Macaulay ring [32]. Thus for commutative noetherian local rings, Auslander-regularity implies Macaulayness. In general, however, Auslander-regularity appears to be a weak form of regularity, whereas Macaulayness is a strong property. For instance, let R be a discrete valuation domain with $\mathfrak{p} := \operatorname{Rad}R$. The R-order

$$\Lambda = \begin{pmatrix} R & \mathfrak{p} & \mathfrak{p}^2 & \mathfrak{p}^2 \\ R & R & \mathfrak{p} & \mathfrak{p} \\ \mathfrak{p} & \mathfrak{p} & R & \mathfrak{p} \\ R & R & R & R \end{pmatrix} \tag{21}$$

has global dimension 2 and is Auslander-regular. Thus $\operatorname{gld}\Lambda \neq \dim R$, i. e. Λ is not weakly regular in the sense of [39].

On the other hand, Λ is a Cohen-Macaulay order, but Λ does not satisfy the Macaulay property.

3. Invertible ideals

Let R be a ring. The importance of regular normal elements $g \in R$ has already been exemplified. Every such g generates an *invertible* ideal $P := Rg = gR$, that is, for a suitable over-ring $S \supset R$ there exists an (R, R)-subbimodule P^{-1} of S such that $PP^{-1} = P^{-1}P = R$. Abstractly, the relationship between P and P^{-1} can be expressed by the natural isomorphisms (cf. [37])

$$P^{-1} = \mathrm{Hom}_R(P, R); \quad P \otimes_R P^{-1} = P^{-1} \otimes_R P = R. \tag{22}$$

In other words, an ideal P of R is invertible if and only if the bimodule ${}_RP_R$ belongs to the Picard group $\mathrm{Pic}(R)$ of R (see [16], §55). In the special case where P is generated by a regular normal element g we speak of a *principal* invertible ideal. The semigroup of invertible ideals in R will be denoted by R°. Notice that R° is lower directed: $P, Q \in R^\circ$ implies $PQ \in R^\circ$.

There is a natural minimal over-ring $\widetilde{R}$ containing all the inverses of invertible ideals in R:

$$\widetilde{R} := \varinjlim\{P^{-1} \mid P \in R^\circ\}. \tag{23}$$

The multiplication in $\widetilde{R}$ is given as follows. For $a : {}_RP \to {}_RR$ and $b : {}_RQ \to {}_RR$ in $\widetilde{R}$ we define

$$a \cdot b : QP \xrightarrow{a} Q \xrightarrow{b} R. \tag{24}$$

Notice that the definition of $\widetilde{R}$ is left-right symmetric: Every $a : {}_RP \to {}_RR$ corresponds to a homomorphism $a^* : P_R \to R_R$ by means of the formula $(\forall x, y \in P)$:

$$a(x) \cdot y = x \cdot a^*(y). \tag{25}$$

Therefore,

$$\mathrm{Hom}_R({}_RP, {}_RR) = \mathrm{Hom}_R(P_R, R_R). \tag{26}$$

Let R_{reg} denote the semigroup of regular normal elements in R. Obviously, this is a left and right denominator set (see, e. g., [30]).

Recall that R is said to be *semilocal* if $R/\operatorname{Rad} R$ is semi-simple (artinian).

For the remainder of this section we collect some facts on invertible ideals, suggesting a generalization of R-regular sequences in commutative algebra, and perhaps also of interest in their own right. For a slightly different presentation and alternative proofs we refer to [37], §1.

PROPOSITION 1. *Let R be a semilocal ring with $J := \operatorname{Rad} R$ and t isomorphism classes of simple left R-modules. Let m be the least common multiple of $\{1, \ldots, t\}$, and $P \in R^\circ$. Then $PJ = JP$, and P^m is principal. Hence $\widetilde{R} = R[R_{\text{reg}}^{-1}]$. If R is commutative, then every $P \in R^\circ$ is principal.*

Proof. Since $P \in \operatorname{Pic}(R)$, the left R-module P is finitely generated projective, and $\operatorname{End}_R(P)^{\text{op}} = R$. By the latter identification, $J = \operatorname{Hom}_R(P, JP) = \operatorname{Hom}_R(P, R) \otimes_R JP = P^{-1}JP$. Hence P commutes with J. The functor

$$P \otimes_R - : R\text{-}\mathbf{Mod} \to R\text{-}\mathbf{Mod}$$

permutes simple R-modules. Thus $P^m \otimes_R -$ fixes the simples. Consequently,

$$R/J \cong P^m \otimes_R R/J = P^m/P^m J = P^m/JP^m,$$

whence ${}_RR \cong P^m$, say, $P^m = Rr$ for some left regular $r \in R$. Since P is invertible, r is also right regular: $ra = 0 \Rightarrow a = 0$. The isomorphism ${}_RR \xrightarrow{\sim} P^m$ is given by the right multiplication $\rho_r : a \mapsto ar$. for every $a \in R$ we infer $\rho_r \rho_a \rho_r^{-1} \in \operatorname{End}_R(P^m)^{\text{op}} = R$, whence $\rho_r \rho_a = \rho_b \rho_r$ for some $b \in R$. Therefore, $ar = rb$, which yields $Rr \subset rR \subset RrR = P^m R = P^m = Rr$. If R is commutative, then $R/J \cong P^m \otimes_R R/J$ holds for $m = 1$ since the simple direct summands of ${}_R(R/J)$ are pairwise non-isomorphic. $\square$

Remark. Generalizing regular elements in commutative rings, invertible ideals have the following pleasant features: Every invertible ideal Q remains invertible modulo a prime ideal $P \not\supset Q$ ([15], Theorem 4.3). If R is a prime ring, then every non-zero normal element generates an invertible ideal. The Krull Principal Ideal theorem generalizes to invertible ideals ([15], Theorem 4.6). If R is noetherian, localization at an invertible prime ideal is classical ([21], Theorem 5.2; [14], Theorem 1.3).

Let M be an R-module. An invertible ideal $P \in R^\circ$ will be called *invertible with respect to* M, or simply M*-invertible* if $Px = 0$ implies $x = 0$ for all $x \in M$. The subsemigroup of M-invertible ideals in R° will be denoted by $R^\circ(M)$. We have the following equivalent characterizations:

PROPOSITION 2. *For* $M \in R\text{-}\mathbf{Mod}$ *and* $P \in R^\circ$, *the following are equivalent:*

(a) $P \in R^\circ(M)$.
(b) *The natural map* $P \otimes_R M \twoheadrightarrow PM$ *is an isomorphism.*
(c) $\mathrm{Hom}_R(R/P, M) = 0$.

Proof. The equivalence (a) $\Leftrightarrow$ (c) is trivial. Condition (c) says that the natural homomorphism $\mathrm{Hom}_R(R, M) \to \mathrm{Hom}_R(P, M)$ is monic, or equivalently, that $M \to P^{-1} \otimes_R M$ is monic. As P is a progenerator, this is tantamount to the injectivity of $P \otimes_R M \to R \otimes_R M$, hence to (b). □

COROLLARY 1. *Let* M_1, M_2 *be submodules of* $M \in R\text{-}\mathbf{Mod}$, *and* $P \in R^\circ(M)$. *Then* $P(M_1 \cap M_2) = PM_1 \cap PM_2$.

Proof. Consider the exact sequence

$$M_1 \cap M_2 \hookrightarrow M \to M/M_1 \oplus M/M_2.$$

By Proposition 2, and by virtue of the isomorphisms $P \otimes_R M/M_i = PM/PM_i$, tensoring with P yields an exact sequence

$$P(M_1 \cap M_2) \hookrightarrow PM \to (PM/PM_1) \oplus (PM/PM_2).$$

Therefore, the exact sequence

$$PM_1 \cap PM_2 \hookrightarrow PM \to (PM/PM_1) \oplus (PM/PM_2)$$

proves the assertion. □

COROLLARY 2. *Let I be an ideal of a ring R, and $M \in R$-**Mod**. If $P \in R^\circ(M)$, then*

$$P \in R^\circ(M/IM) \iff PM \cap IM = PIM. \tag{27a}$$

Proof. The short exact sequence $IM \hookrightarrow M \twoheadrightarrow M/IM$ induces a commutative diagram with an exact first row:

$$\begin{array}{ccccc} P \otimes_R IM & \rightarrowtail & P \otimes_R M & \twoheadrightarrow & P \otimes_R M/IM \\ \downarrow \alpha & & \downarrow \beta & & \downarrow \gamma \\ PIM & \rightarrowtail & PM & \twoheadrightarrow & P(M/IM). \end{array}$$

By assumption, α and β are isomorphisms. Moreover, $P \in R^\circ(M/IM)$ if and only if γ is an isomorphism. This holds if and only if the second row is exact, which is equivalent to $PM \cap IM = PIM$. □

COROLLARY 3. *Let R be a ring, I an ideal of R, and $P \in R^\circ$. Then $P+I/I$ is invertible in R/I if and only if P is invertible with respect to ${}_R(R/I)$ and $(R/I)_R$.*

Proof. Consider the subring $R' := \bigcup_{n=0}^{\infty} P^{-n}$ of $\widetilde{R}$. Suppose first that $P + I/I \in (R/I)^\circ$. Then $P + I/I$ in invertible with respect to ${}_{R/I}(R/I)$ and $(R/I)_{R/I}$. Hence P is invertible with respect to ${}_R(R/I)$ and $(R/I)_R$. Conversely, assume that the latter condition holds. Then it also holds for all P^n, $n \in \mathbb{N}$. By (27a) this implies $P^n \cap I = P^n I = IP^n$ for all $n \in \mathbb{N}$. Hence

$$R'I = \bigcup_{n=0}^{\infty} P^{-n} I = \bigcup_{n=0}^{\infty} IP^{-n} = IR'.$$

Moreover,

$$\begin{aligned} R'I \cap R = (\bigcup P^{-n} I) \cap R = \bigcup (P^{-n} I \cap R) = \bigcup P^{-n} P^n (P^{-n} I \cap R) \\ = \bigcup P^{-n}(I \cap P^n) = \bigcup P^{-n} P^n I = I. \end{aligned}$$

Therefore, we have an embedding

$$R/I \hookrightarrow R'/R'I \tag{28}$$

that maps R/I to $R + R'I/R'I$. The ideal $P + I/I$ is mapped to $P + R'I/R'I$. Obviously, this ideal of $R + R'I/R'I$ has the inverse $P^{-1} + R'I/R'I$. □

For a noetherian ring R, Corollary 3 can be improved by the following

PROPOSITION 3. *Let R be a ring that satisfies the a.c.c. for two-sided ideals. Let I be an ideal and $P \in R^{\circ}$. Then P is invertible with respect to ${}_R(R/I)$ if and only if P is invertible with respect to $(R/I)_R$.*

Proof. Applying (27a) to $M = R$, we only have to show that $P \cap I = PI$ implies $PI = IP$. The assumption gives $IP \subset P \cap I = PI$, and thus $I \subset PIP^{-1} \subset R$. This yields an ascending chain of ideals

$$I \subset PIP^{-1} \subset P^2IP^{-2} \subset \cdots$$

that has to be stationary, say, $P^nIP^{-n} = P^{n+1}IP^{-n-1}$. Hence $I = PIP^{-1}$, i. e. P and I commute. □

4. Cohen-Macaulay rings

The preceding results show that invertible ideals behave rather similar to regular elements in a commutative ring. This motivates the following definition [37].

DEFINITION 2. A noetherian ring R will be called a *Cohen-Macaulay ring* if there are invertible ideals $P_1, \ldots, P_n \subset \operatorname{Rad} R$ such that the partial sums $I_i := P_1 + \ldots + P_i$ satisfy

(1) $I_{i+1}/I_i \in (R/I_i)^{\circ}$ for $i < n$.

(2) R/I_n is of finite length.

We call $\{P_1, \ldots, P_n\}$ a *CM system* for R.

If R is commutative, the definition says that R is semilocal and Macaulay in the sense of Nagata ([32], §25). Cf. also the last paragraph of §2. By [37], Proposition 10 and Theorem 3, we have:

PROPOSITION 4. *Let R be a Cohen-Macaulay ring with CM system $\{P_1, \ldots, P_n\}$. Then every CM system for R has n elements. Moreover, $P_iP_j = P_jP_i$ holds for all pairs i, j. Any permutation of the $\{P_1, \ldots, P_n\}$, or a replacement of P_i by some power P_i^m $(m > 0)$, gives again a CM system.*

If R is a commutative noetherian local domain, an R-algebra Λ is said to be an *R-order* if ${}_R\Lambda$ is finitely generated and torsion-free. By [37], Theorem 4, we have:

PROPOSITION 5. *Let Λ be an order over a noetherian local domain R. Then Λ is Cohen-Macaulay if and only if ${}_R\Lambda$ is a Cohen-Macaulay module in the classical sense.*

We shall see below that for a Cohen-Macaulay ring R, the cardinality of a CM system is equal to the dimension of R, defined as follows. Let R be a noetherian ring. An ideal I of R will be called a *defining ideal* if $I \subset \operatorname{Rad} R$ and the length $l(R/I)$ of ${}_R(R/I)$ is finite. Clearly, this implies that R is semilocal. Hence I contains some power of $\operatorname{Rad} R$, and thus $(R/I)_R$ is also of finite length. As a counterpart to Gelfand-Kirillov dimension, we define the *dimension* of a finitely generated R-module M as follows.

$$\dim M := \inf\{d \in \mathbb{R} \,|\, l(M/I^nM) \leqslant n^d \text{ for almost all } n \in \mathbb{N}\}. \quad (29)$$

In particular, $\dim 0 = -\infty$ and $\dim M = 0$ if and only if $M \neq 0$ and $l(M) < \infty$. Similarly to (2), $\dim M$ does not depend on the choice of I. An ideal I of R is said to have the *left (right) full Artin-Rees property* [29] if for every $M \in R$-**mod** (resp. $M \in$ **mod**-R) and any submodule N there exists an integer $m \in \mathbb{N}$ such that

$$N \cap I^{i+m}M = I^i(N \cap I^mM) \quad (30)$$

holds for all $i \in \mathbb{N}$. Condition (30) can be expressed in terms of the blowup ring

$$B_I(R) := \bigoplus_{i=0}^{\infty} I^it^i \subset R[t]. \quad (31)$$

Accordingly, to any left R-module M there corresponds a graded $B_I(R)$-module $B_I(M) := \bigoplus_{i=0}^{\infty} t^iI^iM$. This gives a functor

$$B_I : R\text{-}\mathbf{Mod} \longrightarrow B_I(R)\text{-}\mathbf{Mod}. \quad (32)$$

PROPOSITION 6. *Let R be left noetherian. An ideal I satisfies the left full Artin-Rees property if and only if the functor (32) preserves finitely presented modules.*

Proof. Let N be a submodule of $M \in R$-**mod**. For the natural projection $p : M \twoheadrightarrow M/N$, consider the induced epimorphism

$$B_I(p) : B_I(M) \twoheadrightarrow B_I(M/N).$$

The kernel of $B_I(p)$ is given by

$$\operatorname{Ker} B_I(p) = \bigoplus_{i=0}^{\infty} t^i(N \cap I^i M).$$

Condition (30) states that $\operatorname{Ker} B_I(p)$ is finitely generated. Since every finitely presented R-module can be given in the form M/N with $M = R^n$, the left full Artin-Rees property implies that (32) preserves finitely presented R-modules. Conversely, this property of B_I implies that $B_I(M)$ and $B_I(M/N)$ are finitely presented. Hence $\operatorname{Ker} B_I(p)$ is finitely generated. □

We will say that an ideal has the *full Artin-Rees property* if it satisfies the left and right full Artin-Rees property. If this holds for a defining ideal I, the dimension (29) behaves nicely (cf. [39], Proposition 1 and 2):

PROPOSITION 7. *Let R be a noetherian ring with a defining ideal that satisfies the full Artin-Rees property. Then the dimension function is exact on short exact sequences $M' \rightarrowtail M \twoheadrightarrow M''$ in R-**mod**, i. e.*

$$\dim M = \sup\{\dim M', \dim M''\}. \tag{33}$$

PROPOSITION 8. *Let R be a noetherian ring. Let $P \in R^\circ$ be contained in a defining ideal I that satisfies the full Artin-Rees property. If P is invertible with respect to $M \in R$-**mod**, then*

$$\dim M = 1 + \dim M/PM.$$

DEFINITION 3. A noetherian ring R will be called a *weak Cohen-Macaulay ring* if there exists a chain of ideals

$$0 = I_0 \subset I_1 \subset \cdots \subset I_n \tag{34}$$

in R such that I_n is a defining ideal having the full Artin-Rees property, and $I_i/I_{i-1} \in (R/I_{i-1})^\circ$ for all $i \in \{1, \dots n\}$. We call (34) a *defining chain* for R.

In contrast to Definition 2, we have inserted the full Artin-Rees property of I_n. The following result, a two-sided version of [37], Proposition 6, shows that this is automatically satisfied for Cohen-Macaulay rings:

PROPOSITION 9. *Let R be a noetherian ring. Any finite sum of pairwise commuting invertible ideals has the full Artin-Rees property.*

COROLLARY. *Every Cohen-Macaulay ring is weakly Cohen-Macaulay.*

In Definition 3, however, the condition on I_n cannot be dropped. For example, let $\mathfrak{g} = kx \oplus ky$ be the solvable Lie algebra with $[x, y] = x$. The ideal I of the universal enveloping algebra $\mathrm{U}(\mathfrak{g})$ generated by x and y is localizable (see [37], Proposition 12). The localization $R := \mathrm{U}(\mathfrak{g})_I$ has a chain of ideals $0 \subset I_1 \subset I_2$ with $I_1 := Rx$ and $I_2 := Rx + Ry$ as in Definition 3, but I_2 has not the full Artin-Rees property. (Take $M = R$ and $N = I_1$ in (30)).

PROPOSITION 10. *Let R be a weak Cohen-Macaulay ring with a defining chain $0 \subset I_1 \subset \cdots \subset I_n$. Then $n = \dim R$.*

Proof. By Proposition 8, $\dim R = 1 + \dim R/I_1$. Moreover, $0 \subset I_2/I_1 \subset \cdots \subset I_n/I_1$ is a defining chain for the weak Cohen-Macaulay ring R/I_1. Therefore, the assertion follows by induction. □

In particular, Proposition 10 implies that the dimension of a Cohen-Macaulay ring is equal to the cardinality of any CM system.

There are many similarities between non-commutative and commutative Cohen-Macaulay rings. Some are collected in the following proposition.

PROPOSITION 11. *Let R be a Cohen-Macaulay ring with CM system $\{P_1, \dots, P_n\}$, and $I := P_1 + \dots + P_n$. The following holds.*

(a) *The Hilbert function of R is given by*

$$l(R/I^{i+1}) = l(R/I) \cdot \binom{n+i}{n}.$$

Moreover, $\bigcap I^i = 0$.

(b) *Every non-zero left ideal of R has dimension n.*

(c) *The grade $j(R/\operatorname{Rad} R)$ is n, and*

$$\operatorname{Ext}^n_R(P_1^{-1} \cdots P_n^{-1} \otimes_R (R/\operatorname{Rad} R), R) = \operatorname{Hom}_R(R/\operatorname{Rad} R, R/I). \tag{35}$$

Proof. (a) follows by [37], Proposition 8, and Corollary 1 of Proposition 6. (b) follows by [37], Proposition 9, and (c) by [37], Proposition 10. □

Remarks. 1. Proposition 11(a) also implies that $n = \dim R$.

2. The invertible ideal $P_1^{-1} \dots P_n^{-1}$ in (35) merely permutes the simple R-modules. Hence, if $I = \operatorname{Rad} R$, Proposition 11(c) gives a version of the AS-Gorenstein property (see §2).

5. Regular rings

In the commutative case, regular local rings form a special class of Cohen-Macaulay rings. Here our two concepts of Cohen-Macaulayness (Definition 2 and 3) coincide. For non-commutative noetherian rings, we consider two corresponding versions of regularity.

DEFINITION 4. A noetherian ring R will be called (strongly) *regular* if R has a CM system $\{P_1, \dots, P_n\}$ such that $P_1 + \dots + P_n = \operatorname{Rad} R$. We call R *weakly regular* if R is a weak Cohen-Macaulay ring such that $\operatorname{gld} R = \dim R$.

By [38], Proposition 1.4, regular rings are weakly regular. For orders over a regular local ring, the weak version coincides with the usual regularity concept (cf., e. g., [34, 18]). Replacing GKdim in Definition 1 by our dimension (29), we call a noetherian semilocal ring R *Macaulay* if

$$j(M) + \dim M = \dim R < \infty \tag{36}$$

holds for every $M \in R$-**mod**. Then we have [39]

THEOREM 4. *Every weakly regular ring is Auslander-regular and Macaulay.*

Example. Let $\mathfrak{g}$ be an n-dimensional nilpotent Lie algebra over k. (Here the field k can be arbitrary.) Then there exists an increasing chain of ideals

$$0 = \mathfrak{g}_0 \subset \mathfrak{g}_1 \subset \cdots \subset \mathfrak{g}_n = \mathfrak{g}$$

with $\dim \mathfrak{g}_i = i$. Let I_i denote the ideal of the universal enveloping algebra $\mathrm{U}(\mathfrak{g})$ generated by $\mathfrak{g}_i$. Then I_i/I_{i-1} is invertible in $\mathrm{U}(\mathfrak{g})/I_{i-1}$. By [43], Theorem 4.3, the ideal I_n is localizable. Let R be the localization of $\mathrm{U}(\mathfrak{g})$ at I_n. Then R is weakly regular.

Now we turn our attention to strongly regular rings. For dimension one there is no difference to the weak case:

PROPOSITION 12. *Every one-dimensional weakly regular ring is regular.*

Proof. Let R be weakly regular. Then R has an invertible defining ideal, and $\operatorname{gld} R = 1$, i. e. R is (left and right) hereditary. In [40] we define the *arithmetic* $\mathcal{A}(R)$ of R as the semigroup of (R,R)-subbimodules I of $\widetilde{R}$ such that $P \subset I \subset P^{-1}$ holds for some $P \in R^{\circ}$. For $I, J \in \mathcal{A}(R)$ the *left* and *right quotients*

$$(I:J)_l := \{a \in \widetilde{R} \,|\, aJ \subset I\}; \qquad (I:J)_r := \{a \in \widetilde{R} \,|\, Ja \subset I\} \quad (37)$$

are again in $\mathcal{A}(R)$. Since R is hereditary, [40], Proposition 1, implies

$$(R:I)_r I = (I:I)_r; \quad I(R:I)_l = (I:I)_l$$

for all $I \in \mathcal{A}(R)$. Therefore, the inclusion $(R:I)_l I \subset R$ gives

$$I \subset (R:(R:I)_l)_r \subset (IR:I(R:I)_l)_r = (I:(I:I)_l)_r \subset (I:R)_r = I.$$

By symmetry, we get

$$(R:(R:I)_r)_l = (R:(R:I)_l)_r = I.$$

Consequently, $\tau I := (R:(R:I)_r)_r$ defines a lattice automorphism

$$\tau : \mathcal{A}(R) \xrightarrow{\sim} \mathcal{A}(R).$$

In particular, τ permutes the primitive ideals, and thus $\tau(\operatorname{Rad} R) = \operatorname{Rad} R$. With $J := \operatorname{Rad} R$, this implies $J = (R : (R : J)_r)_r$, whence $(J : J)_r = (R : J)_r J \subset R$. By symmetry, we get $(J : J)_r = (J : J)_l = R$. Consequently, J is invertible, i. e. R is regular. □

The semiperfect regular rings of dimension one have been characterized by Michler [31]. They are finite products of rings that are Morita equivalent to some

$$R = \begin{pmatrix} V & M & \cdots & M \\ \vdots & \ddots & \ddots & \vdots \\ \vdots & & \ddots & M \\ V & \cdots & \cdots & V \end{pmatrix}, \tag{38}$$

with V a (non-commutative) discrete valuation domain, and $M = \operatorname{Rad} V$.

For higher-dimensional (indecomposable basic) semiperfect regular rings R, there is still a matrix description $R = (V_{ij})$ with fractional ideals V_{ij}. However, the V_{ii} are no longer regular, the V_{ij} are no longer invertible (they are sums of invertible ideals), and in general, the V_{ij} cannot be chosen as ideals $V_{ij} \subset V$. In order to classify such rings R, we have to introduce higher-dimensional discrete valuations

$$v : S \longrightarrow \widehat{F}. \tag{39}$$

Here the free abelian group $F = \mathbb{Z}^n$ is enlarged to a partially orderd semigroup $\widehat{F}$ consisting of formal intersections of elements in F. More precisely, let L be a subgroup of $\mathbb{Z}^n = \mathbb{Z}e_1 \oplus \cdots \oplus \mathbb{Z}e_n$ with $\mathbb{Z}^n/L$ finite. Then $\mathbb{N}^n$ is the positive cone of a partial ordering on $\mathbb{Z}^n$ that induces a partial ordering on L. A subset I of L is said to be a *filter* if $\beta \geqslant \alpha \in I$ implies $\beta \in I$. For two filters I, J, the *sum* $I + J$ is defined as the filter generated by the $\alpha + \beta$ with $\alpha \in I$ and $\beta \in J$. By $\widehat{L}$ we denote the set of finitely generated filters $I \subset L$. Then $\widehat{L}$ is an additive semigroup, and a lattice with respect to $I \wedge J := I \cup J$ and $I \vee J := I \cap J$. If the elements $\alpha \in L$ are identified with the principal filters $[\alpha] := \{\beta \in L \mid \beta \geqslant \alpha\}$, then L becomes a partially ordered subsemigroup of $\widehat{L}$, and every element

$\alpha \in \widehat{L}$ admits a unique representation

$$\alpha = \alpha_1 \wedge \cdots \wedge \alpha_m \tag{40}$$

with pairwise incomparable $\alpha_i \in L$. The empty intersection will be denoted by ∞. It is easily seen that an element $\alpha \in \widehat{L}$ belongs to L if and only if it is invertible with respect to "+". Furthermore,

$$(\alpha \wedge \beta) + \gamma = (\alpha + \gamma) \wedge (\beta + \gamma) \tag{41}$$

holds for all $\alpha, \beta, \gamma \in \widehat{L}$, and $\widehat{L}$ is a distributive lattice (i. e. (41) remains true if "+" is replaced by "$\vee$"). By these rules, when the operations $+, \wedge, \vee$ are applied to expressions (40), the resulting expressions can be put into the same form. In the special case $F = \mathbb{Z}$ we simply have

$$\widehat{\mathbb{Z}} = \mathbb{Z} \cup \{\infty\}. \tag{42}$$

DEFINITION 5. Let $F := \mathbb{Z}^n$ be a free abelian group. For a ring S, a map $v : S \to \widehat{F}$ will be called an *almost regular valuation* if $v(S) \supset mF$ holds for some positive integer m, and the following conditions are satisfied for all $a, b \in S$:

(V1) $v(a) \in F \Rightarrow a \in S^\times \Rightarrow v(a^{-1}) = -v(a)$

(V2) $v(a) = \infty \Leftrightarrow a = 0$

(V3) $v(a + b) \geqslant v(a) \wedge v(b)$

(V4) $v(ab) \geqslant v(a) + v(b)$

(V5) Every $a \in S$ has a decomposition $a = a_1 + \cdots + a_n$ with $a_i \in S^\times$ and $v(a_i) \geqslant v(a)$.

For $n = 1$, the definition coincides with that of a discrete valuation, and then S is a skew-field. In general, we have a local ring

$$V := \{a \in S \mid v(a) \geqslant 0\} \tag{43}$$

which we call the *valuation ring* of v. The subgroup $L := v(S) \cap F$ of F will be called the *value group* of v. The inclusion $L \subset F$ extends to a natural embedding

$$\widehat{L} \hookrightarrow \widehat{F} \tag{44}$$

that respects the operations "+" and "∧". Then

$$v(S) = \widehat{L}. \tag{45}$$

In [38] we give an intrinsic characterization of noetherian rings of the form (43). Every regular local ring arises in this way. We therefore call such noetherian rings (43) *almost regular*. By [38], Proposition 3.4, and Corollary of Proposition 3.3, we have

PROPOSITION 13. *Every almost regular ring is a local Cohen-Macaulay domain.*

Example. Let k be a field, and let V be an n-dimensional k-rational representation of a torus $G = \mathrm{GL}_1(k)^m$. Let $\mathcal{O} = \mathrm{S}(V)^G$ be the invariant ring for the induced action of G on the symmetric algebra $\mathrm{S}(V)$. Equivalently, $\mathcal{O}$ can be given as the ring of regular functions on an affine toric variety [17]. Hochster [23] showed that $\mathcal{O}$ is a semigroup ring $\mathcal{O} = k[L \cap \mathbb{N}^n] \subset k[\mathbb{N}^n] = \mathrm{S}(V)$, where L is a subgroup of $\mathbb{Z}^n$, and $\mathbb{N}^n \subset \mathbb{Z}^n$ is the positive cone with respect to the natural ordering on $\mathbb{Z}^n$. Moreover, he proved that $\mathcal{O}$ is a Cohen-Macaulay ring. If the index of L in $\mathbb{Z}^n$ is finite, then the localization of $\mathcal{O}$ at the ideal generated by $L \cap \mathbb{N}^n$ is almost regular. Essentially, the almost regular rings are non-commutative versions of the local rings obtained in this way.

To any almost regular valuation $v : S \to \widehat{F}$ with valuation ring V and value group $L := v(S) \cap F$, we can associate a regular ring as follows. Let C be any system of representatives of the residue classes in F/L. For $\alpha, \beta \in C$ we define $V_{\alpha,\beta} := V_{\beta-\alpha}$, where

$$V_\gamma := \{a \in S \mid v(a) \geqslant \gamma\} \tag{46}$$

for $\gamma \in \widehat{F}$. Let $N := \mathbb{N}^n$ denote the positive cone in F. If $\gamma \in L \cap N$, then V_γ is an invertible ideal in V. Otherwise, V_γ is of the form $I^{-1}(I_1 + \cdots + I_s)$ with $I, I_1, \ldots, I_s \in V^\circ$. We define

$$R_v := (V_{\alpha,\beta}) \subset \mathrm{M}_r(S), \tag{47}$$

where $r := |C|$. For different choices of the representative system C of residue classes, the ring (47) merely differs by conjugation. Now we can state ([38], Theorem 4.1):

THEOREM 5. *For every almost regular valuation $v : S \to \widehat{F}$ with noetherian valuation ring V, the ring (47) is indecomposable basic semiperfect and regular. Conversely, every indecomposable basic semiperfect regular ring is of this form.*

In the one-dimensional case, the theorem says that an indecomposable basic semiperfect regular ring R can be described in terms of a discrete valuation $v : S \to \mathbb{Z} \cup \{\infty\}$ of a skew-field S. Let V denote the valuation ring, $M := \operatorname{Rad} V$, and let n be the index of $v(S \setminus 0)$ in $\mathbb{Z}$. Then R is conjugate to a ring $(V_{ij}) \subset \mathrm{M}_n(S)$ of the form (38).

For a regular ring (47) and $\gamma \in F$, consider the (R_v, R_v)-bimodule

$$P_\gamma := (V_{\alpha,\beta+\gamma}) \subset \mathrm{M}_r(S). \tag{48}$$

Then $P_\gamma P_{-\gamma} = R_v$. Therefore, the P_γ with $\gamma \in F \cap N$ are invertible ideals of R_v. If $\{e_1, \ldots, e_n\}$ denotes the basis of F that generates the positive cone, $N = \mathbb{N}e_1 \oplus \cdots \oplus \mathbb{N}e_n$, then the results of [38] imply that $\{P_{e_1}, \ldots, P_{e_n}\}$ is a CM system for R_v.

We call the valuation $v : S \to \widehat{F}$ *reduced* if the projections of the value group $L = v(S) \cap F$ to the $\mathbb{Z}e_i$ are surjective. Otherwise, we may assume that $L \subset F_0 := \mathbb{Z}e_1 \oplus \cdots \oplus \mathbb{Z}e_{n-1} \oplus \mathbb{Z}me_n$ with an integer $m > 1$. Let C_0 be a system of representatives for the residue classes in F_0/L. With respect to the reduced valuation

$$v_0 : S \to \widehat{F_0} \tag{49}$$

we obtain a regular ring

$$R_{v_0} = (V_{\alpha,\beta}) \subset \mathrm{M}_{r_0}(S),$$

where $\alpha, \beta \in C_0$ and $r_0 = |C_0|$. In order to get R_v, consider the following system of representatives for the residue classes in F/L:

$$C := C_0 \amalg (C_0 + e_n) \amalg \cdots \amalg (C_0 + (m-1)e_n). \tag{50}$$

Then

$$R_v = \begin{pmatrix} R_{v_0} & P_n & \cdots & P_n \\ \vdots & \ddots & \ddots & \vdots \\ \vdots & & \ddots & P_n \\ R_{v_0} & \cdots & \cdots & R_{v_0} \end{pmatrix} \subset \mathrm{M}_m(R_{v_0}), \tag{51}$$

where P_n denotes the invertible ideal P_{me_n} of R_{v_0}.

In dimension one, the regular rings R_v with a reduced valuation v are local. For dimension two, this is no longer true. We have

PROPOSITION 14. *Let $v : S \to \widehat{F}$ be a reduced almost regular valuation with noetherian valuation ring V and value group $L := v(S) \cap F$, where $F = \mathbb{Z}e_1 \oplus \mathbb{Z}e_2$. If $L \cap \mathbb{Z}e_i = \mathbb{Z}r_i e_i$ for $i \in \{1, 2\}$, then F/L is cyclic of order $r_1 = r_2 =: r$. For $i \in \{0, \ldots, r\}$ consider the ideals*

$$I_i := \{a \in V \mid v(a) \geqslant ie_1\}. \tag{52}$$

Then I_r is invertible, and $R_v = (V_{ij}) \subset M_r(S)$ with

$$V_{ij} = \begin{cases} I_{j-i} & \text{for } i \leqslant j \\ I_r^{-1} I_{r+j-i} & \text{for } i > j. \end{cases} \tag{53}$$

Proof. Since v is reduced, there is an element $me_1 + e_2 \in L$. Hence $r_1 e_2 = r_1(me_1 + e_2) - r_1 me_1 \in L$, and so $r_1 = r_2$. Moreover, we infer $F = \mathbb{Z}e_1 + L$, whence F/L is generated by the residue class of e_1. Therefore, we can choose $C := \{e_1, 2e_1, \ldots, re_1\}$ as a system of representatives of the residue classes in F/L. Consequently, $R_v = (V_{ij})$ with $V_{ij} = \{a \in V \mid v(a) \geqslant (j - i)e_1\}$. Since $re_1 \in L$, the ideal I_r is invertible, whence (53). □

Example. The simplest non-trivial example of a reduced almost regular valuation already shows that the fractional ideals (53) cannot be replaced by ideals in V. In the power series ring $k[[x, y]]$ consider the subring V generated by the quadratic monomials x^2, xy, y^2. If S denotes the localization $V[x^{-2}, y^{-2}]$ of V, and $F := \mathbb{Z}^2$, then S admits a natural almost regular valuation

$$v : S \to \widehat{F}$$

which associates to a power series $a = \sum a_{ij} x^i y^j$ the intersection of the minimal $\binom{i}{j} \in F$ with $a_{ij} \neq 0$. For example,

$$v\Big(\sum_{n=2}^{\infty} \sum_{i=0}^{2n} x^i y^{2n-i}\Big) = \binom{0}{4} \wedge \binom{1}{3} \wedge \binom{2}{2} \wedge \binom{3}{1} \wedge \binom{4}{0}.$$

Obviously, v is reduced, and the valuation ring of v is V. The value group $v(S) \cap F$ has index 2 in F, and the ideals (52) are $I_1 = Vx^2 + Vxy$ and $I_2 = Vx^2$. Hence

$$R_v = \begin{pmatrix} V & Vx^2 + Vxy \\ V + Vx^{-1}y & V \end{pmatrix} \cong \begin{pmatrix} V & Vx + Vy \\ Vx + Vy & V \end{pmatrix}. \quad (54)$$

Notice, however, that $Vx + Vy$ is not an ideal in V, and there is indeed no way to represent R_v as a matrix ring (47) with ideals $V_{\alpha,\beta} \subset V$. Namely, if $R_v \cong \binom{V\,J}{I\,V}$ would be such a representation with ideals $I, J \subset V$, then $IJ = (Vx + Vy)^2 = \operatorname{Rad} V$. But this would imply that the indecomposable projective R_v-modules P_1, P_2 satisfy $P_1 \cong \operatorname{Rad} P_2$ or $P_2 \cong \operatorname{Rad} P_1$, in contrast to (54).

References

[1] K. Ajitabh, S. P. Smith, J. J. Zhang: Auslander-Gorenstein rings, Commun. Algebra 26 (1998), 2159-2180

[2] M. Artin, W. Schelter: Graded algebras of global dimension 3, Adv. Math. 66 (1987), 171-216

[3] M. Artin, J. Tate, M. Van den Bergh: Some algebras associated to automorphisms of elliptic curves, The Grothendieck Festschrift, vol. 1, Boston 1990, pp. 33-85

[4] M. Artin, J. Tate, M. Van den Bergh: Modules over regular algebras of dimension 3, Invent. math. 106 (1991), 335-388

[5] M. Artin, M. Van den Bergh: Twisted homogenous coordinate rings, J. Algebra 133 (1990), 249-271

[6] M. Auslander: Functors and Morphisms determined by Objects, Proc. Conf. Representation Theory, Philadelphia 1976 (Marcel Dekker 1978), pp. 1-244.

[7] M. Auslander, D. A. Buchsbaum: Homological dimension in local rings, Trans. Amer. Math. Soc. 85 (1957), 390-405

[8] J.-E. Björk: Rings of Differential Operators, Amsterdam – Oxford – New York 1979

[9] J.-E. Björk: The Auslander condition on noetherian rings, Sém. Dubreil-Malliavin 1987-1988, LNM 1404 (Springer 1989), 137-173

[10] N. Bourbaki: Algèbre commutative, Chap. I-IX. Hermann, Masson, 1961-1983

[11] K. A. Brown, C. R. Hajarnavis: Homologically homogenous rings, Trans. Amer. Math. Soc. 281 (1984), 197-208

[12] K. A. Brown, C. R. Hajarnavis, A. B. MacEacharn: Rings of finite global dimension integral over their centers, Comm. Alg. 11 (1983), 67-93

[13] W. Bruns, J. Herzog: Cohen-Macaulay rings, Cambridge Univ. Press, 1993

[14] A. W. Chatters, S. M. Ginn: Localization in Hereditary Rings, J. Algebra 22 (1972), 82-88

[15] A. W. Chatters, A. W. Goldie, C. R. Hajarnavis, T. H. Lenagan: Reduced rank in noetherian rings, J. Alg. 61 (1979), 582-589

[16] C. W. Curtis, I. Reiner: Methods of Representation Theory, II, New York – Chichester – Brisbane 1987

[17] V. I. Danilov: The Geometry of Toric Varieties, Uspekhi Mat. Nauk. 33:2 (1978), 85-134 = Russian Math. Surveys 33:2 (1978), 97-154

[18] Yu. A. Drozd, K. W. Roggenkamp: Cohen-Macaulay rings of global dimension two and Krull dimension two, Comm. Alg. 22 (1994), 3297-3329

[19] D. Eisenbud: Commutative algebra with a view toward algebraic geometry, Springer 1995

[20] E. K. Ekström: The Auslander condition on graded and filtered noetherian rings, Sém. Dubreil-Malliavin 1987-1988, Springer LNM 1404 (1989), 220-245

[21] A. W. Goldie: Localization in Non-Commutative Noetherian Rings, J. Algebra 5 (1967), 89-105

[22] A. Grothendieck: Eléments de géométrie algébriques, Chap. IV, Publ. Math. I.H.E.S. 20, 24, 28, 32 (1964-1967)

[23] M. Hochster: Rings of Invariants of Tori, Cohen-Macaulay Rings generated by Monomials, and Polytopes, Ann. Math. 96 (1972), 318-337

[24] A. V. Jategaonkar: Localization in noetherian rings, LMS Lecture Note Series 98, Cambridge 1986

[25] G. Krause: On Fully Left Bounded Left Noetherian Rings, J. Algebra 23 (1972), 88-99

[26] T. Levasseur: Complexe bidualisant en algèbre non-commutative, Sém. Dubreil-Malliavin, LNM 1146 (Springer 1984), 270-287

[27] T. Levasseur: Some properties of noncommutative regular graded rings, Glasgow Math. J. 34 (1992), 277-300

[28] H. Matsumura: Commutative algebra, 2nd ed., London – Amsterdam 1980

[29] J. C. McConnell: Localization in enveloping rings, J. London Math. Soc. 43 (1968), 421-428. Erratum and addendum: J. London Math. Soc. 3 (1971), 409-410

[30] J. C. McConnell, J. C. Robson: Noncommutative Noetherian Rings, Chichester – New York – Brisbane 1987

[31] G. O. Michler: Structure of Semi-Perfect Hereditary Noetherian Rings, J. Alg. 13 (1969), 327-344

[32] M. Nagata: Local rings, NewYork 1962
[33] M. Ramras: Maximal orders over regular local rings of dimension two, Trans. Amer. Math. Soc. 142 (1969), 457-479
[34] I. Reiten, M. Van den Bergh: Two-dimensional tame and maximal orders of finite representation type, Memoirs Amer. Math. Soc. 408, 1989
[35] J.-E. Roos: Bidualité et structure des foncteurs dérivés de $\varprojlim$ dans la catégorie des modules sur un anneau régulier, C. R. Acad. Sci. Paris 254 (1962), 1556-1558
[36] J. J. Rotman: An introduction to homological algebra, Academic Press, New York, 1979
[37] W. Rump: Non-commutative Cohen-Macaulay rings, J. Algebra, to appear
[38] W. Rump: Non-commutative regular rings, Preprint
[39] W. Rump: A class of Auslander-regular Macaulay rings, Preprint
[40] W. Rump: Invertible ideals and non-commutative arithmetics, Preprint
[41] J.-P. Serre: Algèbre Locale Multiplicités, LNM 11 (3ème ed.), 1975
[42] L. Silver: Non-commutative localizations and applications, J. Algebra 7 (1967), 44-76
[43] P. F. Smith: The Artin-Rees property, Sém. d'Algèbre P. Dubreil et M.-P. Malliavin, LNM 924 (1982), pp. 197-240
[44] J. T. Stafford: Auslander-regular algebras and maximal orders, J. London Math. Soc. 50 (1994), 276-292
[45] J. T. Stafford, J. J. Zhang: Homological Properties of (Graded) Noetherian PI Rings, J. Alg. 168 (1994), 988-1026
[46] D. R. Stephenson: Artin-Schelter regular algebras of global dimension three, J. Algebra 183 (1996), 55-73
[47] D. R. Stephenson: Algebras associated to elliptic curves, Trans. AMS 349 (1997), 2317-2340
[48] K. M. Teo: Homological Properties of Fully Bounded Noetherian Rings, J. London Math. Soc. 55 (1997), 37-54
[49] R. Walker: Local rings and normalizing sets of elements, Proc. Amer. Math. Soc. 24 (1972), 27-45

LANGLANDS' PHILOSOPHY AND KOSZUL DUALITY

WOLFGANG SOERGEL*
Universität Freiburg
Mathematisches Institut
Eckertstraße 1
D-79104 Freiburg
Germany

1. Introduction

In this article I will formulate conjectures relating representation theory to geometry and prove them in some cases. I am trying to extend work of Adams-Barbasch-Vogan (ABV92) and some joint work of myself with Beilinson and Ginzburg (BGS96). All this should be regarded as a contribution to (the local case of) Langlands' philosophy.

The localization of Beilinson-Bernstein also establishes a relation between representation theory and geometry. This is however very different from the relation to be investigated in this article: Whereas localization leads to geometry on the group itself (or its flag manifold), the results in (ABV92) to be extended in this article lead to geometry on the Langlands dual group.

The article was around as a preprint for quite some time and owes its premature publication here to the rules governing the summer school where part of its ideas were discussed.

* Partially supported by the TMR network "Algebraic Lie Theory" ERB FMRX-CT97-0100

K.W. Roggenkamp and M. Ştefănescu (eds.), Algebra - Representation Theory, 379–414.

1.1. THE BASIC CONJECTURE

Let G be (only for a short moment) a real reductive Zariski-connected algebraic group. We want to study admissible representations of the associated Lie group $G(\mathbb{R})$ of real points of G. More precisely, we want to study the category $\mathcal{M}(G(\mathbb{R}))$ of smooth admissible representations of $G(\mathbb{R})$. Remark that by a theorem of Casselman-Wallach this is nothing else but the category of Harish-Chandra modules for $G(\mathbb{R})$, modulo the choice of a maximal compact subgroup $K \subset G(\mathbb{R})$. Let $\mathfrak{g}$ be the complexification of the Lie algebra of $G(\mathbb{R})$, $U - \text{mod} - U = U - \text{mod} - U(\mathfrak{g})$ the enveloping algebra of $\mathfrak{g}$ and $Z \subset U - \text{mod} - U$ the center of $U - \text{mod} - U$. Then $\mathcal{M}(G(\mathbb{R}))$ decomposes according to infinitesimal character into

$$\mathcal{M}(G(\mathbb{R})) = \bigoplus_{\chi} \mathcal{M}(G(\mathbb{R}))_{\chi}$$

where χ runs over the set $\operatorname{Max} Z$ of maximal ideals of Z. There are several classifications of the simple objects in $\mathcal{M}(G(\mathbb{R}))$. The most popular ones are by Langlands (Lan73), Vogan-Zuckerman (Vog81), and Beilinson-Bernstein (BB81). In this article I will exploit a version of the first classification scheme, due to Langlands and Shelstad, and more precisely the form given to it by Adams-Barbasch-Vogan (ABV92).

Let me roughly explain its structure. First of all, we have to forget about the real algebraic group G and just remember the inner class of real forms for $G \times_{\mathbb{R}} \mathbb{C}$ given by G. In other words and other notations, we start out with a complex connected reductive algebraic group G endowed with an inner class of real forms. To simplify the exposition let us furthermore choose a finite central subgroup F of G. To these data one associates in a more or less unique way (and to detail this "more or less" takes most of the time):

1. a finite family S_F of antiholomorphic involutions $\delta : G \to G$, all from the fixed inner class.
2. a finite connected algebraic covering $G_F^\vee$ of the dual group $G^\vee$ (i.e. of the complex connected reductive algebraic group $G^\vee$ whose root system is dual to the root system of G).
3. a collection $\{X(\chi)\}_{\chi \in \operatorname{Max} Z}$ of complex algebraic $G^\vee$−varieties.

I will give details in section 1.3. For every antiholomorphic involution $\delta : G \to G$ let $G(\mathbb{R}, \delta)$ denote the fixed point set of the induced map $\delta : G(\mathbb{C}) \to G(\mathbb{C})$, and for any $\chi \in \mathrm{Max}Z$ let $G(\mathbb{R}, \delta)^\wedge_\chi$ denote the set of (isomorphism classes of) simple objects in $\mathcal{M}(G(\mathbb{R}, \delta))_\chi$. Then Adams-Barbasch-Vogan (ABV92) establish a bijection (depending on additional choices) between

(a) the set $\dot{\bigcup}_{\delta \in S_F} G(\mathbb{R}, \delta)^\wedge_\chi$ and

(b) the set of (isomorphism classes of) pairs (Y, τ), where Y is a $G^\vee$-orbit on $X(\chi)$ and τ an irreducible $G^\vee_F$-equivariant local system on Y.

Strictly speaking this classifies the representations of a whole collection of real forms simultaneously rather then representations of a single group at a time. To explain which geometric parameters (Y, τ) belong to which real form $\delta \in S_F$ wouldn't make good reading for this introduction (compare (Vog93), 4.16). Let me mention however that any real form in the given inner class is equivalent to $G(\mathbb{R}, \delta)$ with $\delta \in S_F$ if we choose F big enough, hence the above classification captures all real forms.

One point of setting up the above classification scheme is the very beautiful and miraculous fact that the KL-data on the representation theoretic side are mirrored by the IC-data on the geometric side. This is the essence of Vogan's character duality (Vog82). I conjecture that this duality lifts to the level of categories. To give more details, I have to introduce more notation.

Let H be a complex algebraic group and X a complex algebraic H-variety. Suppose H decomposes X into finitely many orbits. Then we can form a graded $\mathbb{C}$-algebra $\mathrm{Ext}_H(X)$ as follows: Take the equivariant derived category $\mathcal{D}_H(X)$ as defined in (BL92). This is a triangulated $\mathbb{C}$-category containing the category $\mathcal{P}_H(X)$ of equivariant perverse sheaves as a full abelian subcategory. For $\mathcal{F}, \mathcal{G} \in \mathcal{D}_H(X)$ let us denote by $\mathrm{Hom}_H(\mathcal{F}, \mathcal{G})$ the space of homomorphisms in $\mathcal{D}_H(X)$ and put $\mathrm{Hom}^n_H(\mathcal{F}, \mathcal{G}) = \mathrm{Hom}_H(\mathcal{F}, \mathcal{G}[n])$. Since H was supposed to act with finitely many orbits, there are (up to isomorphism) only finitely many simple objects in $\mathcal{P}_H(X)$. Let $\mathcal{L} \in \mathcal{P}_H(X)$ denote the direct sum of "all" simple objects (i.e. take one from each isomorphism class). Then put

$$\mathrm{Ext}_H(X) = \oplus_{n\geq 0}\mathrm{Hom}^n_H(\mathcal{L},\mathcal{L}) = \mathrm{End}^\bullet_H(\mathcal{L}).$$

There is our graded $\mathbb{C}$-algebra. We will refer to it as the "geometric extension algebra".

For any ring R let $R-Mod$ be the category of all left R-modules. For a graded $\mathbb{C}$-algebra $A = \oplus_{n\geq 0}A^n$ define the full subcategory $A - Nil \subset A - Mod$ by

$$A - Nil = \{M \in A - Mod \mid \dim_{\mathbb{C}} M < \infty, A^n M = 0 \text{ for } n \gg 0\}.$$

So objects in $A - Nil$ are not assumed to be graded, but only "continous along the graduation". Recall now the classification scheme from (ABV92). The main objective of this article is to formulate, motivate and discuss the following

BASIC CONJECTURE 1. *There exists an equivalence of categories*

$$\bigoplus_{\delta\in S_F} \mathcal{M}(G(\mathbb{R},\delta))_\chi \cong \mathrm{Ext}_{G^\vee_F}(X(\chi)) - Nil.$$

Remarks:

1. Take again our algebraic group H acting on our variety X with finitely many orbits. Then the simple objects of $\mathrm{Ext}_H(X) - Nil$ are in an obvious bijection with the simple objects of $\mathcal{P}_H(X)$, and these in turn are parametrized by pairs (Y,τ) with Y an H-orbit on X and τ an irreducible H-equivariant local system on Y, c.f. (BL92). Thus we see that the simple objects on the left and on the right of our conjectured equivalence are parametrized by the same set. Certainly our equivalence should identify simple objects which have he same parameter.
2. In this article I will check the conjecture in the following cases: (1) if G is a torus, (2) for arbitrary G and "almost all" central characters, under the additional assumption $F = 1$, (3) for $G = SL_2$, (4) for complex groups, more precisely when $G = G_c \times G_c$ for a semisimple simply connected complex algebraic group G_c and $F = 1$ and the inner class of real forms is such that S_F consists just of one δ and $G(\mathbb{R},\delta) = G_c(\mathbb{C})$, under the additional assumption that the central character χ is integral regular.

3. I will explain in the last section 4 how I expect this conjecture to be related to Vogan's character duality formulas (Vog82; ABV92).
4. It seems reasonable to make analogous conjectures for nonarchimedean fields as well. In these cases $X(\chi)$ should be replaced by spaces of quasiadmissible morphisms of the Weil-Deligne group to the Langlands dual group. I expect that one may identify in special situations the geometric extension algebras with Lusztig's graded affine Hecke algebras.

1.2. THE EASIEST EXAMPLE

To give the reader a more concrete idea of what is going on, let me treat the easiest example. Let us take $G = \mathbb{C}^\times$ endowed with the inner class of real forms consisting of the split form δ so that $G(\mathbb{R}, \delta) = \mathbb{R}^\times$, and choose $F = 1$.

Then S_F will consist of the single element δ, the covering $G_F^\vee \to G^\vee$ will be the identity $\mathbb{C}^\times \to \mathbb{C}^\times$, and $X(\chi)$ will consist just of two points, for every $\chi \in \mathrm{Max}Z$. We find that

$$\begin{aligned} \mathrm{Ext}_{G_F^\vee}(X(\chi)) &= \mathrm{Ext}_{\mathbb{C}^\times}(\text{two points}) \\ &= \mathrm{Ext}_{\mathbb{C}^\times}(\text{point}) \times \mathrm{Ext}_{\mathbb{C}^\times}(\text{point}) \\ &= H^\bullet(B\mathbb{C}^\times; \mathbb{C}) \times H^\bullet(B\mathbb{C}^\times; \mathbb{C}) \\ &= H^\bullet(\mathbb{P}^\infty\mathbb{C}; \mathbb{C}) \times H^\bullet(\mathbb{P}^\infty\mathbb{C}; \mathbb{C}) \\ &= \mathbb{C}[t] \times \mathbb{C}[t] \end{aligned}$$

where $B\mathbb{C}^\times = \mathbb{P}^\infty\mathbb{C}$ is the classifying space of $\mathbb{C}^\times$ and $deg(t) = 2$. On the other hand, $\mathcal{M}(\mathbb{R}^\times)$ is just the category of continuous finite dimensional representations of $\mathbb{R}^\times$. I leave it to the reader to check for every $\chi \in \mathrm{Max}Z$ the equivalence

$$\mathcal{M}(\mathbb{R}^\times)_\chi \cong (\mathbb{C}[t] \times \mathbb{C}[t]) - Nil.$$

This proves our conjecture in this special case.

1.3. THE GEOMETRY OF ADAMS-BARBASCH-VOGAN

I promised at the beginning that given a complex connected reductive algebraic group G, endowed with an inner class of real forms and a finite central subgroup F, I would construct (1) a multiset S_F

of real forms, (2) a finite connected covering $G_F^\vee$ of the dual group and (3) a collection $\{X(\chi)\}_{\chi \in \mathrm{Max} Z}$ of $G^\vee$-varieties.

To fullfill this promise I have to make some choices. In (ABV92) the authors investigate in detail how these choices affect the final outcome.

1. Let us choose an antiholomorphic involution $\gamma : G \to G$ in the given inner class that stabilizes a Borel subgroup. Then form the (real) Lie group ${}^\Gamma G = G \rtimes \Gamma (= G(\mathbb{C}) \rtimes \Gamma$ if we wouldn't abuse notation), where $\Gamma = Gal(\mathbb{C}/\mathbb{R}) = \{1, \gamma\}$ acts on G via this antiholomorphic involution γ. For any $\delta \in {}^\Gamma G - G$ with $\delta^2 \in Z(G)$ the conjugation $int(\delta) : G \to G$ is an antiholomorphic involution, inner to γ. Let $S_F \subset {}^\Gamma G - G$ be a system of representatives for the G-conjugacy classes of $\delta \in {}^\Gamma G - G$ with $\delta^2 \in F$. As explained we may consider S_F as a multiset of antiholomorphic involutions of G. For example $S_1 = H^1(\Gamma; G)$.
   I warn the reader that for a complex algebraic group I quite often write just G for the complex points of G, to be denoted $G(\mathbb{C})$ in more careful notation. I hope this does not lead to confusion.
2. Let us choose in addition $T \subset B \subset G$ a γ-stable maximal torus and Borel subgroup of G. This gives rise to a based root system $(\mathfrak{X}, R, \Delta; \mathfrak{X}^\vee, R^\vee, \Delta^\vee)$ with an involution $\gamma s : \mathfrak{X} \to \mathfrak{X}$ stabilizing R and Δ. In more detail, $\mathfrak{X} = \mathfrak{X}(T) = \{$homomorphisms of complex algebraic groups $\lambda : T \to \mathbb{C}^\times\}$ and $\gamma s : \mathfrak{X} \to \mathfrak{X}$ is characterized by the formula $(\gamma s(\lambda))(t) = \overline{\lambda(\gamma(t))} \quad \forall t \in T$. (The notation γs is explained in more detail in 3.1). Then choose a triple $(G^\vee, T^\vee, \phi^\vee)$ consisting of a connected reductive complex algebraic group $G^\vee$, a maximal torus $T^\vee \subset G^\vee$ and an isomorphism $\phi^\vee : \mathfrak{X}^\vee \to \mathfrak{X}(T^\vee)$ that identifies the root system of $G^\vee$ with the dual of the root system of G.
   Now the finite central subgroups F of G are in one-to-one correspondence with finite quotients of $\mathfrak{X}/\mathbb{Z}R$, via $F \mapsto \mathfrak{X}(F)$. But the finite connected coverings $\pi : H \to G^\vee$ of $G^\vee$ are also in one-to-one correspondence with finite quotients of $\mathfrak{X}/\mathbb{Z}R$, via $H \mapsto \mathfrak{X}/\mathfrak{X}^\vee(\pi^{-1}(T^\vee))$. In this way any finite central subgroup F of G gives rise to a finite connected covering $G_F^\vee$ of $G^\vee$.
3. This is the most subtle part. We start introducing some notations.

Let H be any complex connected reductive algebraic group, $\mathfrak{h} = LieH$ its Lie algebra and $\lambda \in \mathfrak{h}$ a semisimple element. Put

$$\begin{aligned} \mathfrak{h}(\lambda)_n &= \{\mu \in \mathfrak{h} \mid [\lambda, \mu] = n\mu\} \\ \mathfrak{n}(\lambda) &= \oplus_{n=1,2,\ldots}\mathfrak{h}(\lambda)_n \\ e(\lambda) &= exp(2\pi i\lambda) \in H. \end{aligned}$$

The "canonical flat through λ" is the affine subspace $\mathcal{F}(\lambda) = \lambda + \mathfrak{n}(\lambda) \subset \mathfrak{h}$. One proves (see (ABV92)) that the canonical flats partition the set $\mathfrak{h}_s$ of semisimple elements of $\mathfrak{h}$ and even partition each conjugacy class of semisimples. Furthermore the function e is constant on each canonical flat. Let $\mathcal{F}(\mathfrak{h})$ denote the space of all canonical flats (through semisimple elements of $\mathfrak{h}$) and for a semisimple orbit $\mathcal{O} \subset \mathfrak{h}$ put $\mathcal{F}(\mathcal{O}) = \{\Lambda \in \mathcal{F}(\mathfrak{h}) \mid \Lambda \subset \mathcal{O}\}$. Then e is a function $e : \mathcal{F}(\mathfrak{h}) \to H$.

Now recall $G^\vee$ and put $\mathfrak{g}^\vee = LieG^\vee$. We may and will identify $Lie\mathbb{C}^\times = \mathbb{C}$ in such a way that the exponential map of Lie groups corresponds to the usual exponential map. This determines an embedding

$$\mathfrak{X} = \mathrm{Hom}(T, \mathbb{C}^\times) \to \mathrm{Hom}(LieT, Lie\mathbb{C}^\times) = (LieT)^*,$$

hence an isomorphism $\mathfrak{X} \otimes_{\mathbb{Z}} \mathbb{C} = (LieT)^*$. On the other hand $\phi^\vee$ determines an embedding

$$\mathfrak{X} = \mathrm{Hom}(\mathbb{C}^\times, T^\vee) \to \mathrm{Hom}(Lie\mathbb{C}^\times, LieT^\vee) = LieT^\vee,$$

hence an isomorphism $\mathfrak{X} \otimes_{\mathbb{Z}} C = LieT^\vee$.

Let $\mathcal{W}$ be the Weyl group. We have bijections

$$\mathrm{Max}Z = (LieT)^*/\mathcal{W} = (\mathfrak{X}\otimes_{\mathbb{Z}\mathbb{C}})/\mathcal{W} = (LieT^\vee)/\mathcal{W} = \mathfrak{g}_s^\vee/G^\vee$$

induced respectively by the Harish-Chandra homomorphism, our above isomorphisms and the embedding $LieT^\vee \subset \mathfrak{g}_s^\vee$. Denote by $\chi \mapsto \mathcal{O}(\chi)$ their composition.

Now choose a collection $\{x_{\alpha^\vee}\}_{\alpha\in\Delta}$ of isomorphisms $x_{\alpha^\vee} : \mathbb{C} \to U_{\phi^\vee(\alpha^\vee)}$ of the additive group $\mathbb{C}$ onto the simple root subgroups of $G^\vee$. The involution $\gamma s : \mathfrak{X} \to \mathfrak{X}$ has an adjoint $(\gamma s)^\vee : \mathfrak{X}^\vee \to \mathfrak{X}^\vee$. We will abreviate $(\gamma s)^\vee = \gamma : \mathfrak{X}^\vee \to \mathfrak{X}^\vee$. This gives rise to an involutive automorphism $\gamma : G^\vee \to G^\vee$ of

the complex algebraic group $G^\vee$ characterized by $\gamma(T^\vee) = T^\vee$, $\phi^\vee\gamma = \gamma\phi^\vee : \mathfrak{X}^\vee \to \mathfrak{X}(T^\vee)$ and $\gamma \circ x_{\alpha^\vee} = x_{\gamma(\alpha^\vee)} \forall \alpha \in \Delta$. Thus we may form the semidirect product ${}^\Gamma G^\vee = G^\vee \rtimes \Gamma$, the so-called L-group.
Now for any $\chi \in \mathrm{Max}Z$ we define

$$X(\chi) = \{(y, \Lambda) \mid y \in {}^\Gamma G^\vee - G^\vee, \Lambda \in \mathcal{F}(\mathcal{O}(\chi)) \text{ such that } y^2 = e(\Lambda)\}.$$

This is in a natural way a complex algebraic variety on which $G^\vee$ acts via conjugation. For more details see (ABV92). This fullfills my promise.

It is useful to remark, that by (ABV92), (12.11)(e) the component groups of isotropy groups of $G_F^\vee$ acting on $X(\chi)$ are all commutative. Hence all local systems τ appearing in our geometric parameters are of rank one.

1.4. THANKS

I thank David Vogan, Victor Ginsburg, and Valery Lunts for very helpful discussions. I thank the DFG, MPI, and CNRS for financial support, and the latter two for providing a stimulating research atmosphere.

2. Notations and conventions

2.1. GROUPS AND VARIETIES

Let G, H be complex algebraic groups. A holomorphic (resp. antiholomorphic) morphism from G to H is a morphism of group schemes $\phi : G \to H$ such that the diagram

$$\begin{array}{ccc} G & \stackrel{\phi}{\to} & H \\ \downarrow & & \downarrow \\ Spec\mathbb{C} & \to & Spec\mathbb{C} \end{array}$$

with the lower horizontal the identity (resp. the complex conjugation) commutes.

Let G be a complex algebraic group. Two antiholomorphic involutions δ, δ' of G are said to belong to the same inner class if and only if $\delta\delta'$ is an inner automorphism of G, i.e. $\delta = \delta' int(x)$

with $x \in G$. Two antiholomorphic involutions δ, δ' of G are said to be equivalent if and only if they are conjugate by an inner automorphism of G, i.e. $\delta = int(x^{-1})\delta' int(x)$ with $x \in G$. Any two equivalent antiholomorphic involutions belong to the same inner class.

Let A, B be groups and $\sigma : B \to \mathrm{Aut}A$ a group homomorphism. Let us write $(\sigma(b))(a) = a^b \quad \forall a \in A, b \in B$. We define a new group $A \rtimes_\sigma B = A \rtimes B$ as follows: As a set $A \rtimes B = A \times B$ and the multiplication is given by $(a, b)(a_1, b_1) = (a(a_1^b), bb_1)$.

Let G be a complex algebraic group. Then we denote by $LieG$ its Lie algebra. This is a Lie algebra over $\mathbb{C}$. Let H be a Lie group. Then we denote by $Lie_{\mathbb{R}}H$ the Lie algebra of H. This is a Lie algebra over $\mathbb{R}$. We denote its complexification by $Lie_{\mathbb{C}}H = \mathbb{C} \otimes_{\mathbb{R}} Lie_{\mathbb{R}}H$. Let G be a complex algebraic group. Then $LieG = Lie_{\mathbb{R}}(G(\mathbb{C}))$ canonically. This is an isomorphism of Lie algebras over $\mathbb{R}$, where on the left we quietly restrict scalars from $\mathbb{C}$ to $\mathbb{R}$. Let δ be an antiholomorphic involution on G. Then $LieG = Lie_{\mathbb{C}}(G(\mathbb{R}; \delta))$ canonically.

Let X be a complex algebraic variety. We often abuse notation and write X instead of $X(\mathbb{C})$. Sometimes we also abuse notation in the other direction and write for example $\mathbb{C}^\times, \mathbb{C}$ instead of $\mathbb{G}_{m,\mathbb{C}}, \mathbb{G}_{a,\mathbb{C}}$.

2.2. GROTHENDIECK GROUPS

Let $\mathcal{A}$ be an abelian (resp. triangulated) category. Then we let $[\mathcal{A}]$ be the Grothendieck group of $\mathcal{A}$, generated by the objects with the usual relations for all short exact sequences (resp. distinguished triangles). Any object $A \in \mathcal{A}$ gives an element $[A] \in [\mathcal{A}]$.

Let $\mathcal{A}$ be an additive category. Then we let $\langle \mathcal{A} \rangle$ be the *split* Grothendieck group of $\mathcal{A}$, generated by the objects with the usual relations for all *split* short exact sequences. Any object $A \in \mathcal{A}$ gives an element $\langle A \rangle \in \langle \mathcal{A} \rangle$.

2.3. EQUIVARIANT DERIVED CATEGORIES

My basic reference for what follows is (BL92). Let G be a complex algebraic group and X a complex algebraic G-variety. To X we may associate a triangulated $\mathbb{C}$-category $\mathcal{D}_G(X)$, the bounded constructible G-equivariant derived category, called $D^b_{G,c}(X)$ in (BL92), 4.1. We note Hom_G the morphisms in $\mathcal{D}_G(X)$, put $\mathrm{Hom}^\bullet_G = \oplus_n \mathrm{Hom}^n_G$ and use analogous notation End_G, $\mathrm{End}^\bullet_G$ for endomorphisms. If $G = 1$ we come back to our usual triangulated categories $\mathcal{D}_1(X) = D^b_c(X) = \mathcal{D}(X)$ and since Hom_1 looks so stupid, we write $\mathrm{Hom}_\mathcal{D}$ in this case.

I cannot repeat the details of the definition of $\mathcal{D}_G(X)$. To give the reader an idea of what $\mathcal{D}_G(X)$ is like, let me just remark that it embeds as a full triangulated subcategory in the bounded derived category $D^b(Sh(EG \times_G X))$ of sheaves of $\mathbb{C}$-vectorspaces on $EG \times_G X$, where EG is a contractible space with a free right G-action, e.g. the total space of the universal bundle on the classifying space BG of G.

The usual formalism of the six operations of Grothendieck extends to this equivariant situation: For $\mathcal{F}, \mathcal{G} \in \mathcal{D}_G(X)$ we may form $\mathcal{F} \otimes \mathcal{G}, \mathcal{R}Hom(\mathcal{F}, \mathcal{G}) \in \mathcal{D}_G(X)$ and any equivariant morphism $f : X \to Y$ of G-varieties gives rise to adjoint pairs of triangulated functors (f^*, f_*) and $(f_!, f^!)$ relating $\mathcal{D}_G(X)$ and $\mathcal{D}_G(Y)$. The categories $\mathcal{D}_G(X)$ also come equipped with a (perverse) t-structure, and we let $\mathcal{P}_G(X)$ be its core, i.e. the G-equivariant perverse sheaves for the middle perversity, and $H^i : \mathcal{D}_G(X) \to \mathcal{P}_G(X)$ the perverse cohomology functors. Let $\underline{pt} \in \mathcal{P}_G(pt)$ be the constant object $\mathbb{C}$, and put $\underline{X} = p^*\underline{pt} \in \mathcal{D}_G(X)$ for any X, where $p : X \to pt$ is the constant map.

The irreducible objects of $\mathcal{P}_G(X)$ are parametrized by pairs (Y, τ) with Y a G-stable locally closed subset of X and τ an irreducible G-equivariant local system on Y. Namely to such a pair one associates the intersection cohomology complex with twisted coefficients $\mathcal{IC}(Y, \tau) \in \mathcal{P}_G(X)$.

Let $\phi : H \to G$ be a morphism of algebraic groups. Then we may define the restriction functors $Res : \mathcal{D}_G(X) \to \mathcal{D}_H(X)$. They are triangulated and compatible with all our "internal" operations. If ϕ is either (1) a finite covering or (2) the inclusion of a reductive Levi factor, then Res induces isomorphisms $\mathrm{Hom}_G(\mathcal{F}, \mathcal{G}) \to \mathrm{Hom}_H(Res\mathcal{F}, Res\mathcal{G})$ for all $\mathcal{F}, \mathcal{G} \in \mathcal{D}_G(X)$.

2.4. GEOMETRIC EXTENSION ALGEBRAS

Assume now again that X is a H-variety decomposing into finitely many orbits. Remember we defined the graded $\mathbb{C}$-algebra $\mathrm{Ext}_H(X) = \oplus_{n\geq 0}\mathrm{Hom}^n_H(\mathcal{L},\mathcal{L})$ where $\mathcal{L} = \oplus_{(Y,\tau)}\mathcal{IC}(Y,\tau)$ is the sum over all simple objects of $\mathcal{P}_H(X)$, so that Y runs over all H-orbits in X and τ over all irreducible H-equivariant local systems on Y.

Suppose $H \subset G$ is a closed subgroup. Then the "induction equivalence" gives an isomorphism $\mathrm{Ext}_H(X) = \mathrm{Ext}_G(G\times_H X)$.

Suppose H is connected. Then $\mathrm{Ext}_H(pt) = H^\bullet(BH;\mathbb{C})$ is just the cohomology ring of the classifying space of H. If in particular T is a torus, then $\mathrm{Ext}_T(pt) = S((LieT)^*)$ is just the ring of regular functions on its Lie algebra, graded in such a way that $deg(LieT)^* = 2$. More generally, if H is any connected algebraic group, then $\mathrm{Ext}_H(pt) = H^\bullet(BH;\mathbb{C}) = S((LieT)^*)^{\mathcal{W}}$ consists of the invariants of the Weyl group in the regular functions on the Lie algebra of a maximal torus $T\subset G$.

Let N be a complex algebraic group, $H\subset N$ its identity component and $\mathcal{W} = N/H$ the quotient. Then N acts on H by conjugation, hence it acts on $H^\bullet(BH;\mathbb{C})$ and since H is connected this N-action factorizes to give an action of $\mathcal{W}$ on $H^\bullet(BH;\mathbb{C})$. Let us denote this action by $\sigma : \mathcal{W} \to \mathrm{Aut}_{\mathbb{C}}(H^\bullet(BH;\mathbb{C}))$.

Now whenever we are given a $\mathbb{C}$-algebra A together with an action $\sigma : \mathcal{W}\to \mathrm{Aut}_{\mathbb{C}}A$ of a group $\mathcal{W}$ on A by algebra automorphisms, we may form the "twisted group ring" $A^\sigma[\mathcal{W}] = A\otimes_{\mathbb{CC}}[\mathcal{W}]$ with multiplication $(a\otimes w)(b\otimes v) = a(\sigma(w)(b))\otimes wv$. In the above notations, we prove:

THEOREM 1. *Let N be a complex algebraic group, $H\subset N$ its identity component and $\mathcal{W} = N/H$ the quotient. If $\mathcal{W}$ is commutative, then*

$$\mathrm{Ext}_N(pt) = H^\bullet(BH;\mathbb{C})^\sigma[\mathcal{W}].$$

If not these two rings are still Morita-equivalent.

In fact we will prove even a more general theorem to prepare for the calculations concerning SL_2. Namely let $H \hookrightarrow N \twoheadrightarrow \mathcal{W}$ be as above, let X be an N-variety and $\mathcal{L}\in\mathcal{D}_N(X)$ an object in the equivariant derived category. We have the restriction functor $Res : \mathcal{D}_N(X)\to\mathcal{D}_H(X)$. Its right and left adjoints coincide in our situation, so we denote them just by $Ind : \mathcal{D}_H(X) \to \mathcal{D}_N(X)$. See (BL92), 3.7 for the definitions of Ind_*, $Ind_!$. Let me first give the statement I want, before I explain and prove it.

THEOREM 2. $\mathrm{End}^\bullet_N(Ind\ Res\mathcal{L}) = \mathrm{End}^\bullet_H(Res\mathcal{L})^\sigma[\mathcal{W}]$.

Explanation: I ought to explain the action σ of $\mathcal{W}$ on $\mathrm{End}^\bullet_H(Res\mathcal{L})$. Let $E = EN$ denote the universal bundle over the classifying space of N. Let $\pi : E \times_H X \to E \times_N X$ be the projection. We have the following two commutative squares

$$\begin{array}{ccc} \mathcal{D}_N(X) & \hookrightarrow & D^+(Sh(E \times_N X)) \\ {\scriptstyle Res} \downarrow\uparrow {\scriptstyle Ind} & & {\scriptstyle \pi^*} \downarrow\uparrow {\scriptstyle \pi_*} \\ \mathcal{D}_H(X) & \hookrightarrow & D^+(Sh(E \times_H X)) \end{array}$$

where π^*, π_* stand for the functors usually denoted $L\pi^{-1}, R\pi_*$. The horizontal arrows are fully faithful embeddings. In fact there is a right $\mathcal{W}$-action on on $E \times_H X$ given by

$$[w] : E \times_H X \to E \times_H X, (e, x) \mapsto (en, n^{-1}x)$$

for any $n \in N$ representing $w \in \mathcal{W}$. Our $\pi : E \times_H X \to E \times_N X$ is just the quotient by this right $\mathcal{W}$-action.

Let the image of $\mathcal{L}$ by the upper horizontal in the above diagram be denoted $\tilde{\mathcal{L}}$. Thus $\mathrm{End}^\bullet_H(Res\mathcal{L}) = \mathrm{End}^\bullet(\pi^*\tilde{\mathcal{L}})$. But any $w \in \mathcal{W}$ gives

$$\sigma(w) : \mathrm{End}^\bullet(\pi^*\tilde{\mathcal{L}}) \to \mathrm{End}^\bullet([w]^*\pi^*\tilde{\mathcal{L}}) \to \mathrm{End}^\bullet(\pi^*\tilde{\mathcal{L}})$$

where the first map is induced by $[w]^*$, the second by $\pi \circ [w] = \pi$. There is our $\mathcal{W}$-action on $\mathrm{End}^\bullet(\pi^*\tilde{\mathcal{L}})$.

Proof: We procede in four steps.

1. Define a morphism $\mathrm{End}^\bullet_H(Res\mathcal{L}) \to \mathrm{End}^\bullet_N(Ind\ Res\mathcal{L})$.
2. Define a morphism $\mathbb{C}[\mathcal{W}] \to \mathrm{End}^\bullet_N(Ind\ Res\mathcal{L})$.
3. Prove that they are "σ-compatible", i.e. fit together to define a morphism

$$\mathrm{End}^\bullet_H(Res\mathcal{L})^\sigma[\mathcal{W}] \to \mathrm{End}^\bullet_N(Ind\ Res\mathcal{L}).$$

4. Prove that this morphism is an isomorphism.

So let's go to work!

(1) The first step is clear by functoriality of *Ind*. Translating into a different picture, the morphism is

$$\pi_* : \mathrm{End}^\bullet(\pi^*\tilde{\mathcal{L}}) \to \mathrm{End}^\bullet(\pi_*\pi^*\tilde{\mathcal{L}}).$$

(2) Translating via our diagram we see that we need a morphism $\mathbb{C}[\mathcal{W}] \to \mathrm{End}^\bullet(\pi_*\pi^*\tilde{\mathcal{L}})$. Indeed any $w \in \mathcal{W}$ defines a morphism $\langle w \rangle : \pi_*\pi^*\tilde{\mathcal{L}} \to \pi_*\pi^*\tilde{\mathcal{L}}$ as the composition

$$\pi_*\pi^*\tilde{\mathcal{L}} \to \pi_*[w]_*[w]^*\pi^*\tilde{\mathcal{L}} \to \pi_*\pi^*\tilde{\mathcal{L}}$$

where the first map comes from the adjointness transformation $in \to [w]_*[w]^*$ and the second from $\pi \circ [w] = \pi$. One may check that $\langle w \rangle\langle v \rangle = \langle wv \rangle$. Thus we defined a morphism $\mathbb{C}[\mathcal{W}] \to \mathrm{End}(\pi_*\pi^*\tilde{\mathcal{L}})$.

(3) Let $f \in \mathrm{End}^\bullet(\pi^*\tilde{\mathcal{L}})$ and $w \in \mathcal{W}$. We have to establish in $\mathrm{End}^\bullet(\pi_*\pi^*\tilde{\mathcal{L}})$ the equation $\langle w \rangle(\pi_* f) = (\pi_*(\sigma(w)f))\langle w \rangle$. For this we consider the commutative diagram

$$\begin{array}{ccccccc}
\pi_*\pi^*\tilde{\mathcal{L}} & \to & \pi_*[w]_*[w]^*\pi^*\tilde{\mathcal{L}} & \to & \pi_*[w]^*\pi^*\tilde{\mathcal{L}} & \to & \pi_*\pi^*\tilde{\mathcal{L}} \\
{\scriptstyle \pi_* f} \downarrow & & {\scriptstyle \pi_*[w]_*[w]^* f} \downarrow & & {\scriptstyle \pi_*[w]^* f} \downarrow & & {\scriptstyle \pi_*\sigma(w) f} \downarrow \\
\pi_*\pi^*\tilde{\mathcal{L}} & \to & \pi_*[w]_*[w]^*\pi^*\tilde{\mathcal{L}} & \to & \pi_*[w]^*\pi^*\tilde{\mathcal{L}} & \to & \pi_*\pi^*\tilde{\mathcal{L}}
\end{array}$$

which gives us the required equation.

(4) We have to prove that the product

$$\mathrm{End}^\bullet(\pi^*\tilde{\mathcal{L}}) \otimes_{\mathbb{C}\mathbb{C}} [\mathcal{W}] \to \mathrm{End}^\bullet(\pi_*\pi^*\tilde{\mathcal{L}}), \quad f \otimes w \mapsto \pi_* \circ \mathrm{id} \circ \langle w \rangle$$

is an isomorphism or, equivalently, that the product

$$\begin{array}{c}
\mathrm{Hom}^\bullet(\pi^*\tilde{\mathcal{L}}, \pi^*\tilde{\mathcal{L}}) \otimes_{\mathbb{C}\mathbb{C}} [\mathcal{W}] \to \mathrm{Hom}^\bullet(\pi^*\pi_*\pi^*\tilde{\mathcal{L}}, \pi^*\tilde{\mathcal{L}}) \\
f \otimes w \mapsto \pi_* \circ adj \circ \pi^*\langle w \rangle
\end{array}$$

is an isomorphism where *adj* is the map corresponding to the identity id via the adjointness (π^*, π_*). But $\pi^*\pi_*\pi^*\tilde{\mathcal{L}} \cong \oplus_{w \in \mathcal{W}} \pi^*\tilde{\mathcal{L}}$ and under this isomorphism the $\mathcal{W}$-action on the left by the $\pi^*\langle w \rangle$ corresponds to the "regular action" on the right. So indeed we get an isomorphism. *q.e.d.*

Proof of the first theorem: Apply the second theorem to $X = pt$, $\mathcal{L} = \underline{pt}$. *q.e.d.*

3. Examples

3.1. TORI

To establish the conjecture for G a torus T is not to hard. We begin with some definitions. As is well known there is a contravariant equivalence of abelian categories

$$\left\{\begin{array}{c}\text{diagonalizable}\\ \text{complex algebraic}\\ \text{groups}\end{array}\right\} \begin{array}{c}\mathfrak{X}\\ \rightarrow\\ \leftarrow\\ Diag\end{array} \left\{\begin{array}{c}\text{finitely generated}\\ \text{commutative}\\ \text{groups}\end{array}\right\}$$

For morphisms γ of the left (resp. right) category we will often write just γ instead of $\mathfrak{X}(\gamma)$ (resp. $Diag(\gamma)$). The reason that this does not lead us into a catastrophe is that we only have to work with pairwise commuting involutions.

Any diagonalizable complex algebraic group G admits a unique antiholomorphic involution $s = s_G : G \rightarrow G$ such that $\phi(s(g)) = \overline{\phi(g)}\ \forall g \in G, \phi \in \mathfrak{X}(G)$. This antiholomorphic involution gives rise to what one calls the split real form of G. For every (holomorphic or antiholomorphic) morphism of complex diagonalizable groups $f : G \rightarrow H$ we have $f \circ s_G = s_H \circ f$.

Both the holomorphic and antiholomorphic homomorphisms from one complex diagonalizable group to another form an abelian group. We write these groups additively. For example, if $f : G \rightarrow H$ is holomorphic (or antiholomorphic), we have

$$(-f)(g) = f(g)^{-1}\ \forall g \in G.$$

Now let us get at our torus T. It comes with an inner class of real forms, i.e. with an antiholomorphic involution γ. This gives us an involution $\gamma s = \mathfrak{X}(\gamma s)$ on $\mathfrak{X} = \mathfrak{X}(T)$ and thus an involution $\gamma = (\gamma s)^\vee$ on the dual lattice $\mathfrak{X}^\vee$. Let us describe the varieties $X(\chi)$. They are either empty or isomorphic to $X(0)$, which we henceforth abreviate to X. The isotropy group I is the same for all points of X and is in fact the kernel of $(\mathrm{id} - \gamma) : T^\vee \rightarrow T^\vee$. Applying our exact contravariant functor $\mathfrak{X}$ we get $\mathfrak{X}(I) = \mathfrak{X}^\vee/(\mathrm{id} - \gamma)\mathfrak{X}^\vee$. Thus $(LieI)^* = \mathfrak{X}(I) \otimes_{\mathbb{Z}} \mathbb{C} = (\mathfrak{X}^\vee \otimes_{\mathbb{Z}} \mathbb{C})^\gamma$ and $\mathrm{Ext}_{T_F^\vee} X$ is just a cartesian product of say n_g copies of the symmetric algebra $S = S((\mathfrak{X}^\vee \otimes_{\mathbb{Z}} \mathbb{C})^\gamma)$ over the dual of the Lie algebra of the isotropy group.

On the other hand, $T(\mathbb{R},\gamma)$ admits a Cartan decomposition $T(\mathbb{R},\gamma) = K \times A$ with K compact and A a vector group. In fact, A is just the one-component of $T(\mathbb{R},\gamma) \cap T(\mathbb{R},s)$, and this intersection in turn consists just of all fixed points for some anti-holomorphic involution on the kernel of $(\mathrm{id} - s\gamma) : T \to T$. Thus $Lie_{\mathbb{C}}A = Lie(\ker(\mathrm{id} - s\gamma)) = (\mathfrak{X}^{\vee} \otimes_{\mathbb{Z}} \mathbb{C})^{\gamma}$. In other words, the S from above is also the enveloping algebra of $Lie_{\mathbb{C}}A$. So for each $\chi \in \mathrm{Max}Z$ either

$$\bigoplus_{\delta \in S_F} \mathcal{M}(T(\mathbb{R},\delta))_{\chi} \cong \bigoplus_{n_r \text{ copies}} S - Nil$$

for some n_r, or the left hand side vanishes. So to establish the conjecture we just have to show that $X(\chi) \neq \emptyset \Leftrightarrow \mathcal{M}(T(\mathbb{R},\gamma))_{\chi} \neq 0$ and that if both sides do not vanish, then $n_g = n_r$. But this is calculated in detail in (ABV92) and essentially even in (Lan73) (for $F = 1$).

3.2. GENERIC CENTRAL CHARACTER

For most $\chi \in \mathrm{Max}Z$ the situation is very simple. Namely for χ outside a countable union of proper closed subvarieties of $\mathrm{Max}Z$ only the split real form admits representations with infinitesimal character χ, and the irreducible ones are just the principal series induced from a Borel. I now test the conjecture on this example.

Let $(LieT)^*_{ni} = \{\lambda \in (LieT)^* \mid \langle \lambda, \check{\alpha} \rangle \notin \mathbb{Z} \ \forall \alpha \in R\}$ be the set of all "not at all integral" weights and $(\mathrm{Max}Z)_{ni} \subset \mathrm{Max}Z$ its image under the Harish-Chandra homomorphism.

LEMMA 1. *If G is endowed with the inner class of real forms containing the split form and $\chi \in (\mathrm{Max}Z)_{ni}$, then*

$$X(\chi) = G^{\vee}/T^{\vee} \times \{t \in T^{\vee} \mid t^2 = 1\}$$

as a $G^{\vee}$-variety.

Proof: Since G is endowed with the split inner class, ${}^{\Gamma}G^{\vee} = G^{\vee} \times \Gamma$. For $\chi \in (\mathrm{Max}Z)_{ni}$ each element of $\mathcal{O}(\chi)$ is its own canonical flat, hence $X(\chi) = \{(y,\lambda) \mid y \in G^{\vee}, \lambda \in \mathcal{O}(\chi)$ such that $y^2 = exp(2\pi i\lambda)\}$. Let us determine for one $\lambda_0 \in \mathcal{O}(\chi)$ all possible y. Assume to simplify $\lambda_0 \in LieT^{\vee}$. Since χ was supposed to be not at all integral, the centralizer of $exp(2\pi i\lambda_0)$ in $G^\vee$ is just $T^\vee$ and we find that necessarily $y \in T^\vee$. Thus a bijection $\{t \in T^\vee \mid t^2 = 1\} \to \{y \in G^\vee \mid y^2 = exp(2\pi i\lambda_0)\}$, $t \mapsto t\ exp(\pi i\lambda_0)$. But every $\chi \in (\mathrm{Max}Z)_{ni}$ is in particular regular, so $\mathcal{O}(\chi) = G^\vee/T^\vee$ and we find indeed $X(\chi) = G^\vee/T^\vee \times \{t \in T^\vee \mid t^2 = 1\}$ with $G^\vee$ acting only on the first factor. *q.e.d.*

Let us now assume for simplicity that $F = 1$.

LEMMA 2. *Suppose G is endowed with the inner class of real forms containing the split form. Then only one of the $\delta \in S_1$ gives a split form of G.*

Proof: If we identify $S_1 = H^1(\Gamma; G)$, then $\delta \in S_1$ gives the split form if and only if it maps to the identity element in $H^1(\Gamma; G/Z(G))$. However consider the exact sequence

$$H^1(\Gamma; Z(G)) \to H^1(\Gamma; G) \to H^1(\Gamma; G/Z(G))$$

of nonabelian cohomology. Since the first map factors over the cohomology of a split torus $T \subset G$ and $H^1(\Gamma; T) = 1$, the preimage of the identity element in $H^1(\Gamma; G/Z(G))$ is just the identity element in $H^1(\Gamma; G)$. *q.e.d.*

Let $s \in S_1$ represent the split form. For $\chi \in (\mathrm{Max}Z)_{ni}$ we know that $\mathcal{M}(G(\mathbb{R}, \delta))_\chi = 0$ if $\delta \in S_1, \delta \neq s$. So we have for $\chi \in (\mathrm{Max}Z)_{ni}$ just to establish an equivalene of categories

$$\mathcal{M}(G(\mathbb{R}, s))_\chi \cong \mathrm{Ext}_{G^\vee} X(\chi) - Nil.$$

But both sides are just Nil-modules over a Cartesian product of 2^{rankG} copies of the symmetric algebra $S = S(LieT)$ of the Lie algebra of the maximal torus. For the right hand side this follows from $\mathrm{Ext}_{G^\vee}(G^\vee/T^\vee) = \mathrm{Ext}_{T^\vee}(pt) = H^\bullet(BT^\vee; \mathbb{C}) = S(LieT)$ where the grading is determined by $deg(LieT) = 2$. For the left hand side it follows from the fact that for any $\lambda \in (LieT)^*_{ni}$ the functor

$$\mathcal{M}(T(\mathbb{R}, s))_\lambda \to \mathcal{M}(G(\mathbb{R}, s))_{\xi(\lambda)}$$

of extending trivially to a Borel and inducing (with the shift that ensures that unitary representations go to unitary ones) is an equivalence of categories.

3.3. THE CASE $G = SL_2$

There is only one inner class of real forms. Certainly $G^\vee = PGL(2;\mathbb{C})$. Let us compute the $X(\chi)$. We distinguish three cases.

χ **not integral.** Then $X(\chi) = G^\vee/T^\vee \times \{\text{two points}\}$ as we showed already more generally.

χ **singular.** Then $\mathcal{O}(\chi) = \{0\}$ and

$$X(\chi) = \{y \in G^\vee \mid y^2 = 1\}.$$

But any element of finite order is semisimple, hence contained in a maximal torus and in a torus there are precisely two roots of one. The stabilizers in $G^\vee$ of these two are $G^\vee$ and $N_{G^\vee}(T^\vee)$, which we will henceforth abreviate N. Thus $X(\chi) = pt \,\dot{\cup}\, (G^\vee/N)$.

χ **regular integral.** Then $exp(2\pi i\lambda) = 1$ for all $\lambda \in \mathcal{O}(\chi)$, hence $X(\chi)$ is just the Cartesian product of $\mathcal{F}(\mathcal{O}(\chi))$ with $\{y \in G^\vee \mid y^2 = 1\}$. But if $\chi \in \mathrm{Max}Z$ is regular integral, we have quite generally $\mathcal{F}(\mathcal{O}(\chi)) = G^\vee/B^\vee$. Hence in our situation $\{\mathcal{F}(\lambda) \mid \lambda \in \mathcal{O}(\chi)\} = G^\vee/B^\vee$ and

$$X(\chi) = G^\vee/B^\vee \quad \dot{\cup} \quad G^\vee/B^\vee \times G^\vee/N$$

where $G^\vee$ acts diagonally on the second part.

3.3.1. *The subcase $F == 1$*

Let us first check the conjecture for $F = 1$. Then $S_1 = H^1(\Gamma; SL(2,\mathbb{C}))$ consists just of one element δ with $G(\mathbb{R},\delta) = SL(2,\mathbb{R})$. For any $\chi \in \mathrm{Max}Z$ let us abreviate $\mathcal{M}(SL(2,\mathbb{R}))_\chi = \mathcal{M}_\chi$. These categories break down according to K-types into $\mathcal{M}_\chi = \mathcal{M}_\chi^{ev} \oplus \mathcal{M}_\chi^{odd}$. Now we procede case by case.

χ **not integral.** This has already been treated in general.

χ **singular.** It is known that $\mathcal{M}_\chi^{ev}$ is isomorphic to the category of all finite dimensional continuous modules over the completion $Z_\chi^\wedge$ of Z at χ. Via the Harish-Chandra homomorphism we find

$\mathcal{M}_\chi^{ev} \cong S^{\mathcal{W}} - Nil$, with $S = S(LieT)$ the symmetric algebra over $LieT$. But on the other hand

$$\mathrm{Ext}_{G^\vee}(pt) = H^\bullet(BG^\vee; \mathbb{C}) = H^\bullet(BT^\vee; \mathbb{C})^{\mathcal{W}} = S^{\mathcal{W}}$$

so that indeed

$$\mathcal{M}_\chi^{ev} \cong \mathrm{Ext}_{G^\vee}(pt) - Nil.$$

The structure of $\mathcal{M}_\chi^{odd}$ is more complicated. The irreducible objects of this category are the two limits of principal series, and these extend among each other. More precisely the picture is as follows: Choose in $\mathfrak{g}$ a "standard" basis X, H, Y with $[H, X] = 2X$, $[H, Y] = -2Y$, $[X, Y] = H$ and such that $\mathbb{C}H \subset \mathfrak{g}$ is the complexification of the Lie algebra of the compact torus $SO(2) \subset SL(2, \mathbb{R})$. For $M \in \mathcal{M}$, $n \in \mathbb{Z}$ put $M_n = \{v \in M \mid Hv = nv\}$. Then

$$M \mapsto \left\{ M_{-1} \underset{X}{\overset{Y}{\rightleftarrows}} M_{+1} \right\}$$

is an equivalence of $\mathcal{M}_\chi^{odd}$ to finite dimensional representations of the quiver

$$\bullet \underset{\phi}{\overset{\psi}{\rightleftarrows}} \bullet$$

with relations $\phi\psi$ nilpotent. This is not difficult, and follows for example from the more general results in (Kho80). I leave it to the reader to identify the quiver algebra with the twisted group ring $S^\sigma[\mathcal{W}]$, where $\sigma : \mathcal{W} \to \mathrm{Aut}_{\mathbb{C}} S$ is the obvious action. Putting things together, we get $\mathcal{M}_\chi^{odd} \cong S^\sigma[\mathcal{W}] - Nil$.

But on the other hand

$$\begin{aligned} \mathrm{Ext}_{G^\vee}(G^\vee/N) &= H^\bullet(BN; \mathbb{C}) \\ &= H^\bullet(BT^\vee; \mathbb{C})^\sigma[\mathcal{W}] \quad \text{by theorem 1 of section 2.4} \\ &= S^\sigma[\mathcal{W}]. \end{aligned}$$

So indeed $\mathcal{M}_\chi^{odd} \cong \mathrm{Ext}_{G^\vee}(G^\vee/N) - Nil$ and with our previous result we obtain finally $\mathcal{M}_\chi \cong \mathrm{Ext}_{G^\vee} X(\chi) - Nil$ for singular χ.

χ **regular integral.** By the translation principle we may and will henceforth assume $\chi = \mathrm{Ann}_Z\mathbb{C}$ is just the annihilator of the trivial representation. Then $\mathcal{M}_\chi^{odd}$ can be identified with the category of all finite dimensional continuous modules over the completion $Z_\chi^\wedge$ of Z at χ, and via the Harish-Chandra homomorphism we find $\mathcal{M}_\chi^{odd} \cong S - Nil$. But $\mathrm{Ext}_{G^\vee}(G^\vee/B^\vee) = H^\bullet(BB^\vee; pt) = S$, so $\mathcal{M}_\chi^{odd} \cong \mathrm{Ext}_{G^\vee}(G^\vee/B^\vee) - Nil$.

The category $\mathcal{M}_\chi^{ev}$ is much more interesting. Using the preceding notations, we may describe it as follows: The functor

$$M \mapsto \left\{ \quad M_{-2} \;\underset{X}{\overset{Y}{\rightleftarrows}}\; M_0 \;\underset{X}{\overset{Y}{\rightleftarrows}}\; M_{+2} \quad \right\}$$

establishes an equivalence of $\mathcal{M}_\chi^{ev}$ to finite dimensional representations of the quiver

$$\underset{-2}{\bullet} \;\underset{\phi^-}{\overset{\psi^-}{\rightleftarrows}}\; \underset{0}{\bullet} \;\underset{\phi^+}{\overset{\psi^+}{\rightleftarrows}}\; \underset{+2}{\bullet} \qquad (*)$$

with relations $\phi^-\psi^- = \psi^+\phi^+$ and $\psi^+\phi^+$ nilpotent. This is not difficult, and follows for example from the more general results in (Kho80). Now let us attack the geometric extension algebra. Since

$$G^\vee/B^\vee \times G^\vee/N = G^\vee \times_N (G^\vee/B^\vee)$$

we have (using the induction identity) just to understand $\mathrm{Ext}_N(G^\vee/B^\vee)$. There are two N-orbits on $(G^\vee/B^\vee) = \mathbb{P}^1\mathbb{C} = X$, namely $Y = \{0, \infty\}$ and $U = \mathbb{C}^\times$. Here Y supports just one irreducible N-equivariant local system $\underline{Y}$, but U supports two of them, with trivial and nontrivial N-action, say $\underline{U}^+$ and $\underline{U}^-$. The irreducibles in $\mathcal{P}_N(X)$ are correspondingly $i_*\underline{Y}$, $\underline{X}^+[1]$ and $\underline{X}^-[1]$ where $i : Y \to X$ is the inclusion and $\underline{X}^+$ (resp. $\underline{X}^-$) denotes just the constant sheaf $\mathbb{C}$ on X with trivial (resp. nontrivial) N-action. By definition, $\mathrm{Ext}_N X = \mathrm{End}_N^\bullet(i_*\underline{Y} \oplus \underline{X}^+[1] \oplus \underline{X}^-[1])$. But since we only need to know this ring up to Morita equivalence, let us rather compute the equivariant selfextensions of $i_*\underline{Y} \oplus i_*\underline{Y} \oplus \underline{X}^+[1] \oplus \underline{X}^-[1]$. This object has the advantage to be isomorphic to $IndRes(i_*\underline{Y} \oplus \underline{X}^+[1])$ where we consider N as the semidirect product $N = T^\vee \rtimes \mathcal{W}$ and use the notations of 2.4, so we may apply theorem 2 of section 2.4 to $\mathcal{L} = i_*\underline{Y} \oplus \underline{X}^+[1]$. The theorem tells us that we should first compute $\operatorname{End}^\bullet_{T^\vee}(Res\mathcal{L})$ with its $\mathcal{W}$-action. This is not to hard, so I just give the result.

PROPOSITION 1. $\operatorname{End}^\bullet_{T^\vee}(Res\mathcal{L})$ *is the quiver algebra of the quiver*

$$\underset{l}{\bullet}\quad \underset{(ml)}{\overset{(lm)}{\leftrightarrows}}\quad \underset{m}{\bullet}\quad \underset{(rm)}{\overset{(mr)}{\leftrightarrows}}\quad \underset{r}{\bullet} \qquad (**)$$

with relations $(rm)(ml) = 0 = (lm)(mr)$ *where every arrow has degree one. The reflection of the quiver along the vertical symmetry axis induces the looked-for action* σ *of* $\mathcal{W}$ *on the quiver algebra.*

Proof: Left to the reader. Perhaps it would have been more suggestive to call the extremal points 0 and ∞ instead of r and l.

The finite dimensional representations of $\operatorname{Ext}_N X$ can be identified by Morita equivalence with finite dimensional representations of

$$\operatorname{End}^\bullet_N(Ind\ Res\mathcal{L}) = \operatorname{End}^\bullet_{T^\vee}(Res\mathcal{L})^\sigma[\mathcal{W}],$$

and these in turn are just finite dimensional representations of the quiver with relations from the proposition equipped with an "action of $\mathcal{W}$ on the representation compatible with the above action of $\mathcal{W}$ on the quiver". Such an object consists of a diagram of finite dimensional spaces

$$W_l\quad \underset{(ml)}{\overset{(lm)}{\leftrightarrows}}\quad W_m\quad \underset{(rm)}{\overset{(mr)}{\leftrightarrows}}\quad W_r$$

together with isomorphisms $s : W_r \to W_l$, $s : W_m \to W_m$, $s : W_l \to W_r$ such that the relations

$$s^2 = \mathrm{id}, (lm)(mr) = 0 = (rm)(ml), s(mr) = (ml)s, s(lm) = (rm)s$$

are satisfied. We got to identify the category of such objects with the category of finite dimensional representations of the related quiver $(*)$.

This can be done as follows: To a representation

$$M_{-2} \underset{\phi^-}{\overset{\psi^-}{\rightleftarrows}} M_0 \underset{\phi^+}{\overset{\psi^+}{\rightleftarrows}} M_{+2}$$

of the related quiver $(*)$ we associate the representation

$$M_0 \underset{\binom{\phi^+}{\phi^-}}{\overset{(\psi^+,\psi^-)}{\rightleftarrows}} M_{+2} \oplus M_{-2} \underset{(\psi^+,-\psi^-)}{\overset{\binom{\phi^+}{-\phi^-}}{\rightleftarrows}} M_0$$

of the related quiver $(**)$ with $\mathcal{W}$-action given by

$$s = \mathrm{id} : M_0 \to M_0, \qquad s = \begin{pmatrix} \mathrm{id} & 0 \\ 0 & -\mathrm{id} \end{pmatrix} : M_{+2} \oplus M_{-2} \to M_{+2} \oplus M_{-2}.$$

This can be seen to be an equivalence of categories and induces an equivalence of categories $\mathcal{M}_\chi^{ev} \cong \mathrm{Ext}_N X - Nil$. So putting things together we indeed established the desired equivalence of categories $\mathcal{M}_\chi \cong \mathrm{Ext}_{G^\vee} X(\chi) - Nil$ for regular integral χ.

3.3.2. *The subcase $F = Z(G)$*
Now take for F the full center of $SL(2,\mathbb{C})$, i.e. $F = \{+1, -1\}$. Then S_F can be seen to consist of three elements, giving rise to the three real forms $SU(2,0)$, $SU(1,1)$ and $SU(0,2)$. Certainly two of them are just the compact form $SU(2)$ and the middle one is the split form $SL(2,\mathbb{R})$. So on the representation theoretic side of our conjecture only for χ regular integral there are some changes, and we have to consider instead of $\mathcal{M}(SL(2,\mathbb{R}))_\chi = \mathcal{M}_\chi$ the category

$$\bigoplus_{i=-1,0,1} \mathcal{M}(SU(1+i, 1-i))_\chi \cong (\mathbb{C} - Nil) \oplus \mathcal{M}_\chi \oplus (\mathbb{C} - Nil).$$

On the geometric side we get $G_F^\vee = SL(2,\mathbb{C})$. Certainly homomorphisms between objects in our equivariant derived categories do not change when we pass from $G^\vee$ to $G_F^\vee$. However there might be more simple objects in $\mathcal{P}_{G_F^\vee} X(\chi)$ than in $\mathcal{P}_{G^\vee} X(\chi)$, and this happens precisely when the isotropy groups of points of $X(\chi)$ with respect to $\mathcal{P}_{G_F^\vee}$ have more components than the isotropy groups with respect to $\mathcal{P}_{G^\vee}$. In our situation, this happens only for χ

regular integral and for points in the open orbit on $G^\vee/B^\vee \times G^\vee/N \subset X(\chi)$. To verify our conjecture for $F = \{+1, -1\}$ we just have to check that

$$\mathrm{Ext}_{G_F^\vee}(G^\vee/B^\vee \times G^\vee/N) \cong (\mathrm{Ext}_{G^\vee}(G^\vee/B^\vee \times G^\vee/N)) \times \mathbb{C} \times \mathbb{C}.$$

Via the "induction identity" this can be reduced to showing

$$\mathrm{Ext}_{N_F} X \cong (\mathrm{Ext}_N X) \times \mathbb{C} \times \mathbb{C}$$

where $N_F = N_{G_F^\vee}(T^\vee)$. Let $\mathcal{L} \in \mathcal{P}_N X$ be the sum of all simple objects, $\mathcal{L} = i_* \underline{Y} \oplus \underline{X}^+[1] \oplus \underline{X}^-[1]$. In $\mathcal{P}_{N_F} X$ there are an additional two simple objects. In more detail, the isotropy group in N_F of a point in the open orbit $U \subset X$ is just $\mathbb{Z}/4\mathbb{Z}$, thus there are four simple objects $\mathcal{F}_0, \mathcal{F}_1, \mathcal{F}_2, \mathcal{F}_3$ in $\mathcal{P}_{N_F} X$, corresponding to the four irreducible characters on $\mathbb{Z}/4\mathbb{Z}$. Let $j : U \to X$ be the inclusion. Then $j_{!*}\mathcal{F}_0 = \underline{X}^+[1]$, $j_{!*}\mathcal{F}_2 = \underline{X}^-[1]$ and our two new irreducible objects in $\mathcal{P}_{N_F} X$ are $j_!\mathcal{F}_1 = j_*\mathcal{F}_1$ and $j_!\mathcal{F}_3 = j_*\mathcal{F}_3$. I leave it to the reader to check that

$$\mathrm{End}^\bullet_{N_F}(\mathcal{L} \oplus j_*\mathcal{F}_1 \oplus j_*\mathcal{F}_3) \cong \mathrm{End}^\bullet_N(\mathcal{L}) \times \mathbb{C} \times \mathbb{C}.$$

This gives the conjecture for $G = SL(2, \mathbb{R})$, $F = \{+1, -1\}$.

3.4. COMPLEX GROUPS

Here comes one of the main motivations for the conjecture. Namely we check that it correctly predicts the structure of part of the representation theory of $G_c(\mathbb{C})$ for a semisimple connected complex algebraic group, where we consider $G_c(\mathbb{C})$ as a real Lie group. In fact we will only give details for G_c simply connected and assume this from now on. To get in the set-up of the conjecture, choose a split antiholomorphic involution $s : G_c \to G_c$ and put $G = G_c \times G_c$ with the quasisplit antiholomorphic involution

$$\gamma : G \to G, \gamma(x, y) = (sy, sx).$$

Then $G(\mathbb{R}, \gamma) = G_c(\mathbb{C})$ and $1 = H^1(\Gamma; G) = S_1 = \{\gamma\}$. We prove under the above assumptions:

THEOREM 3. *Suppose χ is a regular integral central character. Then there is an equivalence of categories*

$$\mathcal{M}(G(\mathbb{R}, \gamma))_\chi \cong \mathrm{Ext}_{G^\vee} X(\chi) - Nil.$$

Proof[Theorem]: We begin by simplifying the right hand side. Abreviate $G_c^\vee = \mathcal{G}$ and let $\mathcal{B} \subset \mathcal{G}$ be a Borel subgroup.

LEMMA 3. *Suppose χ is regular integral. Then $X(\chi) \cong G^\vee \times_{\mathcal{B}} (\mathcal{G}/\mathcal{B})$ for a suitable embedding $\mathcal{B} \hookrightarrow G^\vee$.*

Proof: Left to the reader.
Next we simplify the right hand side. Let $\mathcal{HC}h$ denote the category of "Harish-Chandra bimodules" for $\mathfrak{g}_c = LieG_c$ investigated in (Soe92). Let $Z_c \subset U-\text{mod}-U(\mathfrak{g}_c)$ be the center and $Z_c^+ = \text{Ann}_{Z_c}\mathbb{C}$ the central annihilator of the trivial representation of $\mathfrak{g}_c$. Define $\mathcal{H} = \{M \in \mathcal{HC}h \mid (Z_c^+)^n M = 0 = M(Z_c^+)^n \text{ for } n \gg 0\}$ the category of Harish-Chandra bimodules with trivial central character from both sides.

LEMMA 4. *For χ regular integral there is an equivalence of categories*

$$\mathcal{M}(G(\mathbb{R},\gamma))_\chi = \mathcal{M}(G_c(\mathbb{C}))_\chi \cong \mathcal{H}.$$

Proof: This is contained in the literature (BG80).
Modulo these two lemmas we see that we need to establish an equivalence of categories $\mathcal{H} \cong \text{Ext}_{\mathcal{B}}(\mathcal{G}/\mathcal{B}) - Nil$. This was a conjecture in (BG86). It was essentially solved in (Soe92), however I did not write down there the topological part. Merely I constructed in a combinatorial way a graded $\mathbb{C}$-algebra A and proved the equivalence of categories $\mathcal{H} \cong A - Nil$. So we just have to establish an isomorphism $A \cong \text{Ext}_{\mathcal{B}}(\mathcal{G}/\mathcal{B})$ of graded $\mathbb{C}$-algebras.

Let $\mathcal{T} \subset \mathcal{B}$ be a maximal torus and $\mathcal{W} = N_{\mathcal{G}}\mathcal{T}/\mathcal{T}$ the Weyl group. Then the irreducible objects of $\mathcal{P}_{\mathcal{B}}(\mathcal{G}/\mathcal{B})$ are just the intersection cohomology complexes $\mathcal{IC}_x$ of the closures of the Schubert varieties $\overline{\mathcal{B}x^{-1}\mathcal{B}/\mathcal{B}}$, for $x \in \mathcal{W}$. Now by definition $\text{Ext}_{\mathcal{B}}(\mathcal{G}/\mathcal{B}) = \text{End}^\bullet_{\mathcal{B}}(\oplus_{x\in\mathcal{W}}\mathcal{IC}_x)$. We got to understand this algebra better.

The $\mathcal{B}$-equivariant cohomology ring of $\mathcal{G}/\mathcal{B}$ is well known to be $\text{End}^\bullet_{\mathcal{B}}(\underline{\mathcal{G}/\mathcal{B}}) = S \otimes_{S^{\mathcal{W}}} S$ where $S = S((Lie\mathcal{T})^*)$ are the regular functions on $Lie\mathcal{T}$, graded in such a way that $(Lie\mathcal{T})^*$ is the degree two part of S. Now equivariant hypercohomology is a functor

$$\mathbb{H}_{\mathcal{B}} = \text{Hom}^\bullet_{\mathcal{B}}(\underline{\mathcal{G}/\mathcal{B}},\ \ ) : \mathcal{D}_{\mathcal{B}}(\mathcal{G}/\mathcal{B}) \to S \otimes_{S^{\mathcal{W}}} S - mod$$

from the equivariant derived category to the category of graded modules over the equivariant cohomology ring.

PROPOSITION 2. *For all $x, y \in \mathcal{W}$, equivariant hypercohomology induces an isomorphism*

$$\mathbb{H}_{\mathcal{B}} : \mathrm{Hom}^{\bullet}_{\mathcal{B}}(\mathcal{IC}_x, \mathcal{IC}_y) \to \mathrm{Hom}_{S\otimes S}(\mathbb{H}_{\mathcal{B}}\mathcal{IC}_x, \mathbb{H}_{\mathcal{B}}\mathcal{IC}_y).$$

Proof: Postponed to the end of this subsection.
In particular we deduce an isomorphism

$$\mathrm{Ext}_{\mathcal{B}}(\mathcal{G}/\mathcal{B}) \cong \mathrm{End}_{S\otimes S}(\oplus_{x\in\mathcal{W}}\mathbb{H}_{\mathcal{B}}\mathcal{IC}_x).$$

But the algebra A was defined as

$$A = \mathrm{End}_{S\otimes S}(\oplus_{x\in\mathcal{W}}\mathbb{B}_x)$$

where the $\mathbb{B}_x$ are certain graded $S \otimes S$-modules described in a combinatorial way in (Soe92). So to prove our theorem we need just

LEMMA 5. *For all $x \in \mathcal{W}$ there is an isomorphism of graded $S \otimes S$-modules $\mathbb{H}_{\mathcal{B}}\mathcal{IC}_x \cong \mathbb{B}_x$.*

Proof: The proof of this lemma is very similar to the proof of its non-equivariant analogon in (Soe90), so I will rather give a sketch of proof.

Let $(\mathcal{W}, \mathcal{S})$ be the Coxeter system associated to $\mathcal{G} \supset \mathcal{B}$ and let $\mathcal{H} = \mathcal{H}(\mathcal{W}, \mathcal{S}) = \oplus_{x\in\mathcal{W}}\mathbb{Z}[t, t^{-1}]\mathbf{T}_x$ be the corresponding Hecke algebra. For $s \in \mathcal{S}$ let $\mathcal{P}_s \supset \mathcal{B}$ be the associated parabolic. Let $\pi_s : \mathcal{G}/\mathcal{B} \to \mathcal{G}/\mathcal{P}_s$ be the projection. One may make $\langle\mathcal{D}_{\mathcal{B}}(\mathcal{G}/\mathcal{B})\rangle$ into a $\mathcal{H}$-module by the prescription $t\langle\mathcal{F}\rangle = \langle\mathcal{F}[-1]\rangle$ and $(\mathbf{T}_s + 1)\langle\mathcal{F}\rangle = \langle\pi_s^*\pi_{s*}\mathcal{F}\rangle$ for all $s \in \mathcal{S}$. If $\{\mathbf{C}'_x\}_{x\in\mathcal{W}}$ is the selfdual basis of the Hecke algebra defined in (KL79), then $\mathbf{C}'_x\langle\mathcal{IC}_e\rangle = \langle\mathcal{IC}_x\rangle$ for all $x \in \mathcal{W}$.

Now let us specify that we choose the identification $H^{\bullet}_{\mathcal{B}}(\mathcal{G}/\mathcal{B}) = S \otimes_{S^{\mathcal{W}}} S$ such that the pullback via $\pi_s : \mathcal{G}/\mathcal{B} \to \mathcal{G}/\mathcal{P}_s$ corresponds to the inclusion $S^s \otimes_{S^{\mathcal{W}}} S \hookrightarrow S \otimes_{S^{\mathcal{W}}} S$ on equivariant cohomology rings. Furthermore, let us identify graded $S \otimes S$-modules with graded S-bimodules in the obvious way. Let $S - mof - S$ denote the category of all graded S-bimodules which are finitely generated as left and as right modules. Then we may interpret $\mathbb{H}_{\mathcal{B}}$ as a functor

$$\mathbb{H}_{\mathcal{B}} : \mathcal{D}_{\mathcal{B}}(\mathcal{G}/\mathcal{B}) \to S - mof - S,$$

and use essentially the same arguments as in (Soe90) to prove $\mathbb{H}_{\mathcal{B}}\pi_s^*\pi_{s*}\mathcal{F} = S \otimes_{S^s} (\mathbb{H}_{\mathcal{B}}\mathcal{F})$ for $s \in \mathcal{S}$, $\mathcal{F} \in \mathcal{D}_{\mathcal{B}}(\mathcal{G}/\mathcal{B})$.

Now recall from (Soe92) how we made $\langle S - mof - S\rangle$ into a $\mathcal{H}$-module. The above formula says then just that $\mathbb{H}_{\mathcal{B}}$ induces a morphism of $\mathcal{H}$-modules $\mathbb{H}_{\mathcal{B}} : \langle \mathcal{D}_{\mathcal{B}}(\mathcal{G}/\mathcal{B})\rangle \to \langle S - mof - S\rangle$. Certainly $\mathbb{H}_{\mathcal{B}}\langle \mathcal{IC}_e\rangle = \langle S\rangle$, thus

$$\mathbb{H}_{\mathcal{B}}\langle \mathcal{IC}_x\rangle = \mathbb{H}_{\mathcal{B}}\mathbf{C}'_x\langle \mathcal{IC}_e\rangle = \mathbf{C}'_x\mathbb{H}_{\mathcal{B}}\langle \mathcal{IC}_e\rangle = \mathbf{C}'_x\langle S\rangle = \langle \mathbb{B}_x\rangle$$

by definition of $\mathbb{B}_x$. *q.e.d.[Lemma and theorem]*

We still have to prove proposition 2. Let us start with some general considerations. Let X be a topological space filtered by closed subspaces $X = X_0 \supset X_1 \supset \ldots \supset X_r = \emptyset$ and let $\mathcal{F} \in D^b(ShX)$ be an object in the bounded derived category of sheaves on X. To calculate its hypercohomology $\mathbb{H}^\bullet\mathcal{F}$ we have to choose an injective resolution $I^\bullet$ of $\mathcal{F}$ and take cohomology of the complex of global sections $\Gamma(I^\bullet)$. Certainly this is filtered by supports, $0 \subset \Gamma_{X_r}(I^\bullet) \subset \ldots \subset \Gamma_{X_0}(I^\bullet) = \Gamma(I^\bullet)$. So its cohomology is the limit of a spectral sequence with E_1-term $E_1^{p,q} = \mathbb{H}^{p+q}_{X_p - X_{p+1}}(\mathcal{F})$ the local hypercohomology of $\mathcal{F}$ along $X_p - X_{p+1}$. Let $j : Y \hookrightarrow X$ be the embedding of a locally closed subset. Then we can write the local hypercohomology as $\mathbb{H}^\bullet_Y(\mathcal{F}) = \mathbb{H}^\bullet v^!\mathcal{F}$, i.e. as the hypercohomology of the derived local cohomology sheaf $v^!\mathcal{F} \in D^+(ShY)$. Let $j_p : X_p - X_{p+1} \hookrightarrow X$ denote the inclusion. We may then write our E_1-term as $E_1^{p,q} = \mathbb{H}^{p+q}j_p^!\mathcal{F}$.

Suppose next that X is a complex algebraic $\mathcal{B}$-variety and that all X_p are closed $\mathcal{B}$-stable subvarieties. Applying the above considerations to $E\mathcal{B} \times_{\mathcal{B}} X$ we find that for $\mathcal{F} \in \mathcal{D}_{\mathcal{B}}(X)$ the equivariant hypercohomology $\mathbb{H}^\bullet_{\mathcal{B}}\mathcal{F}$ is the limit of a spectral sequence with E_1-term $E_1^{p,q} = \mathbb{H}^{p+q}_{\mathcal{B}}j_p^!\mathcal{F}$. In particular, if $\mathcal{G} \in \mathcal{D}_{\mathcal{B}}(X)$ is another object the equivariant extensions $\mathrm{Hom}^\bullet_{\mathcal{B}}(\mathcal{F}, \mathcal{G}) = \mathbb{H}^\bullet_{\mathcal{B}}\mathcal{R}Hom(\mathcal{F}, \mathcal{G})$ are the limit of a spectral sequence with E_1-term

$$E_1^{p,q} = \mathbb{H}^{p+q}_{\mathcal{B}}j_p^!\mathcal{R}Hom(\mathcal{F}, \mathcal{G}) = \mathbb{H}^{p+q}_{\mathcal{B}}\mathcal{R}Hom(j_p^*\mathcal{F}, j_p^!\mathcal{G}).$$

Suppose that X is a $\mathcal{B}$-variety with a $\mathcal{B}$-stable stratification $X = \dot{\bigcup}_{w\in\mathcal{W}}X_w$ into smooth irreducible subvarieties. Put $|w| = \dim X_w$ and let $\mathcal{C}_w \in \mathcal{P}_{\mathcal{B}}(X_w)$ be the constant perverse sheaf $\mathcal{C}_w = \underline{X_w}[|w|]$. Let $\mathcal{IC}_w \in \mathcal{P}_{\mathcal{B}}(X)$ denote the intersection cohomology complex of the closure of X_w, extended by zero to X.

Let $j_w : X_w \hookrightarrow X$ be the inclusions of the strata and suppose there exist isomorphisms

$$j_v^* \mathcal{IC}_w \cong \oplus_\nu n_{v,w}^\nu \mathcal{C}_v[\nu] \qquad (*)$$

in $\mathcal{D}_\mathcal{B}(X_v)$, for all $v, w \in \mathcal{W}$, with suitable integers $n_{v,w}^\nu$ defined by these isomorphisms.

PROPOSITION 3. *Suppose we have in addition to the above assumptions parity vanishing, i.e. (1) $n_{v,w}^\nu = 0$ unless ν has the same parity as $|v| + |w|$ and (2) $H_\mathcal{B}^i(X_w) = 0$ for all w unless i is even. Then*

i.) $\mathbb{H}_\mathcal{B}^\bullet$ induces an injection

$$\mathrm{Hom}_\mathcal{B}^\bullet(\mathcal{IC}_x, \mathcal{IC}_y) \to \mathrm{Hom}_\mathbb{C}(\mathbb{H}_\mathcal{B}^\bullet \mathcal{IC}_x, \mathbb{H}_\mathcal{B}^\bullet \mathcal{IC}_y).$$

ii.) $\dim_\mathbb{C} \mathrm{Hom}_\mathcal{B}^\nu(\mathcal{IC}_x, \mathcal{IC}_y) = \sum_{a+i+j=\nu,\, z \in \mathcal{W}} n_{z,v}^i n_{z,w}^j \dim_\mathbb{C} H_\mathcal{B}^a(X_z)$.

Proof: ii.) Consider the filtration on X given by $X_p = \bigcup_{|x|+p \leq \dim X} X_x$. If we calculate $\mathrm{Hom}_\mathcal{B}^\nu(\mathcal{IC}_x, \mathcal{IC}_y)$ by the associated spectral sequence, we find as E_1-term

$$\begin{aligned} E_1^{p,q} &= \mathbb{H}_\mathcal{B}^{p+q} \mathcal{R}Hom(j_p^* \mathcal{IC}_x, j_p^! \mathcal{IC}_y) \\ &= \oplus_{|z|+p=\dim X} \mathbb{H}_\mathcal{B}^{p+q} \mathcal{R}Hom(j_z^* \mathcal{IC}_x, j_z^! \mathcal{IC}_y) \\ &= \oplus_{|z|+p=\dim X} \mathbb{H}_\mathcal{B}^{p+q} \mathcal{R}Hom(\oplus_i n_{z,x}^i \mathcal{C}_z[i], \oplus_j n_{z,y}^j \mathcal{C}_z[-j]) \\ &= \oplus_{|z|+p=\dim X} \oplus_{i,j} n_{z,x}^i n_{z,y}^j H_\mathcal{B}^{p+q-i-j}(X_z). \end{aligned}$$

By our parity assumptions this vanishes unless $p+q$ has the same parity as $|x| + |y|$. In particular, our spectral sequence degenerates at this E_1-term and ii.) follows.

i.) Also $\mathbb{H}_\mathcal{B}^i \mathcal{IC}_x$ is the limit of a spectral sequence with E_1-term $E_1^{p,q} = \mathbb{H}_\mathcal{B}^{p+q} j_p^! \mathcal{IC}_x$, and again by parity vanishing it degenerates at this term. It is sufficient to show that any nonzero morphism $f : \mathcal{IC}_x \to \mathcal{IC}_y[\nu]$ in $\mathcal{D}_\mathcal{B}(X)$ induces for some p a nonzero morphism $j_p^! f : j_p^! \mathcal{IC}_x \to j_p^! \mathcal{IC}_y[\nu]$. Let $i_p : X_p \hookrightarrow X$ be the inclusion. We may decompose X_p into an open and a closed subset

$$X_{p+1} \overset{i}{\hookrightarrow} X_p \overset{u}{\hookleftarrow} (X_p - X_{p+1}).$$

Then we have a distinguished triangle

$$(i_! i^!, \mathrm{id}, u_* u^*) i_p^! = (i_* i_{p+1}^!, i_p^!, u_* j_p^!)$$

which says that $i_{p+1}^! f = 0$ and $j_p^! f = 0$ imply $i_p^! f = 0$. So if all $j_p^! f = 0$ then $i_0^! f = f = 0$. *q.e.d.*

Proof[Proposition 2]: It will be sufficient to (1) prove injectivity of the map and (2) check that the Dimensions coincide on both sides, where for a graded vector space $V = \oplus V^i$ we understand its Dimension to be the polynomial $\mathrm{Dim}V = \sum t^i \dim_{\mathbb{C}} V^i$. First we check that the preceding proposition applies in our situation, i.e. for the stratification of $X = \mathcal{G}/\mathcal{B}$ by the Bruhat cells $X_w = \mathcal{B}w^{-1}\mathcal{B}/\mathcal{B}$. We need

LEMMA 6. *In $\mathcal{D}_{\mathcal{B}}(X_v)$ the $j_v^*\mathcal{IC}_w$ decompose into the direct sum of their perverse cohomology, i.e. $j_v^*\mathcal{IC}_w \cong \oplus_\nu n_{v,w}^\nu \mathcal{C}_v[\nu]$ for suitable integers $n_{v,w}^\nu$.*

Proof: In the nonequivariant situation this follows from pointwise purity: Namely the $j_v^*\mathcal{IC}_w$ are "pure", hence the sum of their perverse cohomology groups.

We may reduce our equivariant situation to a nonequivariant situation as follows: First choose a finite interval $I \subset \mathbb{Z}$ such that $\mathcal{IC}_w \in \mathcal{D}_{\mathcal{B}}^I(X)$. Then we have to prove the lemma for the $j_v^*\mathcal{IC}_w \in \mathcal{D}_{\mathcal{B}}^I(X_v)$. Now choose a smooth n-acyclic free right $\mathcal{B}$-space M, where n is bigger then the length of I. Then by (BL92) we have a commutative diagram

$$\begin{array}{ccc} \mathcal{D}_{\mathcal{B}}^I(X) & \rightarrow & \mathcal{D}^I(M \times_{\mathcal{B}} X) \\ j_v^* \downarrow & & \downarrow (\mathrm{id}\times j_v)^* \\ \mathcal{D}_{\mathcal{B}}^I(X_v) & \rightarrow & \mathcal{D}^I(M \times_{\mathcal{B}} X_v) \end{array}$$

where the horizontal maps are fully faithful embeddings. But the nature of the singularities is the same for $M \times_{\mathcal{B}} \overline{X_w}$ and $\overline{X_w}$. Hence the intersection cohomology complex of $M \times_{\mathcal{B}} \overline{X_w}$ is also pointwise pure and the lemma follows. *q.e.d.*

Now we may in fact use proposition 3 to show that our map

$$\mathbb{H}_{\mathcal{B}} : \mathrm{Hom}_{\mathcal{B}}^\bullet(\mathcal{IC}_x, \mathcal{IC}_y) \rightarrow \mathrm{Hom}_{S\otimes S}(\mathbb{H}_{\mathcal{B}}\mathcal{IC}_x, \mathbb{H}_{\mathcal{B}}\mathcal{IC}_y).$$

is an injection. So we just have to check that the Dimensions are equal. But

$$\begin{aligned} \mathrm{Dim}\mathrm{Hom}_{\mathcal{B}}^\bullet(\mathcal{IC}_x, \mathcal{IC}_y) &= \mathrm{Dim}\mathrm{Hom}_{\mathcal{D}}^\bullet(\mathcal{IC}_x, \mathcal{IC}_y) \otimes_{\mathbb{C}} S \\ & \qquad \text{by proposition 3} \end{aligned}$$

and

$$\begin{aligned}\mathrm{DimHom}_{S\otimes S}(\mathbb{H}_{\mathcal{B}}\mathcal{IC}_x, \mathbb{H}_{\mathcal{B}}\mathcal{IC}_y) &= \mathrm{DimHom}_{S\otimes S}(\mathbb{B}_x, \mathbb{B}_y) \text{ by} \\ &= \mathrm{DimHom}_{S\otimes S}(\mathbb{B}_x, \mathbb{B}_y) \otimes_{S\mathbb{C}} \otimes_{\mathbb{C}S} \text{ by} \\ &= \mathrm{DimHom}_S(\mathbb{B}_x \otimes_{S\mathbb{C}}, \mathbb{B}_y \otimes_{S\mathbb{C}}) \otimes_{\mathbb{C}S} \text{ by} \\ &= \mathrm{DimHom}_S(\mathbb{H}^{\bullet}\mathcal{IC}_x, \mathbb{H}^{\bullet}\mathcal{IC}_y) \otimes_{\mathbb{C}S} .\end{aligned}$$

But we know already from (Soe90) that

$$\mathrm{Hom}^{\bullet}_{\mathcal{D}}(\mathcal{IC}_x, \mathcal{IC}_y) = \mathrm{Hom}_S(\mathbb{H}^{\bullet}\mathcal{IC}_x, \mathbb{H}^{\bullet}\mathcal{IC}_y).$$

q.e.d. [Proposition 2]

3.5. COMPACT GROUPS

Suppose under the ABV-parametrization some geometric parameter (Y, τ) corresponds to a representation of a compact form of G. Our basic conjecture would say that

$$\mathrm{Hom} j_{G^{\vee}_F}(\mathcal{IC}(Y, \tau), \mathcal{IC}(Y', \tau')[n]) = 0$$

unless $(Y, \tau) = (Y', \tau')$ and $n = 0$. The results of (ABV92) on the duality between IC-data and KL-data tell us already that the above space of homomorphisms vanishes unless $(Y, \tau) = (Y', \tau')$. We will verify in addition that the isotropy group of points in Y is finite. This implies that there are now higher selfextensions of $\mathcal{IC}(Y, \tau)$.

So let $\chi \in \mathrm{Max}Z$ be integral regular. Put $I = \{y \in {}^{\Gamma}G^{\vee} - G^{\vee} \mid y^2 = 1\}$. Clearly $X(\chi) = I \times G^{\vee}/B^{\vee}$ with the diagonal $G^{\vee}$-action. In (ABV92) it is shown that I decomposes into a finite number of $G^{\vee}$-orbits, that these orbits are all closed and that the corresponding isotropy groups are just complexifications of maximal compact subgroups of real forms of $G^{\vee}$. More precisely the real forms appearing here are just all real forms of one inner class of real forms on $G^{\vee}$, and this inner class corresponds to the involution $(-w_{\circ}\gamma)$ of our based root system, where γ is the involution of our based root system coming from the choosen inner class of real forms on G and $w_{\circ}$ is the longest element of the Weyl group.

Hence if G is endowed with the inner class of real forms containing the compact one, then $\gamma = -w_{\circ}$ so that the corresponding inner class of real forms on $G^{\vee}$ contains the split form. In particular some components of $X(\chi)$ are isomorphic to $G^\vee/K \times G^\vee/B^\vee$ where K is the complexification of a maximal compact subgroup of a split form of $G^\vee$. Now the representations of compact forms of G correspond just to open $G^\vee$-orbits on these $G^\vee/K \times G^\vee/B^\vee \subset X(\chi)$. And clearly the isotropy groups of points in these open orbits are finite.

3.6. TEMPERED REPRESENTATIONS

I want to suggest a link of my basic conjecture with the results of A. Guichardet (Gui92). He considers an L-packet of irreducible tempered representations and the category $\mathcal{T}$ of all Harish-Chandra modules with composition factors only from this L-packet. He then proves that this category $\mathcal{T}$ is equivalent to a category $S^\sigma[R] - Nil$, for a suitable polynomial ring S with a suitable action $\sigma : R \to \mathrm{Aut}_{\mathbb{C}S}$ of a suitable finite group R on it.

In the ABV-parametrization a tempered L-packet corresponds to a collection of geometric parameters (Y, τ_i) for a fixed open orbit $Y \subset X(\chi)$. If my basic conjecture is true, there should be an equivalence of categories

$$\mathcal{T} \cong \mathrm{Ext}_{G_F^\vee} X(\chi)/I - Nil$$

where the twosided ideal I is generated by all idempotents which belong to parameters other than the (Y, τ_i). Now restriction gives us a map

$$\mathrm{Ext}_{G_F^\vee} X(\chi) \to \mathrm{Ext}_{G_F^\vee} Y$$

which annihilates I, and by 2.4 the ring $\mathrm{Ext}_{G_F^\vee} Y$ has the form $S^\sigma[R]$. I didn't check however that these S, σ, R are indeed those of Guichardet.

3.7. THE CENTER AND GEOMETRY

To each additive category $\mathcal{A}$ we may associate a commutative ring $Z(\mathcal{A})$, the center of $\mathcal{A}$, as follows: $Z(\mathcal{A})$ consists just of all natural transformations of the identity functor $\mathrm{id} : \mathcal{A} \to \mathcal{A}$ to itself, i.e. $Z(\mathcal{A}) = \mathrm{End}(\mathrm{id})$. Let R be a commutative ring. By an "action of R on $\mathcal{A}$" we mean a ring homomorphism $R \to Z(\mathcal{A})$.

Now consider our basic conjecture. On the representation theoretic side of the conjectured equivalence there acts certainly the completion $Z_\chi^\wedge$ at χ of the center $Z \subset U - \text{mod} - U$. I want to explain what this should correspond to on the geometric side. Remark that the construction of $X(\chi)$ gives us a $G^\vee$-equivariant map $X(\chi) \to \mathcal{F}(\mathcal{O}(\chi))$. This leads to a morphism

$$H^\bullet_{G^\vee}(\mathcal{F}(\mathcal{O}(\chi))) \to \operatorname{Ext}_{G_F^\vee} X(\chi)$$

of the equivariant cohomology ring of $\mathcal{F}(\mathcal{O}(\chi))$ to our geometric extension algebra, and this morphism has central image. It should correspond to the action of $Z_\chi^\wedge$ on the representation theoretic side of the conjectured equivalence via the morphism $H^\bullet_{G^\vee}(\mathcal{F}(\mathcal{O}(\chi))) \to Z_\chi^\wedge$ defined as the composition

$$H^\bullet_{G^\vee}(\mathcal{F}(\mathcal{O}(\chi))) \overset{(1)}{=} H^\bullet_{G^\vee}(\mathcal{O}(\chi)) \overset{(2)}{=} H^\bullet(BG_\lambda^\vee;\mathbb{C}) \overset{(3)}{=} S^{\mathcal{W}_\lambda} \overset{(4)}{\to} (S_0^\wedge)^{\mathcal{W}_\lambda} \overset{(5)}{\to} Z_\chi^\wedge$$

which we will now explain step by step.

1. This is clear since $\mathcal{O}(\chi) \twoheadrightarrow \mathcal{F}(\mathcal{O}(\chi))$ is a fibration with affine spaces as fibres.
2. We choose $\lambda \in \mathcal{O}(\chi) \cap LieT^\vee$ and let $G_\lambda^\vee$ be the isotropy group of λ. Then this is the induction equivalence.
3. The Weyl group of $G_\lambda^\vee$ is the stabilizer $\mathcal{W}_\lambda$ of λ in $\mathcal{W}$. We let $S = S(LieT)$ be the regular functions on the Lie algebra $LieT^\vee$ of the maximal torus $T^\vee \subset G_\lambda^\vee$. For any linear complex algebraic group the cohomology ring of its classifying space can be identified with the regular functions on the Lie algebra of a maximal torus which are invariant under the Weyl group.
4. We let $S_0^\wedge$ be the completion of S at the origin of $LieT^\vee$.
5. We will rather construct the inverse morphism. The Harish-Chandra homomorphism $\xi : Z \to S$, $\xi(Z) = S^{\mathcal{W}}$ certainly induces a morphism between the respective completions $Z_\chi^\wedge \to S_\lambda^\wedge$. If we compose with the isomorphism $S_\lambda^\wedge \to S_0^\wedge$ defined by translation, we get a morphism $Z_\chi^\wedge \to S_0^\wedge$. By some invariant theory this factorizes over an isomorphism $Z_\chi^\wedge \to (S_0^\wedge)^{\mathcal{W}_\lambda}$. The inverse of this isomorphism is the last map of our above composition.

4. Further conjectures

To provide additional motivation for my basic conjecture, I want to explain how it should be related to character duality (Vog82;

ABV92). Let me cite from (ABV92). "It is natural to look for a functorial relationship between a category of representations of strong real forms of G, and one of our geometric categories. We do not know what form such a relationship should take, or how one may hope to establish it directly." I conjecture that the relationship in question should take the form of Koszul duality. However I have no idea how to establish such a relationship directly. I want now to give more details and start by explaining some Koszul duality.

4.1. GENERALITIES ON KOSZUL DUALITY

Let $A = \oplus_{i\geq 0} A_i$ be a graded $\mathbb{C}$-algebra with $A_0 = \oplus_{x\in\mathcal{W}}\mathbb{C}1_x$, where $\mathcal{W}$ is a suitable finite parameter set. Let A-mod be the category of graded A-modules $M = \oplus_i M_i$. We define $M\langle 1\rangle \in A$-mod by $(M\langle 1\rangle)_i = M_{i-1}$. For $M, N \in A$-mod we let $\hom_A(M,N)$ (resp. $\mathrm{Hom}j_A(M,N)$) denote morphisms in A-mod (resp. A-Mod). So $\hom_A(M,N) = \{f \in \mathrm{Hom}_A(M,N) \mid f(M_i) \subset N_i\}$.

Let us form the bounded from above derived category $\mathcal{D} = D^-(A\text{-mod})$. It comes with three automorphisms [1] "shift of a complex", $\langle 1\rangle$ "shift of the grading on all terms of a complex" and $(1/2) = [-1]\langle -1\rangle$. The strange notation comes from the fact that later $(1/2)$ should correspond to a "root of the Tate twist". We will denote morphisms in $\mathcal{D}$ by $\hom_{\mathcal{D}}$.

We consider two t-structures on $\mathcal{D}$. The "representation theoretic" t-structure $(\mathcal{D}^{\leq 0,r}, \mathcal{D}^{\geq 0,r})$ is the obvious t-structure with heart A-mod. The "geometric" t-structure $(\mathcal{D}^{\leq 0,g}, \mathcal{D}^{\geq 0,g})$ is its "Koszul dual". More precisely, $\mathcal{D}^{\leq 0,g}$ (resp. $\mathcal{D}^{\geq 0,g}$) consists of all complexes which are quasiisomorphic to complexes $\ldots \to X^i \to X^{i+1} \to \ldots$ of projective objects X^i of A-mod such that X^i is generated by its parts of degree $\leq -i$ (resp. $\geq -i$).

The twists $\langle 1\rangle$ (resp. $(1/2)$) respect the representation theoretic (resp. geometric) t-structure. Up to twist the simple objects in the hearts of the respective t-structures are the $L_x = A_0 1_x$ (resp. the $L^x = A1_x$). Let $\mathcal{D}_r$ (resp. $\mathcal{D}_g$) be the full triangulated subcategories of $\mathcal{D}$ generated by the L_x (resp. L^x) and their twists. The respective t-structures on $\mathcal{D}$ induce t-structures $(\mathcal{D}_r^{\leq 0}, \mathcal{D}_r^{\geq 0})$ and $(\mathcal{D}_g^{\leq 0}, \mathcal{D}_g^{\geq 0})$ on $\mathcal{D}_r$ and $\mathcal{D}_g$ respectively.

4.2. APPLICATION TO OUR SITUATION

Let us now go back to the situation of our basic conjecture. There we conjectured an equivalence of categories

$$\bigoplus_{\delta \in S_F} \mathcal{M}(G(\mathbb{R}, \delta))_\chi \cong \mathrm{Ext}_{G_F^\vee}(X(\chi)) - Nil.$$

Let us abreviate the left hand side category by $\mathcal{HC}h$ and put $G_F^\vee = \mathcal{G}$, $X(\chi) = X$. Let $\{\mathcal{L}_x\}_{x \in \mathcal{W}}$ and $\{\mathcal{L}^x\}_{x \in \mathcal{W}}$ represent the simple isomorphism classes in $\mathcal{HC}h$ and $\mathcal{P}_{\mathcal{G}}(X)$ respectively, where the indices are choosen in such a way that objects with the same index correspond to each other under the ABV-parametrization $(a) \leftrightarrow (b)$ from the beginning. In particular $\mathcal{W}$ is not at all the Weyl group. Finally put $A = \mathrm{Ext}_{\mathcal{G}} X = \mathrm{End}^\bullet_{\mathcal{G}}(\oplus \mathcal{L}^x)$. This is a graded $\mathbb{C}$-algebra and $A_0 = \oplus_{x \in \mathcal{W}} \mathbb{C}1_x$ in an obvious way. The basic conjecture predicts an equivalence of categories $\mathcal{HC}h \cong A - Nil$.

Now construct to A as explained before the triangulated category $\mathcal{D} = D^-(A\text{-mod})$ with its twists $[1]$, $\langle 1 \rangle$, $(1/2)$ and the two t-structures giving rise to two full triangulated subcategories

$$\mathcal{D}_r \hookrightarrow \mathcal{D} \hookleftarrow \mathcal{D}_g.$$

CONJECTURE 1. *There exists a t-exact functor $v_r : \mathcal{D}_r \to D^b(\mathcal{HC}h)$ together with canonical isomorphisms $v_r(M\langle 1 \rangle) = v_r M$ such that*

i.) $\mathrm{Hom}_{D^b(\mathcal{HC}h)}(v_r M, v_r N) = \oplus_i \mathrm{hom}_{\mathcal{D}}(M, N\langle i \rangle)$ *for all* $M, N \in \mathcal{D}_r$.

ii.) $v_r L_x \cong \mathcal{L}_x \quad \forall x \in \mathcal{W}$.

CONJECTURE 2. *There exists a t-exact functor $v_g : \mathcal{D}_g \to \mathcal{D}_{\mathcal{G}(X)}$ together with canonical isomorphisms $v_g(M(1/2)) = v_g M$ such that*

i.) $\mathrm{Hom}_{\mathcal{G}}(v_g M, v_g N) = \oplus_i \mathrm{hom}_{\mathcal{D}}(M, N(i/2))$ *for all* $M, N \in \mathcal{D}_g$.

ii.) $v_g L^x \cong \mathcal{L}^x \quad \forall x \in \mathcal{W}$.

The conjectures mean that the categories $\mathcal{D}_r$ and $\mathcal{D}_g$ should be "mixed" or "graded" versions of $D^b(\mathcal{HC}h)$ and $\mathcal{D}_{\mathcal{G}}(X)$ respectively, and the functors v_r, v_g just "forget the grading". The next conjectures stipulate the existence of graded versions for "standard objects". Let me explain what I mean. Certainly any $\mathcal{IC}(Y, \tau)$ is the unique simple quotient of the zero'th perverse cohomology of $\mathcal{M}^x = i_!\tau[\dim Y]$, where $i : Y \hookrightarrow X$ is the inclusion of a $\mathcal{G}$-orbit and τ an irreducible $\mathcal{G}$-equivariant local system on Y, hence $\tau \in \mathcal{D}_{\mathcal{G}(Y)}$. On the other hand any $\mathcal{L}_x$ is the socle of a standard object $\mathcal{N}_x \in \mathcal{HC}h$ induced from an irreducible tempered representation.

CONJECTURE 3. *For all $x \in \mathcal{W}$ there exists $M^x \in \mathcal{D}_g$ such that $v_g M^x \cong \mathcal{M}^x$ and $\hom_{\mathcal{D}}(M^x, L^x) \neq 0$. This M^x is unique up to isomorphism.*

CONJECTURE 4. *For all $x \in \mathcal{W}$ there exists $N_x \in \mathcal{D}_r$ such that $v_r N_x \cong \mathcal{N}_x$ and $\hom_{\mathcal{D}}(N_x, L^x) \neq 0$. This N_x is unique up to isomorphism.*

The essential ingredient we would need to get the character duality of Vogan's is the following

CONJECTURE 5. *We have*

$$\hom_{\mathcal{D}}(M^x(i/2), N_y\langle n\rangle) = \begin{cases} \mathbb{C} & x = y, i = n = 0 \\ 0 & \textit{else.} \end{cases}$$

Now comes character duality. We make the Grothendieck groups $[\mathcal{D}_r]$ and $[\mathcal{D}_g]$ into $\mathbb{Z}[t, t^{-1}]$-modules by the prescription $t[M] = [M\langle 1\rangle]$ for $M \in \mathcal{D}_r$, respectively

$$t[M] = -[M(1/2)] = [M\langle -1\rangle]$$

for $M \in \mathcal{D}_g$. We obtain a $\mathbb{Z}[t, t^{-1}]$-bilinear pairing $[\mathcal{D}_g] \times [\mathcal{D}_r] \to \mathbb{Z}[t, t^{-1}]$ between these two Grothendieck groups by the prescription

$$(M, N) \mapsto \langle M, N\rangle = \sum_{\nu, i} (-1)^\nu \dim_{\mathbb{C}} \hom_{\mathcal{D}}(M, N[\nu]\langle -i\rangle) t^i.$$

Certainly $[\mathcal{D}_g]$ is a free $\mathbb{Z}[t, t^{-1}]$-module with two bases $\{[L^x]\}_{x \in \mathcal{W}}$ and $\{[M^x]\}_{x \in \mathcal{W}}$, which we call the simple and the standard basis. Also $[\mathcal{D}_r]$ is a free $\mathbb{Z}[t, t^{-1}]$-module with "simple" basis $\{[L_x]\}_{x \in \mathcal{W}}$ and "standard" basis $\{[N_x]\}_{x \in \mathcal{W}}$. It is clear from the definitions that $\langle L^x, L_y\rangle = \delta_{x,y}$ and if we believe conjecture 5 we find as well $\langle M^x, N_y\rangle = \delta_{x,y}$.

Let now P_r be the transition matrix from the simple basis $\{[L_x]\}$ to the standard basis $\{[N_x]\}$. So P_r is a $\mathcal{W} \times \mathcal{W}$-matrix with coefficients $(P_r)_{x,y} \in \mathbb{Z}[t, t^{-1}]$ given by the equation

$$[N_y] = \sum_x (P_r)_{x,y}[L_x].$$

Similarily we define the geometric base change matrix P_g by the equation

$$[M^z] = \sum_x (P_g)_{u,z}[L^u]$$

in $[\mathcal{D}_g]$. Now our formulas give us

$$\begin{aligned} \delta_{z,y} &= \langle M^z, N_y \rangle \\ &= \textstyle\sum_{u,x} (P_g)_{u,z}(P_r)_{x,y} \langle L^u, L_x \rangle \\ &= (P_g^T P_r)_{z,y} \end{aligned}$$

where the upper index T transposes the matrix. In other words, we find the equation

$$P_g^T P_r = 1$$

of $\mathcal{W} \times \mathcal{W}$-matrices with coefficients in $\mathbb{Z}[t, t^{-1}]$.

This equation is essentially Vogan's character duality. To see this, write $(P_r)_{x,y} = \sum m_r(x, y; i)t^i$ so that $[N_y] = \sum_{x,i} m_r(x, y; i)[L_x\langle i \rangle]$. Applying v_r to both sides, we see that if we set $t = 1$ in P_r then $P_r(1)$ is just the representation theoretic multiplicity matrix, $(P_r(1))_{x,y} = m_r(x, y)$ in the notation of (ABV92), (1.21). On the other hand, write $(P_g)_{u,z} = \sum m_g(u, z; j)t^j$ so that $[M^z] = \sum_{u,j} m_g(u, z; j)(-1)^j[L^u(j/2)]$. Applying v_g to both sides, we see that if we set $t = -1$ in P_g then $P_g(-1)$ is just the intersection cohomology matrix, $(P_g(-1))_{u,z} = m_g(u, z)$ in the notation of (ABV92), (1.22)(d).

To finish, we have to assume some parity vanishing. More precisely, for $x \in \mathcal{W}$ let $d(x)$ denote the dimension of the corresponding $\mathcal{G}$-orbit on X.

CONJECTURE 6. *We have $m_g(u, z; j) = 0$ unless j has the same parity as $d(u) + d(z)$.*

From this we can deduce

$$(P_g(1))_{u,z} = (-1)^{d(u)+d(z)}(P_g(-1))_{u,z} = (-1)^{d(u)+d(z)} m_g(u, z)$$

and thus specializing our equation $P_g^T P_r = 1$ to $t = 1$ gives precisely Vogan's character duality

$$\sum_u (-1)^{d(u)+d(z)} m_g(u,z) m_r(u,x) = \delta_{z,x}$$

as formulated in (ABV92), Corollary 1.25. a).

4.3. MOTIVATION

The first of these conjectures is almost our basic conjecture. For the other conjectures there is very little hard evidence. At least for tori everything works, and for complex groups one can find results in the spirit of the above conjectures in (BGS96).

References

Jeffrey Adams, Dan Barbasch, and David A. Vogan, Jr. *The Langlands classification and irreducible characters for real reductive groups.* Progress in Mathematics. Birkhäuser, 1992.

Alexander A. Beilinson and Joseph N. Bernstein. Localisation de $\mathfrak{g}$-modules. *C. R. Acad. Sci. Paris, Sér. 1*, 292:15–18, 1981.

Joseph N. Bernstein and Sergei I. Gelfand. Tensor products of finite and infinite representations of semisimple Lie algebras. *Compositio Math.*, 41:245–285, 1980.

Alexander A. Beilinson and Victor Ginsburg. Mixed categories, Ext-duality and representations (results and conjectures). Preprint, 1986.

Alexander A. Beilinson, Victor Ginzburg, and Wolfgang Soergel. Koszul duality patterns in representation theory. *JAMS*, 9(2):473–527, 1996.

Joseph N. Bernstein and Valery Lunts. Equivariant sheaves I (preliminary version, third draft). Preprint, 1992.

Alain Guichardet. Géometrie différentielle locale sur le dual tempéré d'un groupe de Lie semi-simple. Preprint, 1992.

S. M. Khoroshkin. Category of Harish-Chandra modules of the group $SU(n,1)$. *Funct. Anal. Appl.*, 14:153–155, 1980.

David Kazhdan and George Lusztig. Representations of Coxeter groups and Hecke algebras. *Inventiones*, 53:191–213, 1979.

Robert P. Langlands. On the classification of irreducible representations of real algebraic groups. Preprint (appeared in: Representation theory and harmonic analysis on semisimple Lie groups, P. J. Sally, Jr. and D. A. Vogan, Jr., editors, AMS 1989), 1973.

Wolfgang Soergel. Kategorie $\mathcal{O}$, perverse Garben und Moduln über den Koinvarianten zur Weylgruppe. *Journal of the AMS*, 3:421–445, 1990.

Wolfgang Soergel. The combinatorics of Harish-Chandra bimodules. *Journal für die reine und angewandte Mathematik*, 429:49–74, 1992.

David A. Vogan, Jr. *Representations of real reductive Lie groups*, volume 15 of *Progress in Mathematics*. Birkhäuser, 1981.

David A. Vogan, Jr. Irreducible characters of semisimple Lie groups: Character-multiplicity duality. *Duke Math. J.*, 49:943–1073, 1982.

David A. Vogan, Jr. The local Langlands conjecture. In Adams et al., editor, *Contemporary Mathematics*, volume 145, pages 417–423. AMS, 1993.

NONCOMMUTATIVE PROJECTIVE GEOMETRY

J. T. STAFFORD (jts@math.lsa.umich.edu)
Department of Mathematics,
University of Michigan,
Ann Arbor, MI 48109, USA.

The aim of the three lectures I gave at Constanta was to introduce the audience to the theory of "Noncommutative Projective Geometry." The material for these talks was taken from the survey article (6) and so this abstract will give a very brief discussion of this material leaving the reader to see (6) for the details.

Noncommutative affine geometry—where one think of the correspondence between a ring A and the associated variety $\operatorname{Spec}(A)$ consisting of (say) all the prime ideals of A—is an old idea. However, it has really not been very successful at the abstract level. In specific cases, of course, variants of this theme have been very successful; just think of quantum groups or Kontsevich's work. In contrast, the analogues of projective geometry seem to have some real depth, and the philosophy of these lectures—and (6)—is to justify this assertion.

So, what is meant by the term "Noncommutative Projective Geometry?" I have two themes in mind: First, and this is my main motivation, to use the techniques of classical algebraic geometry to undersand the structure of noncommutative connected graded k-algebras $R = k \oplus R_1 \oplus R_2 \oplus \cdots$, where k is a base field that will be fixed throughout. Secondly, to understand what one should even mean by noncommutative projective varieties. As phrased, the latter do not exist. However, in the commutative case, one can replace the given variety by the category of sheaves of modules over it. Analogues of this category do exist at an abstract noncommutative level and they turn out to be profitable objects of study. If R is a

K.W. Roggenkamp and M. Ştefănescu (eds.), Algebra - Representation Theory, 415–417.

connected graded ring, as above, then one such category is $\mathrm{qgr}(R)$, the category of noetherian, $\mathbb{Z}$-graded R-modules modulo those of finite length. Moreover, one can go back and forth between the rings and their categories to get a better understanding of both.

The three lectures covered the following four topics:

Geometric Constructions. The key technique for obtaining noncommutative algebras from (classical) commutative schemes is through the *twisted homogeneous coordinate ring* $B(X, \mathcal{L}, \sigma)$ of (3). This is defined as follows: Let σ be an automorphism of a k-variety or scheme X and let $\mathcal{L}$ be a σ-ample invertible sheaf (when X is an irreducible curve, the σ-ample invertible sheafs are just the ample invertible sheafs). Then $B(X, \mathcal{L}, \sigma) = k \oplus \bigoplus_{n>0} B_n$ where $B_n = \mathrm{H}^0(X, \mathcal{L}_n)$ and multiplication is defined in the natural way. The aim of the first lecture is to describe the properties and applications of these algebras. For example Serre's Theorem generalizes:

THEOREM 1. *(3) Let X be a projective scheme with an automorphism σ and let $\mathcal{L}$ be a σ-ample invertible sheaf. Then, $B = B(X, \mathcal{L}, \sigma)$ is a right noetherian ring such that* $\mathrm{qgr}(B)$ *is equivalent to* $\mathrm{coh}(X)$*, the category of coherent sheaves on X. By (5), B is also left noetherian.*

Noncommutative projective curves. Twisted homogeneous coordinate rings can be used to classify noncommutative domains of quadratic growth, as well as to determine the noncommutative analogues of projective curves. For example, one has:

THEOREM 2. *(1) Let $R = k \oplus R_1 \oplus R_2 \oplus \cdots$ be a connected graded domain, generated in degree one, such that* $\dim_k R_n$ *grows linearly. Then, up to a finite dimensional vector space, $R \cong B(X, \mathcal{L}, \sigma)$ for some irreducible curve X. Thus R is noetherian with* $\mathrm{qgr}(R) \sim \mathrm{coh}(X)$.

Homological Results. The aim of this section is to describe the applications of homological techniques to noncommutative graded rings. For example, in surprisingly general circumstances one has an analogue of Serre duality for noncommutative varieties (see (7, 10) or (6, Section 7)).

Noncommutative projective surfaces. This theory is at an early stage of development, but the general picture is beginning to take shape. For example, by (2) and (4), one has a good notion of (the coordinate ring of) noncommutative projective 2-space—these are the Artin-Schelter regular rings of dimension three—reasonable analogues of quadrics (9) and blowing up and blowing down (8) and some rather brave conjectures about the general picture. See (6, Sections 8–13) for the details.

References

1. M. Artin and J. T. Stafford, *Noncommutative graded domains with quadratic growth*, Invent. Math. **122** (1995), 231–276.
2. M. Artin, J. Tate, and M. Van den Bergh, *Some algebras associated to automorphisms of elliptic curves*, The Grothendieck Festschrift, vol. 1, Birkhäuser, 1990, pp. 33–85.
3. M. Artin and M. Van den Bergh, *Twisted homogeneous coordinate rings*, J. Algebra **133** (1990), 249–271.
4. A. I. Bondal and A. E. Polishchuk, *Homological properties of associative algebras: the method of helices*, Russian Acad. Sci. Izv. Math **42** (1994), 219–260.
5. D. S. Keeler, *Criteria for σ-ampleness*, J. Amer. Math. Soc., **13** (2000), 517-532.
6. J. T. Stafford and M. Van den Bergh, *Noncommutative projective curves and surfaces, Bull. Amer. Math. Soc.,* **38** *(2001), 171-216.*
7. M. Van den Bergh, *Existence theorems for dualizing complexes over noncommutative graded and filtered rings*, J. Algebra, **195** (1997), 662-679.
8. M. Van den Bergh, *Blowing-up of noncommutative smooth surfaces*, Mem. Amer. Math. Soc., to appear; math.QA/980911, 1998.
9. M. Van den Bergh, *Noncommutative quadrics*, in preparation.
10. A. Yekutieli and J. J. Zhang, *Serre duality for noncommutative projective schemes*, Proc. Amer. Math. Soc., **125** (1997), 697-707.

BRAID GROUPS AS SELF-EQUIVALENCES OF DERIVED CATEGORIES

ALEXANDER ZIMMERMANN *
(alexander.zimmermann@u-picardie.fr)
Faculté de Mathématiques,
Université de Picardie Jules Verne
33 rue St Leu,
F-80039 Amiens, FRANCE
http://delambre.mathinfo.u-picardie.fr/alex/azim.html

Abstract. We shall report about equivalences of derived categories for module categories and in a parallel setting for coherent sheaves on algebraic varieties. The group of self-equivalences of these derived categories is introduced. The braid group of an affine Dynkin diagram of type $\tilde{A}_2$ maps to the group of auto-equivalences of the derived category of the group ring of the alternating group of degree 4 over any extension of of the 2-adic integers containing a third root of unity. We study the action of the self-equivalence group on the group cohomology, first in general, then for our example and give some consequences.

1. Introduction

Derived categories of algebras become more and more the appropriate working place of representation theorists. In the representation theory of groups this interest arose from various places notably from Broué's conjecture which states in its simplest form that for a big enough field k of characteristic p the derived categories of the principal block of the group ring kG of the finite group G and that of the principal block of the group ring $kN_G(P)$ of the normalizer

* These notes are an extended version of the talk I gave at the Euroconference "Rings, Modules and Representations" in Constanţa in August 2000. I acknowledge support by the CCCI.

K.W. Roggenkamp and M. Ştefănescu (eds.), Algebra - Representation Theory, 419–438.

$N_G(P)$ of an abelian p-Sylow subgroup P of G are equivalent as triangulated categories. Though this is our main motivation at the moment, there are more situations which we would like to recall. There is the development of character sheaves which was pushed very far by Lusztig, Shoji and others to describe generically the representation theory of algebraic groups. Moreover, many relationships between commutative or non commutative geometries to the representation theory of algebras has been discovered. This relationship is made precise and can be formulated by an equivalence between various corresponding derived categories.

In these notes we shall be concerned with consequences of the established or conjectured equivalences. In particular we work on symmetries of the representation theory which are implied by the existence of the derived category in the background. To be more precise for R a commutative ring we shall study the group of self-equivalences of the derived category of an R-algebra which is projective as an R-module. This group often is quite rich as we established in joint work with Raphaël Rouquier in (RZ-96). In fact, braid groups seems to play an important rôle in the structure of these groups. Parallel developments by Khovanov, Lenzing, Meltzer, Miyachi, Polishchuk, Seidel, Thomas, Yekutieli and others also indicate in parts that for nice situations braid group symmetries arise.

In the present note we shall review some of these developments. Furthermore we shall compute the case of the principal block of the group ring of a group with Klein Four Sylow subgroup. There, a generalized braid group of type $\tilde{A}_2$ occurs. Moreover, we explain how these self-equivalences of the derived category of a principal block of a group ring RG act on the cohomology ring $H^*(G, R)$ in a functorial way. Furthermore, we study this action for the case of the Klein Four Sylow subgroup, and in particular for the alternating group $\mathfrak{A}_4$.

2. Generalities on equivalences between derived categories

We recall some facts about equivalences between derived categories as is developed in Rickard (Ric-89, Ric-91, Ric-96) and Keller (Ke-93). For the notations we refer to (KZ-98).

2.1. THE ALGEBRAIC SETUP

Jeremy Rickard proved in his fundamental work a necessary and sufficient criterion for two derived categories of rings to be equivalent. In the situation we study here this criterion is as follows.

THEOREM 1 (Jeremy Rickard (Ric-91), Bernhard Keller (Ke-93)). *Let S be a commutative ring and let A be a finite dimensional S-algebra which is projective as S-module. Then, in case B is another S-algebra and $D^b(A) \simeq D^b(B)$ as triangulated categories, there is a complex $X \in D^b(B \otimes_S A^{op})$ so that $F_X := X \otimes^{\mathbb{L}}_A -$ is a functor inducing an equivalence.*

We call functors *standard* if they are of the form F_X and if they induce an equivalence $D^b(A) \simeq D^b(B)$. In case F_X is a standard equivalence, we call X a *twosided tilting complex.*

(2.1) Recall that an S-algebra A for an integral domain S with field of fractions K is an S-*order* if first A is finitely generated projective as S-module and second $K \otimes_S A$ is a semisimple K-algebra. An S-algebra A is symmetric provided $Hom_S(A, S) \simeq A$ as $A \otimes_S A^{op}$-modules.
Suppose now S is a field and A an arbitrary S-algebra or S is a Dedekind domain and A is a symmetric S-order. Then if $D^b(A) \simeq D^b(B)$, also B is projective as S-module (Z-99a) and a quasi-inverse of F_X is standard as well.

(2.2) We keep the assumption that A is an S-projective S-algebra. We define (RZ-96)

$$TrPic_S(A) := \{ \text{ isomorphism classes } [X] \mid \\ X \in D^b(A \otimes_S A^{op}) \text{ is a twosided tilting complex}\}.$$

Then, $TrPic_S(A)$ is a group under the group law $[X] \cdot [Y] := [X \otimes^{\mathbb{L}}_A Y]$. This group contains the classical Picard group of A (cf (Ba-68)), with it also the group of automorphisms of A modulo inner automorphisms of A, and an infinite cyclic central subgroup generated by shift in degrees.

2.2. THE GEOMETRIC SETUP

There is a geometric analogue to this construction. Let X and Y be algebraic varieties of the same dimension and let $D^b(X)$ be the derived category of bounded complexes of coherent sheaves on X and likewise $D^b(Y)$ be the derived category of bounded complexes of coherent sheaves on Y.

(2.3) Let P be a bounded complex of coherent sheaves on $X \times Y$. Denote by p_X the projection of $X \times Y$ on X and by p_Y the projection of $X \times Y$ on Y. Then, in case

$$F_P(-) := (p_X)_*(P \otimes^{\mathbb{L}}_{\mathcal{O}_{X\times Y}} (p_Y)^* -) : D^b(Y) \longrightarrow D^b(X)$$

is an equivalence of categories, this equivalence is called ((M-81, Brid-99, P-95)) *Fourier-Mukai transformation.* If X is a point, then a Fourier-Mukai transformation is a standard equivalence.

(2.4) Tom Bridgeland gave in (Brid-99) some partial analogue to Rickard's theorem. For this suppose that X and Y are smooth projective varieties of the same dimension and that P is a vector bundle on $X \times Y$. The bundle P is called strongly simple over Y if for each $y \in Y$ the bundle P_y on X is simple and if $Ext^i_X(P_{y_1}, P_{y_2}) = 0$ for any $(y_1, y_2) \in Y \times Y$ and any i.

THEOREM 2 (Tom Bridgeland (Brid-99)). *The functor F_P is fully faithful if and only if P is strongly simple over Y. The functor F_P is an equivalence precisely when one also has $P_y \simeq P_y \otimes \omega_X$ for all $y \in Y$.*

Theorem 2 is some sort of analogue to the theorem of Rickard and Keller in case the twosided tilting complex is just a module. As far as I know there is no general theory like Rickard's for the geometric situation.

3. Some results on braid groups

We recall briefly some facts about braid groups.

(3.1) Let W be a real reflection group generated by orthogonal reflections $\{s_1, \ldots, s_n\} =: S$ on some hyperplanes $\{H_1, \ldots H_n\}$ in

some real euclidean space E. We associate to (W,S) a graph $\Gamma_{(W,S)}$. The vertices of $\Gamma_{(W,S)}$ are the elements of S. There is an edge with weight $m_{i,j}$ between s_i and s_j if and only if s_i and s_j do not commute and the order of s_is_j is $m_{i,j}+2$.
It is well known that W is finite if and only if $\Gamma_{(W,S)}$ is a Dynkin diagram.

(3.2) Now, W acts on the arrangement of hyperplanes and also on the complement. It is a result of Brieskorn (Brie-71) that $\pi_1((\mathbb{C}\otimes_{\mathbb{R}}E\backslash\bigcup_{i=1}^{n}\mathbb{C}\otimes_{\mathbb{R}}H_i)/W) =: Braid(\Gamma_{(W,S)})$ is a group with presentation $Braid(\Gamma_{(W,S)}) = < t_1,\ldots,t_n | \underbrace{t_it_jt_i\ldots}_{m_{i,j}+2 \text{ factors}} = \underbrace{t_jt_it_j\ldots}_{m_{i,j}+2 \text{ factors}} >$.
The case of Dynkin diagram $\Gamma_{(W,S)} = A_e$ is due to Arnold while the other cases are due to Brieskorn.

(3.3) This item was communicated to me by Claudio Procesi. We shall be concerned in particular with the braid group associated to the Dynkin diagram A_2 (which is just a graph with one edge and 2 vertices) and the affine Dynkin diagram $\tilde{A}_2$ (which is the unique graph with three edges and three vertices, each vertex being adjacent to two edges).
Then

$$Br(A_2) = < s,t |\ sts = tst >$$

and

$$Br(\tilde{A}_2) = < s_1,s_2,s_3 |\ s_is_js_i = s_js_is_j \forall\ i,j \in \{1,2,3\} > .$$

The mappings $Br(A_2) \stackrel{\alpha}{\to} Br(\tilde{A}_2)$ defined by $\alpha(s) = s_1$ and $\alpha(t) = s_2$ as well as the mapping $Br(\tilde{A}_2) \stackrel{\beta}{\longrightarrow} Br(A_2)$ given by $\beta(s_1) = s$, $\beta(s_2) = t$ and $\beta(s_3) = t$ are group homomorphisms. We see immediately $\beta\circ\alpha = id_{Br(A_2)}$. Hence, $Br(\tilde{A}_2) = ker\beta \rtimes Br(A_2)$.

(3.4) Independently of the existence of a Weyl group we define for any weighted graph Γ a generalized braid group $Braid(\Gamma)$. The case of complex reflection groups (W,S) is studied by Broué, Malle, Rouquier (BMR-98) and even more involved types of diagrams and associated braid groups have been introduced.

4. Some results on self-equivalences of derived categories

4.1. THE ABELIAN SITUATION

The group $TrPic_S(A)$ was first studied in (RZ-96) (see also (Z-96)). As one of the main results we got the following description.

THEOREM 3 (R. Rouquier and A. Zimmermann (RZ-96)). *Let k be a prime field and let B_e be an indecomposable Brauer tree algebra with no exceptional vertex for a Brauer tree with e edges. Then, there is a group homomorphism $\varphi_e : Braid(A_e) \longrightarrow TrPic_k(B_e)$. Moreover, φ_2 is injective and the image of φ_2 is normal of index 8. For any $e > 1$ the image of a standard generator is infinite cyclic.*

(4.1) Injectivity is proven by identifying

$$Braid(A_2)/Z(Braid(A_2)) \simeq PSL_2(\mathbb{Z})$$

and one then identifies the images of projective indecomposable modules with the matrix coefficients of elements in $PSL_2(\mathbb{Z})$. Actually, let P_+ and P_- be the two indecomposable projective B_2-modules. For $[X] \in TrPic_k(B_2)$ let $T_+ := X \otimes_{B_2} P_+$ and $T_- := X \otimes_{B_2} P_-$. Replace T_+ by the representative in the isomorphism class of T_+ with the smallest k-dimension and likewise for T_-. Then, $T_+ = P_+^{n+} \oplus P_-^{n-}$ and $T_- = P_+^{m+} \oplus P_-^{m-}$ as modules, forgetting the differentials. Then one proves that the image of X in $PSL_2(\mathbb{Z})$ is the matrix $\begin{pmatrix} \pm n_+ & \pm m_+ \\ n_- & m_- \end{pmatrix}$. This description immediately gives the statement.

Surjectivity is proven by using that there are only *two* simple modules up to isomorphism and any complex has *two* ends. For more ample details the reader should consult (RZ-96).

(4.2) There is some analogue for orders over a complete discrete valuation domain. There, so-called Green orders take the rôle of Brauer tree algebras and one gets the same situation but replacing B_e by the corresponding order Λ_e. The interest is that tensoring with the field of fractions gives a very different situation to tensoring obtained by tensoring with the residue field. Some nice arithmetic structure comes into the play. For more ample details the reader might consult (Z-00a).

(4.3) Later, independently, Mike Khovanov and Paul Seidel discovered this homomorphism φ_e.

THEOREM 4 (M. Khovanov,P. Seidel (KS-00, ST-00, HK-00)). *φ_e is injective for all $e > 2$.*

A more general class of algebras, so-called zig-zag algebras, is covered. The proof is very involved and uses sophisticated methods related to Mirror symmetry. In particular one of the key steps is to establish an equivalence of categories between the derived category of the algebra under consideration and some object coming from symplectic geometry. Using Floer cohomology the authors are able to prove injectivity.

(4.4) In the geometrical setting we get some braid group picture as well.

THEOREM 5 (Alexander Polishchuk (P-95)). *Let A be a connected abelian variety of dimension g over a field k endowed with a principal symmetric polarisation. Then, $Braid(A_2)$ acts on $D^b(A)$ by Fourier-Mukai transformations.*

The symmetric polarisation is used to identify the variety with its dual. The construction goes along the following lines. One finds some sort of involution and a complex inducing a Fourier-Mukai transformation. Conjugating the complex by the involution one gets a second complex and these two complexes are the braid group generators.

There is one more example in this direction using Fourier-Mukai transformation actions on derived categories.

THEOREM 6 (Raphaël Rouquier (Rou-98)). *Let G be a semisimple complex algebraic group with Weyl group W and let $\mathcal{B}$ be the variety of Borel subgroups of G. Then, the braid group B_W associated to W acts on $D^b(\mathcal{B})$ by Fourier-Mukai transformations.*

4.2. NON ABELIAN STATEMENTS

(4.5) Amnon Yekutieli is interested in non communicative geometry and was interested for this reason in algebras with finite global dimension.

THEOREM 7 (Jun-Ichi Miyachi and Amnon Yekutieli (MY-99)). *Let A be a hereditary basic finite dimensional k-algebra of finite representation type with quiver Δ. Then, there is a group isomorphism $TrPic_k(A) \simeq Aut_k(\overrightarrow{\mathbb{Z}}\Delta)^{\tau}$.*

Here, $\tau = A^*[-1] \otimes_A^{\mathbb{L}} -$ and $\overrightarrow{\mathbb{Z}}\Delta$ is a quiver obtained by repetition of Δ (Ried-80). By this, Miyachi and Yekutieli get explicitly the case of Δ being a Dynkin diagram. No braid group arises, though the groups coming up are all of a nice algebraic nature.

(4.6) Helmut Lenzing and Hagen Meltzer study all kinds of properties for a rather special class of algebras, so-called canonical algebras. These are algebras of finite global dimension and are defined by quivers and relations. Moreover, these algebras can serve as a model for a 'weighted projective line'. For details the reader may consult Geigle and Lenzing (GL-87). Recently these algebras came into play by the attempt of Happel and Reiten to classify all abelian categories which are derived equivalent to a hereditary one. Canonical algebras play an important rôle in the list of categories they expect to get. Here is one of the main results on the group of self-equivalences of the derived category of these algebras.

THEOREM 8 (Meltzer and Lenzing (LM-00)). *Let X be a weighted projective line of tubular type. Then the group of isomorphism classes of self-equivalences of $D^b(X)$ is isomorphic to the group $Braid(A_2) \rtimes (Pic_0(X) \rtimes AutX)$.*

For details on the definitions we refer to (GL-87). It is interesting to see that also in this case the generators for the braid group part are analogous to the case studied in (RZ-96) for Brauer tree algebras with two simple modules.

(4.7) It is surprising to get the braid group also in this non abelian situation. It seems that this happens since the assumption of the weighted projective line to be tubular has some regularizing effect, and so the situation becomes "almost abelian". In fact for general types the braid groups do not arise and actually the derived category in this case has only relatively few symmetries.

5. Action on the cohomology ring

Let G be a finite group and let R and be a complete discrete valuation ring of characteristic 0 with residue field of characteristic p. In fact in order to establish the theory that follows the assumptions are much too strict on R but the hypotheses on R as above simplify a lot the presentation. Let S be any commutative noetherian ring.

(5.1) In order to study the group $TrPic_R(B_0(RG))$ we shall look for natural modules. Since for any two SG-modules M and N we have

$$Ext^n_{SG}(N,M) \simeq Hom_{D^b(SG)}(N,M[n])$$

it is natural to consider $M = N$ and to look at the subgroup of $TrPic_S(SG)$ that fixes the module M. Then, morally at least, this group should act on $Ext^*_{SG}(M,M)$ in a multiplicative way. Since $H^*(G,S) = Ext^*(S,S)$ is the cohomology ring of the group G with values in S, we are particularly interested in the case $M = S$. Denote

$$HD_S(G) := \{[X] \in TrPic_S(SG) \mid X \otimes_{SG} S \simeq S\}$$

It is reasonable to restrict in case of $S = R$ to the derived category of the principal block. We shall do this without mentioning and hope that this will not produce any confusion.
To get an action of $HD_S(G)$ on $H^*(G,S)$ there are many obstacles on the way. E.g. there are many isomorphisms we choose on the way, the functors are equivalences and not isomorphisms of categories to cite only a few. But, we get the following first result.

THEOREM 9 ((Z-99b)). *Keep the assumptions of this section. Then, the cohomology ring $H^*(G,S)$ is an $S\,HD_S(G)$-module. This action is compatible with the ring structure of $H^*(G,S)$.*
If $S_1 \longrightarrow S_2$ is a ring homomorphism, then this ring homomorphism induces a group homomorphism $HD_{S_1}(G) \longrightarrow HD_{S_2}(G)$

and this way the ring homomorphism $H^*(G, S_1) \longrightarrow H^*(G, S_2)$ is $S_1\ HD_{S_1}(G)$-linear.

The proof is not only a checking of being well defined, but an actual construction of a bigger group acting properly, and then seeing that the extension that had to be considered actually enlarges the group by something in the kernel of the action.

(5.2) What do we know with respect to functoriality of this structure in the first variable? Here we have to face the problem that derived equivalences behave very badly with respect to subrings, quotient rings, extension rings or similar constructions. There is one exception, the so-called splendid tilting complexes introduced by Jeremy Rickard (Ric-96).

DEFINITION 5.1. **(Jeremy Rickard (Ric-96))** Let G and H be finite groups with a common p-Sylow subgroup P and let R be a complete discrete valuation domain of characteristic 0 with residue field of characteristic p or a field of characteristic p. Denote by ΔP a fixed diagonal embedding of P in $G \times H$. Let X be a bounded complex of $B_0(RG) \otimes_R B_0(RH)^{op}$ modules so that

- Each homogeneous component of X is projective as RG-module and projective as RH-module,
- Each homogeneous component of X is a ΔP-projective p-permutation module (that is a direct factor of a permutation module),
- $Hom_{RG}(X, X)$ is *homotopic* to $B_0(RH)$ as complex of bimodules and $Hom_{RH}(X, X)$ is *homotopic* to $B_0(RG)$ as complex of bimodules.

(5.3) I do not know any twosided tilting complex between two principal blocks of group rings which cannot be modified by a Morita equivalence so that the modified complex is splendid.

(5.4) For splendid tilting complexes Rickard proves that the Brauer construction induces equivalences on the local levels. We shall explain this statement below.
Let G be a finite group and let k be a field of characteristic p. Then, for any kG-module M and a p-subgroup U of G define the

$kN_G(U)$-module $M(U) := M^U / \sum_{V<U} Tr_V^U(M^V)$. This construction, the Brauer construction, is functorial and carries over to the homotopy category. It is the right environment to restrict splendid derived equivalences to centralizers of subgroups.

THEOREM 10 (Jeremy Rickard (Ric-96)). *Let X be a splendid tilting complex in $D^b(B_0(kG)\otimes_k B_0(kH))$. Then, for any p-subgroup Q of G the complex $X(\Delta Q)$ is a splendid tilting complex in $D^b(B_0(kC_G(Q)) \otimes_k B_0(kC_H(Q)))$.*

Moreover, Rickard proves that for any splendid tilting complex $\overline{X}$ in $D^b(B_0(kG) \otimes_k B_0(kH))$ there is an, up to isomorphism, unique splendid tilting complex X in $D^b(B_0(RG) \otimes_R B_0(RH))$ with $X \otimes_R k \simeq \overline{X}$.

(5.5) Denote by $[X]$ the isomorphism class of a complex X *in the derived category* and denote by (X) the isomorphism class of a complex X *in the homotopy category*. Note that being homotopy equivalent is stronger than being quasi-isomorphic.

(5.6) It is natural to define (Z-01)

$$SplenPic_R(G) := \{(X) \mid X \text{ is a splendid tilting complex in } D^b(B_0(RG) \otimes_R B_0(RG))\}$$

and similarly for k as coefficient domain. It is readily verified that this in fact is a group under tensor product. It is clear that we get a group homomorphism $SplenPic_R(G) \longrightarrow TrPic_R(B_0(RG))$ by setting $\nu(\ (X)\ ) := [X]$. The same notation is used for k as coefficient domain.
In view of the action on the cohomology ring we define

$$HSplen_R(RG) := \nu^{-1}(HD_R(G)) \cap SplenPic_R(RG)$$

and similarly for k instead of R.

(5.7) The action of $HSplen_R(G)$ on $H^*(G,R)$ is via φ though. So, working up to homotopy equivalence only or working up to quasi-isomorphism does not make any difference for the action on the cohomology ring of the group.

(5.8) Why do we like to work up to homotopy equivalence only instead of up to quasi-isomorphism? The reason is the following result.

PROPOSITION 5.1 ((Z-01)). *Let k be a field of characteristic p, let G be a finite group and let Q be a p-subgroup of G. Then, the Brauer construction provides a group homomorphism* $SplenPic_k(G) \longrightarrow SplenPic_k(C_G(Q))$

If one would work over isomorphism classes in the derived category the Brauer construction is not necessarily well defined. Actually there is no obvious reason why $[B_0(kG)] = [X]$ implies $[B_0(kC_G(Q)))] = [X(\Delta Q)]$.

(5.9) Let us study the obstruction. The equation $[B_0(kG)] = [X]$ implies that there is a quasi-isomorphism $B_0(kG) \overset{s}{\longrightarrow} X$ (and not necessarily in the other direction) since the homology of X is concentrated in degree 0 and is isomorphic to $B_0(kG)$. The mapping cone of s is therefore acyclic (i.e. with homology 0). Form the mapping cone by the standard construction

$$C(s) := (B_0(kG)[1] \oplus X, \begin{pmatrix} d_{B_0(kG)} & s \\ 0 & d_X \end{pmatrix}) \, .$$

Then, it is easily verified that $C(s(\Delta Q)) = (C(s))(\Delta Q)$. In order to define the mapping in Proposition 5.1 in the derived category we have to prove that the Brauer functor applied to an acyclic complex is acyclic. Actually this is slightly stronger than what is needed since not every acyclic complex is the mapping cone of a quasi-isomorphism as above. Nevertheless, we shall give an example showing that it is not correct that the Brauer construction applied to an acyclic complex of p-permutation modules is acyclic.

(5.10) Let k be the field with 2 elements. The sequence

$$0 \longrightarrow k \longrightarrow kC_2 \longrightarrow k \longrightarrow 0$$

coming from the augmentation of the cyclic group of order 2 is exact. Moreover, this exact sequence can be seen as acyclic complex of kC_2-permutation modules with three non zero terms. Applying the Brauer functor $-(C_2)$ to this complex gives

$$(\dots \longrightarrow 0 \longrightarrow k \longrightarrow 0 \longrightarrow k \longrightarrow 0 \longrightarrow \dots)$$

since $kC_2(C_2) = 0$ and $k(C_2) = k$. Obviously, this complex is not acyclic.

THEOREM 11 ((Z-01, Z-00b)). *Let k be a field of characteristic p, let G be a finite group, P a p-Sylow subgroup and let Q be a p-subgroup of G. Then, for any $[X] \in HSplen_k(G)$ so that $X(\Delta Q) \in HSplen_k(C_G(Q))$, we get $HSplen_R(G) \simeq HSplen_k(G)$ and the diagrams*

$$\begin{array}{ccc} H^*(G,k) & \stackrel{X\otimes -}{\longrightarrow} & H^*(G,k) \\ res^G_{C_G(Q)} \downarrow & & res^G_{C_G(Q)} \downarrow \\ H^*(C_G(Q),k) & \stackrel{X(\Delta Q)\otimes -}{\longrightarrow} & H^*(C_G(Q),k) \end{array}$$

and

$$\begin{array}{ccc} H^*(G,k) & \stackrel{X\otimes -}{\longrightarrow} & H^*(G,k) \\ \uparrow trans^G_{C_G(Q)} & & \uparrow trans^G_{C_G(Q)} \\ H^*(C_G(Q),k) & \stackrel{X(\Delta Q)\otimes -}{\longrightarrow} & H^*(C_G(Q),k) \end{array}$$

are commutative. In case P is abelian, an

$$X \in HSplen_k(G) \cap (-(\Delta P))^{-1}(HSplen_k(C_G(P)))$$

acts on $H^*(G,k)$ via automorphisms of P.

(5.11) It should be noted that the first point is not a trivial consequence of Rickard's result that $SplenPic_k(G) \simeq SplenPic_R(G)$. It has to be shown that the unique lift fixes the trivial module as well.
The second relation is essentially proved by first showing that $X(\Delta Q) \otimes_{C_G(Q)} k \simeq X^G(Q)$.
The third item uses the second and then as essential ingredient the theorem of Roggenkamp and Scott on the isomorphism problem for integral group rings of p-groups (RS-87, Rog-92).

6. The alternating group of degree 4

We shall give the group ring $R\mathfrak{A}_4$ for R being a complete discrete valuation ring of characteristic 0 and residue field of characteristic 2, so that R has a third root of unity. For this one may as well use the description in (N-98) of $R\,SL_2(4)$ by Gabriele Nebe using that $SL_2(4) \simeq \mathfrak{A}_5$ and the explicit tilting complex inducing a splendid equivalence between the principal block of $R\,SL_2(4)$ and $R\mathfrak{A}_4$ given by Jeremy Rickard in (Ric-96). Nevertheless, a more direct approach is possible as is shown below.

Let k be a field of characteristic 2 containing a third root of unity and let R be a complete discrete valuation ring of characteristic 0 containing a third root of unity and with residue field k of characteristic 2.

Since $\hat{Z}_2[\zeta_3]$ is a splitting field for $\mathfrak{A}_4$ and all of its subgroups, it is enough to compute $R\mathfrak{A}_4$ for R being the (unique) unramified extension of $\hat{\mathbb{Z}}_2$ of degree 2. Let K be the field of fractions of R.

Since $\mathfrak{A}_4 \simeq (C_2 \times C_2) \rtimes C_3$ K is a splitting field for $\mathfrak{A}_4$. We use a representation $A_4 \simeq < a, b, c : a^2, b^2, c^3, (a,b), a^c = ab, b^c = a >$.

Now, $\mathbb{Q}\mathfrak{A}_4$ is a direct product of a copy of $\mathbb{Q}$, a copy of $\mathbb{Q}[\zeta_3]$ and a 3×3 matrix ring over $\mathbb{Q}$. Moreover, $K \otimes_{\mathbb{Q}} \mathbb{Q}[\zeta_3] \simeq K \times K$.

We are interested, how $R\mathfrak{A}_4$ is embedded into $K\mathfrak{A}_4$. Over R, however, we get 3 indecomposable projective modules, over $\hat{\mathbb{Z}}_2$ we get 2 projective indecomposable modules. Corresponding to the irreducible characters χ_i, $i = 1, 2, 3$ of C_3, we get the idempotents $e_i := \frac{1}{3}\sum_{j=1}^{3} \chi_i(c^j)c^j$ Over $\hat{\mathbb{Z}}_2$ we get the two indecomposable idempotents χ_1 and $\chi_2 + \chi_3$. We know, that for $i \neq j$ $e_i R\mathfrak{A}_4 e_j$ is a non zero fractional R–ideal in K and for $i = j$ we know that $e_i R A_4 e_i$ is an R–order in $K \times K$.

The central idempotent $\delta := \frac{1}{4}(1+a+ab+b)$ in $K\mathfrak{A}_4$ corresponds to the natural projection onto KC_3. This induces a pullback diagram

$$\begin{array}{ccc} R\mathfrak{A}_4 & \longrightarrow & RC_3 \\ \downarrow & & \downarrow \\ RA_4 \cdot (1-\delta) & \longrightarrow & (R/4R)C_3 \end{array}.$$

$\Lambda := R\mathfrak{A}_4 \cdot (1-\delta)$ is an R-order in $\begin{pmatrix} K & K & K \\ K & K & K \\ K & K & K \end{pmatrix}$. We use the induced representation from a one dimensional character of the Klein Four group

$$a \longrightarrow \begin{pmatrix} -1 & 0 & 0 \\ 0 & 1 & 0 \\ 0 & 0 & -1 \end{pmatrix}, b \longrightarrow \begin{pmatrix} -1 & 0 & 0 \\ 0 & -1 & 0 \\ 0 & 0 & 1 \end{pmatrix} \text{ and } c \longrightarrow \begin{pmatrix} 0 & 1 & 0 \\ 0 & 0 & 1 \\ 1 & 0 & 0 \end{pmatrix}.$$

Then, we conjugate by $\begin{pmatrix} 1 & 1 & 1 \\ 1 & \zeta_3^2 & -\zeta_3 \\ 1 & -\zeta_3 & \zeta_3^2 \end{pmatrix}$ to get $\Lambda = \begin{pmatrix} R & 2R & 2R \\ 2R & R & 2R \\ 2R & 2R & R \end{pmatrix}$.

After some computation we get

$$R\mathfrak{A}_4 \simeq \{(a_1, a_2, a_3, (b_{i,j})_{1\le i,j\le 3}) \in R \times R \times R \times \Lambda \mid \\ b_{i,i} - a_i \in 4R\ \forall 1 \le i \le 3\}$$

7. The braid group generators

We keep the hypotheses of the beginning of section 6. The group ring $R\mathfrak{A}_4$ has three projective indecomposable modules P_1, P_2, P_3 up to isomorphism. Denote by P_1 the projective cover of the trivial module. Note that if R does not have a third root of unity, then there are only two projective indecomposable modules. The discussion in that case differs considerably of what we shall develop here.

Denote in order not to overload the notation by A the group ring $R\mathfrak{A}_4$. We denote by S_i the complex

$$\dots \longrightarrow 0 \longrightarrow P_i \otimes P_i^* \longrightarrow A \longrightarrow 0 \longrightarrow \dots$$

and the functors $F_i := S_i \otimes_A -$ for any $i \in \{1, 2, 3\}$. The following result was known to Joe Chuang [1] for k as coefficient domain as action on the simple modules. We shall include a proof for R and this implies at once the result for k.

PROPOSITION 7.1. *$S_i \otimes_A S_j \otimes_A S_i \simeq S_j \otimes_A S_i \otimes_A S_j$ for any pair $(i, j) \in \{1, 2, 3\} \times \{1, 2, 3\}$.*

Proof. It is obvious by the symmetry of $R\mathfrak{A}_4$ that it is enough to prove the statement for $i = 1$ and $j = 2$. We compute

$$S_1 \otimes P_1 \simeq P_1 \longrightarrow 0, \quad S_1 \otimes P_2 \simeq P_1 \longrightarrow P_2 \text{ and } S_1 \otimes P_3 \simeq P_1 \longrightarrow P_3$$

where the right most position for the complexes on the right of the isomorphism is in degree 0. Analogous results hold for S_2 instead of S_1. Then, $S_2 \otimes_A S_1 \otimes_A P_1 \simeq S_2 \otimes_A P_1[1] \simeq P_2 \longrightarrow P_1 \longrightarrow 0$ again with the convention that the right hand side of the complex on the right is meant to be in degree 0.

$$S_2 \otimes_A S_1 \otimes_A P_2 \simeq F_2(cone(P_1 \to P_2)) \simeq cone(F_2(P_1) \to F_2(P_2)) \simeq P_1[1]\ .$$

[1] In September 1999 in a private communication

Likewise

$$S_2 \otimes_A S_1 \otimes_A P_3 \simeq F_2(cone(P_1 \longrightarrow P_3)) \simeq P_2 \longrightarrow P_1 \oplus P_2 \longrightarrow P_3 .$$

Furthermore, $F_1(P_2 \longrightarrow P_1 \longrightarrow 0) \simeq P_2[2]$ and $F_1(P_1[1]) \simeq P_1[2]$. Compute the image of P_3:

$$\begin{aligned} F_1(F_2(F_1(P_3))) &\simeq F_1(cone((P_2 \to P_1) \to (P_2 \to P_3))) \\ &\simeq cone(F_1(P_2 \to P_1) \to F_1(P_2 \to P_3)) \\ &\simeq cone(P_2[1] \to (P_2 \longrightarrow P_1 \oplus P_2 \to P_3)) \\ &\simeq \ldots \to 0 \begin{array}{c} \nearrow P_1 \longrightarrow P_2 \searrow \\ \searrow P_2 \longrightarrow P_1 \nearrow \end{array} P_3 \longrightarrow 0 \longrightarrow \ldots \end{aligned}$$

where in the final complex the differential in degree 2 maps the projective P_1 in degree 2 to the projective P_2 in degree 1 and the projective P_2 in degree 2 to the projective P_1 in degree 1.
In any case the result is invariant under changing indices 1 and 2. By (RZ-96) this proves that $F_1F_2F_1$ differ from $F_2F_1F_2$ only by an automorphism of $R\mathfrak{A}_4$. As is easily seen from the structure of $R\mathfrak{A}_4$ in section 6, apart from the identity there is no R-linear automorphism which fixes each projective indecomposable up to isomorphism. We proved the statement. ∎

COROLLARY 7.1. *Let $Br(\tilde{A}_2)$ be the braid group associated to the affine Dynkin diagram $\tilde{A}_2$. Then, there is a group homomorphism $Br(\tilde{A}_2) \longrightarrow SplenPic_R(\mathfrak{A}_4)$.*

It is clear that the group generated by F_2 and F_3 is the image under the above homomorphism of an Artin braid group on three strings. Moreover, these two self-equivalences stabilize the trivial module.

8. Studying the action on the group cohomology

(8.1) The outer automorphism group of $\mathfrak{A}_4$ is isomorphic to the cyclic group of order 2 since any automorphism of $\mathfrak{A}_4$ induces an automorphism of $\mathfrak{S}_4$ where each automorphism is inner. Moreover (A-M 94, Chapter III Theorem 1.3),

$$H^*(\mathfrak{A}_4, \mathbb{F}_2) = \mathbb{F}_2[A, B](C)/(C^3 + A^2 + B^2 + AB)$$

where $A = f_1^3 + f_2^3$, $B = \zeta f_1^3 + \zeta^2 f_2^3$ and $C = f_1 f_2$. Here, ζ is a primitive third root of unity over $\mathbb{F}_2$. The elements f_1 and f_2 are the two canonical generators of $H^*(C_2^2, \mathbb{F}_2)$ in degree 1.
The group of automorphisms of C_2^2 is $GL_2(2) \simeq \mathfrak{S}_3$. The automorphisms of order 3 are inner in $\mathfrak{A}_4$ and induce hence the identity on the subring $H^*(\mathfrak{A}_4, \mathbb{F}_2)$ of $H^*(C_2^2, \mathbb{F}_2)$. It is an easy and straightforward calculation that the elements of $GL_2(2)$ of order 2 act trivially on degree 2 cohomology and act on degree 3 cohomology as regular representation ($A \longrightarrow A$ and $B \longrightarrow -A - B$).

(8.2) Carlson and Rouquier proved (CR-00) that for a group $G = P \rtimes Q$ where P is an abelian p-group and Q is a cyclic p'-group, the group $StPic_{\mathbb{F}_p}(\mathbb{F}_p G)$ of self-equivalences of Morita type of the stable category is generated by the syzygy operator and the usual Picard group of $\mathbb{F}_p G$. Hence, the action of $HD_{\mathbb{F}_p}(G)$ on $H^*(G, \mathbb{F}_p)$ is via automorphisms of $\mathbb{F}_p G$. In case $G = \mathfrak{A}_4$ and $p = 2$, the action described above is the only possible non trivial action of $HSplen_{\mathbb{F}_2}(\mathfrak{A}_4)$ on $H^*(\mathfrak{A}_4, \mathbb{F}_2)$.

(8.3) Recall that $< F_2, F_3 >$, which is the homomorphic image of an Artin braid group, is in $HSplen_R(R\mathfrak{A}_4)$. By (Z-00b) we conclude in our case of abelian Sylow subgroups that $HSplen_R(\mathfrak{A}_4)$ acts by automorphisms of the Sylow subgroup on the cohomology. Since for a splendid tilting complex X one has $X(\Delta Q) \otimes_{C_G(Q)} R \simeq X^G(Q)$ (where G acts on the right and Q on the left), one sees easily that any element of $< F_2(\Delta C_2^2), F_3(\Delta C_2^2) >$ fixes the trivial module as well. Moreover, it is easy to see that $< F_2, F_3 >$ acts trivially an $H^*(\mathfrak{A}_4, \mathbb{F}_2)$ since this action factors via the action of the group of stable self-equivalences (see (Z-99b)) and there S_2 and S_3 become just isomorphic to $R\mathfrak{A}_4$. So, we proved that $< F_2, F_3 >$ does not contain $Out(\mathfrak{A}_4)$.
Denoting by γ the element (of order 2) in $Out(R\mathfrak{A}_4)$ given by conjugation by an element in $\mathfrak{S}_4 \setminus \mathfrak{A}_4$, we get

$$< F_2, F_3, \gamma > = < F_2, F_3 > \rtimes C_2 \leq HD_R(\mathfrak{A}_4) .$$

(8.4) We observe that $F_2(P_2) \simeq P_2[1]$ and $F_2(P_3) \simeq (P_2 \longrightarrow P_3)$ with homology in degree 0 and 1, and likewise for F_3 with the rôles of P_2 and P_3 interchanged. The very same proof as in (RZ-96,

Theorem 4.6) shows that the morphism

$$Br(A_2) \longrightarrow < F_2, F_3 > \subseteq HSplen_{\mathbb{F}_2}(\mathfrak{A}_4)$$

is injective.

(8.5) Since the symmetric group of degree 3 acts naturally on the three generators of $Br(\tilde{A}_2)$ and by the fact that the relations obey the same symmetry also on $Br(\tilde{A}_2)$, we get

$$\begin{array}{ccccc}
Br(\tilde{A_2}) \rtimes \mathfrak{S}_3 & \simeq & (ker\beta \rtimes Br(A_2)) \rtimes (C_3 \rtimes C_2) & \longrightarrow\!\!\!\!\rightarrow & Br(A_2) \rtimes C_2 \\
\downarrow & & & & \cap\!\!\downarrow \\
< F_1, F_2, F_3 > \rtimes \mathfrak{S}_3 & & & & < F_2, F_3 > \rtimes < \gamma > \\
\cap & & & & \cap \\
TrPic_R(R\mathfrak{A}_4) & & & & HD_R(\mathfrak{A}_4)
\end{array}$$

(8.6) May it be that derived categories "in an abelian situation" have generalized braid group symmetries? I believe that the above observations are not just an accident.

References

A-M 94. A. Adem and R. J. Milgram, Cohomology of finite groups, Springer Verlag 1994

Ba-68. H. Bass, Algebraic K-theory, Benjamin, New York 1968

Brid-99. T. Bridgeland, *Equivalences of triangulated categories and Fourier-Mukai Transformations*, Bull. London Math. Soc. **31** (1999) 25-34.

Brie-71. E. Brieskorn, *Die Fundamentalgruppe des Raumes der regulären Orbits einer endlichen komplexen Spiegelungsgruppe*, Invent. Math. **12** (1971) 37-61.

BMR-98. M. Broué, G. Malle and R. Rouquier, *Complex Reflection Groups, Braid Groups and Hecke Algebras*, Journal reine angew. Mathematik **500** (1998) 127-190.

CR-00. J. F. Carlson and R. Rouquier, *Self-equivalences of stable module categories*, Math. Z. **233** (2000) 165-178.

GL-87. W. Geigel and H. Lenzing, *A class of weighted projective lines arising in representation theory of finite dimensional algebras*, Springer Lecture Notes in Mathematics **1273** (1987) 265-297.

HK-00. R. Huerfano and M. Khovanov, *A category for the adjoint representations*, preprint

Ke-93. B. Keller, *A remark on tilting theory and DG-algebras*, manuscripta mathematatica **79** (1993) 247-253.

KS-00. M. Khovanov and P. Seidel, *Quivers, Floer Cohomology and Braid Group Actions*, preprint

KZ-97. S. König and A. Zimmermann, *Tilting selfinjective algebras and Gorenstein orders*, Oxford Quarterly Journal of Mathematics (2) 48 (1997) 351-361.

KZ-98. S. König and A. Zimmermann, "Derived equivalences for group rings (with Contributions by B. Keller, M. Linckelmann, J. Rickard and R. Rouquier" , Lecture Notes in Math. 1685, Springer Verlag, Berlin 1998.

LM-00. H. Lenzing and H. Meltzer, *The automorphism group of the derived category for a weighted projective line*, Communications in Algebra **28** (2000) 1685-1700.

MY-99. J.-I. Miyachi and A. Yekutieli, *Derived Picard groups of finite dimensional hereditary algebras*, preprint

M-81. S. Mukai, *Duality between $D(X)$ and $D(\hat{X})$ with its application to Picard sheaves,* Nagoya Mathematical Journal **81** (1981) 153-175.

N-98. G. Nebe, *The group ring of $SL_2(p^2)$ over the p-adic integers,* J. Algebra **210** (1998) 593-613.

P-95. A. Polishchuk, *A remark on the Fourier Mukai transform,* Mathematical Research Letters **2** (1995) 193-202.

Ric-89. J. Rickard, *Morita theory for derived categories,* J. London Math. Soc. **39** (1989), 436–456.

Ric-91. J. Rickard, *Derived equivalences as derived functors* J. London Math. Soc. **43** (1991) 37-48.

Ric-96. J. Rickard, *Splendid equivalences: derived categories and permutation modules*, Proc. London Math. Soc. **72** (1996), 331-358.

Ried-80. C. Riedtmann, *Algebren, Darstellungsköcher, Überlagerungen und zurück*, Comm. Math. Helv. **55** (1980) 199-224.

Rog-92. K. Roggenkamp, *Subgroup Rigidity of p-adic group rings (Weiss' argument revisited)*, J. London Mathematical Society **59** (1992) 534-544.

RS-87. K.W. Roggenkamp, L.L. Scott, *Isomorphisms of p-adic group rings*, Annals of Mathematics, **126** (1987) 593–647.

Rou-98. R. Rouquier, *Action du groupe de tresses sur la catégorie dérivée de la variété de drapeaux*, manuscript

RZ-96. R. Rouquier and A. Zimmermann, *Picard groups for derived module categories*, accepted for Proceedings of the London Mathematical Society.

ST-00. P. Seidel and R. Thomas, *Braid group actions on derived categories of coherent sheaves*, preprint

Y-00. A. Yekutieli, *Dualizing complexes, Morita equivalence and the derived Picard group of a ring*, Journal of the London Mathematical Society (2000), **60** (1999) 723-746.

Z-96. A. Zimmermann, *Derived equivalences of orders*, Proceedings of the ICRA VII, Mexico, eds : Bautista, de la Pena, Martinez, Can. Math. Soc. Conf. Proc. **18** (1996) 721–749.

Z-99a. A. Zimmermann, *Tilted symmetric orders are symmetric orders*, Archiv der Mathematik **73** (1999) 15-17.

Z-99b. A. Zimmermann, *Auto-equivalences of derived categories acting on cohomology*, preprint

Z-00a. A. Zimmermann, *Automorphisms of Green orders and their derived categories*, preprint

Z-00b. A. Zimmermann, *Cohomology of groups, abelian Sylow subgroups and splendid equivalences*, Algebra Montpellier Announcements 01-2000

Z-01. A. Zimmermann, *Cohomology of groups and splendid equivalences of derived categories*, to appear in Mathematical Proceedings of the Cambridge Philosophical Society (2001).

MODULES WITH GOOD FILTRATION AND INVARIANT THEORY

A.N.ZUBKOV
(e-mail: zubkov@iitam.omsk.net.ru)
Omsk State Pedagogical University,
Omsk branch of Institute of mathematics (SORAN)
644099, Russia, Omsk,
Tuhachevskogo embankment 14,
chair of algebra

The aim of this talk is to demonstrate some applications of theory of modules with good filtration in the invariant theory of classical groups. For the first time it was done by Donkin (1), (9) to find the generators of invariant ring of several matrices in the modular case. In the characteristic zero case it is well known Procesi-Razmyslov's theorem (2), (3) which asserts that this ring is generated by traces of products of matrices. Procesi supposed that in the case when the ground field is infinity of arbitrary characteristic then this invariant ring is generated not only traces but all another coefficients of characteristic polynomial. The last hypothesis was confirmed in (9).

Besides, Procesi and Razmyslov have found not only generators but all defining relations between them using the language of trace identities. The problem to generalize this second part of their result to the modular case was solved in (4).

In the first section of this notices we remind some definitions and notations from the theory of modules with good filtration. Next, we remind definitions of Akin-Buchsbaum-Weyman filtrations and resolvents and concentrate on the simplest case of invariants of several matrices.

In the second section we give some sketch of the proof of main

K.W. Roggenkamp and M. Ştefănescu (eds.), Algebra - Representation Theory, 439–460.

theorem from (4) to demonstrate how the theory of modules with good filtration is applied in the modular case. This section is finished by some short observation of further results concerning representations of quivers.

In the third section we represent some method to reduce the problem of finding of generators and defining relations between them for another classical groups case to the the same kind problem for so-called "mixed" representations of quivers (see all definitions below).

Finally, in the fourth part we formulate and briefly comment some open problems which seem important from the author's point of view.

1. Some definitions and notations

Let G be a reductive algebraic groupdefined over an algebraically closed field k. Denote by T a maximal torus in G and by $X(T)$ the group of characters of T. Let R be a root system associated to the pair (G, T) and R^+ be a set of positive roots in R. Denote by X^+ the set of dominant weights, i.e.

$$X^+ = \{\nu \mid < \nu, \alpha* > \geq 0 \text{ for all } \alpha \in R^+\},$$

where α^* denotes the coroot assotiated with α.

EXAMPLE 1.1. *Let*

$$G = GL_n(k), T = T_n = \{diag(t_1, \ldots, t_n) \mid t_i \in k, i = 1, \ldots, n\}.$$

Then any character $\nu \in X(T_n)$ has a form $(\nu_1, \ldots, \nu_n) \in Z^n$ with respect to the action $\nu(t) = t_1^{\nu_1} \ldots t_n^{\nu_n}$, where $t = diag(t_1, \ldots, t_n)$. Besides, $X(T_n)^+ = \{\nu \mid \nu_1 \geq \nu_2 \geq \ldots \geq \nu_n\}$.

Let B be a Borel subgroup of G corresponding to the set of negative roots $-R^+$.

For any $\lambda \in X(T)$ denote by $\nabla(\lambda)$ the induced module $Ind_B^G k_\lambda = \{f \in k[G] \mid \forall g \in G, \forall b \in B, f(gb) = \lambda(b)^{-1} f(g)\}$.

DEFINITION 1.1. *A G-module V is said to be a module with good filtration (briefly, module with GF) if there is an ascending chain of submodules*

$$0 = V_0 \subseteq V_1 \subseteq \dots$$

such that $\forall i \geq 1, V_i/V_{i-1} \cong \nabla(\lambda_i)$ *for some* $\lambda_i \in X^+$ *and* $\bigcup_{i\geq 1} V_i = V$.

Dualizing all these definitions we can define coinduced module $\Delta(\lambda) = \nabla(\lambda^*)^*$, where $\lambda^* = -w_0(\lambda)$ and w_0 is the longest element of the Weyl group $W(G,T) = N_G(T)/T$. Replacing in the definition of modules with GF induced modules by coinduced ones we obtain the definition of modules with Weyl filtration (briefly, modules with WF). Finally, a G-module V is called a tilting one if it has both good and Weyl filtrations. Note that if V is finitely dimensional then V is a tilting module iff V and V^* both with GF (or with WF).

EXAMPLE 1.2. *Let* $V = k^n$ *is a standard* $GL_n(k)$*-module. For any* $0 \leq t \leq n$ *the exterior power* $\Lambda^t(V)$ *is isomorphic to* $\nabla(\epsilon_1 + \dots + \epsilon_t)$, *where* $\epsilon_i = (0, \dots, \underbrace{1}_{i}, \dots, 0), i = 1, \dots, n,$ *as a* $GL_n(k)$*-module. In particular, this module with GF. Besides,* $\Lambda^t(V)^* \cong \nabla(-\epsilon_{n-t+1} - \dots - \epsilon_n)$. *Therefore this module is a tilting one.*

List some standard facts about modules with GF.

THEOREM 1.1. *((7), (6)) Let* V *and* W *are modules with GF. Then* $V \otimes W$ *is a module with GF with respect to the diagonal action of* G *too.*

EXAMPLE 1.3. *Let* $\lambda = (\lambda_1, \dots, \lambda_s) \in N^s$. *Denote by* $\Lambda^\lambda(V)$, $V = k^n$ *the tensor product* $\Lambda^{\lambda_1}(V) \otimes \dots \otimes \Lambda^{\lambda_s}(V)$ *considered as a* $GL_n(k)$*-module under the diagonal action. By the previous theorem this module is a tilting one.*

PROPOSITION 1.1.

1. *If* V, W *are modules with GF and* $V \subseteq W$ *then* W/V *with GF too.*

2. If we have a short exact sequence of G-modules

$$0 \to U \to W \to V \to 0$$

such that U with GF then

$$0 \to U^G \to W^G \to V^G \to 0$$

is an exact one too.

DEFINITION 1.2. *Let $N \geq n$ and $G = GL_N(k), T = T_N$. The weight (character) $\lambda = (\lambda_1, \ldots, \lambda_N)$ is called n-restricted if $\lambda_i = 0$ for all $i > n$. The suset of $X = X(T_N)$ consisting of all n-restricted weights denote by $X(N, n)$.*

From now on we identify the group $GL_n(k)$ with the block-diagonal subgroup of $GL_N(k)$ by the rule

$$\forall g \in GL_n(k), g \to \tilde{g}, \text{where } \tilde{g}_{ij} = g_{ij} \text{ iff } 1 \leq i, j \leq n, \text{ otherwise}$$
$$\tilde{g}_{ij} = 0, 1 \text{ if } i \neq j \text{ or } i = j \text{ respectively.}$$

DEFINITION 1.3. *((8)) Under the same conditions as in previous definition we can define the truncation (or Schur, see (8)) functor $Tr_{N,n}$ from the category of polynomial GL_N-modules to the category of polynomial GL_n-ones by the rule $Tr_{N,n}(V) = \oplus_{\lambda \in X(N,n)} V_\lambda$, where $\oplus_{\lambda \in X} V_\lambda$ is the weight decomposition of $GL_N(k)$-module V.*

EXAMPLE 1.4. *$Tr_{N,n}(\Lambda^t(k^N)) = 0$ if $t > n$ otherwise $Tr_{N,n}(\Lambda^t(k^N)) = \Lambda^t(k^n)$.*

More general variant of this definition can be found in (10). The following lemma explains the role of Schur functor.

LEMMA 1.1. *((10), (11)) Let V, W be polynomial $GL(N)$-modules with GF and with WF respectively, $n < N$. The natural restriction $Hom_{GL(N)}(W, V) \to Hom_{GL(n)}(Tr_{N,n}(W), Tr_{N,n}(V))$ is an epimorphism.*

Let V, W be finitely dimensional vector spaces considered as a standard $GL(V)$ and $GL(W)$-modules respectively. A homogeneous component $S^t(V \otimes W)$ of the symmetric algebra $S(V \otimes W)$ has a filtration with quotients $L_\lambda(V) \otimes L_\lambda(W)$ as a $GL(V) \times GL(W)$-module , where $L_\lambda(?)$ is a Shur module corresponding to an ordered partition λ of degree $\mid \lambda \mid = t$ (5). It is well known that $L_\lambda(?) \cong \nabla(\tilde{\lambda})$ as a $GL(?)$-module, where $\tilde{\lambda}$ is the conjugated partition.

Let us remind the definition of this filtration.

First of all, put in order all partition of degree t lexicographically from the left to the right. For any ordered partition $\lambda = (\lambda_1, \lambda_2, \ldots)$ of degree t such that $\lambda_1 \geq \max\{\dim V, \dim W\}$ let us define a pairing

$$\delta_\lambda : \Lambda^\lambda(V) \otimes \Lambda^\lambda(W) \cong$$

$$\Lambda^{\lambda_1}(V) \otimes \Lambda^{\lambda_1}(W) \otimes \Lambda^{\lambda_2}(V) \otimes \Lambda^{\lambda_2}(W) \otimes \ldots \to S^t(V \otimes W)$$

by the rule:

$$(v_{11} \wedge \ldots \wedge v_{1,\lambda_1}) \otimes (w_{11} \wedge \ldots \wedge w_{1,\lambda_1}) \otimes (v_{21} \wedge \ldots \wedge v_{2,\lambda_2}) \otimes (w_{21} \wedge \ldots \wedge w_{2,\lambda_2}) \ldots$$

$$\to \det(v_{1i} \otimes w_{1j})_{i,j=1}^{i,j=\lambda_1} \det(v_{2i} \otimes w_{2j})_{i,j=1}^{i,j=\lambda_2} \ldots$$

Sum of all subspaces $Im\delta_\mu$, where μ runs all partitions are not lexicographically less than λ denote by M_λ. We have a filtration

$$0 \subseteq M_{(t,0,\ldots)} \subseteq \ldots M_{(1,1,\ldots,1)} = S^t(V \otimes W)$$

For any M_λ its predecessor $\dot{M}_\lambda$ in this filtration can be defined as the sum of all $Im\delta_\mu$ such that each partition μ is lexicographically greater than λ. The quotient $M_\lambda/\dot{M}_\lambda$ is isomorphic to $L_\lambda(V) \otimes L_\lambda(W)$ (5).

Besides, any Shur module has a finite resolvent

$$0 \to C_m \to C_{m-1} \to \ldots \to C_1 \to L_\lambda(V) \to 0$$

such that all members of one are direct sums of tensor products of some exterior powers $\Lambda^t(V)$ and $C_1 = \Lambda^\lambda(V)$. In particular, all members of this resolvent are with GF as $GL(V)$-modules.

Denote by $[1, r]$ the initial segment $\{1, \ldots, r\}$ of N and by S_r the symmetric group naturally acting on $[1, r]$. If we have a surjective map $g : [1, r] \to [1, d]$ one can define an Young subgroup S_g consisting of all $\sigma \in S_r$ such that $g \circ \sigma = g$. The fibers of g are called the fibers of the Young subgroup S_g. It is clear that $S_g = S_{g^{-1}(1)} \times \ldots S_{g^{-1}(d)}$. In the general case the fibers of S_g can be not segments because of we do not require that the map g is a monotone function.

2. Procesi-Razmyslov's theorem

The matrix algebra $M_n(k)$ can be considered as a $GL_n(k)$-module with respect to the diagonal adjoint action. It is clear that it is isomorphic to $V^* \otimes V, V = k^n$.

Let us consider a space of m $n \times n$ matrices

$$X_{n,m} = M_n(k)^m = \underbrace{M_n(k) \times \ldots \times M_n(k)}_{m \text{ times}}$$

The group $GL_n(k)$ acts on this variety diagonally by conjugations.

The coordinate algebra of the affine space $X_{n,m}$ is $k[x_{ij}(s) \mid 1 \leq i, j \leq n, 1 \leq s \leq m]$. The general matrices $X_s = (x_{ij}(s)), s = 1, \ldots, m$ can be considered as the matrix coordinate functions on $X_{n,m}$.

Any homogeneous component of the algebra $k[X_{n,m}]$ of degrees $t_1, \ldots, t_m$ in variables $X_1, \ldots, X_m$ respectively can be identified with $S^{t_1}(V \otimes V^*) \otimes \ldots \otimes S^{t_m}(V \otimes V^*)$.

In particular, tensoring Akin-Buchsbaum-Weyman (ABW) filtrations of all tensor factors we obtain some filtration of the $GL_n(k)$-module

$$k[X_{n,m}](t_1, \ldots, t_m) \cong S^{t_1}(V \otimes V^*) \otimes \ldots \otimes S^{t_m}(V \otimes V^*)$$

whose members are indicated by collections of partitions (superpartitions) $\tilde{\lambda} = \{\lambda(1), \ldots, \lambda(m)\}$ such that $\mid \lambda(i) \mid = t_i, i = 1, \ldots, m$. Denote these submodules by $M_{\tilde{\lambda}}$. We have

$$M_{\tilde{\lambda}}/\dot{M}_{\tilde{\lambda}} \cong L_{\tilde{\lambda}}(V) \otimes L_{\tilde{\lambda}}(V^*),$$

where $L_{\tilde{\lambda}}(?) = L_{\lambda(1)}(?) \otimes L_{\lambda(2)}(?) \otimes \ldots$.

Tensoring resolvents of modules $L_{\lambda(i)}(V)$ we obtain some resolvent of $L_{\tilde{\lambda}}(V)$ whose members with GF (see (12) for detailed proof). In particular, we get a short exact sequence

$$0 \to R_{\tilde{\lambda}}(V) \to \Lambda^{\tilde{\lambda}}(V) \to L_{\tilde{\lambda}}(V) \to 0$$

whose members with GF too and

$$\Lambda^{\tilde{\lambda}}(V) = \Lambda^{\lambda(1)}(V) \otimes \Lambda^{\lambda(2)}(V) \otimes \ldots.$$

Notice that the module $L_{\lambda}(V^*)$ with GF because of all members of its resolvent are tilting modules. In particular, the module $L_{\tilde{\lambda}}(V^*)$ has a resolvent whose members with GF too.

Using standard identifications one can obtain a short exact sequence

$$0 \to T_{\tilde{\lambda}}(V)^* \to \Lambda^{\tilde{\lambda}}(V)^* \to L_{\tilde{\lambda}}(V^*) \to 0,$$

where the module $T_{\tilde{\lambda}}(V)$ with WF (see (11) for detailed proof). Combaining we get a short exact sequence

$$0 \to D_{\tilde{\lambda}}(V) \to \Lambda^{\tilde{\lambda}}(V) \otimes \Lambda^{\tilde{\lambda}}(V)^* \to L_{\tilde{\lambda}}(V) \otimes L_{\tilde{\lambda}}(V^*) \to 0,$$

where the module $D_{\tilde{\lambda}}(V) = R_{\tilde{\lambda}}(V) \otimes \Lambda^{\tilde{\lambda}}(V)^* + \Lambda^{\tilde{\lambda}}(V) \otimes T_{\tilde{\lambda}}(V)^*$ has a filtration

$$0 \subseteq R_{\tilde{\lambda}}(V) \otimes T_{\tilde{\lambda}}(V)^* \subseteq D_{\tilde{\lambda}}(V)$$

with the quotients

$$R_{\tilde{\lambda}}(V) \otimes T_{\tilde{\lambda}}(V)^* \text{ and}$$

$$L_{\tilde{\lambda}}(V) \otimes T_{\tilde{\lambda}}(V)^* \oplus R_{\tilde{\lambda}}(V) \otimes L_{\tilde{\lambda}}(V^*)$$

In particular, $D_{\tilde{\lambda}}(V)$ with GF!

For the sake of convenience we will replace the symbol V in all these notations by n, where $n = \dim V$. Sometimes we will omit even n, k or $\tilde{\lambda}$ if it doesn't lead to confusion.

Using the proposition 1.1 we obtain

PROPOSITION 2.1. *((1), (4),(11)) The homogeneous component $k[X_{n,m}](t_1,\ldots,t_m)$ has the filtration $\{M_{\tilde\lambda}(n)^{GL_n(k)}\}$ with quotients $(L_{\tilde\lambda}(n)\otimes L_{\tilde\lambda}(n^*))^{GL_n(k)}$, where $L_{\tilde\lambda}(n^*)$ means the same as $L_{\tilde\lambda}(V^*)$. Besides, for any $\tilde\lambda$ we have a short exact sequence*

$$0\to D_{\tilde\lambda}(n)^{GL_n(k)}\to(\Lambda^{\tilde\lambda}(n)\otimes\Lambda^{\tilde\lambda}(n)^*)^{GL_n(k)}\to(L_{\tilde\lambda}(n)\otimes L_{\tilde\lambda}(n^*))^{GL_n(k)}\to 0,$$

or in slightly different but equivalent form

$$0\to D_{\tilde\lambda}(n)^{GL_n(k)}\to End_{GL_n(k)}(\Lambda^{\tilde\lambda}(n))\to M_{\tilde\lambda}(n)^{GL_n(k)}/\dot M_{\tilde\lambda}(n)^{GL_n(k)}\to 0$$

The next question is how to define the image of any endomorphism $\phi\in \mathrm{End}_k(\Lambda^{\tilde\lambda}(n))$ in $k[X_{n,m}](t_1,\ldots,t_m)$?

LEMMA 2.1. *The image of ϕ equals $tr(\phi\circ(X_1\otimes\ldots\otimes X_m))$, where the operator $X_1\otimes\ldots\otimes X_m$ acts on $\Lambda^{\tilde\lambda}(n)$ by such way that each X_i acts diagonally on $\Lambda^{\lambda(i)}(n), i=1,\ldots,m$.*

EXAMPLE 2.1. *((2)) If $t_1=\ldots=t_m=1$ then the multilinear component $k[X_{n,m}](1,\ldots,1)$ can be identified with $End_k(V^{\otimes m})$ and its invariant subspace $k[X_{n,m}](1,\ldots,1)^{GL_n(k)}$ with $End_{GL_n(k)}(V^{\otimes m})$. It is well known that the algebra $End_{GL_n(k)}(V^{\otimes m})$ is an epimorphic image of the group algebra $k[S_m]$ of the symmetric group S_m. Any $\sigma\in S_m$ acts on the space $V^{\otimes m}$ by the rule*

$$\sigma(v_1\otimes\ldots\otimes v_m)=v_{\sigma^{-1}(1)}\otimes\ldots\otimes v_{\sigma^{-1}(m)}.$$

It can easily be checked that the image of given σ equals $tr(X_a\ldots X_b)\ldots tr(X_c\ldots X_d)$, where $(a\ldots b)\ldots(c\ldots d)$ is the cyclic decomposition of σ. We denote the element $tr(X_a\ldots X_b)\ldots tr(X_c\ldots X_d)$ by $tr(\sigma)$.

Notice that $End_{GL_n(k)}(V^{\otimes m})\cong k[S_m]$ iff $n\geq m$ otherwise the kernel K_n of the epimorhism $k[S_m]\to End_{GL_n(k)}(V^{\otimes m})$ is generated by the element $\sum_{\tau\in S_{n+1}}(-1)^\tau\tau$ (13).

The previous lemma can be formulated in the following equivalent form. Let us introduce new general matrices $Y_1,\ldots,Y_t$, where $t=t_1+\ldots+t_m$. Let us define a map $g:[1,t]\to[1,m]$ by the rule $g(i)=s$ iff $t_1+\ldots+t_{s-1}+1\leq i\leq t_1+\ldots+t_s$. In other words, we devide the segment $[1,t]$ into sequential subsegments of lengths $t_1, \ldots, t_m$ and map any element from given subsegment to its number.

Besides, for any $\tilde{\lambda}$ let us define more refinement map $g_{\tilde{\lambda}} : [1, t] \to [1, k_{\tilde{\lambda}}]$. To be precise, deleting all disjunctive brackets we can consider the superpartition $\tilde{\lambda}$ as an ordinary partition. Then $k_{\tilde{\lambda}}$ is the number of its coordinates. Because of the sum of all coordinates of $\tilde{\lambda}$ equals t we can devide $[1, t]$ into sequential subsegments of the same lengths as coordinates of $\tilde{\lambda}$ being read from the left to the right.

Sometimes we will denote the Young subgroup $S_{g_{\tilde{\lambda}}}$ just by $S_{\tilde{\lambda}}$.

Identify $\mathrm{End}_{GL_n(k)}(\Lambda^{\tilde{\lambda}}(n))$ with a subspace of $\mathrm{End}_{GL_n(k)}(V^{\otimes t})$ by the rule $\phi \to \iota_{\tilde{\lambda}} \phi \pi_{\tilde{\lambda}}$, where $\pi_{\tilde{\lambda}} : V^{\otimes t} \to \Lambda^{\tilde{\lambda}}(n)$ is the canonical projection and $\iota_{\tilde{\lambda}} : \Lambda^{\tilde{\lambda}}(n) \to V^{\otimes t}$ is the tensor product of alternating maps $\Lambda^d(V) \to V^{\otimes d}, v_1 \wedge \ldots \wedge v_d \to \sum_{\tau \in S_d} (-1)^{\tau} \tau(v_1 \otimes \ldots \otimes v_d)$. Up to this identification the image of $\phi \in \mathrm{End}_{GL_n(k)}(\Lambda^{\tilde{\lambda}}(n))$ equals $g(\frac{1}{|S_{\tilde{\lambda}}|} tr(\phi))$, where $tr(\phi)$ is calculated on matrices $Y_1, \ldots, Y_t$ and the map g specializes $Y_i \to X_{g(i)}, i = 1, \ldots, t$. In general the last element is not defined in the modular case but only in the characteristic zero one. Nevertheless, one can find a base of $\mathrm{End}_{GL_n(k)}(\Lambda^{\tilde{\lambda}}(n))$ such that $\chi(\phi, g)$ is defined over Z for any its element ϕ and therefore $\chi(\phi, g)$ can be reduced by any modulo p. Besides, denote by $tr(\phi, g)$ the element $tr(\phi \circ (X_1 \otimes \ldots \otimes X_m))$.

Denote by the same symbol $Tr_{N,n}$ the canonical map $k[X_{N,m}] \to k[X_{n,m}]$ defined by $x_{ij}(s) \to 0$ iff either $i > n$ or $j > n$ otherwise $x_{ij}(s) \to x_{ij}(s)$, $1 \leq i, j \leq N, 1 \leq s \leq m$.

PROPOSITION 2.2. *((1), (4)) The truncation functor $Tr_{N,n}$ induced a commutative diagram*

$$\begin{array}{ccccccccc} 0 \to & D_{\tilde{\lambda}}(N)^{GL_N} & \to & End_{GL_N}(\Lambda^{\tilde{\lambda}}(N)) & \to & M_{\tilde{\lambda}}(N)^{GL_N}/\dot{M}_{\tilde{\lambda}}(N)^{GL_N} & \to 0 \\ & \downarrow & & \downarrow & & \downarrow & \\ 0 \to & D_{\tilde{\lambda}}(n)^{GL_n} & \to & End_{GL_n}(\Lambda^{\tilde{\lambda}}(n)) & \to & M_{\tilde{\lambda}}(n)^{GL_n}/\dot{M}_{\tilde{\lambda}}(n)^{GL_n} & \to 0 \end{array}$$

Here the right vertical map coincides with the map induced by the projection mentioned above and all vertical ones are surjective.

Proof. By the lemma 1.1 $\mathrm{End}_{GL_N}(\Lambda^{\tilde{\lambda}}(N)) \to \mathrm{End}_{GL_n}(\Lambda^{\tilde{\lambda}}(n))$ is surjective. Thus follows surjectivness of the right vertical map. Next, the module $D^{GL}_{\tilde{\lambda}}$ has the filtration

$$0 \subseteq (R_{\tilde{\lambda}} \otimes T^*_{\tilde{\lambda}})^{GL} \subseteq D^{GL}_{\tilde{\lambda}}$$

with the quotients

$$(R_{\tilde{\lambda}} \otimes T^*_{\tilde{\lambda}})^{GL} \cong \mathrm{Hom}_{GL}(T_{\tilde{\lambda}}, R_{\tilde{\lambda}}) \text{ and}$$

$$(L_{\tilde{\lambda}} \otimes T^*_{\tilde{\lambda}} \oplus R_{\tilde{\lambda}} \otimes L_{\tilde{\lambda}}(^*))^{GL} \cong \mathrm{Hom}_{GL}(T_{\tilde{\lambda}}, L_{\tilde{\lambda}}) \oplus \mathrm{Hom}\,(L_{\tilde{\lambda}}(^*)^*, R_{\tilde{\lambda}})$$

Applying the same lemma we see that the left vertical map is surjective too.

Denote by $J_{n,m}$ the invariant algebra $k[X_{n,m}]^{GL_n(k)}$.

COROLLARY 2.1. *$Tr_{N,n}$ induces a surjective homomorphism $J_{N,m} \to J_{n,m}$. Moreover, if $N, n \geq t$ then $J_{N,m}(t) \to J_{n,m}(t)$ is an isomorphism so that we can identify both homogeneous components. Further, if $N \geq t$ then the kernel of this map is spanned by the elements $tr(\phi, g)$, where $\phi \in$ $End_{GL_N(k)}(\Lambda^{\tilde{\lambda}}(N))$ and either $\Lambda^{\tilde{\lambda}}(n) = 0$ or $\phi \mid_{\Lambda^{\tilde{\lambda}}(n)} = 0$.*

Proof. The first assertion is a trivial consequence of the previous proposition. If $N \geq n$ we have a commutative diagram

$$\begin{array}{ccc} \mathrm{End}_{GL_N}(\Lambda^{\tilde{\lambda}}(N)) & \to & \mathrm{End}_{GL_N}(W^{\otimes t}) \\ \downarrow & & \downarrow \\ \mathrm{End}_{GL_n}(\Lambda^{\tilde{\lambda}}(n)) & \to & \mathrm{End}_{GL_n}(V^{\otimes t}), \end{array}$$

where $W = k^N, V = k^n$ and the horizontal arrows are embeddings defined above. Since $\mathrm{End}_{GL_n}(V^{\otimes t}) \cong \mathrm{End}_{GL_N}(W^{\otimes t}) \cong k[S_t]$ if $N \geq n \geq t$ thus follows the second one. Finally, if $z \in M_{\tilde{\lambda}}(N)$ belongs to the kernel of $Tr_{N,n}$ and $\mathrm{End}_{GL_n}(\Lambda^{\tilde{\lambda}}(n)) \neq 0$ then using the commutative diagram from the proof of the previous proposition one can find an endomorphism $\phi \in \mathrm{End}_{GL_N}(\Lambda^{\tilde{\lambda}}(N))$ such that $z - tr(\phi, g) \in \dot{M}_{\tilde{\lambda}}(N)$ and $\phi \mid_{\Lambda^{\tilde{\lambda}}(n)} = 0$. Indeed, if ψ is a preimage of z in $\mathrm{End}_{GL_N}(\Lambda^{\tilde{\lambda}}(N))$ then $\pi = Tr_{N,n}(\psi)$ belongs to $D_{\tilde{\lambda}}(n)^{GL_n}$. Let α is a preimage of π in $D_{\tilde{\lambda}}(N)^{GL_N}$. It is clear that $\phi = \psi - \alpha$ is the required endomorphism.

Going down along our filtration we get either the case when $\Lambda^{\tilde{\lambda}}(n) = 0$ or the final member $\Lambda^{t_1}(N) \otimes \ldots \lambda^{t_m}(N)$ of our filtration where our statement is obvious.

DEFINITION 2.1. *Let us consider an inverse spectrum of invariant algebras* $\{J_{N,m}, Tr_{N,n}, 1 \leq n \leq N\}$. *The "restricted" invesre limit of one is called a free invariant algebra depending of* m *(matrix) variables. To be precise, we consider "homogeneous" parts of this spectrum, i.e.* $\{J_{N,m}(t), Tr_{N,n}, 1 \leq n \leq N\}, t = 0, 1, \ldots$ *and define* t*-th homogeneous component of this algebra as the inverse limit of* t*-th "homogeneous" part of this spectrum.*

Denote the free invariant algebra depending of m *(matrix) variables by* J_m. *It is not hard to prove that* J_m *is canonically embedded into* J_M, *where* $M \geq m$. *Therefore, we can define a free invariant algebra* $J = \varinjlim J_m$ *depending of countable set of (matrix) variables.*

Notice that by the corollary 2.1 $J_m(t)$ can be identified with $J_{n,m}(t)$ for any $n \geq t$. In particular, we can define a "formal" i-th coefficient of characteristic polynomial belonging to J_m as an element $\sigma_{i,n}(X), n \geq i$, where X is a general $n \times n$ matrice and $\sigma_{i,n}$ is the i-th coefficient of the characteristc polynomial $\chi_n(t) = \det(X - t \cdot E_n)$. Denote it just by $\sigma_i(X)$.

This notation is well defined because of for any $N \geq n \geq i$ $Tr_{N,n}(\sigma_{i,N}) = \sigma_{i,n}$. By the same way we can define "formal" invariants $\sigma_i(X_{s_1} \ldots X_{s_l}) \in J_m, i \geq 1, 1 \leq s_1, \ldots, s_l \leq m$.

Donkin proved that any algebra J_m is a free polynomial one with free generators $\sigma_i(X_{s_1} \ldots X_{s_l})$, where the sequence $s_1, \ldots, s_l$ has no any proper period, i.e. there is not any proper divisor d of l such that $s_i = s_{i+d}, 1 \leq i \leq l - d$ (1).

Denote by $I_{n,m}$ the kernel of the epimorphism $J_m \to J_{n,m}$. Let $T_m = J_m < X_1, \ldots, X_m >$ be a free associative algebra over the ring J_m. The case $m = \infty$ is admissible and all statements below can be reformulated easily for this case.

The ideal $I_{n,m}$ or $I_n = I_{n,\infty}$ is stable under "formal" substitutions $X_i \to F_i$, where $F_i \in \sum_{1 \leq i \leq m} X_i T_m$. In spite of $I_{n,m}(I_n)$ is not T-ideal in rigorous sense of this term we use one in these lectures (compare with (3)).

THEOREM 2.1. *The ideal* I_n *as well as all ideals* $I_{n,m}, m \geq 1$ *is generated by the elements* $\sigma_d(X), d = n + 1, n + 2, \ldots$ *as a* T*-ideal.*

We won't give a complete proof of this theorem but only some sketch of one. Fix some homogeneous component of $J(t)$ and let us consider any element $tr(\phi, g) \in J(t)\bigcap I_n$ such that $\phi \in \mathrm{End}_{GL_N}(\Lambda^{\tilde{\lambda}}(N)), N >> t$ and $\phi \mid_{\mathrm{End}_{GL_n}(\Lambda^{\tilde{\lambda}}(n))} = 0$ (another case can be reduced to this one (4)).

The algebra $k[S_t]$ can be identified with the "multilinear" component of $S^t(E \otimes E)$, where E is a t-dimensional vector space with a base $e_1, \ldots, e_n$, by the rule $\sigma \to \prod_{i=1}^{i=t} e_{\sigma(i)} \otimes e_i$.

Using the same ABW filtration for $S^t(E \otimes E)$ one can notice that the ideal K_n has a filtration consisting of $M_\lambda \bigcap k[S_t]$ such that $\lambda_1 \geq n+1$ (13). On the other hand, the space $\mathrm{End}_{GL_N}(\Lambda^{\tilde{\lambda}}(N))$ is identified with the subspace $R_{\tilde{\lambda}}$ of $k[S_t] = \mathrm{End}_{GL_N}(W^{\otimes t}), W = k^N$, consisting of all $\phi \in k[S_t]$ such that $\forall \sigma \in S_{\tilde{\lambda}}, \sigma\phi = \phi\sigma = (-1)^\sigma \phi$.

The kernel of the epimorphism

$$\mathrm{End}_{GL_N}(\Lambda^{\tilde{\lambda}}(N)) \to \mathrm{End}_{GL_n}(\Lambda^{\tilde{\lambda}}(n))$$

coincides with $R_{\tilde{\lambda}} \bigcap K_n$. To use the theory of modules with GF we notice that $R_{\tilde{\lambda}}$ equals $S^t(E \otimes E)^{GL(\tilde{\lambda}) \times GL(\tilde{\lambda}), \det \times \det}$, where $GL(\tilde{\lambda})$ is a subgroup of GL_t consisting of all block-diagonal matrices whose blocks have the same sizes as coordinates of $\tilde{\lambda}$ considered as a ordinary partition (see above). In particular $R_{\tilde{\lambda}} \bigcap K_n$ has a filtration $\{M_\mu^{GL(\tilde{\lambda}) \times GL(\tilde{\lambda}), \det \times \det}\}$, where μ runs over all partitions of t such that $\mu_1 \geq n+1$. Since any GL_t-module with GF remains with GF as a $GL(\tilde{\lambda})$-module we should compute only suitable generators of $(M_\mu / \dot{M}_\mu)^{GL(\tilde{\lambda}) \times GL(\tilde{\lambda}), \det \times \det}$ and interpret ones in $\mathrm{End}_{GL_N}(\Lambda^{\tilde{\lambda}}(N))$. To do it one can use the resolvent of the quotient $M_\mu / \dot{M}_\mu = L_\mu \otimes L_\mu$, i.e. just compute the generators of the space $(\Lambda^\mu \otimes \Lambda^\mu)^{GL(\tilde{\lambda}) \times GL(\tilde{\lambda}), \det \times \det}$. The last space can be identified with $(S_t/S_\mu \times S_t/S_\mu)^{S_{\tilde{\lambda}} \times S_{\tilde{\lambda}}, \mathrm{sign} \times \mathrm{sign}}$, where $S_{\tilde{\lambda}}$ acts on the set of left cosets S_t/S_μ by the left multiplication.

After doing all these computations and identifications one can obtain some set of generators of $I_n \bigcap J(t)$ defined over Z. But more useful way is to compute the generators of bigger spaces $(S_t/S_\mu \times S_t/S_\mu)^{\mathrm{diag}(S_{\tilde{\lambda}} \times S_{\tilde{\lambda}})}$. In this case we obtain the generators $tr(\phi_{\sigma_1, \sigma_2}, g)$, where

$$\phi_{\sigma_1, \sigma_2} = \sum_{x \in S_{\tilde{\lambda}}/(S_{\tilde{\lambda}} \cap S_\mu^{\sigma_1} \cap S_\mu^{\sigma_2})} \sum_{\tau \in S_\mu} (-1)^\tau x \sigma_1 \tau \sigma_2^{-1} x^{-1} \in R_{\tilde{\lambda}} \bigcap K_n$$

Up to some partial linearizations, glueings of matrix variables and rearranging of ones we can suppose that $\sigma_1 = 1$ and $S_{\bar{\lambda}} = S_g = S_\mu \cap S_\mu^\sigma$. In particular, we have to consider only the elements $tr(\phi_\sigma, g)$, where

$$\phi_\sigma = \sum_{\tau \in S_\mu} (-1)^\tau \tau \sigma^{-1}$$

and $\sigma = \sigma_1^{-1}\sigma_2$

The next step is to prove that all elements $\chi(\phi_\sigma, g)$ can be obtained from the elements $\sigma_{n+1}, \sigma_{n+2}, \ldots$ by the same operations – linearizations, glueings and admissible substitutions mentioned above. Moreover, we can use only those operations among them which are defined over Z. In particular, the elements $\chi(\phi_\sigma, g)$ are defined over Z too, i.e. it can be reduced by any modulo p. Besides, the equation $\chi(\phi_\sigma, g) = tr(\phi_\sigma, g)$ is valid over arbitrary field k!

This step is most difficult part of the proof and even short observation of one would require a lot of time. Besides, the technique used in has pure combinatorial character which does not interfere with our aim to show some applications of the theory of modules with GF in invariant theory. By this reason we omit it and refer interested reader to (4) and (11) (the last article contains some corrections of the previous one).

Notice that the proof of this theorem gives another proof of Donkin's result (1), (9).

Finally, not long ago author has generalized this theorem for quivers (11).

Let us remind some definitions and notations. A quiver is a quadruple $Q = (V, A, h, t)$, where V is a vertex set and A is an edge set of Q. Let the maps $h, t : A \to V$ associate each edge $a \in A$ its origin $h(a) \in V$ and its end $t(a) \in V$. We enumerate elements of the vertex set as $V = \{1, \ldots, n\}$. Let us consider a collection of vector spaces $E_1, \ldots, E_n$. Let $\dim E_1 = d_1, \ldots, \dim E_n = d_n$. Denote by $\bar{d}$ the vector $(d_1, \ldots, d_n)$. This vector is called a dimensional vector. Denote by $GL(\bar{d})$ the group $GL_{d_1}(k) \times \ldots \times GL_{d_n}(k)$. A representations spacc of quiver Q of dimension $\bar{d}$ is the space $R(Q, \bar{d}) = \prod_{a \in A} \operatorname{Hom}_k(E_{h(a)}, E_{t(a)})$. The group $GL(\bar{d})$ acts on $R(Q, \bar{d})$ by the rule:

$$g \cdot (y_a)_{a\in A} = (g_{t(a)} y_a g_{h(a)}^{-1})_{a\in A}$$

$$g = (g_1, \ldots, g_n) \in GL(\bar{d})$$

For example, if Q has one vertex and m loops are incidenced to this vertex then the representations space of this quiver is isomorphic to a space of m $n \times n$-matrices with respect to the diagonal action of the group $GL_n(k)$.

A coordinate ring of the affine variety $R(Q, \bar{d})$ is isomorphic to $K[y_{ij}(a) \mid 1 \leq j \leq d_{h(a)}, 1 \leq i \leq d_{t(a)}, a \in A]$. For any $a \in A$ denote by $Y_{\bar{d}}(a)$ the general matrice $(y_{ij}(a))_{1\leq j\leq d_{h(a)}, 1\leq i\leq d_{t(a)}}$. The action of $GL(\bar{d})$ on $R(Q, \bar{k})$ induces the action on its coordinate ring by the rule $Y_{\bar{d}}(a) \to g_{t(a)}^{-1} Y_{\bar{d}}(a) g_{h(a)}, a \in A$. We omit the lower index $\bar{d}$ if it doesn't lead to confusion. For example, we write just $Y(a)$ instead of $Y_{\bar{d}}(a)$.

Professor Donkin proved that the algebra $J_{\bar{d}}(Q) = K[R(Q, \bar{d})]^{GL(\bar{d})}$ is generated by the elements $\sigma_j(Y(a_r) \ldots Y(a_1))$, where $1 \leq j \leq d$, $d = \max\{d_1, \ldots, d_n\}$ and the edges $a_1, \ldots, a_r$ form an oriented cycle (or path), i.e. $t(a_1) = h(a_2), \ldots, t(a_{r-1}) = h(a_r), t(a_r) = h(a_1)$ (16).

For the first time this theorem was proved in the characteristc zero case (15), (14). It has very interesting applications for describing of a local structure of the quotient variety $R(Q, \bar{d})/GL(\bar{d})$ (14).

By the same way as above one can define a free invariant algebra $J(Q)$ of given quiver Q and a canonical epimorhpism $J(Q) \to J_{\bar{d}}(Q)$. The kernel of this epimorphism is a T-ideal with respect to admissible formal substitutions saving the geometrical structure of Q. Denote this ideal by $I_{\bar{d}}(Q)$.

THEOREM 2.2. *((11)) The ideal $I_{\bar{d}}(Q)$ is generated as a T-ideal by the elements $\sigma_j(Y(a_r) \ldots Y(a_1))$, where $j > d_{h(a_1)} = d_{t(a_r)}$.*

Notice that both theorems describing relations of matrix invariants or invariants of representations of quivers have an equivalent form on the concomitants language.

DEFINITION 2.2. *A polynomial map $\phi : M_n(k)^m \to M_n(k)$ is said to be an concomitant if it is $GL_n(k)$-equivariant one.*

It is clear that the set consisting of all concomitants is an algebra with respect to the matrix multiplication. Using Donkin's result it isn't hard to prove that this algebra is isomorphic to the $J_{n,m}$-subalgebra of $M_n(k[X_{n,m}])$ which is generated by the matrices $X_1, \ldots, X_m$. Denote this algebra by $T_{n,m}$.

By the same way as above one can define a free concomitant algebra T_m as a restricted inverse limit of all algebras $T_{n,m}$. It can easily be checked that $T_m = J_m < X_1, \ldots, X_m >$ (17). Denote by D_m the kernel of the canonical epimorphism $T_m \to T_{n,m}$. It is a T-ideal with respect to the same admissible substitutions as above. Denote by $\chi_n(X)$ a n-th Cayley-Hamilton identity, i.e. the polynomial $X^n - \sigma_1(X)X^{n-1} + \ldots + (-1)^n\sigma_n(X)$.

THEOREM 2.3. *((4)) The ideal D_m is generated by $\chi_n(X), \chi_{n+1}(X), \ldots$.*

If the characteristic of k is zero we obtain Procesi-Razmyslov's theorem that all trace identities of $n \times n$ matrices are followed from n-th Cayley-Hamilton one.

Finally, this theorem remains true for any quiver case after necessary reformulations (11). Slightly changing the notations from this article one can prove that a free concomitant algebra $T(Q)$ of given quiver $Q = (V, A, h, t)$ is isomorphic to a path $J(Q)$-algebra of Q. To be precise, it is the $J(Q)$-algebra with generators $X_i, i = 1, \ldots, n, X_a, a \in A$ and defining relations

$$\forall a, b \in A, \forall i \in V, X_i^2 = X_i, \sum_{1 \leq i \leq n} X_i = 1,$$

$$X_a X_b = 0 \text{ iff } h(a) \neq t(b), X_i X_a = X_a (X_a X_i = X_a) \text{ if}$$

$$i = t(a)(i = h(a)) \text{ otherwise } X_i X_a (X_a X_i) \text{ equal zero.}$$

For given dimensional vector $\bar{d}$ the corresponding concomitant algebra $T_{\bar{d}}(Q)$ is a $J_{\bar{d}}(Q)$-subalgebra of the algebra $M_{d_1+\ldots+d_n}(J_{\bar{d}}(Q))$ which is generated by the operators $Y_i', i \in V$ and $Y(a)', a \in A$, where $Y_i' \mid_{E_j} = \delta_{ij} id_{E_j}, Y(a)' \mid_{E_j} = \delta_{h(a),j} Y(a)$.

It is clear that we have a canonical epimorphism $T(Q) \to T_{\bar{d}}(Q)$ defined as $\forall i \in V, \forall a \in A, X_i \to Y_i', X_a \to Y(a)'$. Denote its kernel by $D_{\bar{d}}(Q)$.

THEOREM 2.4. *((11)) The ideal $D_{\bar{d}}(Q)$ is generated as a T-ideal by the elements $\chi_m(X_{a_l}\dots X_{a_1})$ such that $(a_1,\dots,a_l)$ is some cycle in Q and $d_{t(a_l)} = d_{h(a_1)} \leq m$.*

Notice that in the characteristic zero case this theorem was proved in (19).

3. Another classical groups invariants and relative problems

Let G_n be a one of classical groups O_n, SO_n or Sp_n, where in the orthogonal group case we suppose additionally that *chark* $\neq 2$ and in the symplectic group case n is an even number. This group acts on the space $X_{n,m} = M_n(k)^m$ by the same way, i.e. diagonally by conjugations.

As above one can ask what are the generators of the algebra $k[X_{n,m}]^{G_n}$ and what are the defining relations between them?

In this section we describe some method to reduce these problems for special type representations of quivers – so-called "mixed" ones.

DEFINITION 3.1. *((18)) Let G be a reductive group acting on an affine variety X. We call X a good variety iff $k[X]$ is a G-module with GF. Besides, if Y is a G-subvarity of X then it is called a good one iff its defining ideal $I_Y = \{f \in k[X] \mid f(Y) = 0\}$ with GF as a G-submodule of $k[X]$.*

Keeping in mind the standard properties of modules with GF we see that Y is a good variety too and the canonical epimorphism $k[X] \to k[Y]$ induces the epimorphism $k[X]^G \to k[Y]^G$.

The following lemma is a variant of Frobenius reciprocity.

LEMMA 3.1. *Let A be a algebraic G-algebra and H is a closed subgroup of G. Then we have an isomorphism*

$$(k[A] \otimes k[G/H])^G \to k[A]^H$$

which is defined by the rule $a \otimes f \to af(eH)$..

LEMMA 3.2. *((20)) If $G_n = O_n$ then the quotient GL_n/G_n is isomorphic to the space S_n of all non-degenerative symmetric $n \times n$ matrices with respect to the action of GL_n by the rule $\forall A \in S, \forall g \in GL_n, A^g = gAg^t$. If $G_n = Sp_n$ then GL_n/G_n is isomorphic to the space C_n of all non-degenerative skew-symmetric $n \times n$ matrices with respect to the same action of GL_n.*

Combining these lemmas we see that to compute the generators of $K[X_{n,m}]^{G_n}$ we must find the generators of the algebra $(K[X_{n,m}] \otimes K[L])^{GL_n(k)}$, where L is either

$$S_n = \{A \in GL_n(k) \mid A = A^t\}$$

or

$$C_n = \{A = (a_{ij}) \in GL_n(k) \mid \forall i \neq j, a_{ij} = -a_{ji}, \forall i, a_{ii} = 0\}$$

and then "specialize" L to E_n (to J in the symplectic case, where J is the matrix of the corresponding skew-symmetric bilinear form (20)).

LEMMA 3.3. *The variety $X_{n,m} \times L$ is a good subvariety of $X_{n,m} \times GL_n(k)$, where $GL_n(k)$ acts on the last factor by the same way as on L. In particular, we have the epimorphism*

$$k[X_{n,m} \times GL_n(k)]^{GL_n(k)} \to k[X_{n,m} \times L]^{GL_n(k)}.$$

All one need to prove this lemma is to prove that $k[\mathrm{M}_n(k)]$ is a good variety with respect to the same action of $GL_n(k)$ as on L and the variety L is a good subvariety of one. The first assertion can be proved by the same way as above using ABW filtration of $k[\mathrm{M}_n(k)]$. The next one is a consecuence of Donkin's result (18) asserting that if Y is a complete intersection in X and a minimal system of generators of its defining ideal I_Y spans a G-stable vector subspace E such that the symmetric algebra $S(E)$ is a G-module with GF then Y is a good subvariety of X.

The next step is to identify the variety $X_{n,m} \times GL_n(k)$ with some closed subvariety Z of the variety $X_{n,m} \times \mathrm{M}_n(k)^2$ by the rule

$$(y_1, \ldots, y_m, x) \to (y_1, \ldots, y_m, x^{-1}, x), x \in GL_n(k), (y_1, \ldots, y_m) \in X_{n,m}.$$

The variety Z is defined by the equation $xy = E_n$, where x, y are the matrix coordinates of $\mathrm{M}_n(k)^2$ and E_n is a matrix unit. It is clear that Z is a complete intersection. The group $GL_n(k)$ acts on $\mathrm{M}_n(k)^2$ by the rule $(x, y)^g = ((g^t)^{-1}xg^{-1}, gyg^t)$. Repeating arguments mentioned above one can prove that Z is a good subvariety of $X_{n,m} \times \mathrm{M}_n(k)^2$. Therefore, it remains to find the generators of $k[X_{n,m} \times \mathrm{M}_n(k)^2]^{GL_n(k)}$ and specialize them subsequently to $k[X_{n,m}]^{GL_n(k)}$.

Now, notice that the $GL_n(k)$-variety $X_{n,m} \times \mathrm{M}_n(k)^2$ can be considered as a representation space of the quiver Q consisting of two vertexes, m loops outgoing from one vertex and two arrows between vertex and another one which have opposite directions. The dimensional vector is $\bar{n} = (n, n)$. The difference with ordinary case is although the first vertex is occupied by $V = k^n$ but the second one by V^*. Besides, we replace the group $GL(\bar{n})(k) = GL_n(k) \times GL_n(k)$ by its diagonal subgroup.

Generalizing this example we set the following

DEFINITION 3.2. *Let Q be a quiver and $\bar{d}$ be a dimensional vector. Let us devide the set of vertexes of Q into some subsets, say $V = \bigsqcup_{i=1}^{i=l} V_i$. Our dimensional vector $\bar{d}$ is said to be compatible with this subdivision of V iff $\forall i \forall k, s \in V_i \; d_k = d_s$. In other words, all coordinates of the vector $\bar{d}$ with numbers from the same V_i are equal one to another. Mark some coordinates of $\bar{d}$ by the upper star $*$. Denote by $R_{mixed}(Q, \bar{d})$ the space $\prod_{a \in A} Hom_k(W_{h(a)}, W_{t(a)})$, where $\forall i, \dim W_i = d_i$ and $W_i = E_i^*$ if d_i is equipped by $*$ otherwise $W_i = E_i$. Decompose the group $GL(\bar{d})$ into blocks by the rule*

$$GL(\bar{d}) = \times_{i=1}^{i=l}(\times_{s \in V_i} GL_{d_s}(k)).$$

Denote by H the group $\times_{i=1}^{i=l} diag((\times_{s \in V_i} GL_{d_s}(k)))$. The representation space $R_{mixed}(Q, \bar{d})$ with the action of the group H is called a "mixed" representation space of the quiver Q of (marked) dimension $\bar{d}$.

We formulate

PROBLEM 1. *What are the generators and the defining relations between them for the ring $k[R(Q, \bar{d})]^H$?*

The answer for the first question can be obtained by the same way as in (20). For the sake of simplicity I demonstrate it on the quiver with two vertexes mentioned above. As usual denote the general matrices corresponding to its m loops by $X_1, \ldots, X_m$ and by A, B the matrices corresponding to the arrow outgoing from the first vertex to the second one and from the second to the first respectively.

Then for given dimension $\bar{n} = (n, n^*)$ the invariant algebra of "mixed" representations of this quiver is generated by the elements $\sigma_i(Y_{s_1} \ldots Y_{s_l}), i = 1, \ldots, n$, where any

$$Y_j \in \{X_1, X_1^t, \ldots, X_m, X_m^t, A, A^t, B, B^t\}$$

and the product $Y_{s_1} \ldots Y_{s_l}$ sutisfies the following conditions:

1. If $Y_i = X_?$ then a next matrice must be A, A^t or $X_?$;
2. If $Y_i = A$ then a next matrice must be B, B^t or $X_?^t$;
3. If $Y_i = B$ then a next matrice must be A, A^t or $X_?$.

The cases $Y_i = X_?^t, A^t$, or B^t are obvious and can be obtained by transposing of previous ones.

As for the second question my hypothesis is one has to add to the standard identities as $\sigma_?(??)$ something like polynomials $F_{k,n+1}, 2k \leq n+1$ from (2). To be definite, the Procesi's polynomials are just some complete polarizations of ours. We omit the explicit description of the last ones because of too serious combinatorial difficults.

The positive solution of this problem would allow us to describe relations between invariants of another classical groups. Moreover, one can define "mixed" representations of quivers for all these groups too just replace some factors $GL_?()$ in $GL(\bar{?})$ by $O_?$ or $\mathrm{Sp}_?$. It doesn't mean any real extending of previous definition because of up to using Frobenius reciprocity it will be ordinary "mixed" case again.

Finally, notice that the case $G_n = \mathrm{SO}_n$, where n is an even number requires to describe not only invariants but semiinvariants of "mixed" representations too. Recently, author in collaboration with M.Domokos have described semiinvariants of ordinary representations of quivers (12).

4. Open problems

Let us begin with the problem already mentioned above.

PROBLEM 2. *What are the generators of semiinvariant ring of "mixed" representations of any quiver?*

In his remarcable article (18) Donkin proved that Zarisski closure of any orbit $A^{GL_n(k)}, A \in M_n(k)$ is a good subvariety of $M_n(k)$.

PROBLEM 3. *Is it true that Zarisski closure of any orbit* $A^{GL_n(k)} A \in X_{n,m}, m \geq 2$ *is a good subvariety of* $X_{n,m}$*?*

It is obvious that this problem can be formulated for quiver case too. More generally, let us formulate

PROBLEM 4. *Is it true that Zarisski closure of any orbit* $Gx, x \in X$*, where* X *is a good affine* G*-varity, is a good subvariety too?*

As for the problem 3 my expectations are very optimistic but as for the fourth one I leave it without any comments. I think that even there is any countrexample for the last hypothesis it won't be trivial.

PROBLEM 5. *Is* $J_{n,m}$ *a Cohen-Macaulay ring?*

In the characteristic zero case it is well known that the answer is yes for any reductive group G and its module V. But in the modular case the answer is yes for any G-module V iff G is linearly reductive group, i.e. if G is a finite extension of some torus (21).

PROBLEM 6. *Is* $k[X]^G$ *a Cohen-Macaulay ring for any good* G*-variety* X*?*

Anything I know is the following partial result.

THEOREM 4.1. *((22) For* $1 \leq m \leq 4$ *the algebra* $J_{2,m}$ *is a Cohen-Macaulay one.*

Acknowledgment

I would like to thank Klaus Roggenkamp and Laslo Marki for their supporting of my visit and inviting me to give these lectures. I thank Mirela Stefanescu for her kind advising and helping me with all these things concerning travel conditions, accomodation and etc.

References

1. S.Donkin. Invariant functions on matrices. *Math.Proc.Cambridge Phil.Soc.* **113(1992), no. 23, pp. 23–43.**
2. C.Procesi. The invariant theory of $n \times n$-matrices. *Adv.in Math.***19(1976), pp. 306–381.**
3. Y.P.Razmyslov. Trace identities of matrix algebras over fields of characteristic zero.*Izv.AN USSR. ser.math.* **38(1974), no 4, pp. 723–756.** (russian)
4. A.N.Zubkov. On generalization Procesi-Razmyslov's theorem. *Algebra i Logica* **35(1996), no. 4, pp. 433–457.**(russian)
5. K.Akin, D.A.Buchsbaum and J.Weyman. Schur functors and Schur complexes. *Adv. in Math.* **44(1982), pp. 207–278.**
6. O.Mathieu. Filtrations of G-modules. *Ann.Scient.Ec. Norm. Sup(2)* **no. 23(1990), pp.625–644.**
7. S.Donkin. Rational representations of algebraic groups: tensor products and filtrations. *Lect.Notes in Math.***114, (1985).**
8. J.A.Green. Polynomial representations of GL_n. *Lecture Notes in Math.***830, Springer-Verlag, Berlin/Heidelberg/New York, (1980).**
9. S.Donkin*. Invariants of several matrices. *Invent.Math.* **110(1992), pp. 389–401.**
10. S.Donkin On tilting modules for algebraic groups. *Math.Z.* **212(1993), pp .39–60.**
11. A.N.Zubkov. Procesi-Razmyslov's theorem for quivers. *Fundamental'nay i Prikladnay matematika* **to appear.** (russian)
12. M.Domokos, A.N.Zubkov. Semiinvariants of quivers as determinants. *Transformation Groups* **to appear.**
13. C. de Concini and C.Procesi. A characteristic free approach to invariant theory. *Adv. in Math.* **21(1976), pp.330-354.**
14. L. Le Bruyn and C. Procesi. Semisimple representations of quivers. *Trans. Amer. Math. Soc.* **317(1990), pp. 585–598.**
15. Lieven Le Bruyn* and C.Procesi. Etale local structure of matrix invariants and concomitants *Lect.Notes.Math.* **127(1987), pp. 143-175.**
16. S. Donkin. Polynomial invariants of representations of quivers. *Comment. Math. Helvetici* **69(1994), pp. 137–141.**

17. A.N.Zubkov. Endomorphisms of tensor products of exterior powers and Procesi hypothesis. *Commun. in Algebra* **22(1994), no. 15, pp. 6385-6399.**
18. S. Donkin. The normality of conjugacy classes of matrices. *Inv. Math.* **101 (1990), pp. 717–736.**
19. M. Domokos. Invariants of quivers and wreath products, *Comm. Alg.* **26(1998), pp. 2807–2819.**
20. A. N. Zubkov, Adjoint action invariants of classical groups. *Algebra i Logika* **38(1999), no. 5, pp. 549–584.** (russian)
21. G.Kemper. A Characterization of Linearly Reductive Groups by their Invariants. *Transformation Groups* **to appear.**
22. M.Domokos*, S.G.Kuz'min and A.N.Zubkov. Representations of invariant rings. **in preparetion.**